Theory of Plastic Instability

Henryk Petryk

Theory of Plastic Instability

Selected Works of Henryk Petryk

Springer

Henryk Petryk
Institute of Fundamental Technological
Research
Polish Academy of Sciences
Warsaw, Poland

ISBN 978-3-031-98222-4 ISBN 978-3-031-98223-1 (eBook)
https://doi.org/10.1007/978-3-031-98223-1

This Springer imprint is published by the registered company Springer Nature Switzerland AG
The registered company address is: Gewerbestrasse 11, 6330 Cham, Switzerland

to my wife

Foreword

It is now more than forty years since Professor Henryk Petryk published his first paper concerned with the theory of plastic instability, thus initiating a series of contributions in which the theory has been gradually developed into its current mature state. A selection of those works forms the main part of this book. Furthermore, it is now fifty years since Henryk Petryk decided to pursue a scientific career and joined the Institute of Fundamental Technological Research (IPPT), Polish Academy of Sciences in Warsaw, an institution he has been associated with ever since. It is thus a perfect occasion to briefly recall the most important facts from his professional and personal life.

Two facts from an early stage of Henryk Petryk's life are worth mentioning as they match our image of Henryk as a scientist and a brilliant thinker. First, as a high school student, he received a bronze medal at the International Mathematical Olympiad in 1968 in Moscow. Second, he was a tournament chess player, reaching the category of national master in 1980 and the FIDE rating of 2370.

Henryk graduated from the Warsaw University of Technology (Politechnika Warszawska), Department of Mechanics and Technology in 1974 and soon after he enrolled in the doctoral school at IPPT. It is also since 1974 that Henryk has been married to Ewa; they have three children.

In 1978 Henryk Petryk defended at IPPT his PhD thesis which was concerned with non-uniqueness of solutions of a steady-state flow of a rigid-perfectly plastic material. Problems related to non-uniqueness of solutions in plasticity remained his leading research interest in the following years.

After completing his PhD, Henryk Petryk focused on the stability and bifurcation problems in rate-independent elastoplastic solids. This area of research became the foundation for several key contributions that earned him widespread recognition within the scientific community. His landmark work involved examining stability conditions and the corresponding energy criteria, as well as exploring post-critical deformation in elastoplastic solids — particularly in the context of incrementally nonlinear responses and material heterogeneity. His extensive body of work on plastic instability includes over 40 publications, most of them single-authored.

The concept of incremental energy minimization was employed in a novel approach to tackle the long-standing challenge of selecting active slip systems and deformation banding in rate-independent crystal plasticity. Building on this, an extended version of the incremental energy minimization method was introduced to model the evolution of martensitic microstructures, accounting for interface energy to predict the associated size effects. Size effects in plasticity have driven his recent research aimed at developing a new class of gradient-enhanced crystal plasticity models with a natural length scale.

Other notable contributions include a generalization of time differentiation and spatial averaging rules for materials with moving boundaries and interfaces, analysis of divergence and flutter instabilities in elastoplastic materials, extremal principles in non-equilibrium thermodynamics, modelling of laminated microstructures in shape memory alloys, and micromechanical mean-field modelling of elastoplastic and viscoelastic heterogeneous materials.

As a confirmation of his high reputation in the Polish and international mechanics community, Henryk Petryk has received a number of distinctions and awards. He was awarded the Mercator Professor fellowship in Germany (2001) and was elected a member of EUROMECH Council (2007-2012), a member of the Polish Academy of Sciences (2010), and the Secretary-General of IUTAM (2016-2022), to mention just a few. Since 1999, he has been the Editor-in-Chief of Archives of Mechanics, an international journal published since 1949.

The contribution to the theory of plastic instability is definitely the flagship achievement of Henryk Petryk. This book summarizes the main steps in the development of the theory of plastic instability. Thirteen original works of Henryk Petryk have been selected and are accompanied by an extended introduction by the author that provides a general context and a guide through the individual contributions. These contributions date back to 1982 with the formulation of a general energy approach to plastic instability, which has been thoroughly investigated in subsequent works and became known as incremental energy minimization. The embedded criterion of choice between non-unique solutions led to a novel computational approach with automatic branch switching by eliminating unstable paths. Recently, in 2020, the approach was extended to incremental problems of non-potential type, thus removing the inherent requirement of symmetry of the original incremental energy minimization framework. We believe this book will be a valuable reference for a wide audience, in particular, for those interested various aspects of plastic instability. The derived criteria can be implemented in numerical simulation codes for the detection of bifurcations and localization phenomena in structural calculations and for the analysis of postcritical behaviour.

December 2024

Davide Bigoni
Samuel Forest
Stanisław Stupkiewicz

Preface

The theory of plastic instability is a vital topic. The term 'plastic instability' refers to instability in plastically deformed solids of any kind. It can manifest itself in various ways, including phenomena such as buckling, necking, snap-through, strain localization, shear banding, local deformation patterns and resulting microstructures in plastically deformed metals and alloys as well as in geomaterials. The concept of instability of equilibrium is too narrow to account for such a variety of phenomena in inelastic solids. This book shows that they can be investigated in a consistent manner using a single energy criterion of path instability. The incremental energy minimization presented in its different variants can be directly applied in numerical studies of plastic instability phenomena.

This monograph consists of thirteen selected works of Henryk Petryk, previously published in various books and journals, accompanied by a new unifying introduction by the author. The selection has been made in such a way that the subject matter of the book as a whole is coherent and the individual chapters complement each other. In this way, the theory of plastic instability is presented in successive stages of its development up to its present mature state.

Acknowledgements

The author expresses his gratitude to Professor Klaus Thermann from the University of Dortmund for long-term intensive collaboration in the field of plastic instability. His insightful and precise work in developing incremental energy minimization towards a novel computational approach to incrementally nonlinear elastoplastic solids resulted in a series of joint papers listed in the 'About the Author' section. In article [20] the energy criterion of path stability was employed for the first time in finite element computations. As demonstrated in this and other joint papers, cf. [21, 24, 25, 27, 29], the incremental energy minimization embedded in the finite element method enables automatic branch switching upon detection of plastic instability, even many times along a path when the deformation pattern changes in consecutive time steps.

The author thanks also Dr. Michał Kursa for his significant contribution to the elaboration of algorithms and for performing numerical simulations of single crystal plasticity. Joint articles [35, 36, 41] dealt with the incremental minimization of deformation work in the multimode plasticity framework under the above-mentioned symmetry constraint, and further work [38, 39] extended the previous approach to the actual nonsymmetric slip-system interaction matrix at finite strain by applying incremental energy quasi-minimization. Thanks also go to Dr. Maciej Ryś for developing the algorithms and for finite element simulations of spontaneous formation of deformation patterns in single crystals in [41], and also in [37].

The author would also like to express gratitude to Professor Stanisław Stupkiewicz from IPPT for many years of close cooperation in the field of martensitic phase transformation, which, although physically different from plastic instability, also deals with rate-independent dissipative processes and the inherent non-uniqueness of solutions. In joint articles [33, 34], the incremental energy minimization approach was used to select a rate-independent evolution path in the transformation of martensitic laminates. More recent joint work has incorporated the phase-field method for finite element modelling of the evolving transformation pattern, which in the time-discrete formulation is similar in spirit to the incremental energy minimization or quasi-minimization while taking into account rate-dependent dissipation.

The cooperation and contributions of all co-authors of joint works related to plastic instability, whose names can be found in the reference list in the 'About the Author' section, are highly appreciated. The author is especially indebted to Professors Davide Bigoni, Samuel Forest and the late Quoc Son Nguyen for many inspiring discussions on plastic instability issues. It is impossible to list here all those to whom the author is grateful in other areas.

Finally, the author would like to acknowledge gratefully the continuous support from his home institution, Institute of Fundamental Technological Research (IPPT), Polish Academy of Sciences, Warsaw, Poland, over the last fifty years. The extremely favourable research atmosphere at IPPT encouraged free scientific work on the concepts described in this book.

Warsaw, Poland
December 2024

Henryk Petryk

Contents

About the Author

Henryk Petryk is Professor at the Institute of Fundamental Technological Research (IPPT), Polish Academy of Sciences, Warsaw, Poland. He was awarded the title of Professor in 1995 and elected a member of the Polish Academy of Sciences in 2010. His research interests focus on the mechanics of solids, mainly on theory of plasticity, bifurcation and instability, variational methods, micromechanics of materials, evolution of microstructures in metal crystals and shape memory alloys, internal length scales and size effects. He published over a hundred papers, most of them on fundamental problems of solid mechanics, 40 as the sole author. He was invited as visiting professor in Germany, France, Japan, Italy and China. His scientific work was accompanied by editorial and organizational activities. For example, he has been Editor-in-Chief of the international journal Archives of Mechanics since 1999, and Secretary-General of IUTAM (International Union of Theoretical and Applied Mechanics) from 2016 to 2022.

Henryk Petryk's principal research topic has been the theory of plastic instability. The successive stages of its development are presented in this book, which comprises the following selected works of the author, enumerated as book chapters.

1 H.Petryk. Theory of bifurcation and instability in time-independent plasticity. In CISM Courses and Lectures No. 327, Udine 1991 (Ed. Q.S.Nguyen), pp. 95-152. Springer-Verlag, Wien 1993.
2 H.Petryk. Stability and constitutive inequalities in plasticity. In CISM Courses and Lectures No. 336, Udine 1992 (Ed. W.Muschik), pp. 259-329. Springer-Verlag, Wien 1993.
3 H.Petryk. Theory of material instability in incrementally nonlinear plasticity. In CISM Courses and Lectures No. 414, Udine 1999 (Ed. H.Petryk), pp.261-331. Springer-Verlag, Wien 2000.
4 H.Petryk. A consistent energy approach to defining stability of plastic deformation processes. In Stability in the Mechanics of Continua, Proc. IUTAM Symp., Nuembrecht 1981 (Ed. F.H.Schroeder), pp. 262-272. Springer-Verlag, Berlin Heidelberg 1982.
5 H.Petryk. On energy criteria of plastic instability. In Plastic Instability, Proc. Considère Memorial, Paris 1985, pp. 215-226. Presses de l'ecole nationale des ponts et chaussées, Paris 1985.
6 H.Petryk. The energy criteria of instability in time-independent inelastic solids. Archives of Mechanics 43, 519-545, 1991.
7 H.Petryk. Material instability and strain-rate discontinuities in incrementally nonlinear continua. Journal of the Mechanics and Physics of Solids 40, 1227-1250, 1992.
8 H.Petryk. On stability of time-independent materials at finite strain. In: Finite Inelastic Deformations - Theory and Applications, Proc. IUTAM Symp. Hannover 1991 (Ed. D.Besdo and E.Stein), pp. 189-196. Springer-Verlag, Berlin Heidelberg 1992.
9 H.Petryk. Plastic instability: Criteria and computational approaches. Archives of Computational Methods in Engineering 4, 111-151, 1997.
10 H.Petryk. On the micro-macro transition and hardening moduli in plasticity. In: Proc. IUTAM Symposium on Micro- and Macrostructural Aspects of Thermoplasticity, Bochum 1997 (Ed. O.T. Bruhns and E. Stein), pp. 219-230. Kluwer, Dordrecht 1999.
11 H.Petryk. General conditions for uniqueness in materials with multiple mechanisms of inelastic deformation. Journal of the Mechanics and Physics of Solids 48, 367-396, 2000.
12 H.Petryk. Incremental energy minimization in dissipative solids. Comptes Rendus Mécanique 331, 469-474, 2003.
13 H.Petryk. A quasi-extremal energy principle for non-potential problems in rate-independent plasticity. Journal of the Mechanics and Physics of Solids 136, 10369, 2020.

The following list comprises further publications by the author which are related to plastic instability but not included in this volume. These should not be regarded as citations; the relevant bibliographies (several hundred titles in total) can be found in the works listed.

[14] H.Petryk and K.Thermann. Second-order bifurcation in elastic-plastic solids. Journal of the Mechanics and Physics of Solids 33, 577-593, 1985.
[15] H.Petryk. On the onset of instability in elastic-plastic solids. In Plasticity Today: Modelling, Methods and Applications, Proc. Symp. Udine 1983 (Ed. A.Sawczuk and G.Bianchi), pp. 429-447. Elsevier, London 1985.
[16] H.Petryk. On stability and symmetry conditions in time-independent plasticity. Archives of Mechanics 37, 503-520, 1985.
[17] H.Petryk. On constitutive inequalities and bifurcation in elastic-plastic solids with a yield-surface vertex. Journal of the Mechanics and Physics of Solids 37, 265-291, 1989.
[18] Q.S.Nguyen and H.Petryk. A constitutive inequality for time-independent dissipative solids. Comptes Rendus de l'Académie des Sciences, Série II 312, 7-12, 1991.
[19] H.Petryk. On the second-order work in plasticity. Archives of Mechanics 43, 377-397, 1991.
[20] H.Petryk and K.Thermann. On discretized plasticity problems with bifurcations. International Journal of Solids and Structures 29, 745-765, 1992.
[21] H.Petryk and K.Thermann. On post-bifurcation paths in plastic deformation processes. In Computational Plasticity, Proc. 3rd Int. Conf. Barcelona 1992 (Ed. D.R.J.Owen, E.Onate, E. Hinton), vol I, 593-604. Pineridge Press, Swansea 1992.

[22] H.Petryk. Some constitutive inequalities for incrementally nonlinear models in plasticity. In Modern Approaches to Plasticity (Ed. D. Kolymbas), pp. 59-70, Elsevier, Amsterdam 1993.
[23] H.Petryk. Thermodynamic stability of equilibrium in plasticity. Journal of Non-Equilibrium Thermodynamics 20, 132-149, 1995.
[24] H.Petryk and K.Thermann. On plastic strain localisation in the non-elliptic range under plane stress. In: Computational Plasticity, Proc. 4th Int. Conf. Barcelona 1995 (Ed. D.R.J.Owen, E.Onate, E. Hinton), vol I, pp. 647-658. Pineridge Press, Swansea 1995.
[25] H.Petryk and K.Thermann. Post-critical plastic deformation of biaxially stretched sheets. International Journal of Solids and Structures 32, 689-705, 1996.
[26] H.Petryk. Instability of plastic deformation processes. In: Theoretical and Applied Mechanics 1996, Proc. XIX Int. Congr. Theor. Appl. Mech., Kyoto 1996 (Ed. T. Tatsumi et. al.), pp. 497-516. Elsevier, Amsterdam 1997.
[27] H.Petryk and K.Thermann. Post-critical deformation pattern in plane strain plastic flow with yield-surface vertex effect. International Journal of Mechanical Sciences 42, 2133-2146, 2000.
[28] D.Bigoni and H.Petryk. A note on divergence and flutter instabilities in elastic-plastic materials. International Journal of Solids and Structures 39, 911-926, 2002.
[29] H.Petryk and K.Thermann. Post-critical plastic deformation in incrementally nonlinear materials. Journal of the Mechanics and Physics of Solids 50, 925-954, 2002.
[30] H.Petryk. Second-order work and dissipation on indirect paths. Comptes Rendus Mécanique 330, 121-126, 2002.
[31] H.Petryk. Thermodynamic conditions for stability in materials with rate-independent dissipation, Philosophical Transactions of the Royal Society A 363, 2479-2515, 2005.
[32] H.Petryk. Instability and energy minimization in solids: formation of microstructures. In Proc. Int. Symp. CMDS 11, Paris 2007, (Ed. D.Jeulin and S.Forest), pp. 113-118. Ecole de Mines, Paris 2008.
[33] H.Petryk and S.Stupkiewicz. Interfacial energy and dissipation in martensitic phase transformations. Part I: Theory. Journal of the Mechanics and Physics of Solids 58, 390–408, 2010.
[34] H.Petryk, S.Stupkiewicz, G.Maciejewski. Interfacial energy and dissipation in martensitic phase transformations. Part II: Size effects in pseudoelasticity. Journal of the Mechanics and Physics of Solids 58, 373–389, 2010.
[35] H.Petryk and M.Kursa. The energy criterion for deformation banding in ductile single crystals. Journal of the Mechanics and Physics of Solids 61, 1854–1875, 2013.
[36] H.Petryk and M.Kursa. Incremental work minimization algorithm for rate-independent plasticity of single crystals. International Journal for Numerical Methods in Engineering 104, 157–184, 2015.
[37] M.Ryś, S.Forest, H.Petryk. A micromorphic crystal plasticity model with the gradient-enhanced incremental hardening law, International Journal of Plasticity 128, 102655, 2020.
[38] H.Petryk and M.Kursa. Crystal plasticity algorithm based on the quasi-extremal energy principle. International Journal for Numerical Methods in Engineering 123, 3285-3316, 2022.
[39] H.Petryk and M.Kursa. Energy approach to the selection of deformation pattern and active slip systems in single crystals. European Journal of Mechanics A-Solids 104, 105040, 2024.
[40] H.Petryk. On thermodynamic extremal principles in gradient plasticity with energetic forces. Mathematics and Mechanics of Solids 30(2), 230-246, 2025.
[41] M.Ryś, M.Kursa, H.Petryk. Spontaneous emergence of deformation bands in single-crystal plasticity simulations at small strain. Computational Mechanics 75, 547-574, 2025.

Introduction

About the book

This book aims to present the theory of plastic instability in successive stages of its development by collecting in one volume selected works of the author, gradually published in different books and journals. The selection was made so that the subject matter of the entire book was coherent and its individual chapters complemented each other. The last chapter of the book and other recent publications indicate that the theory of plastic instability remains relevant and continues to evolve today.

The term 'plastic instability' is a shorthand expression for the instability in plastically deformed solids (not plastics). The primary context of its use has changed over time. Several decades ago, this term was often used in connection with buckling or necking phenomena in the plastic range, mainly under uniaxial or biaxial loading, and also in connection with dynamic phenomena beyond a peak on a load-displacement curve. It became frequently used when studying an abrupt change of uniform or smoothly distributed deformation into a localized mode, in the form of shear bands in solid bodies or localized necking in thin sheets. Nowadays, plastic instability increasingly refers to material instability related to the formation of local deformation patterns and resulting microstructures in plastically deformed metals and alloys. It is well known that the concept of instability of equilibrium is too narrow to account for such a variety of instability phenomena in the plastic range. In this book, instability is understood in a broader sense and refers to a quasi-static deformation process (or path), including the above mentioned phenomena as particular symptoms of plastic instability. The concept of path instability is not limited to bifurcation understood as the non-uniqueness of the incremental solution, but is closely related to it. We will see that all the mentioned observable instabilities of the dynamic, geometric, local or material type, which appear to be quite different from each other, can be investigated in a unified way by using a single energy criterion to predict their occurrence.

The first three chapters of this book contain notes from lecture series conducted by the author during advanced courses at the International Centre for Mechanical Sciences (CISM) in Udine, Italy. These lectures were intended to present the theory of plastic instability from three different, complementary perspectives. Chapter 1, with due references to the original works, covers the basic elements of the theory of bifurcation and stability in time-independent solids established by Rodney Hill. It also includes the extension towards a unified energy approach developed by the author. The thermodynamic aspects of this energy approach are presented in Chapter 2, where the original concept of stability of a rate-independent inelastic deformation process was extended to stability in the thermodynamic sense. Chapter 3 specifies the results presented in the first two chapters for the plastic instability of material type. Within the framework of the theory presented in Chapters 1 to 3 and in most of the remaining chapters, it is possible to retrieve the evolution equations for the system by minimizing the increment in global energy, which thus represents an incremental potential. The extension of the theory to non-potential problems is presented in Chapter 13, so that Chapters 1, 2, 3 and 13 can be considered the core of the book. Chapters 4 to 13 are arranged chronologically.

The individual Chapters of this book are reprinted from the first thirteen references listed in the 'About the Author' section, together with their separate bibliographies. Other works by the author, also listed in the 'About the Author' section but not included in this book, are occasionally referred to in this Introduction using numbers in square brackets to facilitate further reading. Numbers in parentheses below indicate the formulae given in this Introduction.

The energy condition for path stability: Incremental energy minimization

The article in which the energy approach to plastic instability in a general framework was first proposed is included in Chapter 4 of this book. This article was originally published in 1982, and the basic concepts, developed further in the other chapters, can already be found there. It was also shown that the proposed energy approach is consistent with Hill's theory of bifurcation and instability for conventional elastic-plastic continua that obey the normality flow rule at finite strain. This was essential because Hill's theory, recapitulated in Chapter 1, was fully developed at that time and provided a firm basis for the study of a class of plastic instability problems, under the general constitutive assumption of the existence of a velocity-gradient potential. The consistency of the energy approach with Hill's theory was subsequently demonstrated in Chapters 5 and 6 under the same general assumption. In this way in a unified energy interpretation of the functionals and the conditions for uniqueness and stability given by Hill was obtained. Moreover, new perspectives were opening up, which are discussed below.

The essence of the developed energy approach can be expressed in a concise form by the following variational inequality

$$\Delta E \geq \Delta E^0 \qquad (1)$$

to be satisfied to *second* order in $\Delta t > 0$ for all kinematically admissible deformation modes used to determine ΔE for $\Delta\lambda \neq 0$, under the symmetry constraint discussed later. In the basic formulation, which can be extended, it is assumed that any constitutive relation used to explore (1) does not depend in any way on natural time t; it is also called time-independent or rate-independent interchangeably. A superscript ^{0} is used to distinguish a quantity in a fundamental, quasi-static and isothermal deformation process, i.e. the solution path whose stability is investigated, from the corresponding quantity in a perturbed process without a superscript. Here, a prefix Δ denotes an increment corresponding to a time increment Δt, and λ is an external control parameter that varies with t in a prescribed manner. A word of caution is needed at this point: in Hill's notation, a prefix Δ had a completely different meaning, referring to the difference of a pair of increments, and that convention was also adopted in Chapters 1, 2, 4 and 5. Therefore, proper interpretation of the related formulae is required when reading these chapters.

ΔE denotes the amount of energy which has to be supplied from external sources to the mechanical *system* consisting of the deformed body and the loading device in order to produce quasi-statically a *virtual* deformation increment, in general with the help of additional perturbing forces. If the mechanical system is extended by an ambient heat reservoir, then ΔE can be identified with a change in the *internal energy* E of the resulting compound thermodynamic system, provided that any rate-dependent dissipation contribution to it (e.g. viscous of thermal) is disregarded, see Chapter 2. This compound thermodynamic system {deforming body + loading device + heat reservoir} has an important property that it is *isolated* when $\lambda = \text{const}$, which gives inequality (1) a clear thermodynamic meaning.

If the fundamental solution path is known (often trivially as a uniform deformation path) then inequality (1) represents the postulated energy criterion for path stability. In the verbal formulation quoted from Chapter 5: 'In the criterion it is assumed that a sufficiently small deformation increment in a stable fundamental process requires less energy to be supplied from external sources to the system consisting of the body and loading device than any other kinematically admissible deformation increment'. In the special case $\Delta\lambda = 0$ we have $\Delta E^0 = 0$, and then inequality (1) reduces to the energy condition for stability of a single equilibrium state.

Inequality (1) can also be rewritten as

$$\Delta E \rightarrow \min \; (= \Delta E^0) \tag{1'}$$

under the same conditions as in (1). This is known as the postulate of *incremental energy minimization*, see Chapter 12, which is just another formulation of the solution property (1) introduced in Chapter 4. Either (1′) or (1) means that a fundamental solution path corresponding to an unknown ΔE^0 that satisfies (1) can be determined by minimizing ΔE. In other words, the energy criterion (1) and the incremental energy minimization (1′) are merely different ways for expressing the same concept. In numerical calculations, the asymptotic condition 'to second order in $\Delta t > 0$' can be omitted in order to determine an approximate incremental solution for a small but finite time step Δt.

The proposed energy approach, based on (1) or (1′), may be regarded as an extension of the classical energy minimization for elastic conservative systems to a class of incrementally nonlinear dissipative solids. The former was designed to eliminate unstable *equilibrium states*, whereas the latter is designed to eliminate unstable *de-*

formation paths. It thus becomes evident that the proposed energy approach brought added value to Hill's theory. First, if there is more than one incremental solution to a given problem – which has been a long-standing issue in rate-independent plasticity – then condition (1) provides a criterion of choice among them in the form of a minimum of ΔE, which was not present in bifurcation theory alone. Second, a variational method for determining the solution is offered that is based on minimization as in (1′), and not only on the stationarity property as in Hill's theory after the primary bifurcation.

Limitation: the incremental potential

There is a limitation similar to that required for the stationarity property just mentioned. In short, ΔE must act as an incremental potential. To make this more clear, a more detailed representation of energy functional ΔE is needed. According to Chapter 4, we have

$$\Delta E = \Delta W + \Delta\Omega. \tag{2}$$

Here, ΔW is the increment of work (evaluated up to second order in Δt) that has to be supplied to the deformed body in order to produce quasi-statically a *virtual* deformation increment, including the work expended by additional perturbing forces. In general, ΔW depends on the deformation path in the plastic range. $\Delta\Omega$ is the increment in potential energy Ω of the external loading device, which is assumed conservative, so that Ω depends on λ and a displacement field, but not on the displacement path in time. It is emphasized that an increment in $(-\Omega)$ is *not* equal to the work transmitted to the deformed body, unless done by given forces independent of λ.

The reduction of the incremental energy functional to the sum (2) of the increments of deformation work and external potential was the key to establishing the consistency of the proposed energy approach with Hill's theory of bifurcation and stability. Chronologically, this was first done in the article included as Chapter 4 and expanded in Chapter 5 by providing a series of hierarchical conditions necessary to satisfy (1). Along a sufficiently regular, kinematically admissible deformation path, the energy increment ΔE was developed into a Taylor series with respect to Δt,

$$\Delta E = \dot{E}\Delta t + \tfrac{1}{2}\ddot{E}(\Delta t)^2 \quad \text{to second order in } \Delta t, \tag{3}$$

where a superimposed dot means a forward rate with respect to t, taken here in a given initial state at some $\lambda = \lambda^0$ corresponding to $\Delta t = 0$. As a consequence of (2) and the virtual work principle in the equilibrium state at λ^0, the first-order term in (3) is independent of the virtual velocity field denoted by $\tilde{\mathbf{v}}$ and equal to a displacement rate field. Fulfilment of (1) therefore requires that the second term, proportional to $\ddot{E}$ and dependent smoothly on $\tilde{\mathbf{v}}$, be stationary with respect to $\tilde{\mathbf{v}}$ at a fundamental solution $\tilde{\mathbf{v}}^0$ in velocities.

In the classical associative plasticity framework, this requires no further constitutive assumption. However, for a broader class of elastic-plastic continua, where in a given state of the material the stress rate is a nonlinear function of the strain rate (as illustrated in Figure 4 of Chapter 1), a more general constitutive assumption is needed to accept (1). This is provided by the constitutive rate-potential introduced by Hill in 1959 for a wide range of time-independent materials,

$$\dot{\mathbf{S}} = \partial U/\partial \dot{\mathbf{F}}, \qquad U(\dot{\mathbf{F}}) = \tfrac{1}{2}\dot{\mathbf{S}}(\dot{\mathbf{F}})\cdot\dot{\mathbf{F}} \tag{4}$$

in the updated notation, where $\dot{\mathbf{S}}$ is the Piola stress rate equal to the transpose of the nominal stress rate, $\dot{\mathbf{F}}$ is the deformation gradient rate equal to velocity gradient in the reference configuration, the rate always means the one-sided material time derivative in a forward sense, and a central dot denotes full contraction. In the assumed absence of higher gradient terms or couple stresses, and without disturbing the constitutive relation itself (which will be allowed later), the potential $U(\dot{\mathbf{F}})$ plays a central role. Its existence implies that the fourth-order tangent stiffness moduli tensor (dependent on the direction of $\dot{\mathbf{F}} \neq \mathbf{0}$, not its magnitude) possesses a diagonal symmetry. The constitutive relationship (4) encompasses the incremental effects of changes in geometry of a material element, which can be of importance for predicting plastic instability regardless of whether the current strain is small or not.

Along a deformation path used to obtain (3), ΔW takes the form of a volume integral in the reference configuration, which consists of two parts:

$$\Delta W = \int (\mathbf{S}\cdot\Delta\mathbf{F} + U(\dot{\mathbf{F}})(\Delta t)^2)\, \mathrm{d}V \ \text{ to second order in } \Delta t. \tag{5}$$

In a given state of mechanical equilibrium, the first term above (where $\mathbf{S}$ is the current Piola stress at $\lambda = \lambda^0$ and the deformation gradient increment $\Delta\mathbf{F}$ is calculated with second-order accuracy) is again cancelled by the corresponding non-constant term of $\Delta\Omega$ in the virtual work principle. By substituting the second part of (5) into (2) and rearranging it using (4), it was shown that (1) or (1′) is consistent with the basic variational statements of Hill's theory. Moreover, it was also concluded from (1) that $\ddot{E}$ must be minimized not only when the condition sufficient for uniqueness holds, as in Hill's theory, but at every stage along a *stable path* of quasi-static deformation. Since not every solution corresponds to a minimum of $\ddot{E}$, this is the selection criterion for the post-bifurcation solution path under assumption (4), which was missing in the bifurcation theory alone, see Chapter 1.

Conditions for plastic instability

Hill's bifurcation and stability theory, which was based on the assumptions of conservative external loading and constitutive potential (4) for a time-independent material, was given an energy interpretation in Chapters 4 and 5, and was also supplemented with (1) as a criterion for path stability under the same assumptions. At that time, in 1985, the question remained open whether the negation of (1), i.e.

$$\Delta E < \Delta E^0 \tag{6}$$

for *some* virtual deformation mode, implies *instability* of the fundamental solution path in a physically meaningful sense under the same assumptions as in (1). Assuming the decomposition (3), with the first part independent of the velocity field, inequality (6) can be equivalently written as

$$\ddot{E} < \ddot{E}^0 \tag{6'}$$

for some kinematically admissible velocity field used to determine $\ddot{E}$ at $\Delta\lambda \neq 0$. A related open question was whether the fulfilment of the condition

$$\Delta E < 0 \quad \text{at} \quad \lambda = \text{const} \tag{7}$$

for some virtual deformation mode implies instability of equilibrium in a dynamic sense. These questions received generally positive answers in the early 1990s in the first approximation similar to that in linear stability analysis but retaining nonlinearity of the incremental constitutive law. The first approximation can be defined here by extending the nonlinear constitutive relationship (4) from the rates to small but finite increments, provided that the latter follow a straight segment in the displacement space. For instance, this means that elastic and elastoplastic stiffness moduli remain unaltered in a neighbourhood of the equilibrium state at λ^0.

In Chapter 6 the following conclusions were proven under the assumptions of conservative external loading and constitutive potential (4):

(i) If an equilibrium state of a spatially discretized system is such that (7) is satisfied, then it is unstable in the dynamic sense, in the first approximation.

(ii) If a fundamental quasi-static deformation process of a spatially discretized system is such that (6') is satisfied while the traversed equilibrium states are directionally stable in the sense of positive definiteness of $\ddot{E}$ for λ = const, then in every such state there exists the secondary incremental solution that minimizes ΔE to second order at $\Delta\lambda \neq 0$.

(iii) Under the same assumptions as in (ii) above but for a continuous system, the deformation process is unstable for persistent disturbances, in the sense of sensitivity of the incremental deformation to arbitrarily small perturbing forces.

In short, if (6) or (6′) is met for a spatially discretized system then

- either the equilibrium states traversed along the fundamental solution path associated with ΔE^0 are unstable (in a dynamic sense as in (i) above), or
- the fundamental path is unstable (in a quasi-static sense as in (ii) above), or
- the limit case of neutral stability of equilibrium is reached (when $\ddot{E}$ becomes just positive semi-definite for λ = const).

In this way, the status of incremental energy minimization (1) or (1′), introduced as a postulate under the same general assumptions as in Hill's theory of bifurcation and stability, was strengthened to the condition that is *necessary* for stability of the fundamental quasi-static solution path. Hence, (6) or (6') has received the status of a condition *sufficient for plastic instability*. The theoretical results obtained in relation to the energy criterion of plastic instability are summarized in Figure 7 of Chapter 1.

Consequences for computational procedures

The above results have consequences for computational procedures for determining a quasi-static solution path in rate-independent plasticity; a review of the state of the art as of 1995 regarding plastic instability can be found in Chapter 9. The common practice of solving plasticity problems has been to move forward incrementally without verifying whether (6) has been met or not. However, in many cases of finite plastic deformation it is quite probable that the solution path determined in this way will be physically meaningless due to its instability. A remedy can be clearly identified in light of the theory outlined above: that is, to eliminate the solution paths along which (6) or (6′) holds. It is not difficult to check the condition sufficient to state that (6) or (6′) has occurred in a numerical procedure for a spatially discretized problem. All that is needed is to check the sign of the smallest eigenvalue of the tangent stiffness matrix associated with the computed nonzero field of velocities. This matrix is symmetric on account of the assumptions of conservative loading and constitutive potential (4) for a time-independent material. If its smallest eigenvalue is negative, this implies the fulfilment of the condition (6′), which in turn is sufficient for instability of the fundamental path, see Chapters 1 and 6. In the range (ii) above, where non-uniqueness persists along a post-critical deformation path, a selection criterion becomes unavoidable. It is then logical to abandon the unstable continuation of the fundamental path in further calculations and instead to look for an alternative branching solution using (1′). The secondary solution, as an incremental energy minimizer, exists as long as the equilibrium states passed are directionally stable, and must correspond to a tangent stiffness moduli field different from that in the fundamental solution. This is the essence of the computational procedure for automatic branch switching proposed jointly with K. Thermann in 1992 in article [20]. In short, the path is followed by applying incremental energy minimization at each time step by abandoning the unstable incremental solutions that correspond to a saddle point of the energy increment.

This procedure is fully justified if, in order to enter the path instability range characterized by (6′), it is *necessary* for the smallest eigenvalue of the tangent stiffness matrix associated with the calculated velocity field to become negative. This converse implication holds for classical associative plasticity, but also for a class of incrementally nonlinear models illustrated in Figure 4 of Chapter 1. In article [17] it was proven that this is true if the following constitutive inequality is satisfied

$$\dot{\mathbf{S}}^0 \cdot \dot{\mathbf{F}} - \dot{\mathbf{S}} \cdot \dot{\mathbf{F}}^0 \geq 0 \quad \text{for all } \dot{\mathbf{F}}. \tag{8}$$

Inequality (8) was derived from micromechanical considerations and expresses the lack of abrupt unloading at the micro-level at a regular point along the fundamental path. Here, $\dot{\mathbf{S}}^0$ and $\dot{\mathbf{F}}^0$ are the rates taken along the fundamental path, and $\dot{\mathbf{S}}$ is constitutively related to $\dot{\mathbf{F}}$ through (4). Inequality (8) was also shown to hold for a class of so-called generalized standard materials [18]. The consequences of (8) are included in Figure 7 of Chapter 1.

Material instability

Of particular interest is the plastic instability of the material type, associated with deformation patterning and microstructure formation independent of the external loading conditions. It is referred to as material instability and can be investigated as follows. Within the plastically deformed body, consider a material element, say G, with a volume sufficiently small so that body forces may be considered negligible. Two distinct types of material instability are identified:

- Instability of the fundamental deformation process within the element if the displacements over the element boundary ∂G are unperturbed.
- Instability of equilibrium within the element if the displacements over the element boundary ∂G are zero.

The former case still corresponds to (6) and the latter case to (7), as the respective instability condition applicable to the material element under consideration. However, the effect of $\Delta\Omega$ is now eliminated due to fully kinematic constraints over the element boundary and negligible body forces. The conditions (6) and (7) specified for material instability can be written jointly as

$$\Delta W < \Delta W^0 \quad \text{subject to } \mathbf{v} = \mathbf{v}^0 \text{ over } \partial G \tag{9}$$

to second order in Δt for some virtual velocity field $\tilde{\mathbf{v}}$ used to determine ΔW. To investigate instability of equilibrium, we substitute $\mathbf{v}^0 = \mathbf{0}$ and $\Delta W^0 = 0$.

A systematic analysis of theoretical aspects of material instability is presented in Chapter 7 for a general class of incrementally nonlinear time-independent solids which admit a constitutive potential (4). Various specialized versions of condition (9), along with their mutual relationships, are summarized in diagram (7.21) of Chapter 7. Conditions for propagation of strain-rate discontinuities and formation of shear bands are also investigated. The example of the instability ranges is extended in Chapter 8, where the theoretical results concerning material instability are presented in a condensed form.

A synthesis of the energy-based theory of plastic instability of various types was provided in reference [26], presented in a sectional lecture at the IUTAM Congress in Kyoto in 1996. This encompassed novel aspects of material instability, such as a closed-form formula for the incipient volume fraction of the localization zone consisting of any number of parallel planar bands, and the analysis of the associated post-critical behaviour. Along with the relation to another approach based on initial imperfections, it complemented the theory of plastic instability that relies on the energy criterion (6) under the rate-independent constitutive relationship (4).

In Chapter 3, a broader view on the theory of material instability in incrementally nonlinear solids is presented under more general boundary conditions for a material element that those in (9) and using constitutive relations other than (4). The analysis includes the thermodynamic framework and multimode plasticity, which are discussed below. In addition to instability of uniform straining, the analysis covers instability with respect to internal structural rearrangements as well as instability under deformation-sensitive loading at the material element boundary. The analysis is preceded by considering an extremely simple, one-dimensional example which – perhaps surprisingly – can illustrate many aspects of the general theory of material instability.

Thermodynamic framework

The applicability of the incremental energy minimization (1) or (1′) is not limited to the constitutive law in the form (4), which relates the stress rate to the total strain rate, but can be extended to constitutive relations stated differently. Quite often the stress rate in a given state is expressed as a function of the strain rate and additional rate variables such as plastic multipliers or the rate of internal variables. In such cases, assuming isothermal rate-independent deformation as before, but now in the framework of classical thermodynamics, the incremental deformation work ΔW in (2) is split into two parts,

$$\Delta W = \Delta\Phi + \Delta\mathcal{D}, \tag{10}$$

where $\Delta\Phi$ is the increment of the Helmholtz free energy of the deformed body, and $\Delta\mathcal{D}$ is the dissipation increment which is a time integral of a global *rate-independent dissipation function*. Accordingly, an increment in the total internal energy E of the compound system {deforming body + loading device + heat reservoir} takes the form

$$\Delta E = \Delta\Phi + \Delta\mathcal{D} + \Delta\Omega. \tag{11}$$

Recall that the compound system is chosen so that it is isolated in the thermodynamic sense if λ = const.

The background to this decomposition is presented in Chapter 2. Note that the symbol Δ has a different meaning there when discussing stability of a quasi-static process than here, therefore it may be easier to read the related Chapter 12 first. The postulate of incremental energy minimization has the same form (1) or (1′) as before and with the same assumptions. However, its implementation becomes different:

- First, ΔE now depends not only on the incremental displacement or velocity field, but also on the fields of other incremental variables, such as plastic multipliers or the increment or rate of internal variables.
- Second, the minimized functional is no longer smooth because of unilateral constraints imposed on the rate of internal variables by the second law of thermodynamics.
- Third, the restriction that ΔE acts as an incremental potential is now imposed on $\Delta\mathcal{D}$ in the form of a symmetry constraint on the state derivative of the dissipation function. For further details, see Chapter 2 and Chapter 12.

Nevertheless, after introducing decomposition (10), the principle remains the same as before and under analogous assumptions. Under the symmetry constraint mentioned above, the minimization of a virtual energy increment (11) still represents a physically reasonable criterion for selecting an incremental solution corresponding to a stable quasi-static deformation path, as first proposed in Chapter 4. Furthermore, in the case of non-uniqueness, (1′) still represents a variational method to determine the solution, eliminating those that are less favourable energetically. Specific features of incremental energy minimization in gradient plasticity theory are discussed in [40].

Multimode plasticity

The advantage of representing the energy increment in the form (11) is its direct applicability to multimode plasticity. This may be accomplished with or without explicit reference to internal state variables, see Chapters 2 and 3. The characteristic feature of multimode plasticity is the existence of multiple yield surfaces (also called loading surfaces) in the stress or strain space, which are associated with different internal modes of plastic deformation; a typical example is crystal plasticity. The surfaces intersect at the current loading point, forming a corner or vertex. In contrast to the classical plasticity models, the direction of the plastic part of strain-rate is then not predetermined at the loading point, but depends on relative intensity of different plastic deformation modes, measured by the so-called plastic multipliers denoted here by $\dot{\gamma}_K$.

Increments $\Delta\Phi$ and $\Delta\mathcal{D}$ can be determined by time integration to second order of their rate expressions transformed to the following form

$$\dot{\Phi} = \int \mathbf{S}\cdot\dot{\mathbf{F}} - \pi_K\dot{\gamma}_K)\,\mathrm{d}V, \qquad \dot{\mathcal{D}} = \int \pi_K^{\mathrm{c}}\dot{\gamma}_K\,\mathrm{d}V, \tag{12}$$

where π_K and π_K^{c} are thermodynamic driving forces and their state-dependent critical values (thresholds), respectively, associated with multipliers $\dot{\gamma}_K$, and the summation convention is used for repeated subscripts. Upon substituting these expressions into (10), the rate of work supplied to deformed body in a perturbed deformation process becomes

$$\dot{W} = \int \mathbf{S}\cdot\dot{\mathbf{F}} - f_K\dot{\gamma}_K)\,\mathrm{d}V, \qquad f_K = \pi_K - \pi_K^{\mathrm{c}}\,. \tag{13}$$

This is a non-standard expression for the *virtual* work rate. The difference f_K between the actual and critical thermodynamic forces is treated as a yield function (also called loading function). To induce a *virtual* $\dot{\gamma}_K$, a nonzero value of f_K can be compensated by an additional set of perturbing forces *acting within* a material element and applied by a disturbing agency; this can be seen as disturbing the constitutive relation itself. As is usually the case, the product $f_K\dot{\gamma}_K$ vanishes along any solution path, and then (13) reduces to the standard formula. Further details can be found in Chapters 2 and 3.

In Chapter 10, it is shown how the structure of incremental constitutive relationships of multimode associative plasticity transmits from a mesoscopic level to a discretized macro-element. Then, by invoking (13) and under the symmetry constraint needed for ΔE to act as an incremental potential, which now takes the form $\partial\dot{f}_K/\partial\dot{\gamma}_L = \partial\dot{f}_L/\partial\dot{\gamma}_K$ for active modes of plastic deformation, the energy condition (1) for path stability is specified in the cases when either strain or stress is macroscopically controlled.

An important question is under which circumstances the values of plastic multipliers are uniquely determined by the current loading conditions at a material point.

General conditions for uniqueness of the material response in multimode plasticity (applicable also to other rate-independent inelastic phenomena and partial kinematic constraints, *without* invoking the above-mentioned symmetry constraint) are given in Chapter 11. As an application to predicting the onset of plastic instability, lower and upper bounds are established for the instant of primary bifurcation in a band along a smooth straining path. In contrast to the commonly used approach based on the condition of ellipticity loss for the tangent stiffness moduli, the analysis here is performed differently and takes into account any possible unloading, without relying on the linearized relationship between the stress and strain rates. Unlike the commonly used approach, this analysis also covers the onset of a laminate deformation pattern corresponding to an abrupt unloading.

A synthesis of the obtained results regarding material stability or instability in multimode plasticity is presented in Chapter 3.

Extension to non-potential incremental problems

The theory of plastic instability, which postulates incremental energy minimization in the form (1′) or (1), equivalently, was established in the 1980s and 1990s and is described in Chapters 1 to 12. It remains valid under the general assumption that the global energy increment ΔE (up to second order) acts as an incremental potential from which the governing incremental equations and inequalities can be retrieved. It was far from evident how the theory might be generalized to cases where this assumption was not valid. The author had spent several years searching the literature for a suitable existing remedy, but without success. No extremum principle for a general class of non-potential plasticity problems was found which would provide, in analogy to incremental energy minimization, both an exact solution to the rate problem and a criterion of choice between non-unique solutions. Finally, in 2019, a stability theory was developed that corresponds to this based on (1) but is applicable to non-potential problems, as presented in Chapter 13.

In order to achieve this goal within the thermodynamic framework for multimode plasticity, the following concepts were introduced simultaneously:

- The rates of the internal variables are related to the plastic multipliers by an arbitrary state-dependent matrix, thus encompassing non-associative plasticity even though the vector $(\dot{\gamma}_L)$ still points to the outward cone normal to the intersecting surfaces of f_K = const in π_L-space. The previous symmetry constraint on the interaction matrix $(\partial \dot{f}_K / \partial \dot{\gamma}_L)$ is no longer imposed, thus encompassing, for instance, the geometrically exact theory of crystal plasticity at finite deformation.
- A virtual energy increment (11) is considered on a special 'dog-leg' path consisting of two short straight segments of the overall deformation increment, which can be scaled down proportionally to its length as $\Delta\lambda \to 0$. The first segment corresponds to the thought solution, while the second segment represents a *sudden* perturbation of the first; this is a key idea that can also be justified physically. The energy increment is evaluated along this special path up to second order and compared to a direct incremental solution with the same $\Delta\lambda \neq 0$.

- The usual optimality principle (1′) is replaced by a novel 'quasi-extremal' principle, in which the objective function depends not only on the variables subject to variation, but also on an unknown optimal solution, which is introduced as a parameter.

As a consequence, the following incremental energy *quasi-minimization* is obtained, in which minimization (1′) is replaced by

$$\Delta x^0 = \arg\min_{\Delta x}(\Delta E(\Delta x) + \Delta x \cdot C \Delta x^0). \qquad (1'')$$

This is to be satisfied to second order in Δt for all kinematically admissible deformation modes Δx for $\Delta\lambda \neq 0$, including all nonnegative $\Delta\gamma_K$, *without* imposing any additional symmetry constraint as previously for (1′). Here, Δx^0 symbolizes a discretized solution, Δx any virtual increment as an alternative to Δx^0, $\Delta E(\Delta x)$ denotes ΔE evaluated on a perturbed *straight* path from 0 to Δx, and $C = (A - A^T)/2$ is a skew-symmetric part of the global matrix A analogous to the local matrix that defines the constitutive relation between $(\Delta \mathbf{S}, \Delta\gamma_K)$ and $(\Delta \mathbf{F}, \Delta\gamma_L)$. The only distinction between (1″) and (1′) is the last term; besides, (1″) can be used in an analogous manner to iteratively determine a solution Δx^0. Incremental energy minimization can be retrieved as a special case of (1″) if $A = A^T$ and $C = 0$. For a continuum problem, it suffices to define matrix A locally. Further details can be found in Chapter 13. For additional insight into numerical applications in crystal plasticity, see also recent references [38, 39].

It is crucial to note that Δx^0 determined from (1″) still tends in the limit as $\Delta\lambda \to 0$ to an exact solution of the given rate problem of non-potential type, which covers non-associative plasticity as well as nonsymmetric interactions between plastic deformation modes. Moreover, this solution exhibits the additional stability property: perturbing forces, if present, produce a non-negative scalar product with a deviation from Δx^0 they cause. Accordingly, if there are multiple solutions, which is common in rate-independent plasticity, then incremental energy quasi-minimization can provide a criterion of choice among them, analogously to the potential case. However, further work is needed to explore the consequences of failure of (1″), which are currently less clear than those identified for potential problems. In this context, see Chapter 11, article [28], and the book by Bigoni (2012) cited in Chapter 13.

It can be concluded, also by looking at the recent literature, that the theory of plastic instability represents a vital topic and is still evolving.

Warsaw, Poland
December 2024

Henryk Petryk

Chapter 1
Theory of Bifurcation and Instability in Time-Independent Plasticity

Abstract The theory is developed for constitutive rate equations which are arbitrarily nonlinear and can thus describe the important effect of formation of a yield-surface vertex. Basic elements of Hill's theory of bifurcation and stability in time-independent plastic solids are presented. In particular, the condition sufficient for uniqueness of a solution to the first-order rate boundary value problem, the stationary and minimum principles for velocities, the concept of a comparison solid and the primary bifurcation point are discussed. Distinction is emphasized between the conditions for uniqueness and for stability of equilibrium, and between the bifurcation point and an eigenstate. Several recent extensions of Hill's theory are next presented. It is shown how the property of the comparison solid required in the uniqueness criterion can be weakened and justified then on micromechanical grounds, without the need of complete specification of the macroscopic constitutive law. The question of non-uniqueness and instability in the post-critical range is examined, and respective theorems are formulated for discretized systems and for a certain class of continuous systems. The energy interpretation of the basic functionals and conditions in the bifurcation theory is given. Finally, a unified approach to various bifurcation and instability problems is presented which is based on the concept of instability of a deformation process and on the relevant energy criterion.

Introduction

These lecture notes are aimed at presenting in a systematic way certain general results obtained in the area of uniqueness and stability, or bifurcation and instability, of quasi-static solutions for time-independent inelastic solids. The word "general" does not mean, of course, that the theory is universal and can be applied to any problem. For instance, the assumptions adopted exclude problems involving fracture

[0] Reprinted from: CISM Courses and Lectures No. 327, Udine 1991 (Ed. Q.S.Nguyen), pp. 95-152, Springer-Verlag, Wien 1993. H.Petryk, Theory of bifurcation and instability in time-independent plasticity. Copyright © 1993 by CISM, Udine.

H. Petryk, *Theory of Plastic Instability*,
https://doi.org/10.1007/978-3-031-98223-1_1

or strong discontinuities in velocities, and most of the conclusions will be obtained for conservative loading and symmetric moduli only. The class of constitutive laws and problems under consideration is, however, still very broad, and the theory can be applied to a variety of phenomena as buckling, necking, snap-through, etc.

Constitutive framework is given in Sect. 1.1 and discussed further in Sect. 1.3, with the emphasis on the effect of changes in geometry and on incremental nonlinearity of the constitutive law. Fundamental results concerning uniqueness and bifurcation in velocities and stability of equilibrium, due to R.Hill, are presented in Sect. 1.2. Later sections complement Hill's theory and are mainly based on the present author's own work during last several years. The classical assumption of a smooth yield surface is nowadays known to be an oversimplification in bifurcation problems, and for practical applications it is of interest whether the bifurcation criterion may be linearized also at a yield-surface vertex whose formation is predicted by micromechanical studies; this problem is discussed in Sect. 1.3. Stability of equilibrium and incremental uniqueness on post-critical paths where Hill's sufficiency conditions no longer apply are examined in Sect. 1.4. Sect. 1.5 shows that the basic conditions from the bifurcation theory can be given an energy interpretation by expressing them in terms of the second time derivative of the total energy functional. Finally, in Sect. 1.6 a unified approach to bifurcation and stability problems is presented which is based on the concept of instability of a quasi-static deformation process. The concept itself is discussed first, and then the energy criterion of instability of a process is introduced. It is demonstrated how various critical points connected with bifurcation, loss of stability of equilibrium or material instabilities are generated by the energy criterion as particular cases of the onset of instability of a deformation process.

No attempt is made here to review the related vast literature devoted to particular problems or specific material models, or to compare the theory with experimental or numerical results. References to the literature are intended primarily to indicate where theoretical results quoted below were obtained, and not how they relate to other investigations. The contents of the lectures decide that many important contributions but not directly to the general theory described below are not mentioned.

1.1 Preliminaries

1.1.1 Notation. Deformation Gradient and Nominal Stress

Lower case Latin subscripts varying from 1 to 3 denote vector or tensor components on a fixed orthogonal triad of unit vectors $\mathbf{a}^1$, $\mathbf{a}^2$, $\mathbf{a}^3$, and the usual summation convention will be adopted for repeated indices. Boldface letters will symbolize vectors or tensors themselves which often will be identified with a collection of their components, just to simplify the terminology. A dot over a symbol of a quantity denotes the material time derivative understood as a *forward* rate, i.e. the right-hand derivative with respect to a scalar time-like parameter θ.

Two configurations of a continuous deformable material body are considered: the actual one is variable and is called *the current configuration* while the other is fixed and called *the reference configuration.* In many cases it is convenient to assume that these two configurations *momentarily* coincide. Denote by ξ_i and x_i the rectangular Cartesian coordinates of a material point in the reference and current configurations, respectively, on the common basis $\mathbf{a}^i$. Infinitesimal elements of volume and of surface area in the *reference* configuration are denoted by $d\xi$ and da, respectively. Displacement and velocity components are $u_i = x_i - \xi_i$ and $v_i \equiv \dot{x}_i \equiv \partial x_i/\partial\theta^+$, respectively. A spatial field will be distinguished from its value at some point by superimposing a tilde over the symbol.

The first weak variation (Gateaux differential) of a functional J at its argument value $\tilde{\mathbf{v}}$ in the direction $\tilde{\mathbf{w}}$ is denoted by

$$\delta J(\tilde{\mathbf{v}}, \tilde{\mathbf{w}}) \equiv \frac{d}{d\gamma} J(\tilde{\mathbf{v}} + \gamma\tilde{\mathbf{w}})|_{\gamma=0}\,, \tag{1.1}$$

where $\tilde{\mathbf{w}} = \delta\tilde{\mathbf{v}}$ stands for an admissible variation of $\tilde{\mathbf{v}}$ and γ is a real variable.

If the mapping $x_i(\xi_j)$ is differentiable then the formula $F_{ij} = \partial x_i/\partial\xi_j \equiv x_{i,j}$ defines *the deformation gradient*; henceforth the partial differentiation with respect to the rectangular Cartesian coordinates ξ_i (but *not* x_i) is denoted by the corresponding index preceded by a comma. The stress measure which is work-conjugate to F_{ij} is *the nominal stress* N_{ij} , that is,

$$N_{ij}\,\dot{F}_{ji} \;=\; \text{the rate of deformation work per unit reference volume}\,. \tag{1.2}$$

This may be regarded as a definition of N_{ij} ; equivalently, N_{ij} are obtained by decomposing the load vector dT_j (acting in the current configuration) on an infinitesimal surface element in the reference configuration, viz.

$$dT_j \;= n_i\,N_{ij}\,da, \tag{1.3}$$

where n_i and da denote the unit normal to the element and the element area, both taken in the reference configuration. Generally, the nominal stress is not symmetric. It is connected with the Cauchy stress $\boldsymbol{\sigma}$ and the Kirchhoff stress $\boldsymbol{\tau}$ by the formula

$$F_{ji}\,N_{ik} \;= \det(F_{pq})\,\sigma_{jk} \;= \tau_{jk}\,; \tag{1.4}$$

the transpose of $\mathbf{N}$ is also called the first Piola-Kirchhoff stress. Of course, if the reference and current configurations coincide then $F_{ij} = \delta_{ij}$ (δ_{ij} is the Kronecker symbol) and N_{ij} reduces to σ_{ij} .

The moment balance at the assumed absence of body moments requires the Cauchy stress to be symmetric, which is equivalent to

$$F_{ji}\,N_{ik} \;= F_{ki}\,N_{ij}\,. \tag{1.5}$$

The deformation gradient and nominal stress as particular measures of deformation and stress are directly connected with typical experimental measurements.

Consider in the reference configuration a unit cube of the edges defined by the triad of base vectors $\mathbf{a}^i$ (Fig. 1.1). The cube is thought as an element of a specimen in an initial state. After a (macroscopically) *uniform* deformation the cube becomes a parallelepiped with the edges $\mathbf{b}^i$ such that $(\mathbf{b}^i)_j = F_{jk}(\mathbf{a}^i)_k = F_{ji}$; the former equality follows from the definition of F_{ij} and the latter from $(\mathbf{a}^i)_k = \delta_{ik}$. It becomes evident that measurements of the components of $\mathbf{b}^i$ provide a direct way to determine the deformation gradient in an experiment.

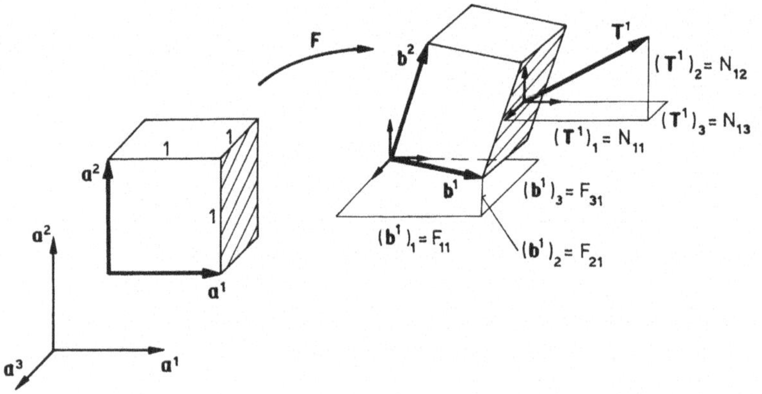

Fig. 1.1 Deformation gradient and nominal stress

The surface tractions are assumed to be (macroscopically) uniformly distributed on each of the parallelepiped faces. The resultant load vector on the parallelepiped face in the current configuration which in the *reference* configuration had the outward normal $\mathbf{a}^i$ is denoted by $\mathbf{T}^i$. In agreement with (1.3), the nominal stress N_{ij} can be directly defined as the j-th component of the load vector $\mathbf{T}^i$ (Fig. 1.1); the components can be measured experimentally. Thus

$$F_{ji} = (\mathbf{b}^i)_j \quad N_{ij} = (\mathbf{T}^i)_j \tag{1.6}$$

where $\mathbf{b}^i$ and $\mathbf{T}^i$ have a clear mechanical interpretation.

Remark Since the basis $\mathbf{a}^i$ has been assumed fixed and common for all configurations, it is not necessary here to distinguish between components connected with the reference or current configurations. However, it should be remembered that both the deformation gradient and nominal stress are in fact based partly on each of the two configurations (they are so-called two-point tensors). This becomes essential when the question of objectivity (or frame-indifference) is examined which, however, is not discussed here.

1.1.2 Incremental Effect of Changes in Geometry

The theory where changes in geometry of a deformed material body are disregarded is customarily called the small strain theory. Unfortunately, the terminology may suggest that the converse implication is valid, that is, that all changes in geometry may be disregarded provided only that the strains are known to be small. That supposition is evidently incorrect in a study of buckling of slender structures but might seem to be justified for solid bodies where the rotations are of order of the strains. However, even then the supposition is generally incorrect if the bifurcation and stability phenomena are examined, as R.Hill clearly demonstrated in the cycle of papers in 1957-1959 [1, 2, 3, 4, 5]. In the analysis of such phenomena the *incremental* effect of changes in geometry has to be taken into account. This can be done by using various variables but Hill has shown that the resulting formulae have particularly concise form when expressed in terms of the rates of the deformation gradient and nominal stress, $\dot{F}_{ij}$ and $\dot{N}_{ij}$. The rate $\dot{F}_{ij}$ of the deformation gradient is, of course, nothing else than the velocity gradient $v_{i,j}$ evaluated in the *reference* configuration. The Eulerian strain rate $\dot{\varepsilon}_{ij} = \frac{1}{2}(\partial v_i/\partial x_j + \partial v_j/\partial x_i)$ and material spin $\omega_{ij} = \frac{1}{2}(\partial v_i/\partial x_j - \partial v_j/\partial x_i)$ can be expressed in terms of $\dot{F}_{ij}$ on substituting $\partial v_i/\partial x_j = \dot{F}_{ik}F^{-1}_{kj}$, where $\mathbf{F}^{-1}$ denotes the inverse of $\mathbf{F}$.

To illustrate the incremental effect of change in geometry, choose the reference configuration to coincide momentarily with the current configuration so that $x_i(\boldsymbol{\xi}) = \xi_i$, $F_{ij} = \delta_{ij}$ and the current strains and rotations are exactly zero. In that special case $\dot{F}_{ij}$ coincides with the spatial velocity gradient $\partial v_i/\partial x_j$. Consider the cube discussed above under a current stress $N_{ij} = \sigma_{ij}$ and suppose that $v_{i,j}$ and constitutively related $\dot{\sigma}_{ij}$ are known. From (1.4) it follows that in the reference configuration we have

$$\dot{N}_{ij} = \dot{\sigma}_{ij} + v_{k,k}\,\sigma_{ij} - v_{i,k}\,\sigma_{kj}\,. \tag{1.7}$$

From (1.6) we obtain that the components $(\dot{\mathbf{T}}^i)_j$ of the rate of the load vector $\mathbf{T}^i$ on the cube face are given by the right-hand expression in (1.7); in the small strain theory the last two terms would be absent. This does not necessarily mean that it is these two terms which represent precisely the effect of changes in geometry. For, $\dot{\sigma}_{ij}$ is influenced by the material spin and can thus hardly be regarded as a proper generalization of the stress-rate from the small strain theory which is related constitutively to $\dot{\varepsilon}_{ij}$ and not to ω_{ij} . Rather, an objective stress flux should be regarded as such a generalization but then a problem arises which (if any) of the many proposals known from the literature is the most appropriate. Leaving this problem aside, consider one of the most popular, the Zaremba–Jaumann flux of Kirchhoff stress, defined by

$$\mathring{\tau}_{ij} = \dot{\tau}_{ij} + \tau_{ik}\,\omega_{kj} - \omega_{ik}\,\tau_{kj}\,. \tag{1.8}$$

Substitution of (1.8) into (1.7) at $F_{ij} = \delta_{ij}$ yields

$$\dot{N}_{ij} = \mathring{\tau}_{ij} - \dot{\varepsilon}_{ik}\,\sigma_{kj} - \sigma_{ik}\,\omega_{kj}\,. \tag{1.9}$$

If the relationship $\overset{\circ}{\tau}_{ij}(\dot{\varepsilon})$ is taken of the same form as the incremental constitutive relationship from the small strain theory then the last two terms in (1.9) (multiplied by $d\theta$) represent the effect of changes in geometry on the increment $d\mathbf{T}^i$ of the load vector associated with a given deformation increment. It is clear that if stress-rates are of order of strain-rate × stress, that is, if the instantaneous stiffness moduli are of order of the stresses, then that incremental geometric effect is *in general* not negligible. It is recalled that one arrives at that conclusion despite the fact that instantaneous strains and rotations are exactly zero.

Exercise: Examine the effect of changes in geometry on the load increment in pure tension and simple shear.

1.1.3 Constitutive Framework for Time-Independent Inelastic Solids

A general assumption is made that mechanical properties of the material in isothermal deformation do not depend in any way on a natural time. Under restriction to piecewise smooth deformation paths, it is also assumed that the stresses vary in time in a continuous and piecewise smooth manner. It must be noted, however, that that natural assumption excludes rigid-plastic solids. The constitutive relationship between the (right-hand) rates of stress and strain, no matter in which variables it is originally formulated to ensure objectivity of the material description, can be expressed as

$$\dot{N}_{ij} = \dot{N}_{ij}(\dot{\mathbf{F}}, \mathcal{H}), \tag{1.10}$$

where the parameter $\mathcal{H}$ symbolizes the influence of the deformation history prior to the current instant, including the current stress and deformation. For instance, $\mathcal{H}$ may be taken as a set $\{F_{ij}\ , \sigma_{ij}\ , \alpha_K\}$, where α_K are internal state variables; arbitrary "hardening" or "softening" characteristics are allowed. If the current state of the material is regarded as known then the symbolic parameter $\mathcal{H}$ will be omitted for simplicity. At a given state, the function $\dot{N}_{ij}(\dot{\mathbf{F}})$ is assumed to be single-valued, continuous and piecewise smooth (with a possible singularity at $\dot{F}_{ij} = 0$). The following function

$$U(\dot{\mathbf{F}}, \mathcal{H}) = \frac{1}{2}\dot{N}_{ij}(\dot{\mathbf{F}}, \mathcal{H})\, \dot{F}_{ji} \tag{1.11}$$

will play a fundamental role in the theory of bifurcation and instability.

For the material to be time-independent, the dependence of $\dot{N}_{ij}$ on $\dot{F}_{kl}$ must be positively homogeneous of degree one, i.e.

$$\dot{N}_{ij}(\gamma\dot{\mathbf{F}}, \mathcal{H}) = \gamma\dot{N}_{ij}(\dot{\mathbf{F}}, \mathcal{H}) \ \text{ for every } \gamma > 0 \tag{1.12}$$

but otherwise may be arbitrarily nonlinear (piecewise linear in particular cases). Consequently, $U(\dot{\mathbf{F}})$ is homogeneous of degree two. By the Euler theorem on homogeneous functions, at every point of differentiability of $\dot{\mathbf{N}}(\dot{\mathbf{F}})$ the equations (1.10) can be written down in the form

$$\dot{N}_{ij} = C_{ijkl}(\dot{\mathbf{F}}, \mathcal{H})\, \dot{F}_{lk}\,, \quad C_{ijkl} \equiv \partial \dot{N}_{ij} / \partial \dot{F}_{lk}\,, \tag{1.13}$$

where, at a given state, C_{ijkl} depend only on the direction of strain rate; the dependence of C_{ijkl} on $\dot{\mathbf{F}}$ may be discontinuous and nonlinear. Later, of primary importance will be *the tangent moduli* $C_{ijkl}(\dot{\mathbf{F}}^0, \mathcal{H})$ evaluated at the actual velocity gradient $\dot{\mathbf{F}}^0$ directed tangentially to a given deformation path.

Suppose that at a certain given state the incremental constitutive relationship is specified in the form

$$\mathring{\tau}_{ij} = L_{ijkl}(\dot{\boldsymbol{\varepsilon}})\, \dot{\varepsilon}_{kl} \tag{1.14}$$

e.g. as an adaptation of a constitutive relationship from the small strain theory. We assume that $L_{ijkl} = \partial \mathring{\tau}_{ij} / \partial \dot{\varepsilon}_{kl}$ and $L_{ijkl} = L_{ijlk}$. Then the moduli $C_{ijkl}(\dot{\mathbf{F}})$ can be found from the following relationship

$$\det(F^{-1}_{rs})\, F_{ip}\, F_{kq}\, C_{pjql}(\dot{\mathbf{F}}) = L_{ijkl}(\dot{\boldsymbol{\varepsilon}}) - \frac{1}{2}(\sigma_{jk}\, \delta_{il} - \sigma_{ik}\, \delta_{jl} + \sigma_{il}\, \delta_{jk} + \sigma_{jl}\, \delta_{ik})\,, \tag{1.15}$$

where $\boldsymbol{\sigma}$ and $\mathbf{F}$ are the current Cauchy stress and deformation gradient, and $\dot{\varepsilon}_{ij} = \frac{1}{2}(\dot{F}_{ik} F^{-1}_{kj} + \dot{F}_{jk} F^{-1}_{ki})$. The derivation of (1.15) is omitted here; all needed transformations are given in [6]. The stress-dependent right-hand terms in (1.15) are associated with the *incremental* effect of changes in geometry discussed in the preceding subsection, while the left-hand multiplicative term dependent on the deformation gradient represents the influence of a *finite* deformation between the reference and current configurations.

Following Hill [5, 6], we shall later introduce an additional assumption that the constitutive rate equations (1.10) admit a potential, viz.

$$\dot{N}_{ij} = \frac{\partial U}{\partial \dot{F}_{ji}}\,, \qquad C_{ijkl} = \frac{\partial^2 U}{\partial \dot{F}_{ji} \partial \dot{F}_{lk}}\,, \tag{1.16}$$

where U necessarily satisfies (1.11) on account of the homogeneity of the constitutive equations and is continuously differentiable by the assumed continuity of $\dot{N}_{ij}(\dot{\mathbf{F}})$. Of course, (1.16) implies that

$$C_{ijkl} = C_{klij} \tag{1.17}$$

whenever C_{ijkl} are well defined, and conversely, differentiation of (1.11) by $\dot{F}_{ji}$ shows that (1.17) implies (1.16). Existence of the potential and diagonal symmetry of the moduli are preserved when (1.2) is expressed in terms of rates of any *work-conjugate* pair of stress and strain measures [6]. In particular, we have (1.17) if and only if the moduli in (1.14) satisfy $L_{ijkl} = L_{klij}$. For elastic materials, this symmetry property is a consequence of existence of a strain energy potential.

Example: Classical elastic-plastic solids

For the classical models of incremental elastic-plastic response, the moduli in the constitutive rate equations (1.14) have the following form (cf. [7])

$$L_{ijkl} = \begin{cases} L^{\mathrm{p}}_{ijkl} & \text{if} \quad \lambda_{ij}\dot{\varepsilon}_{ij} > 0 \quad \text{(loading)} \\ L^{\mathrm{e}}_{ijkl} & \text{if} \quad \lambda_{ij}\dot{\varepsilon}_{ij} < 0 \quad \text{(unloading)} \end{cases} \tag{1.18}$$

with

$$L^{\mathrm{p}}_{ijkl} = L^{\mathrm{e}}_{ijkl} - g^{-1}\lambda^{\mathrm{p}}_{ij}\lambda_{kl} \,. \tag{1.19}$$

λ_{ij} is the (outward) normal to the current smooth yield surface in the *strain* space, g is a positive scalar parameter, and $\lambda^{\mathrm{p}}_{ij}$ defines the direction of the so-called plastic parts $\dot{\varepsilon}^{\mathrm{p}}_{ij}$ and $\overset{\circ}{\tau}{}^{\mathrm{p}}_{ij}$ of the strain-rate and stress-rate when $\lambda_{ij}\,\dot{\varepsilon}_{ij} > 0$, viz.

$$\overset{\circ}{\tau}{}^{\mathrm{p}}_{ij} = \overset{\circ}{\tau}_{ij} - L^{\mathrm{e}}_{ijkl}\dot{\varepsilon}_{kl} = (-g^{-1}\lambda_{kl}\dot{\varepsilon}_{kl})\lambda^{\mathrm{p}}_{ij} \,, \tag{1.20}$$

$$\dot{\varepsilon}^{\mathrm{p}}_{ij} = \dot{\varepsilon}_{ij} - M^{\mathrm{e}}_{ijkl}\overset{\circ}{\tau}_{kl} = (g^{-1}\,\lambda_{pq}\dot{\varepsilon}_{pq})M^{\mathrm{e}}_{ijkl}\lambda^{\mathrm{p}}_{kl} \tag{1.21}$$

and L^{e}_{ijkl} and M^{e}_{ijkl} are elastic moduli and compliancies, respectively, with the assumed symmetry $L^{\mathrm{e}}_{ijkl} = L^{\mathrm{e}}_{klij}$. It can be seen that $L^{\mathrm{p}}_{ijkl} = L^{\mathrm{p}}_{klij}$ if and only if $\lambda^{\mathrm{p}}_{ij}$ has the direction of λ_{ij} , that is, if the normality flow rule holds. Consequently, the normality flow rule is here necessary and sufficient for existence of the constitutive potential (1.16). Note that no distinction has been made between "hardening" and "softening" materials which themselves are not measure-invariant concepts [7].

Example: Elastic-plastic materials with a discrete set of internal plastic deformation mechanisms

Consider constitutive rate equations for crystals deformed plastically by multislip, or more generally, for solids with a finite number N of internal mechanisms of rate-independent inelastic deformation. The constitutive framework for such materials at finite strain under the assumption of the "normality structure" was given by Hill and Rice [8] and Sewell [9], generalizing earlier theories of Koiter [10], Mandel [11] and Hill [12]. It is assumed that

$$\dot{\gamma}_K \geq 0\,, \quad f_K \leq 0\,, \quad f_K\,\dot{\gamma}_K = 0 \quad \text{(no sum)}\,, \tag{1.22}$$

$$\lambda_{Kij} = \partial f_K / \partial \varepsilon_{ij} \,, \tag{1.23}$$

$$\overset{\circ}{\tau}_{ij} = L^{\mathrm{e}}_{ijkl}\,\dot{\varepsilon}_{kl} - \lambda_{Kkl}\,\dot{\gamma}_K \,, \tag{1.24}$$

$$\dot{f}_K = \lambda_{Kkl}\,\dot{\varepsilon}_{kl} - g_{KL}\,\dot{\gamma}_L \,. \tag{1.25}$$

Upper case lower indices varying from 1 to N refer to quantities related to a specific plastic deformation mechanism; the summation convention is adopted here for those indices except when an indication "no sum" appears. $\dot{\gamma}_K$ denotes a scalar measure of the rate of activity of the K-th mechanism and $f_K(\varepsilon_{ij})$ is the respective smooth yield function in strain space, functionally dependent on the deformation history, where ε_{ij} is a strain measure based on the current configuration as reference. For instance, in crystals $\dot{\gamma}_K$ may stand for the rate of shearing on K-th slip system, with $f_K = \tau_K - \tau^{\mathrm{c}}_K$, where τ_K is the *generalized* resolved shear stress on that system and τ^{c}_K is its critical value. If $f_K = 0$ then (1.22) implies $\dot{f}_K \leq 0$ and $\dot{f}_K\,\dot{\gamma}_K = 0$ (no sum) while λ_{Kij} defined by (1.23) represents a normal to the K-th yield surface in strain space, directed outward from the elastic domain. To ensure that $\tau_{ij}(\dot{\varepsilon}_{kl})$ is a single-valued function, one assumes that [8, 9]

$$(g_{KL}) \quad \text{is positive definite}\,, \tag{1.26}$$

where (g_{KL}) is the matrix for *potentially active* mechanisms. Then one easily shows that the constitutive rate equations can be written down in the potential form (1.16) provided that

$$L^{\mathrm{e}}_{ijkl} = L^{\mathrm{e}}_{klij}\,, \qquad g_{KL} = g_{LK}\,. \tag{1.27}$$

In fact, this follows from the explicit expression for the moduli

$$L_{ijkl}(\dot{\boldsymbol{\varepsilon}}) = L^{\mathrm{e}}_{ijkl} - \bar{g}^{-1}_{KL}(\dot{\boldsymbol{\varepsilon}})\,\lambda_{Kij}\,\lambda_{Lkl} \tag{1.28}$$

where the submatrix $\bar{g}_{KL}(\dot{\boldsymbol{\varepsilon}}, \mathcal{H})$ of $g_{KL}(\mathcal{H})$ corresponds only to the mechanisms which are active for that strain-rate $\dot{\boldsymbol{\varepsilon}}$, and $\bar{g}^{-1}_{KL}$ is its inversion.

For N = 1 we recover the constitutive rate equations for the classical elastic-plastic solids obeying the normality flow rule.

Remark The relationship between the first-order rates of stress and strain is nonlinear but homogeneous. This is not so for higher-order rates. Let $\mathcal{H}$ be identified with $\{F_{ij}\,, \sigma_{ij}\,, \alpha_K\}$ and suppose that the constitutive function $\dot{N}_{ij} = \dot{N}_{ij}(\dot{F}_{kl}\,, F_{kl}\,, \sigma_{kl}\,, \alpha_K)$ is "sufficiently smooth" in its arguments. Then straightforward time differentiation of $\dot{N}_{ij}$ followed by substitution of (1.13) yields

$$\ddot{N}_{ij} = C_{ijkl}(\dot{\mathbf{F}}, \mathcal{H})\ddot{F}_{lk} + \frac{\partial \dot{N}_{ij}}{\partial F_{kl}}(\dot{\mathbf{F}}, \mathcal{H})\dot{F}_{kl} + \frac{\partial \dot{N}_{ij}}{\partial \sigma_{kl}}(\dot{\mathbf{F}}, \mathcal{H})\dot{\sigma}_{kl} + \frac{\partial \dot{N}_{ij}}{\partial \alpha_K}(\dot{\mathbf{F}}, \mathcal{H})\dot{\alpha}_K\,. \tag{1.29}$$

In a time-independent material, $\dot{\alpha}_K$ is a homogeneous function of degree one of the strain-rate but we may assume that $\dot{\alpha}_K$ is independent of higher-order rates of the deformation gradient. It follows that all right-hand terms in (1.29), except the leading one, are independent of $\ddot{F}_{lk}$. Now, let a difference between the quantities corresponding to two different continuations of the deformation process be denoted by a prefix Δ. Then from (1.29) we obtain

$$\Delta\ddot{N}_{ij} = C_{ijkl}(\dot{\mathbf{F}}, \mathcal{H})\,\Delta\ddot{F}_{lk} \quad \text{if } \Delta\dot{F}_{ij} = 0\,. \tag{1.30}$$

That procedure could formally be repeated to draw a conclusion that the constitutive relationship between the *differences* of the stress- and deformation-rates of any order is linear with the same coefficients defined as the actual moduli (1.13) provided that there is no difference in all lower-order rates. This observation [13] is essential for studying *regular* higher-order bifurcations. However, the underlying regularity assumptions should always be carefully verified for the problem at hand since they need not always be satisfied. For instance, (1.30) is invalid in the case of so-called "neutral" loading in the classical elastic-plastic solids [14].

1.1.4 Incremental Boundary Value Problem

We are concerned with isothermal deformations of a time-independent solid body subject to a quasi-static loading program. The body may be inhomogeneous but any material property is always assumed to vary at least piecewise smoothly with place. Suppose that the body occupied in a given reference configuration a spatial domain V bounded by a piecewise smooth surface S. Let $\lambda = \lambda(\theta)$ denote a scalar loading parameter which varies continuously in time (infinitely slowly with respect to a natural time); it will be convenient to keep the parameters θ and λ as being distinct. An initial equilibrium state of the body at a certain value of λ is regarded as known. We shall assume that on a nonzero part S_u of S the displacements $u_j = \bar{u}_j(\boldsymbol{\xi}, \lambda)$ are controlled so that rigid-body motions at fixed λ are kinematically excluded. The part S_u might be taken different for different components in order to include the mixed boundary conditions but is regarded as known and independent of λ.

The standard boundary value problem is posed by assuming that on a part S_T of S, complementary to S_u, the nominal surface tractions $T_j = \overline{T}_j(\boldsymbol{\xi}, \lambda)$ (per unit reference area) are controlled, while in V the nominal body forces $b_j = \bar{b}_j(\boldsymbol{\xi}, \lambda)$ (per unit reference volume) are controlled; note that λ need not be a load multiplier. The loading functions $\bar{u}_j, \overline{T}_j, \bar{b}_j$ are assumed to be at least piecewise smooth with respect to their arguments and to exhibit no strong discontinuities other than across material surfaces or lines. In a more general case, T_j and b_j need not be controlled directly but may depend on the actual deformation process.

Suppose that a theoretical quasi-static response of the body to the loading program is known, at least in principle, up to a current equilibrium state of the body at a certain value of λ. A convenient starting point to formulate the incremental boundary value problem is the principle of virtual power, which on account of (1.2) and of the *nominal* character of T_j and b_j can be written down in the form

$$\int_V N_{ij} w_{j,i}\, d\xi = \int_V b_j w_j\, d\xi + \int_{S_T} T_j w_j\, da \quad \text{for every } \tilde{\mathbf{w}} \in \mathcal{W}. \tag{1.31}$$

$\mathcal{W}$ denotes the linear space of admissible variations $\delta\tilde{\mathbf{v}} \equiv \tilde{\mathbf{w}}$ of velocities at a fixed value of the loading parameter λ; the fields from $\mathcal{W}$ are defined in the reference configuration and vanish on the part S_u of the body surface S where displacements and velocities have prescribed values. Different regularity restrictions could be imposed here; if not stated otherwise, we shall assume that $\mathcal{W}$ consists of all continuous and piecewise continuously twice differentiable vector fields on $\overline{V} = V \cup S$ which vanish on S_u . Since S_u has been assumed to be independent of λ, $\mathcal{W}$ is also independent of λ. It is well known that fulfillment of the principle of virtual power in the form (1.31) is equivalent to fulfillment of the standard conditions of equilibrium in the current configuration.

Consider now the first-order rate boundary value problem, formulated at the current equilibrium state. Admissible velocity fields $\tilde{\mathbf{v}}$ must take on S_u the prescribed values $\bar{v}_j = (\partial \bar{u}_j / \partial \lambda)\, \dot{\lambda}$; moreover, $\tilde{\mathbf{v}}$ are assumed to be continuous and piecewise continuously twice differentiable. Let $\tilde{\mathbf{v}}^k$ be such a fixed field; the class $\mathcal{V}$ of kinematically admissible velocity fields is defined by

$$\mathcal{V} = \{\tilde{\mathbf{v}} : \tilde{\mathbf{v}} = \tilde{\mathbf{v}}^k + \tilde{\mathbf{w}}\,,\ \mathbf{v}^k = \bar{\mathbf{v}} \text{ on } S_u,\ \tilde{\mathbf{w}} \in \mathcal{W}\}\,. \tag{1.32}$$

From a mathematical point of view it would be natural to take $\tilde{v}_j$ from a wider space, known as the Sobolev space $H^1(V)$, but we prefer to deal with solutions in the classical sense. Note that any functional defined on $\mathcal{V}$ can equivalently be regarded as a certain other functional defined on the *linear* space $\mathcal{W}$.

Suppose that there are no moving strong discontinuities in **N**, **b** and **T**. Since the integration domains V and S_T are fixed, differentiation of (1.31) with respect to a time-like parameter θ at *fixed* w_j yields

$$\int_V \dot{N}_{ij} w_{j,i}\, d\xi = \int_V \dot{b}_j w_j\, d\xi + \int_{S_T} \dot{T}_j w_j\, da \quad \text{for every } \tilde{\mathbf{w}} \in \mathcal{W}\,. \tag{1.33}$$

Assuming that inertia forces are neglected, a solution to the first-order rate boundary value problem, or simpler: *a solution in velocities*, or *a first-order solution*, is defined as a velocity field $\tilde{\mathbf{v}} \in \mathcal{V}$ such that the stress-rate field related constitutively to $\tilde{\mathbf{v}}$ through (1.10) satisfies the variational equality (1.33). By the standard transformation with the help of Green's theorem, (1.33) is written equivalently as

$$\begin{aligned} \int_V (\dot{N}_{ij,i} + \dot{b}_j) w_j\, d\xi &+ \int_{S_D} [\dot{N}_{ij}] n_i w_j\, da \\ &+ \int_{S_T} (\dot{T}_j - \dot{N}_{ij} \nu_i) w_j\, da = 0 \quad \text{for every } \tilde{\mathbf{w}} \in \mathcal{W}\,, \end{aligned} \tag{1.34}$$

where ν_i denotes the unit outward normal to the body surface, S_D stands for the internal surface(s) of discontinuity in the stress-rate, n_i is the unit normal to S_D , and [ ] denotes a jump across S_D with the former term of the difference taken from the side of S_D into which n_i is pointed. Hence, under the assumed regularity restrictions, fulfillment of (1.33) is equivalent to fulfillment of the set of conditions of continuing equilibrium in the form

$$\dot{N}_{ij,i} + \dot{b}_j = 0 \quad \text{in } V \backslash S_D\,, \quad [\dot{N}_{ij}] n_i = 0 \quad \text{on } S_D\,, \quad \dot{N}_{ij} \nu_i = \dot{T}_j \quad \text{on } S_T\,. \tag{1.35}$$

Higher-order rate problems can in principle be formulated in an analogous way by successive differentiation of (1.33) with respect to time provided the problem is assumed *a priori* to be "sufficiently regular". Also "regular" moving surfaces of discontinuity in stress-rates can be allowed for by using the differentiation rules for integrals defined on varying domains (e.g. [15]) and the respective compatibility conditions on the interfaces (e.g. [16]). We shall not trouble to write down explicitly the resulting formulae. However, it must be remarked that such a formal analysis is much less satisfactory than in the case of the first-order problem since higher-order rate fields may be discontinuous and have other singularities absent in velocity fields.

1.2 Elements of Hill's Theory of Bifurcation and Stability

1.2.1 Condition for Uniqueness of First-Order Solution

The question of uniqueness of a response of the material body to the quasi-static loading program is at the first instance reduced to the problem of uniqueness of a solution in velocities. As a starting point we can adopt the rate form (1.33) of the principle of virtual power. On substituting (1.32) we can rewrite (1.33) equivalently as

$$\int_V \dot{N}_{ij}(\Delta v_j)_{,i}\, d\xi = \int_V \dot{b}_j \Delta v_j\, d\xi + \int_{S_T} \dot{T}_j \Delta v_j\, da \quad \text{for every } \tilde{v}_j^* \in \mathcal{V},\ \Delta v_j = v_j^* - v_j\,. \tag{1.36}$$

If $\tilde{v}_j^*$ is another first-order solution then the equality in (1.33) must hold also when the fields $\tilde{v}_j$ and $\tilde{v}_j^*$ are interchanged, that is, if $\dot{N}_{ij}$, $\dot{b}_j$ and $\dot{T}_j$ are replaced by $\dot{N}_{ij}^*$, $\dot{b}_j^*$ and $\dot{T}_j^*$ which correspond to the velocity field $\tilde{v}_j^*$. By subtracting these two equalities, we obtain the identity

$$\int_V \Delta\dot{N}_{ij}(\Delta v_j)_{,i}\, d\xi = \int_V \Delta\dot{b}_j \Delta v_j\, d\xi + \int_{S_T} \Delta\dot{T}_j \Delta v_j\, da \tag{1.37}$$

valid for a pair of first-order *solutions*, where the differences of corresponding quantities are denoted by the prefix Δ.

Hence [5, 17], *uniqueness of a first-order solution is ensured* (i.e. bifurcation in velocities is excluded) *when*

$$\int_V \Delta\dot{N}_{ij}(\Delta v_j)_{,i}\, d\xi > \int_V \Delta\dot{b}_j \Delta v_j\, d\xi + \int_{S_T} \Delta\dot{T}_j \Delta v_j\, da \tag{1.38}$$

for every pair of kinematically admissible and distinct velocity fields from $\mathcal{V}$. This is the most general form of the uniqueness criterion for velocities; in particular cases it can further be specified. It may be remarked that (1.38) ensures uniqueness also in a wider class of generalized solutions to the first-order rate boundary value problem (in the sense of (1.33)) not necessarily belonging to $\mathcal{V}$. An alternative general formulation of the uniqueness criterion for solids obeying the maximum dissipation principle has been given by Nguyen [18] but it is not discussed here.

Further discussion will for simplicity be limited to the standard boundary value problem, i.e. to controlled nominal loads; an extension to configuration-dependent *conservative* loading will be indicated later. In the standard case, $\dot{b}_j = \dot{\bar{b}}_j = (\partial\bar{b}_j/\partial\lambda)\dot{\lambda}$ in V and $\dot{T}_j = \dot{\bar{T}}_j = (\partial\bar{T}_j/\partial\lambda)\dot{\lambda}$ on S_T are prescribed independently of the actual velocity field, so that $\Delta\dot{b}_j$ and $\Delta\dot{T}_j$ vanish and the uniqueness criterion (1.38) reduces to [5]

$$\int_V \Delta\dot{N}_{ij}\Delta v_{j,i}\, d\xi > 0 \tag{1.39}$$

for every pair of distinct velocity fields from $\mathcal{V}$.

It is worth pointing out that the criterion (1.39) is applicable also to a certain wider class of surface loading than the standard one. Namely, suppose that $\dot{b}_j$ are prescribed while $\dot{T}_j$ depend on the actual velocity field in such a manner that

$$\int_{S_T} \Delta\dot{T}_j \Delta v_j \, da \leq 0 \qquad (1.40)$$

for all pairs of admissible velocity fields from $\mathcal{V}$. It is clear that (1.39) contradicts (1.37) also for that kind of loading. For instance, (1.40) holds for the unilateral frictionless contact with a rigid tool if the actual contact pressure happens to vanish over the contact surface (e.g. [6]).

Let the current configuration be taken as reference; then from (1.9) we obtain

$$\Delta\dot{N}_{ij}\Delta\dot{F}_{ji} = \Delta\mathring{\tau}_{ij}\Delta\dot{\varepsilon}_{ij} - \sigma_{kj}\,(\Delta\dot{\varepsilon}_{ji}\,\Delta\dot{\varepsilon}_{ik} + 2\Delta\omega_{ji}\Delta\dot{\varepsilon}_{ik} + \Delta\omega_{ji}\Delta\omega_{ik})\,. \qquad (1.41)$$

On substituting (1.41), the condition (1.39) takes the form

$$\int_V \Delta\mathring{\tau}_{ij}\Delta\dot{\varepsilon}_{ij}\,d\xi - \int_V \sigma_{kj}(\Delta\dot{\varepsilon}_{ji}\Delta\dot{\varepsilon}_{ik} + 2\Delta\omega_{ji}\Delta\dot{\varepsilon}_{ik} + \Delta\omega_{ji}\,\Delta\omega_{ik})\,d\xi > 0\,. \qquad (1.42)$$

If the constitutive relationship $\mathring{\tau}_{ij}(\dot{\boldsymbol{\varepsilon}})$ is defined as a generalization of an incremental constitutive law from the small strain theory then the second integral in (1.42) which is linearly dependent on the current Cauchy stresses represents the effect of changes in geometry on the uniqueness criterion. It is evident that in general the effect is not negligible.

Example: A pointwise condition sufficient for uniqueness in the case of predominantly tensile stresses [19]

Suppose that uniqueness is examined at the state of predominantly tensile stress, that is, when at each point of the body the principal Cauchy stresses σ_i satisfy the inequalities

$$\sigma_1 + \sigma_2 > 0, \qquad \sigma_2 + \sigma_3 > 0, \qquad \sigma_3 + \sigma_1 > 0. \qquad (1.43)$$

Then, for given $\Delta\dot{\varepsilon}_{ij}$, the right-hand integrand in (1.42) attains a maximum value when

$$\Delta\omega_{pr} = \Delta\dot{\varepsilon}_{pr}\frac{\sigma_p - \sigma_r}{\sigma_p + \sigma_r}, \qquad \text{(no sum)} \qquad (1.44)$$

where the components are taken on the principal axes of $\boldsymbol{\sigma}$. Fulfillment of the inequality (1.42) is thus ensured if

$$\Delta\mathring{\tau}_{ij}\Delta\dot{\varepsilon}_{ij} > \sum_{r=1,2,3} \sigma_r(\Delta\dot{\varepsilon}_{rr})^2 + \sum_{\substack{r,p=1,2,3\\ r\neq p}} \frac{\sigma_p^2 + \sigma_r^2}{\sigma_p + \sigma_r}(\Delta\dot{\varepsilon}_{pr})^2 \qquad (1.45)$$

holds inside the body. It follows that fulfillment of the pointwise condition (1.45) in the body is sufficient for uniqueness in velocities.

Similarly, if at uniaxial tension we have

$$\Delta\mathring{\tau}_{ij}\Delta\dot{\varepsilon}_{ij} > \sigma_1((\Delta\dot{\varepsilon}_{11})^2 + 2(\Delta\dot{\varepsilon}_{12})^2 + 2(\Delta\dot{\varepsilon}_{13})^2) \tag{1.46}$$

then uniqueness of the solution in velocities is ensured with accuracy to rigid body rotations about the tensile axis.

1.2.2 Stationarity and Minimum Principles for Velocities

Consider the standard boundary value problem where $\dot{b}_j$ in V and $\dot{T}_j$ on S_T are prescribed, and suppose that the constitutive rate equations (1.10) admit a velocity-gradient potential (1.16). Then [5] *a velocity field* $\tilde{\mathbf{v}} \in \mathcal{V}$ *is a first-order solution if and only if it assigns to the functional*

$$J(\tilde{\mathbf{v}}) = \int_V (U(\nabla\mathbf{v}) - \dot{\bar{b}}_j v_j)\, d\xi - \int_{S_T} \dot{\bar{T}}_j v_j \, da\,, \qquad \tilde{\mathbf{v}} \in \mathcal{V}\,, \tag{1.47}$$

a stationary value, in the sense that its first weak (Gateaux) variation vanishes in any direction, viz. (cf. (1.1))

$$\delta J(\tilde{\mathbf{v}}, \tilde{\mathbf{w}}) = 0 \quad \text{for every } \tilde{\mathbf{w}} \in \mathcal{W}\,, \tag{1.48}$$

where $\tilde{\mathbf{w}} = \delta\tilde{\mathbf{v}}$ stands for an admissible variation of $\tilde{\mathbf{v}}$. Equivalence between (1.48) and (1.33) is immediate since from (1.16) we have $\delta U(\nabla\mathbf{v}, \nabla\mathbf{w}) = \dot{N}_{ij}(\nabla\mathbf{v})w_{j,i}$.

Suppose now that the uniqueness condition (1.39) holds. Observe that (1.39) can be rewritten as

$$\delta J(\tilde{\mathbf{v}}^*, \tilde{\mathbf{v}}^* - \tilde{\mathbf{v}}) - \delta J(\tilde{\mathbf{v}}, \tilde{\mathbf{v}}^* - \tilde{\mathbf{v}}) > 0 \quad \text{for every } \tilde{\mathbf{v}}^*, \tilde{\mathbf{v}} \in \mathcal{V}, \quad \tilde{\mathbf{v}}^* \neq \tilde{\mathbf{v}}\,, \tag{1.49}$$

which is nothing else than the condition of strict convexity of the functional (1.47) in $\mathcal{V}$. It follows [5, 17, 6] that *in the range where* (1.39) *holds, the unique solution* $\tilde{\mathbf{v}}^0$ *assigns to the functional* $J(\tilde{\mathbf{v}})$ *a strict and absolute minimum value*, viz.

$$J(\tilde{\mathbf{v}}) > J(\tilde{\mathbf{v}}^0) \quad \text{for every } \tilde{\mathbf{v}} \in \mathcal{V}\,, \quad \tilde{\mathbf{v}} \neq \tilde{\mathbf{v}}^0\,. \tag{1.50}$$

1.2.3 Incrementally Linear Comparison Solid

The uniqueness criterion (1.39) may be difficult to be applied directly to practical problems. Hill [4, 5] has proposed the following "linearizing device" to make it more easy to handle, although in general at the cost of overestimating the possibility of bifurcations. Consider a hypothetical material body of the same configuration and Cauchy stress distribution under the current loading as those for the actual inelastic body, but of a different incremental constitutive law. Namely, at each material point the hypothetical solid is assumed to be incrementally *linear*, with the constitutive relationship

$$\dot{N}^L_{ij} = C^L_{ijkl}(\mathcal{H})\dot{F}_{lk}\,. \tag{1.51}$$

The moduli C^L_{ijkl} are dependent on place in general; the moduli field $\tilde{C}^L_{ijkl}$ may be discontinuous, for instance, across the elastic-plastic interface.

Suppose that C^L_{ijkl} are defined in such a way that

$$\Delta\dot{N}_{ij}\Delta\dot{F}_{ji} \geq C^L_{ijkl}\Delta\dot{F}_{ji}\Delta\dot{F}_{lk} \tag{1.52}$$

for every pair of velocity gradients whose difference $\Delta\dot{\mathbf{F}}$ generates the difference $\Delta\dot{\mathbf{N}}$ of the stress-rates which are first found from (1.10) for each velocity gradient separately and then subtracted. The incrementally linear solid of the instantaneous moduli C^L_{ijkl} satisfying (1.52) is called the *comparison solid*. Since diagonally antisymmetric part of C^L_{ijkl} gives no contribution to the right-hand side of (1.52), henceforth we assume that $C^L_{ijkl} = C^L_{klij}$. If the potential (1.16) exists then (1.52) means that the *difference* $U - U^L$ of the two potentials is a convex function of $\dot{\mathbf{F}}$, where $U^L(\dot{\mathbf{F}}) \equiv \frac{1}{2}C^L_{ijkl}\dot{F}_{ji}\dot{F}_{lk}$ is the quadratic potential for the comparison solid. (1.52) is then called the *relative convexity property.*

Define the quadratic functional

$$I^L(\tilde{\mathbf{w}}) = \int_V C^L_{ijkl}w_{j,i}w_{l,k}\,d\xi\,, \qquad \tilde{\mathbf{w}} \in \mathcal{W}\,. \tag{1.53}$$

If (1.52) holds then the integrand in (1.39) is not smaller in value that the integrand in (1.53) with $w_{j,i} = \Delta v_{j,i}$. It follows [5] that *fulfillment of the condition*

$$I^L(\tilde{\mathbf{w}}) > 0 \quad \text{for every } \tilde{\mathbf{w}} \in \mathcal{W}\,, \quad \tilde{\mathbf{w}} \neq \tilde{\mathbf{0}}\,, \tag{1.54}$$

excludes bifurcation in velocities in the actual incrementally nonlinear solid.[1]

The problem lies in optimal choice of the comparison solid, in the sense of making the difference between the integrands in (1.39) and (1.53) as small as possible. In certain cases (see the examples below) this can be decided by fulfillment of (1.52) for C^L_{ijkl} equal to the moduli from a specific constitutive cone for the actual nonlinear law, where the "stiffness" of the material is the lowest. This means that we have equality in (1.52) whenever both velocity gradients lie within that cone. If (1.16) holds then this is equivalent to saying that the convex function $U - U^L$ vanishes in that cone. Of course, in an elastic zone in V where U is quadratic, $U^L = U$ trivially.

From (1.9) or (1.41) it can be seen that (1.52) is equivalent to

$$\Delta\mathring{\tau}_{ij}\Delta\dot{\varepsilon}_{ij} \geq L^L_{ijkl}\Delta\dot{\varepsilon}_{ij}\Delta\dot{\varepsilon}_{kl} \tag{1.55}$$

[1] Fulfillment of (1.54) ensures also uniqueness of a solution in quasi-static accelerations to a *regular second-order* rate boundary value problem [14].

where L^L_{ijkl} are related to C^L_{ijkl} by the formula analogous to (1.15). This observation can be extended to an incremental constitutive relationship expressed in terms of the rates of any pair of work-conjugate measures of strain and stress [6].

Examples of the comparison solid for elastoplastic models

As shown by Hill [4], for the classical elastic-plastic solids obeying the normality flow rule (i.e. $\lambda^{\mathrm{p}}_{ij} = \lambda_{ij}$ in (1.19)), the relative convexity property (1.55) is satisfied if we define L^L_{ijkl} as

$$L^L_{ijkl} = L^{\mathrm{p}}_{ijkl} \tag{1.56}$$

at any material element stressed to the current yield point; of course, L^L_{ijkl} in the elastic zone can be taken as the actual elastic moduli. This means that in the plastic zone the comparison solid is defined by the moduli from the loading branch.

For the classical elastic-plastic solids but without normality, Raniecki [20, 21] has shown that (1.55) is satisfied when in the plastic zone we define

$$L^L_{ijkl} = L^{\mathrm{e}}_{ijkl} - \frac{1}{4gr}(\lambda^{\mathrm{p}}_{ij} + r\lambda_{ij})(\lambda^{\mathrm{p}}_{kl} + r\lambda_{kl}), \qquad r > 0. \tag{1.57}$$

Contrary to the above case with normality, this comparison material does not follow from linearization of the constitutive relationship about a principal (loading) solution.

For the material model defined by (1.22) - (1.27) with the moduli (1.28), Sewell [9] has proved that (1.55) is ensured if the comparison moduli are taken from the so-called total loading branch, that is, if

$$L^L_{ijkl} = L^{\mathrm{e}}_{ijkl} - g^{-1}_{KL}\lambda_{Kij}\lambda_{Lkl}, \tag{1.58}$$

where the subscripts K, L run over all potentially active mechanisms of plastic deformation.

Remark There is an open problem in which circumstances (1.54) excludes also higher-order bifurcations. This delicate question has been discussed when (1.30) holds [13, 22] or need not hold [14] but always under certain simplifying assumptions (sometimes introduced tacitly) concerning regularity of the bifurcating branch. Since the latter is unknown, no definite conclusion of analogous generality as in the first-order case is available at present, to the writer's knowledge.

1.2.4 Primary Bifurcation Point

The discussion above was concerned with conditions *sufficient for uniqueness* of a solution in velocities. Now, we will discuss a condition sufficient for bifurcation. Consider the critical instant beyond which the condition (1.54) fails for the first time along a given deformation path. Generally, that critical instant determines *a lower bound* to the first value of θ at which bifurcation in velocities is possible. Below we shall specify the circumstances in which the critical instant coincides with the primary bifurcation point.

It is natural to assume [2] that at the critical instant the functional (1.53) is just positive semi-definite on $\mathcal{W}$, in the sense that

$$\begin{cases} I^L(\tilde{\mathbf{w}}) \geq 0 & \text{for every} \quad \tilde{\mathbf{w}} \in \mathcal{W}, \\ I^L(\tilde{\mathbf{w}}^*) = 0 & \text{for some} \quad \tilde{\mathbf{w}}^* \in \mathcal{W}, \ \tilde{\mathbf{w}}^* \neq \tilde{0}. \end{cases} \tag{1.59}$$

I^L must be stationary at the minimum point $\tilde{\mathbf{w}} = \tilde{\mathbf{w}}^*$, viz.

$$\delta I^L(\tilde{\mathbf{w}}^*, \tilde{\mathbf{w}}) = 0 \quad \text{for every } \tilde{\mathbf{w}} \in \mathcal{W}. \tag{1.60}$$

This is equivalent to saying that (cf. the equivalence between (1.35) and (1.48)) $\tilde{\mathbf{w}}^*$ is an eigenmode which satisfies the homogeneous equations

$$\begin{aligned} (C^L_{ijkl} w_{l,k})_{,i} &= 0 \text{ in } V, & [C^L_{ijkl}\, w_{l,k}]n_i &= 0 \text{ on } S_D\,, \\ C^L_{ijkl} w_{l,k} \nu_i &= 0 \text{ on } S_T\,, & w_j &= 0 \text{ on } S_u\,. \end{aligned} \tag{1.61}$$

On substituting (1.51) we may interpret (1.61) as the equilibrium equations for $\dot{N}^L_{ij}$ under dead loading. This means that the solution in velocities for the *comparison* solid, if exists, is nonunique for any actual loading $\dot{\bar{b}}_j$, $\dot{\bar{T}}_j$. For, on account of linearity of the constitutive rate equations (1.51) for the comparison material, the eigenmode $\tilde{\mathbf{w}}^*$ with an arbitrary scalar multiplier γ can be added to a solution $\tilde{\mathbf{v}}$ to the non-homogeneous problem (1.35), with $\dot{N}_{ij}$ replaced by $\dot{N}^L_{ij}$, to generate another solution $\tilde{\mathbf{v}} + \gamma\tilde{\mathbf{w}}$ [5].

Now, as indicated above, the moduli C^L_{ijkl} within the body can be equal to the moduli from a specific constitutive cone for the actual nonlinear law. For certain incremental loadings (but not always), among the solutions $\tilde{\mathbf{v}}+\gamma\tilde{\mathbf{w}}$ for the comparison solid there can be those which involve only strain rates from that cone. It is evident that those fields $\tilde{\mathbf{v}} + \gamma\tilde{\mathbf{w}}$ are solutions not only for the comparison material but also for the actual solid. In those circumstances, a solution to the actual first-order rate boundary value problem is nonunique [23]; the critical instant under consideration is hence called a point of bifurcation in velocities. If C^L_{ijkl} satisfy (1.52) then this is the *primary* bifurcation in velocities. On the basis of Hill's theory, that primary bifurcation point was extensively studied analytically or numerically in the literature, starting from [24]. Note that infinitely many solutions in velocities can exist, corresponding to a single eigenmode $\tilde{\mathbf{w}}^*$ and to an interval of the multiplier γ. This does not mean, however, that infinitely many post-bifurcation branches are initiated in that way at the single equilibrium state under consideration. For, higher-order conditions of continuing equilibrium usually eliminate most of the first-order solutions, for instance, all of them except three: the fundamental one and the two associated with

[2] It is worth pointing out that (1.59) need not always hold at the critical instant, even if the comparison moduli vary smoothly along the deformation path.

the limits of the interval of γ within which the involved strain-rates are directed into the constitutive cone where C^L_{ijkl} apply [25, 26]. The critical point can also correspond not to a "true" bifurcation of the deformation path but rather to a so-called limit point; cf. the discussion of eigenstates.

If $\tilde{C}^L_{ijkl}$ do not coincide with the actual moduli field then the critical instant precedes in general the primary bifurcation point which therefore cannot be determined by examining the incrementally linear comparison material only.

Example: Column buckling

Consider a perfect column (Fig. 1.2) subject to an increasing compressive load $P(\theta)$. The column is assumed not to buckle in the elastic range but otherwise to be sufficiently slender to adopt the simplest assumptions of the beam theory (for a more exact analysis, see [24]). Consequently, on using (1.15) with $F_{ij} = \delta_{ij}$ and $\sigma_{ij} = \sigma_1 \delta_{1i} \delta_{1j}$ and substituting (1.56), the integrand in (1.53) in the curent configuration taken as reference is approximated by

$$\begin{aligned} C^L_{ijkl} w_{j,i} w_{l,k} &\approx L^{\mathrm{p}} (w_{1,1})^2 + \sigma_1 (w_{2,1})^2 \\ &= L^{\mathrm{p}} x_2^2 (w'')^2 + \sigma_1 (w')^2, \end{aligned} \tag{1.62}$$

where L^{p} is the current uniaxial tangent modulus, $w(x_1)$ is the deflection mode of

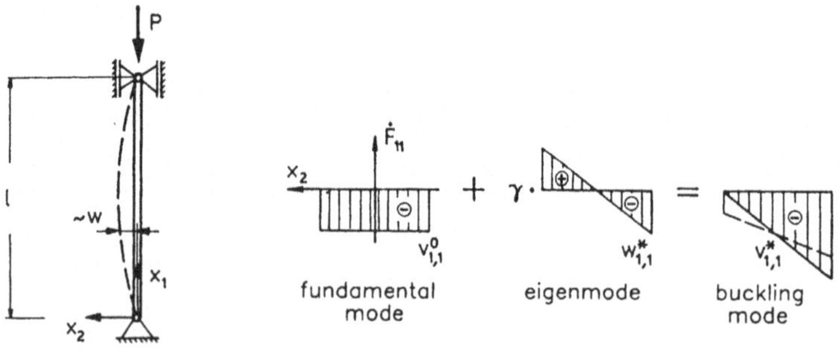

Fig. 1.2 Mode superposition at the primary bifurcation point for a column under increasing compressive load

the column axis and a prime denotes differentiation by x_1. Volume integration of (1.62) yields the sufficiency condition for uniqueness in the form

$$I^L(\tilde{\mathbf{w}}) = \int_0^l (L^{\mathrm{p}} I_c (w'')^2 - P(w')^2)\, dx_1 > 0 \qquad \text{for every } w(x_1), \tag{1.63}$$

where $I_c = \int_A x_2^2 \, da$ and A is the column cross-section. For a uniform column with the boundary conditions as indicated in Fig. 1.2, and at a tangent modulus *continuously varying* with θ, the standard method of the calculus of variation yields the critical load and eigenmode in the classical form

$$P_{\mathrm{T}} = \frac{\pi^2 L^{\mathrm{p}} I_c}{l^2}, \qquad w^*(x_1) = \sin\frac{\pi x_1}{l}. \tag{1.64}$$

This critical load P_{T} (the tangent modulus load) corresponds to the primary bifurcation point discussed above. The eigenmode with a multiplier from some interval can be superimposed, at *increasing* P, on the fundamental mode of uniform compression to produce another solution which describes incipient buckling without abrupt unloading *at* the instant of bifurcation (Fig. 1.2). Higher-order conditions of continuing equilibrium imply that the actual value of the multiplier corresponds to the neutral loading at an outer point of some cross-section (cf. [25]).

1.2.5 Directional Stability of Equilibrium

On a path of quasi-static deformation of the body, consider a single point which represents an equilibrium state. We will now examine stability of that state at a fixed value of the loading parameter λ. Accordingly, admissible velocities vanish on S_u and belong thus to the space $\mathcal{W}$. We assume the standard case of *dead* loading where the *nominal* surface tractions T_j and body forces b_j have fixed values.

Stability *of equilibrium* is essentially a dynamic concept. Roughly speaking, an equilibrium configuration of the body is said to be stable if the distance from that configuration in any dynamic motion caused by a disturbance can be made as small as we please if a measure of the disturbance itself is sufficiently small. That definition has to be made precise by specifying the class of disturbances under consideration and the measures of the distance from the examined equilibrium state and of the disturbance strength.

Path-dependence of an inelastic material creates difficulties in obtaining a condition sufficient for stability. There are also mathematical problems which are met when the continuum problem is to be examined rigorously. Therefore, the reasonable first step is to study a particular kind of stability when only *direct* paths of departure from the equilibrium configuration are taken into account. This means that a perturbed motion is *a priori* assumed to be such that variations of the "direction" of the velocity field along the path are negligible. Stability in that restricted sense is called here *directional stability*.

Along an admissible *direct* path starting from the given equilibrium state, the work of deformation in the body can be expressed as

$$W = \int_V N_{ij}\Delta u_{j,i}\, d\xi + \frac{1}{2}\int_V \Delta N_{ij}\Delta u_{j,i}\, d\xi + o((\Delta\theta)^2), \tag{1.65}$$

where N_{ij} stands for the initial nominal stresses at the equilibrium state, $\Delta\theta$ is a small increment of a time-like parameter θ which measures the "length" of the path, Δu_j are current displacements reached along the path and ΔN_{ij} are the respective stress increments. Since Δu_j vanish on S_u , from (1.31) we obtain that the former integral in (1.65) is exactly equal to the work W^{load} done by the dead loads. Hence, the work difference reads

$$W - W^{\text{load}} = \frac{1}{2}(\Delta\theta)^2 \int_V \dot{N}_{ij}\dot{u}_{j,i}\, d\xi + o((\Delta\theta)^2)\,, \tag{1.66}$$

where the rates are taken with respect to θ at the equilibrium state. It follows that if

$$\int_V \dot{N}_{ij}(\nabla\mathbf{w})w_{j,i}\, d\xi > 0 \quad \text{for every } \tilde{\mathbf{w}} \in \mathcal{W},\ \tilde{\mathbf{w}} \neq \tilde{\mathbf{0}} \tag{1.67}$$

then any movement from the equilibrium configuration in any direction requires some additional energy to be supplied to the system from external sources. *In that sense* (1.67) *is a condition sufficient for directional stability of equilibrium* [4]. On substituting (1.11) the condition (1.67) can be rewritten in a more compact form as

$$I(\tilde{\mathbf{w}}) > 0 \quad \text{for every } \tilde{\mathbf{w}} \in \mathcal{W},\ \tilde{\mathbf{w}} \neq \tilde{\mathbf{0}}\,, \tag{1.68}$$

where

$$I(\tilde{\mathbf{w}}) \equiv \int_V U(\nabla\mathbf{w})\, d\xi\,, \qquad \tilde{\mathbf{w}} \in \mathcal{W}\,. \tag{1.69}$$

A slightly stronger condition will be introduced later.

It is pointed out that (1.68) cannot be regarded as being sufficient for stability of equilibrium in a dynamic sense for arbitrarily circuitous paths unless further assumptions are introduced [4, 5, 27, 28, 53].

Compare now the obtained criteria for uniqueness and for stability of equilibrium. It is evident that if $\bar{v}_j = 0$ on S_u then (1.68) is a consequence of (1.39) since then the zero velocity field is admissible as a particular member of the pair considered in (1.39). If $\bar{v}_j \neq 0$ then the connection is slightly more complicated. Let $\Delta\tilde{\mathbf{v}} = \gamma\tilde{\mathbf{w}}$ with $\tilde{\mathbf{w}}$ and one member $\tilde{\mathbf{v}}$ fixed and γ being a positive number. If γ increases unboundedly then the integral in (1.39) divided by γ^2 tends to the integral in (1.68), by continuity and homogeneity of the relationship $\dot{N}_{ij}(\dot{F}_{lk})$. If (1.39) holds for all pairs then the limit value is never negative; however, in certain cases it may be equal to zero. With the exception of such singular points, (1.68) is a consequence of (1.39). The converse need *not* be true, so that in general the criteria for uniqueness and for stability of equilibrium are distinct and *bifurcation may take place before the loss of stability of equilibrium* [29, 4, 27].

1.2.6 Primary Eigenstate

Suppose that the constitutive rate equations (1.10) admit a potential (1.16) and consider the critical instant when the condition (1.68) fails for the first time along a given deformation path. Similarly as is Sect. 1.2.4 above, suppose that at the critical instant the functional (1.69) is just positive semi-definite on $\mathcal{W}$, in the sense that

$$\begin{cases} I(\tilde{\mathbf{w}}) \geq 0 & \text{for every } \tilde{\mathbf{w}} \in \mathcal{W}, \\ I(\tilde{\mathbf{w}}^*) = 0 & \text{for some } \tilde{\mathbf{w}}^* \in \mathcal{W},\ \tilde{\mathbf{w}}^* \neq \tilde{0}. \end{cases} \tag{1.70}$$

This implies that I is stationary at its minimum point $\tilde{\mathbf{w}} = \tilde{\mathbf{w}}^*$, viz.

$$\delta I(\tilde{\mathbf{w}}^*, \tilde{\mathbf{w}}) = 0 \quad \text{for every } \tilde{\mathbf{w}} \in \mathcal{W}. \tag{1.71}$$

Under dead loading and for $\bar{v}_j = 0$ on S_u, the functional J defined in (1.47) coincides with the functional I so that (1.71) coincides then with (1.48). It follows that $\tilde{\mathbf{w}}^*$ satisfying (1.70) (with any positive multiplier) is a non-trivial solution to the *homogeneous* (but generally nonlinear) first-order rate boundary value problem, obtained e.g. by setting $\dot{\lambda} = 0$. The body configuration at which such a nontrivial solution exists is called *an eigenstate*, and the condition (1.70) defines *the primary eigenstate* [27, 19]. It may be remarked that the method of determining the primary bifurcation point discussed in Sect. 1.2.4 reduces in effect to finding the primary eigenstate for the incrementally linear *comparison* material.

Exercise: Examine conditions for the eigenstate at uniform uniaxial tension under prescribed nominal surface tractions.

Example: Column buckling (continued)

For the perfectly straight column discussed at the end of Sect. 1.2.4, the integrand in the stability condition (1.68) takes the form analogous to (1.62) but with one important distinction: at the zone where $\nabla\mathbf{w}$ corresponds to unloading (i.e. where $w_{1,1} < 0$), the tangent (plastic) modulus must be replaced by an elastic modulus. The critical load corresponding to the primary eigenstate (1.70), i.e. to the possibility of incipient buckling at *fixed* load, is called the reduced modulus load and denoted by P_R. Since the elastic modulus is greater that L^p, we have $P_R > P_T$. The task of determining the value of P_R is not undertaken here.

In the *secondary* post-bifurcation process of buckling, or for an imperfect column, the primary eigenstate is attained when the *controlled* load P reaches its maximum (limit) value along the loading path. The equilibrium states beyond that stage can be regarded as unstable; this question will be examined later on in more detail.

Illustrative diagram

The basic results of Hill's theory quoted in this section are illustrated in Fig. 1.3 on a diagram. The critical values of the loading parameter are denoted by λ_T and λ_R, in analogy to the critical loads P_T and P_R in the example of column buckling, but the picture is now quite general and pertains to many other particular problems. It must be noted, however, that some modifications may be necessary in certain cases. For instance, bifurcation at the tangent modulus load may be "smooth" and not in velocities as indicated in the figure, the values of λ_T and λ_R may coincide with each other or else they may be "jumped over" when the tangent modulus along the deformation path suffers a discontinuous drop, etc. Such cases will be briefly discussed in the other sections.

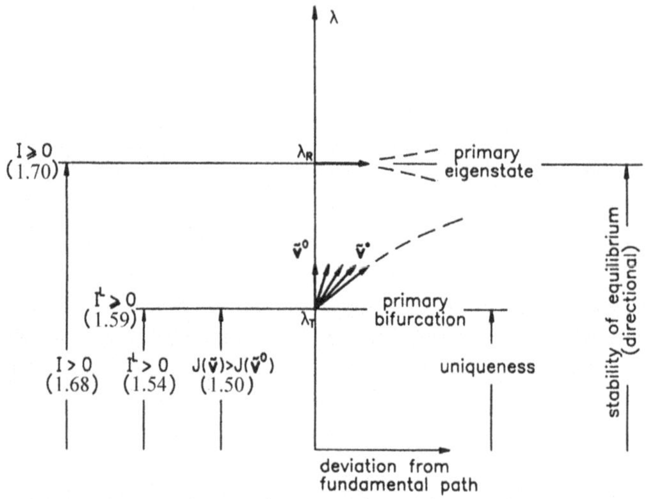

Fig. 1.3 Schematic illustration of the basic results of Hill's theory. It is assumed that the relative convexity property (1.52) holds and C^L_{ijkl} coincide with the current tangent moduli

1.3 Linearization of Bifurcation Problem at a Yield-Surface Vertex

1.3.1 Structure of Constitutive Laws for Polycrystals

The bifurcation problem in velocities is inherently nonlinear for an incrementally nonlinear constitutive law. However, we have seen that in certain circumstances the *primary* bifurcation point can be found, or a lower bound to that point can be determined, from a linearized problem formulated for the comparison material. That approach is justified if the relative convexity property (1.52) has been proved for a specific material model. Within the class of classical elastic-plastic solids obeying the normality flow rule relative to a smooth yield surface the property holds automatically, and the linearization method is thus based on a firm theoretical foundation. Nevertheless, it has turned out that the obtained bifurcation stresses are frequently unrealistically high for structural metals.

The discrepancy with experiment can, at least qualitatively, be attributed to the effect of formation of a yield-surface vertex. Consider a representative heterogeneous element of a polycrystalline elastic-plastic material. If all rheological effects are disregarded then an elastic domain in the stress or strain space is bounded by a yield surface defined as an inner envelope of individual yield surfaces corresponding to a large number of internal mechanisms of plastic deformation. Even if the virgin yield surface may be regarded as smooth, during plastic flow many mechanisms are simultaneously activated so that the respective individual yield surfaces must go through the current loading point. Since there is no reason for distinct mechanisms to have the same loading/unloading macroscopic criterion, the individual surfaces intersect each other and a corner on the global yield surface is inevitably formed [30] (Fig. 1.4a). [3]

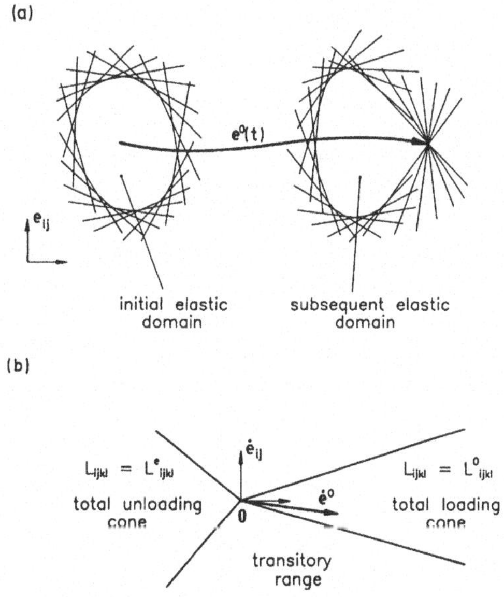

Fig. 1.4 Qualitative behaviour of elastic-plastic polycrystals:
(a) formation of a vertex on the theoretical yield surface at the current loading point,
(b) structure of the incremental constitutive law

This is connected with a more complex structure of the incremental constitutive law than the classical one, as shown schematically in Fig. 1.4b. The strain-rate space (in fact, the set of strain-rate directions) is divided into three constitutive domains: the cone of fully active loading, the cone of total unloading and the intermediate range of partial unloading [30]; the former cone may degenerate to a single ray. Of

[3] Such theoretically sharp corners need not be easy to be detected experimentally [31].

course, the term "fully active" or "total" loading concerns only those mechanisms of plastic flow which are stressed to their current yield point. It is essential that the direction of the plastic component of the strain rate is *not* fixed but depends on the strain-rate direction, also within the total loading cone. This is associated with significant reduction of the incremental stiffness of the material against a change of the straining direction in comparison with the classical elastic-plastic model [32]. As a first approximation, the incremental constitutive relationship in the total loading cone may be assumed integrable along a deformation path which at each instant is directed inside the total loading cone, which justifies the use of the moduli from the deformation theory of plasticity within that cone. Specific models of that type have been developed [33].

In view of the indicated complexity of the whole incremental constitutive law for structural metals, the linearization of the bifurcation problem by using only the "total loading" moduli becomes especially attractive. However, (1.52) has the status of a mathematical condition which may or may not be satisfied if these moduli are substituted in place of C^L_{ijkl}. Fulfillment of (1.52) depends on the actual constitutive relationship in the range of partial unloading whose specification is necessarily connected with introducing some arbitrary assumptions. The usual reference to Sewell's result [9] that the total loading moduli in the class of models described by (1.22)÷(1.28) do satisfy (1.52) is therefore not quite satisfactory, especially due to the assumption (1.26) adopted for large N. We show below how the linearization of the bifurcation problem for velocities can be justified by appealing to another constitutive inequality recently derived from micromechanical considerations. The essence of that approach is that qualitative properties of the material are specified *at a micro-level*, without specifying the macroscopic constitutive law (1.10) itself. Before presenting that approach, we discuss below a modification of Hill's theory in the case when a fundamental solution in velocities is known, which is the common case in practice.

1.3.2 Tangent Comparison Solid

The theory developed by Hill does not require any solution in velocities to be known in advance. In practice, a fundamental solution is usually known or even trivial. As mentioned by Hill [6], the uniqueness condition (1.38) can be weakened if one solution in velocities, say $\tilde{\mathbf{v}}^0$, is known. For, we can then take $\tilde{\mathbf{v}}^0$ as a fixed member of the pair of possible solutions in (1.37). Denote the respective rates of the nominal stress and deformation gradient by $\dot{N}^0_{ij}$ and $\dot{F}^0_{ij}$. It is then clear that under controlled nominal loads the solution $\tilde{\mathbf{v}}^0$ is unique when

$$\int_V (\dot{N}_{ij} - \dot{N}^0_{ij})(\dot{F}_{ji} - \dot{F}^0_{ji})\, d\xi > 0 \quad \text{for every } \tilde{\mathbf{v}} \in \mathcal{V},\ \tilde{\mathbf{v}} \neq \tilde{\mathbf{v}}^0 . \tag{1.72}$$

In comparison with (1.39) the distinction is that now only *one* field is variable; the condition (1.72) is thus generally weaker than (1.39) and more easy to handle.

At a certain stage of the fundamental deformation process, in analogy to (1.53), consider the following quadratic functional

$$I^0(\tilde{\mathbf{w}}) = \frac{1}{2}\int_V C^0_{ijkl} w_{j,i} w_{l,k}\, d\xi\,, \qquad \tilde{\mathbf{w}} \in \mathcal{W}\,, \tag{1.73}$$

based on the actual *tangent moduli*

$$C^0_{ijkl} = C_{ijkl}(\dot{\mathbf{F}}^0, \mathcal{H})\,, \qquad \dot{F}^0_{ij} = v^0_{i,j}\,, \tag{1.74}$$

which correspond to the fundamental *forward* increment of the deformation. The integral in (1.73) is well defined if the tangent moduli themselves are well defined almost everywhere in V, i.e. except possibly in a region of zero volume only (on the elastic-plastic interface, for instance). This is always assumed below whenever the functional I^0 appears in considerations.

As we have seen in preceding section, the inequality

$$I^0(\tilde{\mathbf{w}}) > 0 \quad \text{for every } \tilde{\mathbf{w}} \in \mathcal{W},\ \tilde{\mathbf{w}} \neq \tilde{\mathbf{0}} \tag{1.75}$$

excludes bifurcation in velocities if (1.52) is satisfied within the body for $\tilde{C}^L_{ijkl} = \tilde{C}^0_{ijkl}$. If the fundamental solution $\tilde{\mathbf{v}}^0$ is known then the property (1.52) can be replaced by a less restrictive requirement. Namely, suppose that the following inequality is satisfied [34]

$$(\dot{N}_{ij}(\dot{\mathbf{F}}) - \dot{N}^0_{ij})(\dot{F}_{ji} - \dot{F}^0_{ji}) \geq C^0_{ijkl}(\dot{F}_{ji} - \dot{F}^0_{ji})(\dot{F}_{lk} - \dot{F}^0_{lk}) \quad \text{for every } \dot{\mathbf{F}} \tag{1.76}$$

which can be equivalently rewritten as

$$(\dot{N}_{ij}\ (\dot{\mathbf{F}}) - C^0_{ijkl}\dot{F}_{lk})(\dot{F}_{ji} - \dot{F}^0_{ji}) \geq 0 \quad \text{for every } \dot{\mathbf{F}}\,. \tag{1.77}$$

In comparison with (1.52), one pair of the stress- and deformation-rates is now fixed, so that (1.76) is indeed a weaker restriction then (1.52) with $C^L_{ijkl} = C^0_{ijkl}$. Nevertheless, (1.76) *ensures that the first-order solution* $\tilde{\mathbf{v}}^0$ *is unique if* (1.75) *holds*. For, substitution of (1.76) into (1.75) yields the sufficiency condition (1.72). The incrementally linear comparison solid with the instantaneous moduli C^0_{ijkl} satisfying (1.76) will be called the *tangent comparison solid*. Henceforth up to the end of this section we assume that the incremental constitutive law admits a potential (1.16) so that $C^0_{ijkl} = C^0_{klij}$.

That straightforward modification of Hill's theory brings definite advantages. In Sect. 1.3.3 it will be shown that (1.76) is not only a mathematical condition less restrictive than (1.52) but can also be derived from micromechanical considerations. The difference between the two constitutive inequalities is illustrated graphically in Fig. 1.5, where U^0 is the quadratic potential for the tangent comparison solid, defined by $U^0(\dot{\mathbf{F}}) = \frac{1}{2}C^0_{ijkl}\dot{F}_{ji}\dot{F}_{lk}$.

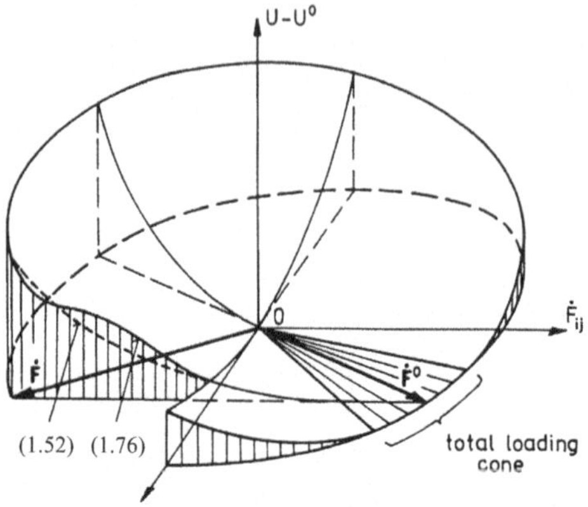

Fig. 1.5 Difference between the relative convexity property (1.52) and its weakened form (1.76). The former requires that $U - U^0$ is a convex function along any ray in $\dot{\mathbf{F}}$-space while the latter requires $U - U^0$ to be only non-decreasing along a ray with the origin at $\dot{\mathbf{F}}^0$.

It is evident that (1.76) implies

$$U(\dot{\mathbf{F}}) \geq U^0(\dot{\mathbf{F}}) \qquad \text{for every } \dot{\mathbf{F}}\,, \tag{1.78}$$

e.g. when $|\dot{\mathbf{F}}^0|/|\dot{\mathbf{F}}|$ approaches 0 . If (1.78) holds almost everywhere in the body then (1.75) implies fulfillment of the condition (1.68) of directional stability of equilibrium.

A fundamental distinction between (1.72) and (1.39) is that (1.72) does *not* ensure convexity of the functional $J(\tilde{\mathbf{v}})$ defined by (1.47). It turns out, however, that (1.72) is still *sufficient for the minimum principle* (1.50) *to be valid* [35]. In particular, (1.75) with (1.76) ensure that the unique solution $\tilde{\mathbf{v}}^0$ assigns to J an absolute minimum value in $\mathcal{V}$. To demonstrate this directly, one can use the identity

$$J(\tilde{\mathbf{v}}) - J(\tilde{\mathbf{v}}^0) = I^0(\tilde{\mathbf{v}} - \tilde{\mathbf{v}}^0) + \int_V (U(\dot{\mathbf{F}}) - U^0(\dot{\mathbf{F}}))\, d\xi\,, \qquad \tilde{\mathbf{v}} \in \mathcal{V} \tag{1.79}$$

valid under the assumptions that $\tilde{\mathbf{v}}^0$ is a solution and $C^0_{ijkl} = C^0_{klij}$. Substitution of (1.75) and (1.78) into (1.79) yields (1.50), although J need not be convex.

Comparison with the standard theorem of the calculus of variations suggests that the functional I^0 has the following interpretation

$$I^0(\tilde{\mathbf{w}}) = \delta^2 J(\tilde{\mathbf{v}}^0, \tilde{\mathbf{w}})\,, \tag{1.80}$$

where

$$\delta^2 J(\tilde{\mathbf{v}}^0, \tilde{\mathbf{w}}) \equiv \frac{1}{2} \lim_{\gamma \to 0} (\delta J(\tilde{\mathbf{v}}^0 + \gamma \tilde{\mathbf{w}}, \tilde{\mathbf{w}}) - \delta J(\tilde{\mathbf{v}}^0, \tilde{\mathbf{w}}))/\gamma \tag{1.81}$$

is the second weak variation of J at $\tilde{\mathbf{v}}^0$ in the direction $\tilde{\mathbf{w}}$. In fact, although the usual regularity assumptions of the theorem are not satisfied, a rigorous proof of (1.80) is available [36]. For validity of (1.80) it is essential that the constitutive potential U is (almost everywhere in the body) twice continuously differentiable at $\dot{\mathbf{F}}^0 = \nabla\mathbf{v}^0$, as it has been assumed above by requiring I^0 to be well defined. It is worth mentioning that if (1.80) and (1.78) hold then $J(\tilde{\mathbf{v}}) - J(\tilde{\mathbf{v}}^0) \geq 0$ is equivalent to $I^0(\tilde{\mathbf{v}} - \tilde{\mathbf{v}}^0) \geq 0$, by (1.79).

If the tangent moduli C^0_{ijkl} vary smoothly along a deformation path then a critical stage can be reached when for some nonzero $\tilde{\mathbf{w}}^* \in \mathcal{W}$ we have

$$I^0(\tilde{\mathbf{w}}^*) = 0 \leq I^0(\tilde{\mathbf{w}}) \quad \text{and} \quad \delta I^0(\tilde{\mathbf{w}}^*, \tilde{\mathbf{w}}) = 0 \quad \text{for every } \tilde{\mathbf{w}} \in \mathcal{W} \tag{1.82}$$

as a particular case of (1.59) and (1.60) when $\frac{1}{2}I^L = I^0$. Then $\tilde{\mathbf{w}}^*$ is an eigenmode for the tangent comparison solid. As a continuation of the discussion from preceding section, the following typical cases of the critical stage can be distinguished.

(i) The condition (1.68) of directional stability of equilibrium still holds, and there exists a constant c *independent of place* such that at each point of the body at which the moduli C^0_{ijkl} associated with $\tilde{\mathbf{v}}^0$ are well defined, we have

$$C_{ijkl}(\dot{\mathbf{F}}) = C^0_{ijkl} \quad \text{if } |\dot{\mathbf{F}} - \dot{\mathbf{F}}^0| < c\,. \tag{1.83}$$

Then the velocity field $\tilde{\mathbf{v}}^* = \tilde{\mathbf{v}}^0 + \gamma\tilde{\mathbf{w}}^*$ also corresponds to the moduli field C^0_{ijkl} provided $|\gamma|$ is sufficiently small. We have thus the situation discussed in Sect. 1.2.4: the primary bifurcation point has been reached, and $\tilde{\mathbf{v}}^*(\gamma)$ represents an infinite family of velocity solutions to the actual first-order rate boundary value problem. It is clear that $J(\tilde{\mathbf{v}}^*) = J(\tilde{\mathbf{v}}^0)$, as can be seen e.g. from (1.48) or from (1.79).

(ii) The condition (1.68) holds but no constant c satisfying (1.83) can be found. The latter case is obvious if the incremental constitutive law is thoroughly nonlinear but is possible also for the classical elastic-plastic solids. Then in general no bifurcation in velocities can take place at the critical instant but rather a smooth bifurcation can be inferred. More precisely, if (1.30) is valid almost everywhere within the body then the eigenmode $\tilde{\mathbf{w}}^*$ with an *arbitrary* multiplier can be added to the fundamental solution in quasi-static *accelerations* to generate another solution to the (linear) second-order rate boundary value problem [14]. The actual order of path bifurcation follows from an analysis of the post-bifurcation expansion and may be different from two [22].

(iii) The condition (1.82) is met simultaneously with the condition (1.70) for the primary eigenstate discussed in Sect. 1.2.6. This take place if the "direction" of the eigenmode $\tilde{\mathbf{w}}^*$ corresponds to the tangent moduli field, in the sense that $C_{ijkl}(\nabla\mathbf{w}^*, \mathcal{H}) = C^0_{ijkl}$ almost everywhere in the body. In typical circumstances, that case occurs along a smooth path when the loading function $\lambda(\theta)$ reaches an analytic extremum point (called also a limit point if λ is a multiplier of controlled loads). For, at that point $\dot{\lambda}$ vanishes so that the fundamental solution $\tilde{\mathbf{v}}^0$, if exists and is finite for the "time" scale defined by a choice of θ, represents itself an eigenmode $\tilde{\mathbf{w}}^*$ satisfying both (1.82) and (1.71). This conclusion remains valid also for the configuration-dependent loading discussed in Sect. 1.5.

1.3.3 A Constitutive Inequality and Uniqueness Criterion

Suppose that the constitutive relationship (1.10) is intended to describe adequately the incremental elastoplastic behaviour of a polycrystalline metal at the macro-level. As discussed in Sect. 1.3.1, the assumption of a smooth yield surface is an oversimplification, at least in bifurcation problems. Suppose thus that (1.10) represents a nonlinear relationship at a yield surface vertex. We may assume that time-independent constitutive relations at the level of a single grain have the known structure (1.22)÷(1.25) which results from the piecewise linearity and normality postulates, commonly accepted for ductile metal crystals deformed by multislip [8]. The symmetry conditions (1.27) are also assumed but fulfillment of (1.26) is not necessary here so that nonuniqueness of the constitutive response *at the micro-level* is not excluded. Under these restrictions on qualitative properties of the material at the micro-level and with the help of the averaging theorems due to Hill [37], it has been shown in [38] that the macroscopic constitutive relationship (1.10) should satisfy the following constitutive inequality

$$\int \dot{F}^*_{ji}\, dN_{ij} \geq \int \dot{N}^*_{ij}\, dF_{ji}\,; \tag{1.84}$$

F and **N** can be replaced by any other pair of work-conjugate measures of strain and stress. The inequality (1.84) is to hold for all segments of every piecewise-smooth deformation path while the starred rates are virtual and may vary arbitrarily along the path. It is understood that $\dot{\mathbf{N}}^*$ is related to $\dot{\mathbf{F}}^*$, similarly as $d\mathbf{N}$ to $d\mathbf{F}$, by the currently valid incremental constitutive law which varies along the path but is unaffected in any way by variations of the starred rates (so that the starred (virtual) and unstarred (actual) rates in (1.84) cannot be interchanged). Validity of (1.84) has recently been examined in [39] for a class of materials obeying the maximum dissipation principle and extended, in particular, to an elastic material with a number of interacting planar Griffith's micro-cracks. Interpretation of (1.84) in terms of the second-order work of deformation is given in [40].

Exercise: Verify validity of (1.84) for the classical elastoplastic solids obeying the normality flow rule relative to a smooth yield surface.

The following two consequences [38] of (1.84) are of special interest for the theory of bifurcation and stability. The first is that (1.84) *implies existence of the constitutive potential* (1.16) (under the *right-hand* continuity restriction on the tangent moduli). The second (immediate) consequence of (1.84) is that along a *regular* deformation path with the actual rates $\dot{\mathbf{F}}^0$ and $\dot{\mathbf{N}}^0$, (1.84) implies (and is implied by)

$$\dot{N}^0_{ij}\dot{F}^*_{ji} - \dot{N}^*_{ij}\dot{F}^0_{ji} \geq 0 \quad \text{for every } \dot{\mathbf{F}}^*,\ \dot{N}^*_{ij} = \dot{N}_{ij}(\dot{\mathbf{F}}^*, \mathcal{H})\,. \tag{1.85}$$

A deformation path (for a material element) is called regular if it consists only of regular points at which the current constitutive relationship $\dot{N}_{ij}(\dot{\mathbf{F}})$ along the path does not change discontinuously.[4] As discussed in [38], (1.85) may be regarded as a mathematical formulation of the physical condition that the actual rates $\dot{\mathbf{F}}^0$ and $\dot{\mathbf{N}}^0$ do not correspond to abrupt unloading at the micro-level. The set of $\dot{\mathbf{F}}^0$ satisfying (1.85) at the current state may thus be identified with the non-unloading range; this is either the closure of the total loading cone discussed above or degenerates to a single ray in the strain-rate space. From (1.84) it can be deduced that the non-unloading range should always vary with the deformation in such a manner that *it contains the actual rate of strain* (or stress) *at almost every instant*, except at singular points. At a singular point, typically at a sufficiently "sharp" corner on the strain path, the forward rate of strain falls *momentarily* outside the non-unloading range, and (1.85) does not hold at that point. Such points are excluded on a regular path. This does not mean that along a regular path there is no unloading, but rather that the unloading at the micro-level takes place gradually (i.e. smoothly at the macro-level).

The left-hand expression in (1.85) is of the form of Hill's "bilinear invariant" [37]. This results in invariance of the constitutive inequality (1.85) and (1.84) under transformation to another work-conjugate pair of stress and strain measures, possibly under a change of the reference configuration. For instance, (1.85) is equivalent to formulation of the physical condition that the actual rates $\dot{\mathbf{F}}^0$ and $\dot{\mathbf{N}}^0$ do not correspond to abrupt unloading at the micro-level. The set of $\dot{\mathbf{F}}^0$ satisfying (1.85) at the current state may thus be identified with the non-unloading range; this is either the closure of the total loading cone discussed above or degenerates to a single ray in the strain-rate space. From (1.84) it can be deduced that the non-unloading range should always vary with the deformation in such a manner that *it contains the actual rate of strain* (or stress) *at almost every instant*, except at singular points. At a singular point, typically at a sufficiently "sharp" corner on the strain path, the forward rate of strain falls *momentarily* outside the non-unloading range, and (1.85) does not hold at that point. Such points are excluded on a regular path. This does not mean that along a regular path there is no unloading, but rather that the unloading at the micro-level takes place gradually (i.e. smoothly at the macro-level).

The left-hand expression in (1.85) is of the form of Hill's "bilinear invariant" [37]. This results in invariance of the constitutive inequality (1.85) and (1.84) under transformation to another work-conjugate pair of stress and strain measures, possibly under a change of the reference configuration. For instance, (1.85) is equivalent to

$$\overset{\circ}{\tau}{}^0_{ij}\dot{\varepsilon}_{ij} - \overset{\circ}{\tau}_{ij}\dot{\varepsilon}^0_{ij} \geq 0 \quad \text{for every } \dot{\varepsilon}_{ij}\,. \tag{1.86}$$

Significance of (1.85) for the bifurcation theory results from the following lemma proved in [38]: If *the potential* (1.16) *exists and the tangent moduli* C^0_{ijkl} *are well defined then* (1.85) *is equivalent to* (1.76). This gives a clear physical meaning to the inequality (1.76) as a condition that $\dot{\mathbf{F}}^0$ does not induce an abrupt unloading at the micro-level. Since existence of the potential (1.16) is a consequence of (1.84), the condition (1.76) which justifies the use of (1.75) as the uniqueness criterion turns

[4] In the sense that the virtual stress-rate $\dot{N}^*_{ij}$ constitutively related to any *fixed* $\dot{\mathbf{F}}^*$ varies continuously along the path.

out to be a consequence of (1.84) provided only that $\dot{\mathbf{F}}^0$, $\dot{\mathbf{N}}^0$ correspond to a regular deformation path and that the tangent moduli C^0_{ijkl} are well defined.

A process of deformation of a finite body may be called *regular* if the set of material points at which the current constitutive relationship $\dot{N}_{ij}(\dot{\mathbf{F}})$ changes discontinuously or the tangent moduli are not well defined is of zero volume at every instant. The uniqueness criterion can then be formulated as follows [38]:

In a regular deformation process under controlled nominal loads, fulfillment of (1.75) *excludes bifurcation in velocities provided the constitutive inequality* (1.84) *holds.*

This uniqueness criterion provides another justification for the common procedure of excluding bifurcation in velocities on the basis of (1.75) only. The previous justification based on (1.52) required the material model to be specified since (1.52) had the status of a mathematical condition which needed verification. The essence of the above result is that the actual incremental constitutive law need not be fully specified to apply the criterion (1.75) since (1.84) may be assumed *a priori* as a constitutive inequality derived from micromechanical considerations. As discussed in Sect. 1.3.1, the incremental constitutive law intended to model adequately the real behaviour of structural polycrystalline metals may be difficult to be determined experimentally.

In the uniqueness range defined by the conditions in the above criterion, both the minimum property (1.50) and the condition (1.68) of directional stability of equilibrium are satisfied. This follows from (1.76) or (1.78) which, as indicated above, in a regular process must hold almost everywhere in the body for (1.84) to be valid.

A method of verification of (1.86) for phenomenological models of elastic-plastic behaviour at a yield-surface vertex has been proposed in [38].

1.4 Nonuniqueness and Instability in Post-Critical Range

1.4.1 Post-Critical Inequalities

We have discussed conditions sufficient for uniqueness of a solution in velocities and for directional stability of equilibrium. Then, the critical stage of the deformation has been examined when one of those conditions fails for the first time. Circumstances have been established in which the solution in velocities is nonunique at the critical instant, either at varying loading (the primary bifurcation point) or at constant loading (the primary eigenstate).

The question now arises: what can be said about uniqueness or stability *beyond* the critical point of primary bifurcation or primary eigenstate, respectively? This question is essential for complete understanding of the bifurcation phenomenon, nevertheless, it is difficult to find in the literature any definite answer or even a discussion on that point. Following [36], we will now consider this problem (and also later on in Sect. 1.6) under the assumption that the constitutive rate equations admit the potential (1.16).

It is possible that the condition (1.75) is again satisfied along the *secondary* post-bifurcation branch. For instance, in elastic-plastic solids with a smooth yield surface, local elastic unloading starts on that branch immediately beyond the bifurcation point so that the elastoplastic tangent moduli in (1.73) are replaced gradually by "stiffer" elastic moduli [25, 26]. The situation is similar when the effect of formation of a yield-surface vertex is taken into account, with the difference that the "stiffer" moduli are initially taken from the range of *partial* unloading before total elastic unloading occurs anywhere. Now, if (1.84) holds and the secondary post-bifurcation segment of the deformation process is regular then fulfillment of (1.75) excludes bifurcation in velocities as stated in Sect. 1.3.3, implying also the minimum property (1.50) and the condition (1.68) of directional stability of equilibrium. In other words, in those circumstances there is no qualitative difference between the fundamental pre-bifurcation path and the secondary post-bifurcation path as far as incremental uniqueness and stability are concerned.

The situation is generally different on the *fundamental* post-bifurcation branch. If the quadratic form $I^0(\tilde{\mathbf{w}})$ is positive definite before the critical point and positive semi-definite at the critical point then, if the deformation path if *smoothly* continued, we will usually have

$$I^0(\tilde{\mathbf{w}}) < 0 \quad \text{for some } \tilde{\mathbf{w}} \in \mathcal{W} \tag{1.87}$$

beyond that point. Then (1.72) fails for $\tilde{\mathbf{v}}^* = \tilde{\mathbf{v}}^0 + \gamma\tilde{\mathbf{w}}$ with $|\gamma|$ sufficiently small, and uniqueness of the solution $\tilde{\mathbf{v}}^0$ is not guaranteed. However, it is not evident whether uniqueness itself is lost since (1.72) is merely sufficient for uniqueness in velocities.

Since nonnegativeness of the second variation is necessary for a minimum, from (1.80) we obtain that (1.87) implies

$$J(\tilde{\mathbf{v}}) < J(\tilde{\mathbf{v}}^0) \quad \text{for some } \tilde{\mathbf{v}} \in \mathcal{V}, \tag{1.88}$$

so that the minimum principle (1.50) necessarily fails in the range of (1.87).

On the other hand, for incrementally nonlinear solids (1.87) does not contradict (1.68) so that, as pointed out in Sect. 1.2, stability of equilibrium can be preserved also on the fundamental post-bifurcation branch. Hence, in typical circumstances (but not always), along that branch a deformation range can be expected where (1.87) and (1.88) are satisfied simultaneously with (1.68). Only at a later stage, usually beyond the primary eigenstate, we will have

$$I(\tilde{\mathbf{w}}) < 0 \quad \text{for some } \tilde{\mathbf{w}} \in \mathcal{W}. \tag{1.89}$$

The condition (1.89) has been introduced by Hill and Sewell [27] as a *part* of a set of sufficiency conditions for instability of equilibrium in a dynamic sense. Although (1.89) means that a spontaneous departure from the equilibrium state is *energetically* possible (cf. the derivation of (1.68)), it is not evident whether such a departure can actually take place under arbitrarily small disturbances.

The following two questions:

- Is (1.89) *sufficient* for instability of equilibrium in a dynamic sense?
- Is (1.87) *sufficient* for non-uniqueness of the fundamental solution $\tilde{\mathbf{v}}^0$?

are addressed below under an additional assumption that the continuum problem has been spatially *discretized*. This means that up to the end of Sect. 1.4, admissible variations of velocities (or of displacements) are restricted to the form

$$\mathbf{w}(\boldsymbol{\xi}) = \sum_{\alpha=1}^{M} w_\alpha \, \boldsymbol{\phi}_\alpha(\boldsymbol{\xi}), \tag{1.90}$$

where $\boldsymbol{\phi}_\alpha$ are given continuous and piecewise smooth "shape functions" vanishing on S_u and w_α are arbitrary scalar coefficients. Their number M is finite so that the fields (1.90) form a finite-dimensional space, for simplicity denoted still by $\mathcal{W}$ as in the continuum problem. The definition of a solution, (1.33) with (1.32), is accordingly changed following the standard discretization procedure. It will later become clear that the above assumption has been introduced for mathematical reasons; the conclusions will pertain to a number of continuous problems as well.

1.4.2 Instability of Equilibrium

Throughout this subsection the loading parameter λ is regarded as fixed, t denotes a natural time, and inertia forces are taken into account. Let $\Delta\mathbf{N}(t) = \mathbf{N}(t) - \mathbf{N}(0)$ denote a stress increment along a deformation path starting from an equilibrium state. By subtracting the virtual work expressions in the initial and current instants, the (discretized) equations of motion can be written down in the form

$$\int_V (\Delta N_{ij} w_{j,i} + \bar{\rho} a_j w_j) \, d\xi = 0 \quad \text{for every } \tilde{\mathbf{w}} \in \mathcal{W}, \tag{1.91}$$

where $\bar{\rho}$ is the material density in the reference configuration and $\mathbf{a} = \ddot{\mathbf{u}}$ denote accelerations (time derivatives are now taken with respect to t).

We will examine the possibility of departure from an equilibrium configuration $\tilde{\mathbf{u}}^0$ along a *direct* path of deformation which, by definition, can be approximated by a straight path

$$\tilde{u}_j(t) = \tilde{u}_j^0 + B(t)\tilde{w}_j, \quad B(0) = 0, \quad \dot{B} \geq 0, \tag{1.92}$$

where $\tilde{\mathbf{w}} \in \mathcal{W}$ is fixed. We introduce a constitutive approximation that along straight (direct) paths and for small B the rates in (1.10) can be replaced by small increments of N_{ij} and F_{ij} with a negligible error, so that

$$\Delta N_{ij}(t) = B(t)\dot{N}_{ij}(\nabla \mathbf{w}) \tag{1.93}$$

where $\dot{N}_{ij}(\cdot)$ denotes the constitutive function at the considered equilibrium state. Of course, the approximation (1.93) would be unacceptable for arbitrarily circuitous paths.

Since $\mathcal{W}$ is now finite-dimensional and U depends continuously on $\dot{\mathbf{F}}$, the functional (1.69) is continuous in $\mathcal{W}$ and, when constrained to a level set of the function

$$\Phi(\tilde{\mathbf{w}}) \equiv \frac{1}{2}\int_V \bar{\rho} w_j w_j \, d\xi \,, \tag{1.94}$$

attains a minimum at some $\tilde{\mathbf{w}}^* \in \mathcal{W}$. By the method of Lagrangian multipliers, there is a number μ such that

$$\delta I(\tilde{\mathbf{w}}^*, \tilde{\mathbf{w}}) = \mu \delta\Phi(\tilde{\mathbf{w}}^*, \tilde{\mathbf{w}}) \quad \text{for every } \tilde{\mathbf{w}} \in \mathcal{W}\,. \tag{1.95}$$

At the equilibrium state and under the assumption that (1.16) holds, this reads

$$\int_V \dot{N}_{ij}(\nabla\mathbf{w}^*) w_{j,i} \, d\xi = \mu \int_V \bar{\rho} w_j^* w_j \, d\xi \quad \text{for every } \tilde{\mathbf{w}} \in \mathcal{W}\,. \tag{1.96}$$

On multiplying both sides of (1.96) by B and using (1.93) we obtain that the stresses along the deformation path (1.92) satisfy (within the approximations involved in (1.93)) the discretized equations of motion (1.91) provided in (1.92) we take $\tilde{w}_j = \tilde{w}_j^*$ and $B(t)$ such that

$$\ddot{B} = -\mu B\,. \tag{1.97}$$

Clearly, the sign of μ decides whether there is a tendency to decrease or to increase the speed of departure from equilibrium.

By homogeneity of degree two of I and Φ, substitution of $\tilde{\mathbf{w}} = \tilde{\mathbf{w}}^*$ into (1.95) yields $I(\tilde{\mathbf{w}}^*) = \mu\Phi(\tilde{\mathbf{w}}^*)$ and

$$\mu = \min_{\substack{\tilde{\mathbf{w}} \in \mathcal{W} \\ \tilde{\mathbf{w}} \neq \mathbf{0}}} \frac{I(\tilde{\mathbf{w}})}{\Phi(\tilde{\mathbf{w}})}\,. \tag{1.98}$$

Suppose now that (1.89) holds; then it is clear that $\mu < 0$. As an admissible function $B(t)$ in (1.92) satisfying (1.97) we can thus take

$$B(t) = (\varepsilon/\kappa)\sinh(\kappa t)\,, \qquad \kappa = (-\mu)^{1/2}\,, \qquad \varepsilon > 0 \tag{1.99}$$

which corresponds to a free inertial motion starting from the equilibrium state at $t = 0$ with initial velocities εw_j^* . Evidently, a given small finite distance from the initial equilibrium configuration is exceeded for t sufficiently large no matter how small ε is. We have thus proved the following theorem on instability of equilibrium (in the first approximation) [36]: *If the discretized system is considered then* (1.89) *implies instability of equilibrium in the dynamic sense for vanishingly small initial disturbances, under the approximation* (1.93).

The method used here to demonstrate dynamic instability of equilibrium of an inelastic time-independent system is an extension of the well known "kinetic" method for linear elastic systems where μ is defined as a square of the lowest natural frequency of vibrations. Of course, for inelastic solids μ defined by (1.98) has no longer this special interpretation.

The obtained result indicates that the condition (1.68), usually treated as a condition sufficient for stability of equilibrium, may be interpreted (with a sign $\geq$) as a condition necessary for stability of equilibrium in the dynamic sense. The most essential assumptions needed to demonstrate validity of that conclusion appear to be the following: conservativeness of the loading and existence of the potential (1.16). The approximations (1.90) and (1.93) have been primarily used to simplify the mathematical proof.

1.4.3 Continuous Range of Bifurcation Points

Consider now a process of quasi-static deformation at varying λ. Since U is a continuously differentiable function of $\dot{\mathbf{F}}$, J is differentiable in the class $\mathcal{V}$ which has been assumed now to be finite dimensional. A minimizer of J in $\mathcal{V}$ must satisfy (1.48) and is thus automatically a (discretized) solution. In general, J may be unbounded from below so that an absolute minimum need not be attained. However, this cannot happen if the equilibrium state at which the discretized rate-problem is formulated is directionally stable in the sense of (1.68). In fact, one can prove the following existence theorem [36]: *If* (1.68) *holds then the discretized rate-problem has a solution which assigns to the functional J its absolute minimum value in $\mathcal{V}$.*

Details of the proof are omitted here; they can be found in [36], and also in [41] in a different version. It is recalled that the above theorem is valid for a broad class of incrementally nonlinear constitutive laws expressible in the potential form (1.16); an essential assumption is that the incremental loading is conservative (not necessarily configuration-insensitive as assumed above).

In the uniqueness range when (1.72) holds, the minimum principle (1.50) is valid and the solution $\tilde{\mathbf{v}}^0$ coincides with the solution guaranteed by the above existence theorem. But if (1.88) holds then the guaranteed solution assigns to J a lower value than $\tilde{\mathbf{v}}^0$ and must thus be distinct from $\tilde{\mathbf{v}}^0$. In turn, if $\tilde{\mathbf{v}}^0$ does minimize J in $\mathcal{V}$ but the minimum is attained also at some other $\tilde{\mathbf{v}}^* \in \mathcal{V}$ then $\tilde{\mathbf{v}}^*$ represents another first-order solution distinct from $\tilde{\mathbf{v}}^0$. Hence, *for uniqueness of a solution $\tilde{\mathbf{v}}^0$ to the discretized rate-problem formulated at a (directionally) stable equilibrium state it is* necessary *that $\tilde{\mathbf{v}}^0$ satisfies the minimum principle* (1.50).

If the tangent moduli C^0_{ijkl} corresponding to $\tilde{\mathbf{v}}^0$ are well defined almost everywhere in V then (1.80) is valid. Since nonnegativeness of the second variation is necessary for a minimum, a necessary condition for $\tilde{\mathbf{v}}^0$ to be a minimizer of J in $\mathcal{V}$ is that

$$I^0(\tilde{\mathbf{w}}) = \frac{1}{2} \sum_{\alpha,\beta=1}^{M} K^0_{\alpha\beta} w_\alpha w_\beta \geq 0 \quad \text{for every } w_\alpha\,, \tag{1.100}$$

where

$$K^0_{\alpha\beta} = \int_V C^0_{ijkl}\phi_{\alpha j,i}\phi_{\beta l,k}\, d\xi \tag{1.101}$$

is the tangent stiffness matrix associated with the solution $\tilde{\mathbf{v}}^0$. As stated above, on account of incremental nonlinearity of the constitutive law, (1.68) can be satisfied along a deformation path also when (1.100) fails, that is, when (1.87) holds. We arrive thus at the following conclusion [36, 41]:

For a discretized system, there is a bifurcation in velocities at every point on a fundamental solution path along which (1.87) *holds simultaneously with* (1.68).

A fundamental distinction can thus be observed between the spectrum of bifurcation points for a typical elastic structure and for the incrementally nonlinear discretized system. In the former case the tangent stiffness matrix (independent of the actual incremental solution) must be singular at a bifurcation point implying that such points form a *discrete* set along a deformation path (of course, provided that trivial bifurcations related e.g. to a rigid-body rotation about a symmetry axis have been eliminated). In the latter case, the above conclusion means that *the bifurcation points form in general a continuous range along a deformation path*; of course, the actual tangent stiffness matrix $K^0_{\alpha\beta}$ (dependent on the actual incremental solution $\tilde{\mathbf{v}}^0$) need *not* be singular in that range. This result, known for the Shanley column [29] for many years, at the present generality seems to be proved only recently as indicated above. Along a smooth deformation path, the continuous range of bifurcations in velocities can be entered at the primary bifurcation point at which (1.75) fails and (1.82) is met (so that $K^0_{\alpha\beta}$ *is* momentarily singular), and terminated at the primary eigenstate at which (1.68) fails and (1.70) is met. This conclusion is *not* necessarily limited to discretized systems; it can remain valid also for many continuous systems under appropriate regularity assumptions which, however, need not always be satisfied. If the primary bifurcation is induced by a discontinuous drop of the tangent moduli then the continuous range of bifurcations can be entered even if the tangent stiffness matrix $K^0_{\alpha\beta}$ is not singular at *any* instant.

The previously discussed bifurcation at an eigenstate for the incrementally linear comparison material turns out thus to be merely a special kind of bifurcation in incrementally nonlinear solids. Contrary to that special case, in the range (1.87) the secondary solution $\tilde{\mathbf{v}}^*$ which minimizes $J(\mathbf{v})$ in $\mathcal{V}$ must correspond to a moduli field different from the fundamental moduli field $\tilde{C}^0_{ijkl}$. For, if the moduli fields were the same then straightforward transformations with the help of symmetry of the moduli would yield $J(\tilde{\mathbf{v}}^*) - J(\mathbf{v}^0) = I^0(\tilde{\mathbf{v}}^* - \tilde{\mathbf{v}}^0) = 0$ which would contradict (1.88).

Illustrative diagram

The results presented in this section are illustrated schematically in a diagram in Fig. 1.6. Comparison with Fig. 1.3 shows how the present conclusions complement Hill's theory which has not dealt with the problems of non-uniqueness and instability in a post-critical range. A picture similar to that in Fig. 1.6 was sometimes in the literature assumed as granted on the basis of an analysis of simplest systems as the Shanley column, but a general proof was lacking. The proof given in [36] can be extended to a class of continuous systems under certain requirements of mathematical nature; cf. the discussion in Sect. 1.6. We will see later that a deformation path in

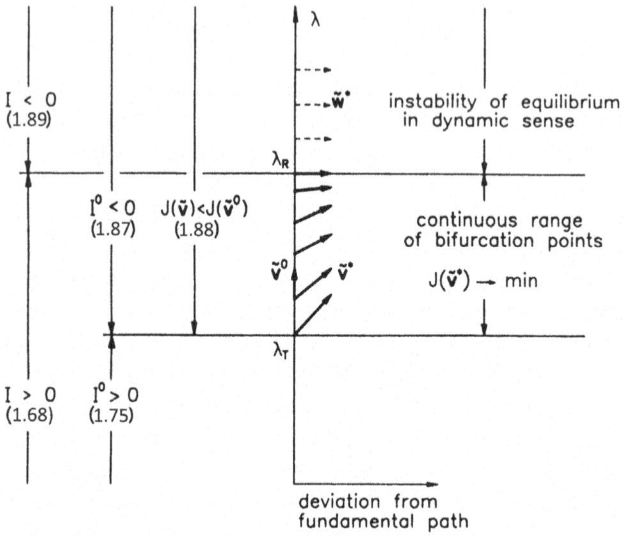

Fig. 1.6 Schematic illustration of the results of Sect. 1.4 for a discretized problem

the range of (1.88) (typically, beyond the primary bifurcation point) can be regarded as unstable, in the sense of instability of a quasi-static deformation *process*.

The question: what happens if the critical value λ_T of the loading parameter λ suffers a discontinuous drop along the deformation path due to a sudden decrease of the tangent modulus, so that the range $\lambda > \lambda_T$ is entered at some other critical value $\lambda_D > \lambda_T$, can now be given a general answer, at least for discretized systems. If $\lambda_T < \lambda_D < \lambda_R$ then a bifurcation in velocities takes place at the critical value λ_D provided λ is further increasing. If $\lambda_D > \lambda_R$ then the equilibrium configuration at the critical instant is unstable (at least in the first approximation) so that a dynamic departure from that configuration becomes possible at constant λ.

1.5 Energy Interpretations in the Bifurcation Theory

1.5.1 The Energy Functional

The work of deformation in the body can be written down as

$$W = \int_V \int N_{ij}\, dF_{ji}\, d\xi \,, \tag{1.102}$$

where the stresses are determined pointwise by integration of the constitutive rate equations (1.10) along the deformation path. A potential energy of the loading device which applies nominal surface tractions $\overline{T}_j$ and nominal body forces $\bar{b}_j$ independently of the body configuration can be expressed as

$$\Omega = \Omega(\tilde{\mathbf{u}}, \lambda) = -\int_V \bar{b}_j u_j \, d\xi - \int_{S_T} \overline{T}_j \, u_j \, da \,. \tag{1.103}$$

Introduce *the energy functional* [42, 34]

$$E = W + \Omega \tag{1.104}$$

defined for any kinematically admissible deformation process at varying or fixed λ. In general, E is a functional of the deformation history due to path-dependence of W. An increment of the value of E can be interpreted as the amount of energy which has to be supplied from external sources to the mechanical system consisting of the deformed body *and* the loading device in order to produce quasi-statically a deformation increment, generally with the help of additional perturbing forces. It is emphasized that an increment of the value of $(-\Omega)$ is generally *not* equal to the work done by the controlled loads unless $\bar{b}_j$ and $\overline{T}_j$ are constant in time.

In the cases of an elastic support or fluid-pressure loading, the nominal surface tractions T_j on S_T are not only functions of ξ_k and λ but depend also on the actual displacements or their surface gradients. For one- or two-dimensional idealizations of the solid body, generalized body forces dependent in a similar way on the displacement field are induced by lateral surface tractions; other examples can also be given. Suppose thus (cf. [17, 6]) that the incremental loading consists not only of a controllable part (distinguished by a bar), as it was assumed above, but also of a deformation-sensitive part, that is

$$\dot{T}_j = \dot{\overline{T}}_j + f_j(\tilde{\mathbf{v}}) \,, \qquad \dot{b}_j = \dot{\bar{b}}_j + g_j(\tilde{\mathbf{v}}) \,, \tag{1.105}$$

where $f_j(\tilde{\mathbf{v}})$ and $g_j(\tilde{\mathbf{v}})$ are linear homogeneous expressions in the velocity v_l and its gradient $v_{l,k}$ at the considered material point. Coefficients in the expressions may depend piecewise continuously on ξ_k and sufficiently smoothly on λ and $\tilde{\mathbf{u}}$. The energy functional (1.104) can still be defined provided the loading is *conservative* in an overall sense (cf. [17, 43]). This means that for a *fixed* value of the loading parameter λ, the total work done by the body forces $b_j = b_j(\boldsymbol{\xi}, \lambda, \tilde{\mathbf{u}})$ and surface tractions $T_j = T_j(\boldsymbol{\xi}, \lambda, \tilde{\mathbf{u}})$ in any virtual motion compatible with the kinematic constraints and leading from a configuration $\tilde{\mathbf{u}}$ to any sufficiently close configuration $\tilde{\mathbf{u}}^*$ is assumed to be *path-independent*, viz.

$$\int_V \int_{\mathbf{u}}^{\mathbf{u}^*} b_j du_j \, d\xi + \int_{S_T} \int_{\mathbf{u}}^{\mathbf{u}^*} T_j du_j \, da = \Omega(\tilde{\mathbf{u}}, \lambda) - \Omega(\tilde{\mathbf{u}}^*, \lambda) \,, \qquad \lambda = \text{const.} \tag{1.106}$$

The functional $\Omega(\tilde{\mathbf{u}}, \lambda)$ is defined to within an additive function of λ which may be chosen arbitrarily. For a physically appropriate choice of that function, Ω can be identified with the potential energy of the loading device and substituted into (1.104).

At a given body configuration $\tilde{\mathbf{u}}$ at a certain λ, introduce the bilinear functional Q of velocities (or their variations) and the respective quadratic functional R, defined by

$$Q(\tilde{\mathbf{v}}, \tilde{\mathbf{v}}^*) = \int_V g_j(\tilde{\mathbf{v}}) v_j^* \, d\xi + \int_{S_T} f_j(\tilde{\mathbf{v}}) v_j^* \, da \,, \quad R(\tilde{\mathbf{v}}) = -\frac{1}{2} Q(\tilde{\mathbf{v}}, \tilde{\mathbf{v}}) \,. \tag{1.107}$$

It can be shown [36] that (1.106) implies

$$Q(\tilde{\mathbf{v}}, \tilde{\mathbf{v}}^*) = Q(\tilde{\mathbf{v}}^*, \tilde{\mathbf{v}}) \quad \text{for every } \tilde{\mathbf{v}}, \tilde{\mathbf{v}}^* \in \mathcal{V} \cup \mathcal{W} \,; \tag{1.108}$$

this is an extension of Hill's [17] "self-adjointness" condition for surface loading to the loading (1.105).

By the trapezoid rule of quadrature we obtain, instead of (1.103), the following expression for the difference between the values of Ω in two body configurations close to each other, correct to second-order:

$$\Omega(\tilde{\mathbf{u}} + \tilde{\mathbf{w}}, \lambda) - \Omega(\tilde{\mathbf{u}}, \lambda) = -\int_V b_j w_j \, d\xi - \int_{S_T} T_j w_j \, da + R(\tilde{\mathbf{w}}) + \ldots, \tag{1.109}$$

where b_j , T_j and R are evaluated at $(\tilde{\mathbf{u}}, \lambda)$.

From (1.105), (1.107) and (1.108) we obtain the equality

$$\int_V \dot{b}_j w_j \, d\xi + \int_{S_T} \dot{T}_j w_j \, da = \int_V \dot{\bar{b}}_j w_j \, d\xi + \int_{S_T} \dot{\bar{T}}_j w_j \, da - \delta R(\tilde{\mathbf{v}}, \tilde{\mathbf{w}}) \tag{1.110}$$

valid for every $\tilde{\mathbf{v}} \in \mathcal{V}$ and $\tilde{\mathbf{w}} \in \mathcal{W}$. By substituting (1.110) into (1.33) it can be shown that the conditions and theorems from the bifurcation theory formulated for controlled nominal loads remain valid for the configuration-dependent conservative loading provided the expressions for J, $\frac{1}{2}I^L$, I^0, I are modified simply by adding the quadratic functional R [17]. Details are omitted here.

Exercise: Derive an explicit expression for $R(\tilde{\mathbf{w}})$ in the cases of elastic foundation and fluid-pressure loading.

1.5.2 Basic Identity

Consider first the case of controlled nominal loads when (1.103) applies. The first-order rate of the energy functional E at a given state of the body is expressed as the affine (linear but inhomogeneous) functional of a velocity field

$$\dot{E}(\tilde{\mathbf{v}}) = \int_V N_{ij} v_{j,i} \, d\xi - \int_V \bar{b}_j v_j \, d\xi - \int_{S_T} \bar{T}_j v_j \, da - \int_V \dot{\bar{b}}_j u_j \, d\xi - \int_{S_T} \dot{\bar{T}}_j u_j \, da \,. \tag{1.111}$$

The first term equals to $\dot{W}$ and the remaining contribute to $\dot{\Omega}$. Since the value of the last two terms in (1.111) is prescribed independently of $\tilde{\mathbf{v}}$, it can be seen by reference to the virtual power principle (1.31) that the considered state is in equilibrium if and only if the value of $\dot{E}$ is independent of $\tilde{\mathbf{v}} \in \mathcal{V}$. The equilibrium conditions can thus be written down as

$$\dot{E}(\tilde{\mathbf{v}}) = \text{ const. in } \mathcal{V}\,. \tag{1.112}$$

Consider now the second-order rate of E evaluated at an equilibrium state. In calculating the second time derivative of the deformation work W one should take into account the possibility of velocity gradient discontinuities which move relative to the material. This can be done by using the transport theorem and kinematic compatibility conditions [5]. It turns out that velocity-gradient discontinuities introduce no extra term in the expression for $\ddot{W}$ *evaluated at an equilibrium state*, which reads[6]

$$\ddot{W} = \int_V (\dot{N}_{ij}\, v_{j,i} + \bar{b}_j \dot{v}_j)\, d\xi + \int_{S_u} N_{ij} \nu_i \dot{v}_j \, da + \int_{S_T} \overline{T}_j \dot{v}_j \, da\,. \tag{1.113}$$

To show validity of (1.113) at the absence of such discontinuities, it suffices to differentiate the first term in (1.111) by θ, use the Green theorem and substitute the equilibrium conditions for N_{ij}.

Calculation of $\ddot{\Omega}$ in the case of configuration-independent loading is straightforward if there are no moving discontinuities in $\dot{\bar{b}}_j$ and $\dot{\overline{T}}_j$[7]. By adding the expression for $\ddot{\Omega}$ to that in (1.113), the integrals of $\bar{b}_j \dot{v}_j$ over V and of $\overline{T}_j \dot{v}_j$ over S_T cancel and we obtain

$$\begin{aligned}\ddot{E} = &\int_V (\dot{N}_{ij} v_{j,i} - 2\dot{\bar{b}}_j v_j)\, d\xi - 2\int_{S_T} \dot{\overline{T}}_j v_j \, da \\ &+ \int_{S_u} N_{ij} \nu_i \ddot{u}_j \, da - \int_V \ddot{\bar{b}}_j u_j \, d\xi - \int_{S_T} \ddot{\overline{T}}_j u_j \, da\,.\end{aligned} \tag{1.114}$$

The last three terms in (1.114) have prescribed values so that $\ddot{E}$ at a given equilibrium state is a functional of velocities only. Comparison of (1.114) and (1.47) yields finally the following identity [42, 34]

$$\frac{1}{2}\ddot{E}(\tilde{\mathbf{v}}^*) - \frac{1}{2}\ddot{E}(\tilde{\mathbf{v}}) = J(\tilde{\mathbf{v}}^*) - J(\tilde{\mathbf{v}}) \quad \text{for every } \tilde{\mathbf{v}}, \tilde{\mathbf{v}}^* \in \mathcal{V}\,. \tag{1.115}$$

Note that the potentiality condition (1.16) is not necessary for validity of (1.115).

Validity of (1.115) can be extended to the case of configuration-dependent conservative loading (1.105) provided the functional J is appropriately modified. Time

[5] There is a misprint in the derivation given in [34]: in eq. (18) there should be $\Delta\ddot{W}$ instead of $\ddot{W}$.

[6] It can be remarked that $\dot{v}_j$ on S in (1.113) need not be equal to the respective limit value on the boundary of the acceleration field $\dot{v}_j$ in V.

[7] If such discontinuities are present then certain extra terms appear in the expression for $\ddot{\Omega}$. However, those terms are independent of the actual continuation of deformation process and have thus no influence on the final result (1.115).

differentiation of (1.109) with $\dot{u}_j = v_j$ and $\dot{w}_j = v_j^* - v_j$, followed by substitution of (1.105), (1.107) and (1.108) shows (cf. [34]) that the contribution of the configuration-sensitive part of the incremental loading to the left-hand side of (1.115) is equal to

$$-Q(\tilde{\mathbf{v}}, \tilde{\mathbf{v}}^* - \tilde{\mathbf{v}}) + R(\tilde{\mathbf{v}}^* - \tilde{\mathbf{v}}) = R(\tilde{\mathbf{v}}^*) - R(\tilde{\mathbf{v}})\,. \tag{1.116}$$

Hence, (1.115) remains valid if the velocity functional J is defined, instead of (1.47), by

$$J(\tilde{\mathbf{v}}) = \int_V (U(\nabla\mathbf{v}) - \dot{\bar{b}}_j v_j)\, d\xi - \int_{S_T} \dot{\bar{T}}_j v_j \, da + R(\tilde{\mathbf{v}})\,. \tag{1.117}$$

This is exactly the modification needed in the bifurcation theory, as mentioned above.

1.5.3 Energy Interpretations

Henceforth we shall always assume that the constitutive rate equations admit the potential (1.16). Then the first-order rate equations of continuing equilibrium can be given the variational formulation (1.48), with J defined either by (1.47) for controlled nominal loads or by (1.117) for the configuration-dependent conservative loading discussed above. Now, by (1.115), the stationarity principle (1.48) has the following energy interpretation:

$$\delta\ddot{E}(\tilde{\mathbf{v}}, \tilde{\mathbf{w}}) = 0 \quad \text{for every } \tilde{\mathbf{w}} \in \mathcal{W}\,. \tag{1.118}$$

On account of (1.112), this may be expressed in words as follows: any solution in velocities corresponds to a stationary value of the increment of E evaluated with accuracy to second order terms.

In the uniqueness range where the minimum property (1.50) holds, substitution of (1.115) shows that the unique solution $\mathbf{v}^0$ strictly minimizes the value of $\ddot{E}$ among all kinematically admissible velocity fields, that is,

$$\ddot{E}(\tilde{\mathbf{v}}) > \ddot{E}(\tilde{\mathbf{v}}^0) \quad \text{for every } \tilde{\mathbf{v}} \in \mathcal{V},\ \tilde{\mathbf{v}} \neq \tilde{\mathbf{v}}^0\,. \tag{1.119}$$

From (1.80) and (1.115) we obtain that the quadratic functional (1.73) based on the tangent moduli, if it is well defined, has the following energy interpretation

$$I^0(\tilde{\mathbf{w}}) = \frac{1}{2}\delta^2\ddot{E}(\tilde{\mathbf{v}}^0, \tilde{\mathbf{w}})\,. \tag{1.120}$$

This remains valid in the case of the configuration-dependent conservative loading if the quadratic functional $R(\tilde{\mathbf{w}})$ is added to the expression for I^0.

If the constitutive inequality (1.84) holds and $\tilde{\mathbf{v}}^0$ corresponds to a regular deformation process then the inequality

$$\delta^2\ddot{E}(\tilde{\mathbf{v}}^0, \tilde{\mathbf{w}}) > 0 \quad \text{for every } \tilde{\mathbf{w}} \in \mathcal{W},\ \tilde{\mathbf{w}} \neq \mathbf{0}\,, \tag{1.121}$$

as being equivalent to (1.75), is sufficient for uniqueness of $\tilde{\mathbf{v}}^0$ and also for (1.119). In other words, in those circumstances, if the actual solution $\mathbf{v}^0$ corresponds to a proper relative minimum of $\ddot{E}(\tilde{\mathbf{v}})$ then existence of other stationarity points of $\ddot{E}(\mathbf{v})$ is excluded and the minimum turns out to be absolute in $\mathcal{V}$.

Consider now the instant of primary bifurcation in velocities when (1.82) is met while (1.83) is valid, still under the assumption that (1.76) is satisfied. As already discussed, the bifurcation solutions $\tilde{\mathbf{v}}^* = \tilde{\mathbf{v}}^0 + \gamma\tilde{\mathbf{w}}^*$ correspond then to the same moduli field $\tilde{C}^0_{ijkl}$ and assign to the functional J the same value as the fundamental solution $\tilde{\mathbf{v}}^0$. It is clear from (1.79) and (1.78) that no field from $\mathcal{V}$ can correspond to a lower value of J. Hence, by (1.115), at the critical point under consideration we have

$$\ddot{E}(\tilde{\mathbf{v}}^*) = \ddot{E}(\tilde{\mathbf{v}}^0) \leq \ddot{E}(\tilde{\mathbf{v}}) \quad \text{for every } \tilde{\mathbf{v}} \in \mathcal{V}. \tag{1.122}$$

Conversely, any kinematically admissible velocity field $\tilde{\mathbf{v}}^*$ satisfying (1.122) assigns to the functional J a stationary value and represents thus another solution to the first-order problem. We have thus the following energy interpretation of the typical primary bifurcation in velocities: the minimum value of the second-order increment of the energy functional (1.104) is attained not solely for the fundamental velocity solution but also for some other deformation mode [44, 34].

Beyond that critical instant, the following condition

$$\delta^2\ddot{E}(\tilde{\mathbf{v}}^0, \tilde{\mathbf{w}}) < 0 \quad \text{for some } \tilde{\mathbf{w}} \in \mathcal{W}, \tag{1.123}$$

equivalent to (1.87), is usually fulfilled on the regular *fundamental* path (but this is not a rule on a *secondary* post-bifurcation path). In that range the fundamental solution in velocities can no longer minimize $\ddot{E}$ (since non-negativeness of the second variation is necessary for a minimum) and *may* be nonunique. In fact, it is nonunique if a minimum of $\ddot{E}$ is actually attained in $\mathcal{V}$; according to the existence theorem from Sect. 1.4.3, this is ensured for discretized systems so long as the equilibrium is directionally stable, i.e. if (1.124) holds.

The condition (1.68) of directional stability of equilibrium, contrary to those in the bifurcation theory, has been derived from the energy balance and requires thus no additional energy interpretation. However, for the sake of comparison, we express it explicitly in terms of the energy functional E. If λ is fixed then an increment of E along any kinematically admissible path is equal to the deformation work minus the work of the given loads and coincides thus with the expression in (1.66). We thus have $I(\tilde{\mathbf{w}}) = \frac{1}{2}\ddot{E}(\tilde{\mathbf{w}})$ for $\tilde{\mathbf{w}} \in \mathcal{W}$, and the condition (1.68) can be equivalently rewritten as

$$\ddot{E}(\tilde{\mathbf{w}}) > 0 \quad \text{at } \lambda = \text{const} \quad \text{for every } \tilde{\mathbf{w}} \in \mathcal{W},\ \tilde{\mathbf{w}} \neq \tilde{\mathbf{0}}. \tag{1.124}$$

Obviously, this equivalence remains valid for the configuration-dependent conservative loading provided $R(\tilde{\mathbf{w}})$ is added to the expression (1.69) for $I(\tilde{\mathbf{w}})$.

Illustrative diagram

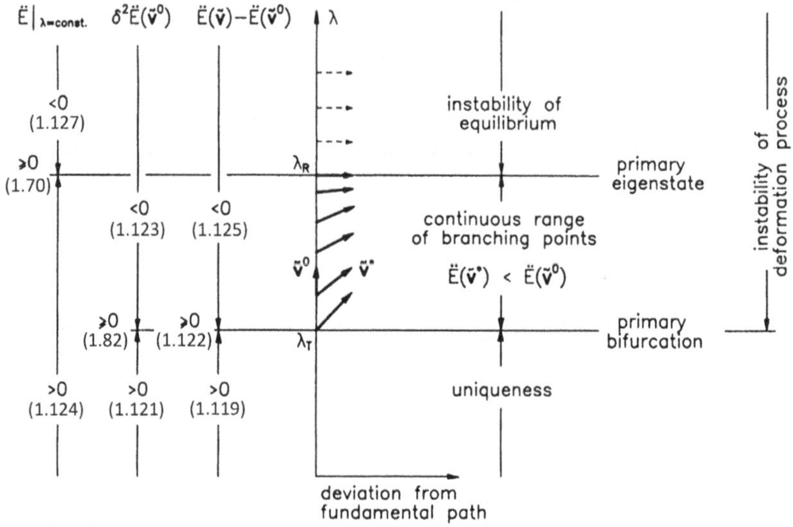

Fig. 1.7 Energy interpretation of the conditions for bifurcation and instability. It is assumed that the constitutive inequality (1.84) holds and that the deformation process is sufficiently regular (and spatially discretized when necessary).

The schematic diagram in Fig. 1.7 shows a summary of the results presented in the preceding sections which are now interpreted in terms of the second time derivative of the energy functional E defined by (1.104). A discussion of the inequalities (1.125) and (1.127) as conditions sufficient for instability of a deformation process and of an equilibrium state, respectively, will be given in Sect. 1.6. Note the role played in the diagram by the assumed constitutive inequality (1.84) which along a regular deformation path is equivalent to (1.76) and implies equivalence between (1.123) and (1.125) below.

A modification of the diagram is needed if at the critical instant $\lambda = \lambda_T$ at which (1.82) holds by assumption, the minimum in (1.119) is still strict, i.e. the condition (1.122) assumed in the diagram is not fulfilled. This implies that bifurcation *at* λ_T cannot be in velocities but rather can occur "smoothly", in a higher-order rate of displacements. As already mentioned, this is a rule when the incremental response of the material is "thoroughly" nonlinear but is possible also in the classical elastic-plastic solids. Another possible modification has been indicated in the discussion of Fig. 1.6; further possibilities will be considered in Sect. 1.6.

No matter what happens at λ_T, the path branching *between* λ_T and λ_R indicated in the diagram can occur through a finite difference in velocities. For discretized systems and for a certain class of continuous systems, branching can be identified with bifurcation, but as discussed below, for a general continuous system the branching may take place under application of arbitrarily small perturbing forces.

1.6 Unified Approach to Bifurcation and Instability Problems

1.6.1 Various Forms of Instability

In Hill's theory outlined in Sect. 1.2 the term "stability" concerns exclusively an equilibrium state at a fixed value of the loading parameter. This is sufficient for examining conditions under which dynamic "snap-through" phenomena may appear. On the other hand, in the engineering language it is common to speak about "instabilities" also when discussing such phenomena as buckling at increasing load or necking in bars or sheets at increasing elongation. The basic tool in analysis of such "geometric" instabilities is not the theory of stability of equilibrium but the bifurcation theory, complemented by studies of post-bifurcation behaviour and imperfection sensitivity (cf. [25, 45, 26, 46, 47, 48, 49] and the references quoted therein). Initiation of shear band formation (and also of localized necking in thin sheets) is studied by using a similar approach. However, in the latter case the theory was developed to some extent independently (see [50] and the references therein) since the bifurcation theory formulated for a finite body with well-defined boundary conditions did not apply without modifications to such "material" instabilities (especially in inhomogeneous bodies). Of still another kind are local instabilities at the body surface which manifest themselves in the form of surface undulations (cf. [51]).

The question arises whether all those observable instabilities of "dynamic", "geometric", "local" or "material" kind, apparently quite distinct from each other, can be investigated in a unified way by using a *single* criterion for predicting their appearance. We will show below, following [34, 36], that an affirmative answer can be given under the general assumptions already adopted in presentation of the bifurcation theory, by extending the energy criterion of instability of equilibrium to quasi-static deformation *processes*. The essence of the unified approach lies in finding a simple and general condition which could help to solve, at least partially, the following problem: what deformation processes *cannot* be practically realized in a physical system.

1.6.2 Instability of a Quasi-Static Deformation Process

Consider a deformation process which represents a theoretical quasi-static solution to an initial - boundary value problem formulated for an inelastic continuous body. Real material properties and boundary conditions are assumed to be adequately modeled. Now, the observable "instabilities" of various kind mentioned above clearly demonstrate that apparently quite correct solutions at sufficiently advanced deformations can be no longer in accord with experimental observations. This means that certain deformation processes, for instance, of macroscopically uniform deformations, are not practically realizable in a post-critical range. One of possible explanations is that this is related to an instability of the deformation *process* (or *path,* in an equivalent terminology) beyond a certain critical stage.

Traditionally, stability of a quasi-static deformation path is tacitly identified with stability of equilibrium states traversed by the path. However, for path-dependent and incrementally nonlinear solids distinction must be made between these two kinds of instability. Stability of equilibrium means that the effect of disturbing influences becomes negligible when the strength of a disturbance is vanishingly small while the value of the loading parameter is kept fixed. Now, if that final effect is very small but nonzero and if a quasi-static loading program is continued, there is no guarantee that the distance between the *subsequent* configurations along the perturbed and fundamental paths at the same loading levels will remain small; note that continuous dependence of this distance on initial data is necessarily broken e.g. at a bifurcation point. This observation alone indicates that stability of an equilibrium state (at constant loading) and of a deformation process (at varying loading) are essentially different concepts. The distinction is even more clear when the effect of application of infinitesimal perturbing forces is studied. Depending on whether such forces are applied at constant loading or at varying loading, different branches of the incrementally nonlinear constitutive law can be activated. In the former case the straining "direction" is defined by the disturbance itself while in the latter it is primarily determined by the fundamental velocity field; for elastic-plastic solids the "stiffness" in the latter case can be much smaller.

The problem arises how to define stability of a deformation process. A natural approach is to identify the process with a quasi-static motion and to apply concepts of the general theory of stability of motion. The sense of the term "stability" lies, roughly speaking, in resistance to small disturbing influences (called also *disturbances*) which should be carefully distinguished from geometric or material *imperfections* : the former perturb the motion of a given material system while the latter change the system itself. There is a vast literature on various theories of stability of motion, in particular, on Lyapunov's theory where *initial* disturbances are considered. In reality, any material body is subject to perpetual disturbing influences *during* a deformation process; in the general theory of stability of motion such influences are called *persistent disturbances* (cf. e.g. [52]). They can be represented, for instance, by small perturbing forces acting in addition to the prescribed loads. The idea of stability for persistent disturbances is not less natural than that for initial disturbances; moreover, in the case of path-dependent solids (where material properties at a perturbed configuration of the body are dependent on the way how this configuration has been reached) it is more easily applicable.

We shall adopt the following concept of instability of a deformation process for persistent disturbances: If application of infinitesimal perturbing forces in any finite interval of time can cause finite deviations from a theoretical, unperturbed deformation process then such a process is regarded as being unstable and, consequently, practically unrealizable in a physical system.

This is still not a precise definition of instability since the measures of the distance and of perturbing forces have not been specified. While it is not difficult to formulate some mathematically rigorous definition of stability or instability, the problem is to ensure that it is physically sound and that it yields practically applicable instability criteria from which the onset of observable instabilities could be found. Therefore, we use another approach: we will first select a certain qualitative property of post-critical solutions as a *possible candidate* for the general condition sufficient for instability of a deformation process, and then examine its connection with the concept of instability for persistent disturbances mentioned above.

1.6.3 The Energy Criterion of Instability of a Process

The fundamental assumptions are the same as in discussing the energy interpretations in the bifurcation theory:

- the constitutive rate equations (1.10) admit the potential (1.16),
- the incremental loading (1.105) is conservative in the overall sense (1.106).

As we have seen in the previous section, along a typical regular deformation path the value of $\ddot{E}(\tilde{\mathbf{v}})$ in $\mathcal{V}$ is minimized by the fundamental solution $\tilde{\mathbf{v}}^0$ up to a certain critical stage and not beyond that stage. This motivates introduction of the following definition: *A quasi-static deformation process is said to be unstable in the energy sense at some stage of the deformation if the respective velocity solution $\tilde{\mathbf{v}}^0$ does not minimize the value of $\ddot{E}$ in $\mathcal{V}$, that is, if*

$$\ddot{E}(\tilde{\mathbf{v}}) < \ddot{E}(\tilde{\mathbf{v}}^0) \quad \text{for some } \tilde{\mathbf{v}} \in \mathcal{V}\,. \tag{1.125}$$

Equivalently, a necessary condition for stability of the fundamental process in the energy sense is that at every stage

$$\ddot{E}(\tilde{\mathbf{v}}) \geq \ddot{E}(\tilde{\mathbf{v}}^0) \quad \text{for every } \tilde{\mathbf{v}} \in \mathcal{V}\,. \tag{1.126}$$

This criterion was proposed [42, 34] as an intuitive criterion of instability of a plastic deformation process, and its justification has been given more recently [36]. Since at an equilibrium state the value of $\dot{E}(\tilde{\mathbf{v}})$ is independent of $\tilde{\mathbf{v}}$, the condition (1.125) (or (1.126)) can be written down in an equivalent incremental form as $\Delta E < \Delta E^0$ (or $\Delta E \geq \Delta E^0$), where ΔE^0 and $\Delta E \equiv (\Delta\theta)\dot{E} + \frac{1}{2}(\Delta\theta)^2\ddot{E}$ are the increments of E evaluated *to second order* along the fundamental path and along any kinematically admissible branching path, respectively.

We shall discuss below the question whether fulfillment of (1.125) does correspond to instability of the deformation process against application of vanishingly small perturbing forces. Two cases are to be distinguished, depending on whether the equilibrium state is directionally stable or not.

Instability of equilibrium

In the special cases when λ= const or when U is quadratic, (1.125) reduces to

$$\ddot{E}(\tilde{\mathbf{w}}) < 0 \quad \text{at } \lambda = \text{const} \quad \text{for every } \tilde{\mathbf{w}} \in \mathcal{W}, \tag{1.127}$$

which is the energy condition of instability of *equilibrium*, equivalent to (1.89). In general, we only have the implication [34]:

$$(1.127) \Rightarrow (1.125)\,. \tag{1.128}$$

This is in agreement with the observation that instability of equilibrium is a narrower concept than instability of a quasi-static deformation process.

To prove the implication (1.128), consider the quantity

$$\frac{1}{\gamma^2}(J(\tilde{\mathbf{v}}^0 + \gamma\tilde{\mathbf{w}}) - J(\tilde{\mathbf{v}}^0)) \tag{1.129}$$

and put $\gamma \to \infty$ while $\tilde{\mathbf{w}} \in \mathcal{W}$ is held fixed. In the case of controlled nominal loads, from (1.47) it is clear that only the volume integral of $U(\nabla\mathbf{v} + \gamma\nabla\mathbf{w})/\gamma^2$ can contribute to a non-zero limit value of (1.129). On account of homogeneity and continuity of $U(\dot{\mathbf{F}})$, the pointwise limit value of this integrand at each point of differentiability of the field $\tilde{\mathbf{w}}$ is equal to $U(\nabla\mathbf{w})$. The limit value of the expression (1.129) as $\gamma \to \infty$ is thus equal to $I(\tilde{\mathbf{w}})$. It follows that if (1.89) holds for some $\tilde{\mathbf{w}}$ then (1.88) is satisfied for $\tilde{\mathbf{v}} = \tilde{\mathbf{v}}^0 + \gamma\tilde{\mathbf{w}}$ for γ sufficiently large. Since (1.89) and (1.88) are equivalent to (1.127) and (1.125), respectively, the implication (1.128) has been proved. Evidently, the implication remains valid for the configuration-sensitive incremental loading (1.105) since adding to $J(\tilde{\mathbf{v}})$ and $I(\tilde{\mathbf{v}})$ of the quadratic term $R(\tilde{\mathbf{v}})$ and $R(\tilde{\mathbf{w}})$, respectively, has no influence on the above argument.

We have shown in Sect. 1.4.2 that fulfillment of (1.89), and hence of (1.127), implies instability of the current equilibrium state in a dynamic sense for vanishingly small disturbances of *initial* velocities, at least in the first approximation and for a discretized system. The argument can easily be modified to extend its validity to the case of persistent disturbances represented by vanishingly small perturbing forces. Let the function $B(t)$ be defined not by (1.99) but by

$$B(t) = \begin{cases} (\varepsilon/\kappa)(\cosh(\kappa t) - 1) & \text{if} \quad 0 \leq t \leq \tau, \\ (\varepsilon/\kappa)(\cosh(\kappa t) - \cosh(\kappa(t - \tau))) & \text{if} \quad t \geq \tau, \end{cases} \tag{1.130}$$

for some $\tau > 0$. This corresponds to imposing a system of perturbing forces in an initial time interval $[0, \tau)$ at unperturbed initial velocities at $t = 0$. For $t < \tau$ we have $\ddot{B} = \kappa^2 B + \varepsilon\kappa$, and by an analogous argument as in Sect. 1.4.2 we obtain that the perturbing forces corresponding to the path (1.92) with $\tilde{\mathbf{w}} = \tilde{\mathbf{w}}^*$ are constant in time in the interval $[0, \tau)$ and can be made as small as we please if ε is taken to be sufficiently small. For $t \geq \tau$ the function (1.130) corresponds to an inertial motion free of perturbing forces since (1.97) is satisfied. As before, a finite distance from equilibrium is reached at t sufficiently large no matter how small ε and τ are. Hence, under the same approximations as previously, we arrive at the following conclusion [36]: *If* (1.127) *holds then a finite distance from the equilibrium configuration can be reached at* $\lambda = const.$ *in a dynamic motion caused by arbitrarily small perturbing forces acting in an arbitrarily short time interval.*

The above demonstration of an instability of equilibrium associated with (1.127) remains valid for the continuum problem provided the absolute minimum of $I(\tilde{\mathbf{w}})$ on the hypersurface $\Phi(\tilde{\mathbf{w}}) = \text{const.}$ in $\mathcal{W}$ is actually attained. The latter requirement need not always be satisfied if $\mathcal{W}$ is infinite-dimensional (the functional $I(\tilde{\mathbf{w}})$ may be unbounded from below on that hypersurface). However, from a physical point of view it looks unlikely that stability of equilibrium in a dynamic sense can be lost just due to discretization of the system (of course, discretization could have a *stabilizing* effect due to elimination of certain instability modes). The approximation (1.93) seems also to have no essential destabilizing effect since the deviation from the equilibrium configuration, albeit finite, has been assumed small (the assumption of a direct path of departure from equilibrium might have again a *stabilizing* effect, e.g. due to elimination of cycling instability modes – cf. [53]).

Instability of deformation process

Consider now a fundamental process of quasi-static deformation at varying λ and assume that the traversed equilibrium states are directionally stable, in the sense of (1.124). Denote the actual velocity solution by $\tilde{\mathbf{v}}^0$ and consider a range of the deformation where (1.125) is satisfied.

Let us discuss first a discretized system. At any equilibrium state the conditions (1.124) and (1.125) are equivalent to (1.68) and (1.88), respectively. Hence, from the conclusions of Sect. 1.4.3 it follows that at any instant in the fundamental process under consideration there is another first-order solution $\tilde{\mathbf{v}}^* \neq \tilde{\mathbf{v}}^0$ which minimizes $\ddot{E}(\tilde{\mathbf{v}})$ in $\mathcal{V}$. We may assume that for the same loading program there exists a bifurcating branch of quasi-static deformation at *varying* λ initiated with the displacement-rate field $\tilde{\mathbf{v}}^*$ (for at least one minimizer if the minimum is not strict). The bifurcating branch deviates from the fundamental one on a finite distance after some increment of θ, while the bifurcation is possible at *every* point of the fundamental path. This is nothing else as a special case of the instability of a process for persistent disturbances which are represented here by *zero* body forces and are thought merely as an agency which selects the actual solution in velocities. If the velocity solutions at a bifurcation point have comparable chances to determine the actual continuation of the deformation process then any segment of the fundamental deformation path, no matter how short, has zero probability to be followed due to existence of infinitely many alternatives. Moreover, the secondary continuations of the deformation process are energetically preferable, according to the minimum property of $\tilde{\mathbf{v}}^*$ and the interpretation of an increment of E. From a physical point of view, this must be interpreted as an instability of the fundamental deformation process. The instability has fundamentally different character than the instability of equilibrium discussed above: now the fundamental *path* can be left *quasi-statically* under *varying* λ, while previously a departure from an equilibrium *state* took place *dynamically* at *fixed* λ. However, in *both* cases the fundamental path may be treated as unrealizable in a physical system.

The above demonstration of instability of the deformation process, associated with (1.125), remains valid for the continuum problem provided the absolute minimum of $J(\tilde{\mathbf{v}})$ in $\mathcal{V}$ is actually attained. This need not be the case when the functional space $\mathcal{W}$ is infinite-dimensional (even if $\mathcal{W}$ were enlarged to be a complete space). The demonstration of instability then breaks down since existence of a continuous range of bifurcations cannot be definitely concluded. Following [36] it will be indicated below that in a general case the instability can rather be interpreted as *sensitivity of the incremental solution to vanishingly small perturbing forces*, in accord with the concept of instability for persistent disturbances.

Measures of the distance between two configurations of a continuous body are not equivalent to each other. Taking into account that rigid body displacements at fixed λ have been excluded, we will use the measure

$$\| \tilde{\mathbf{w}} \| \equiv \left(\int_V w_{j,i} w_{j,i} \, d\xi \right)^{1/2} \tag{1.131}$$

which is a norm on the linear space $\mathcal{W}$.

For the discretized problem the condition (1.68) ensured that the distance from equilibrium attainable on a straight (direct) path vanished with a vanishing input of the energy supplied by a disturbance. The same condition in an infinite-dimensional space does not ensure this rigorously since $I(\tilde{\mathbf{w}})$ can approach zero in a non-trivial way without achieving it. It is thus reasonable to assume the condition of directional stability of equilibrium in a slightly stronger form than (1.124), viz.

$$\ddot{E}(\tilde{\mathbf{w}}) \geq \alpha \|\tilde{\mathbf{w}}\|^2 \quad \text{at } \lambda = \text{const.} \quad \text{for every } \tilde{\mathbf{w}} \in \mathcal{W}, \tag{1.132}$$

where α is a positive parameter, independent of place, which may be arbitrarily small. For a discretized problem, (1.132) is equivalent to (1.124), and for the continuum problem, the equivalence can be expected to be only violated at certain critical instants.

If (1.132) holds then it can be proved that

$$\ddot{E}(\tilde{\mathbf{v}}) \to +\infty \quad \text{when} \quad \|\tilde{\mathbf{v}}\| \to \infty, \; \tilde{\mathbf{v}} \in \mathcal{V} \tag{1.133}$$

and that $\ddot{E}(\tilde{\mathbf{v}})$ is bounded from below in $\mathcal{V}$. Since an absolute minimum of $\ddot{E}(\tilde{\mathbf{v}})$ in $\mathcal{V}$ need not in general be attained, we modify the argument from Sect. 1.4.3 and examine a minimizing sequence $\{\tilde{\mathbf{v}}_n\} \in \mathcal{V}$, $n = 1, 2, \ldots$, rather than a minimizer itself. For instance, the minimizing sequence can be determined by the Ritz method and the velocity fields $\tilde{\mathbf{v}}_n$ can be solutions to a sequence of discretized rate-problems.

It can be shown (a detailed proof given in [36] is not brief and is omitted here) that the minimizing sequence has the following property: the velocity fields $\tilde{\mathbf{v}}_n$ are solutions to the perturbed first-order rate boundary value problems, such that for any fixed material subdomain $G \subseteq V$ we have

$$|\dot{\mathbf{P}}^*(G, \tilde{\mathbf{v}}_n)| \leq \varepsilon_n \to 0 \quad \text{as } n \to \infty\,, \tag{1.134}$$

where $\dot{\mathbf{P}}^*(G, \tilde{\mathbf{v}}_n)$ is the rate of the resultant perturbing force acting on G. If (1.132) holds simultaneously with (1.125) then the distance between velocity fields $\tilde{\mathbf{v}}_n$ from the minimizing sequence and the fundamental solution $\tilde{\mathbf{v}}^0$, when measured in the sense of (1.131), remains finite as the infimum of $J(\tilde{\mathbf{v}})$ is approached at $n \to \infty$, viz.

$$C_2 > \|\tilde{\mathbf{v}}_n - \tilde{\mathbf{v}}^0\| > C_1 \quad \text{for } n > K \tag{1.135}$$

for some positive constants C_1, C_2 and K. Hence, we arrive at the following instability theorem [36]:

If (1.132) *holds and the fundamental process is unstable in the energy sense* (1.125) *then finite deviations from the fundamental velocity field, measured by* (1.131), *can be caused by vanishingly small perturbations of the equations of continuing equilibrium, in the sense of* (1.134).[8]

Let (1.132) and (1.125) hold along the fundamental path of quasi-static deformation at varying λ in a certain interval of θ. As indicated above, at each equilibrium state of the body within that interval there exists a sequence of perturbed velocity fields $\tilde{\mathbf{v}}_n$ satisfying (1.135) while the rates of perturbing forces tend to zero in the sense of (1.134). Consider a branching path initiated with a displacement rate field $\tilde{\mathbf{v}}_n$; we assume that if the *rates* of perturbing forces are kept fixed then variations of the velocity field along any such path in a small time interval can be neglected. According to (1.135), the measure (1.131) of the distance between body configurations in the fundamental and perturbed processes grows in time with a *finite* rate and achieves a given small but finite value after a small increment of θ. This distance can be reached at perturbing forces as small as we please if n is taken sufficiently large, with the initial instant of their application being arbitrarily chosen from the considered time interval.

This is the final interpretation of the instability of a quasi-static deformation process implied by (1.125) when each of the traversed equilibrium states is (directionally) stable. This corresponds to the instability of a process for arbitrarily small *persistent* disturbances when the distance is defined by (1.131) and the perturbing forces are measured in the sense of (1.134). However, it must be noted that the transition from the above theorem formulated in terms of the *rates* to the final statement expressed in *increments,* although it may seem intuitively obvious, is not mathematically rigorous in view of the additional assumption made. In turn, from a physical point of view the conclusion about instability is strengthened by the fact that the initiation of a branching path with a velocity field $\tilde{\mathbf{v}}_n$ for n sufficiently large is energetically preferable to further continuation of the fundamental process. This follows from the interpretation of $\ddot{E}(\tilde{\mathbf{v}})$ and from the inequality $\ddot{E}(\tilde{\mathbf{v}}_n) < \ddot{E}(\mathbf{v}^0)$ which is implied by (1.125) for n sufficiently large.

---

[8] In an equivalent formulation, if (1.132) and (1.125) hold simultaneously then there is $\varepsilon > 0$ such that for every $\delta > 0$, however small, there is $\tilde{\mathbf{v}}_\delta \in \mathcal{V}$ satisfying $|\dot{\mathbf{P}}^*(G, \tilde{\mathbf{v}}_\delta)| < \delta$ for every fixed $G \subseteq V$ and such that $\|\tilde{\mathbf{v}}_\delta - \tilde{\mathbf{v}}^0\| \geq \varepsilon$.

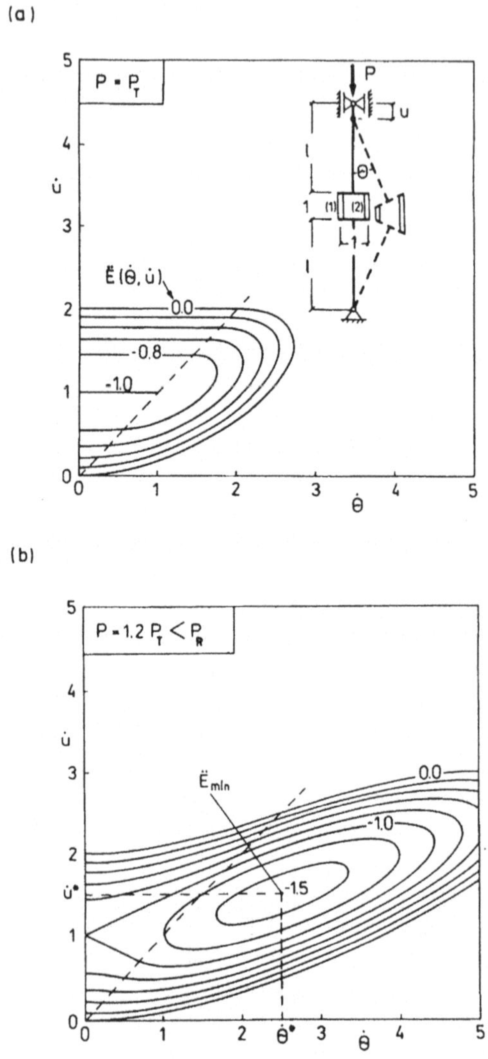

Fig. 1.8 Representation of the functional $\ddot{E}(\tilde{\mathbf{v}})$ for the Shanley column (a) at the tangent modulus load, (b) between the tangent and reduced modulus loads. Numerical values correspond to $\dot{P}/2L^{\mathrm{p}} = 1$, $\ddot{P} = 0$, $L^{\mathrm{p}}/L^{\mathrm{e}} = 0.5$, $L^{\mathrm{p}} = 0.5$. Contours of $\ddot{E} > 0$ are not shown.

Example: The Shanley column

The meaning and range of application of the instability condition (1.125) have already been illustrated on a schematic diagram in Fig. 1.7 for a general system. We will examine now the classical Shanley model of an elastic-plastic column with a central two-flange hinge [23] and give, following [36], an energy interpretation to the well known results. The model has two degrees of freedom: the rotation angle Θ of rigid arms and the relative vertical displacement u of the end points (Fig. 1.8a); $\tilde{\mathbf{v}}$ can thus be identified with $(\dot{\Theta}, \dot{u})$. We will examine stability of the fundamental deformation process $\Theta \equiv 0$ when the vertical compressive load $P \equiv \lambda(t)$ is a given increasing function of time, so that $\Omega = -Pu$. At a certain stage of the fundamental process, the rate of the force carried by an elastoplastic flange (K), $K = 1$ or 2, is given by $\dot{P}_K = L(\dot{e}_K)\dot{e}_K$ (no sum), where $\dot{e}_K$ is the shortening rate of the flange (taken positive for compression). $L(\dot{e})$ is equal to L^{p} (current tangent modulus) or to L^{e} (elastic modulus) for $\dot{e} > 0$ (loading) or $\dot{e} < 0$ (unloading), respectively, with $L^{\mathrm{e}} > L^{\mathrm{p}} > 0$. At an unbuckled configuration with the current dimensions as shown in Fig. 1.8a we have $\dot{e}_1 = \dot{u} + \dot{\Theta}$, $\dot{e}_2 = \dot{u} - \dot{\Theta}$ while $\ddot{u} = \frac{1}{2}(\ddot{e}_1 + \ddot{e}_2) + 2l\dot{\Theta}^2$ (all variables are treated as nondimensional).

Assume for simplicity that P increases with a constant rate so that $\ddot{P} = 0$. Elementary calculations show that the second time derivative of the energy functional (1.104) at an unbuckled equilibrium state reads

$$\ddot{E}(\dot{\Theta}, \dot{u}) = L(\dot{e}_1)\dot{e}_1^2 + L(\dot{e}_2)\dot{e}_2^2 - 2\dot{P}\dot{u} - 2Pl\dot{\Theta}^2. \tag{1.136}$$

The graph of $\ddot{E}(\dot{\Theta}, \dot{u})$ consists of four quadrics joined smoothly along the lines $\dot{\Theta} = \pm\dot{u}$. The uniqueness range (1.121) corresponds to $P < P_{\mathrm{T}}$, where $P_{\mathrm{T}} = L^{\mathrm{p}}/l$ is the tangent modulus load, and in this range the expression (1.136) is minimized by the fundamental continuation $\dot{\Theta}^0 = 0$, $\dot{u}^0 = \dot{P}/2L^{\mathrm{p}}$. If L^{p} decreases continuously along the fundamental path then the uniqueness range terminates at $P = P_{\mathrm{T}}$; this is the primary bifurcation point at which an absolute minimum of $\ddot{E}(\dot{\Theta}, \dot{u})$ is attained at any $\tilde{\mathbf{v}}$ from the "fan" of velocity solutions: $\dot{u} = \dot{u}^0$, $|\dot{\Theta}| \le \dot{u}^0$ (cf. Fig. 1.8a). However, if L^{p} and thus also P_{T} decrease discontinuously then the range $P > P_{\mathrm{T}}$ need not be preceded by $P = P_{\mathrm{T}}$. Within the range $P_{\mathrm{T}} < P < P_{\mathrm{R}}$, where $P_{\mathrm{R}} = 2L^{\mathrm{e}}L^{\mathrm{p}}/l(L^{\mathrm{e}} + L^{\mathrm{p}})$ is the reduced modulus load, the fundamental solution $(\dot{\Theta}^0, \dot{u}^0)$ becomes a saddle point of $\ddot{E}$ so that (1.125) is met while the condition (1.124) of stability of equilibrium is still satisfied. In accord with the theorem from Sect. 1.4.3, along this segment of the fundamental path the absolute minimum value of $\ddot{E}$ is attained at the secondary solution points (cf. Fig. 1.8b) which are easily found to be

$$\dot{\Theta}^* = \pm\frac{\dot{P}}{2L^{\mathrm{p}}}z, \quad \dot{u}^* = \frac{\dot{P}}{2L^{\mathrm{p}}}(1 + (P/P_{\mathrm{T}} - 1)z), \quad \text{where } z \equiv \frac{1 - L^{\mathrm{p}}/L^{\mathrm{e}}}{2(1 - P/P_{\mathrm{R}})}. \tag{1.137}$$

As P is increasing, the column can start to buckle with these rates at *any* point along the segment, without any perturbing forces: this is the present interpretation of the path instability associated with (1.125). Buckling initiated with $(\dot{u}^*, \dot{\Theta}^*)$ requires incrementally less energy to be supplied to the *system* than the fundamental solution, i.e. $\Delta E^* < \Delta E^0$. If $P(t)$ is a gravitational force coming from a mass supplied with a prescribed rate then $\Delta E^* - \Delta E^0$ can be identified with the decrease of the work needed to transport the mass; note that $\dot{u}^* > \dot{u}^0$.

For $P > P_{\mathrm{R}}$ the condition (1.125) is satisfied simultaneously with (1.127), and $\ddot{E}$ is unbounded from below. In this range, instability of the fundamental process is interpreted as instability of the traversed equilibrium states at fixed P in the dynamic sense discussed above.

1.6.4 Applications of the Energy Criterion

1.6.4.1 Unified Treatment of Various Critical Points

Once the general criterion (1.125) for instability of a deformation process has been formulated and justified, it can be applied in a consistent manner to a variety of instability problems mentioned in Sect. 1.6.1. The range of unstable deformations where (1.125) holds can be determined and rejected as apparently unrealizable in a physical system. The initial point of that range represents the critical point on the deformation path, henceforth referred to as $\mathbb{E}$, beyond which some symptoms of instability can be expected. The typical critical points from Hill's theory (i.e. the primary bifurcation point and the primary eigenstate) as well as the critical point associated with shear band bifurcation can be generated in that way from the single condition (1.125) just as *particular cases* of the onset of path instability in the energy sense. An additional advantage of the present energy approach, besides of a unified treatment of various kinds of instability, is that it provides also information about the post-critical behaviour of the system, namely, indicates which post-critical paths cannot be followed in practice and why. Usually, a numerical analysis of development of initial imperfections is adopted to obtain that kind of additional information, and in simplest cases the predictions are the same (cf. e.g. [26, 48]). However, such agreement cannot always be expected since numerical results may be strongly influenced by the form and magnitude of the assumed imperfections, and certain instabilities may remain undetected in the course of numerical computations. Of course, the initial imperfection approach has other advantages as e.g. assessment of the influence of geometrical inaccuracies on the buckling loads.

Suppose that the fundamental deformation process is regular in the sense defined in Sect. 1.3.3, and that the micromechanically based constitutive inequality (1.84) holds. This means that the actual actual velocity solution $\tilde{\mathbf{v}}^0$ corresponds to the rates $\dot{F}^0_{ij}$, $\dot{N}^0_{ij}$ of deformation and stress which satisfy the inequality (1.76). (It is recalled that (1.76) is a weaker restriction than the relative convexity property (1.52) with $C^L_{ijkl} = C_{ijkl}(\dot{\mathbf{F}}^0)$). Then (1.125) is *equivalent* to (1.87) (cf. Fig. 1.7), as follows from (1.115), (1.80), (1.79) and (1.78). For a discretized problem, this can be reformulated as follows: the fundamental process becomes unstable in the energy sense (1.125) exactly when the tangent stiffness matrix (1.101) associated with $\tilde{\mathbf{v}}^0$ becomes indefinite. A number of possibilities now arises depending on the properties of the functionals $I^0(\tilde{\mathbf{w}})$, $J(\tilde{\mathbf{v}})$ and $I(\tilde{\mathbf{w}})$, where $\tilde{\mathbf{w}} \in \mathcal{W}$ and $\tilde{\mathbf{v}} \in \mathcal{V}$, at the critical instant $\mathbb{E}$. Equivalently, one can discuss properties of $\delta^2\ddot{E}(\tilde{\mathbf{v}}^0, \tilde{\mathbf{w}})$, $\ddot{E}(\tilde{\mathbf{v}})$ and $\ddot{E}(\tilde{\mathbf{w}})$; in the discussion below we use the triple $I^0(\tilde{\mathbf{w}})$, $\ddot{E}(\tilde{\mathbf{v}})$ and $I(\tilde{\mathbf{w}})$. The list of possibilities considered below is not intended to be complete.

(i) $\mathbb{E}$ as the primary bifurcation point

Suppose that at $\mathbb{E}$ the condition (1.122) is satisfied simultaneously with (1.68). Then, as already explained in the preceding section, $\mathbb{E}$ coincides with the "usual" primary bifurcation point (in velocities) found from the linearized criterion (1.82). The energy criterion provides additional information that the fundamental post-bifurcation path becomes unstable in the sense discussed in the preceding section and may thus be rejected as a physically unrealistic solution. Stability of the secondary post-bifurcation branches can also be investigated by using (1.125) provided that $\tilde{\mathbf{v}}^0$ changes its previous meaning and denotes the velocity solution along the actually examined branch. If a stable secondary branch exists then it can replace the fundamental one beyond $\mathbb{E}$, which in most typical circumstances can correspond to an observable "geometric" instability.

A smooth bifurcation is found at $\mathbb{E}$ when the condition (1.82) is satisfied but (1.119) still holds. Then bifurcation in velocities (which would require fulfillment of (1.122)) is impossible but the eigenmode $\tilde{\mathbf{w}}^*$ can be added to the fundamental solution in quasi-static *accelerations* to generate another second-order solution, as discussed in Sect. 1.3.2.

$\mathbb{E}$ can correspond also to a primary bifurcation point at which (1.87) is met and which is *not* an eigenstate for the tangent comparison material. This can happen if at $\mathbb{E}$ the tangent moduli C^0_{ijkl} suffer a jump related to an instantaneous drop of the incremental "stiffness" of the material. Although C^0_{ijkl} are discontinuous at $\mathbb{E}$ with respect to θ, they can be still well defined at $\mathbb{E}$ since they relate the *forward* rates of strain and stress. Suppose that at such $\mathbb{E}$ there exists a global minimizer $\tilde{\mathbf{v}}^*$ of $\ddot{E}(\tilde{\mathbf{v}})$ in $\mathcal{V}$, that is,

$$\ddot{E}(\tilde{\mathbf{v}}^0) > \ddot{E}(\tilde{\mathbf{v}}^*) \leq \ddot{E}(\tilde{\mathbf{v}}) \quad \text{for every } \tilde{\mathbf{v}} \in \mathcal{V}. \tag{1.138}$$

By the stationarity principle (1.118), $\tilde{\mathbf{v}}^*$ is a solution so that at $\mathbb{E}$ we have bifurcation in velocities. Existence of that bifurcation cannot be stated by consideration of the incrementally linear comparison solid only. For discretized problems, fulfillment of (1.138) is ensured when (1.87) is satisfied simultaneously with (1.68), as follows from (1.115) and the theorem from Sect. 1.3.3.

Example: Bifurcation under plane strain tension at a non-singular tangent stiffness matrix

The example is taken from the paper [41]. A rectangular specimen is subject to plane strain tension; the material model is from the class proposed by Christoffersen and Hutchinson [33] which describes an incrementally nonlinear response at a yield-surface vertex. The hardening rate is assumed to be initially sufficiently high to ensure uniqueness of a homogeneous solution. Bifurcation is induced by a subsequent discontinuous drop of the current equivalent hardening modulus L^{p} to a value such that the current loading parameter λ, identified e.g. with elongation of the specimen, falls inside the range between λ_{T} and λ_{R} (cf. Fig. 1.7). The bifurcation solutions visualized in Fig. 1.9 have been found by minimizing the velocity functional (1.47), which is equivalent to minimizing $\ddot{E}(\tilde{\mathbf{v}})$, with the help of the finite element method. The velocity field corresponding to Fig. 1.9a describes a diffuse necking

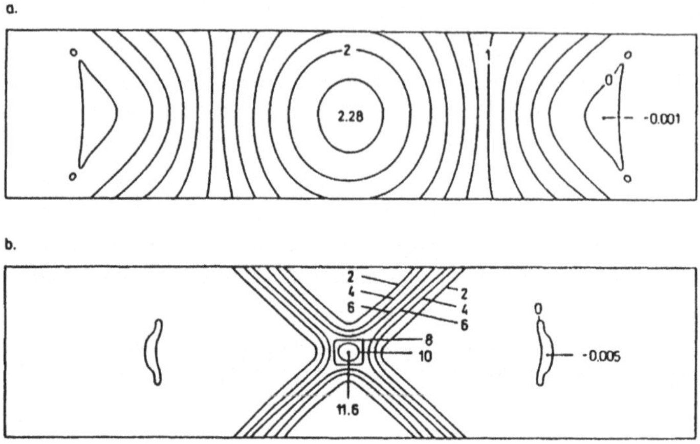

Fig. 1.9 Distribution of maximum principal strain rate in bifurcation solutions at plane strain tension of a rectangular specimen for (a) $L^p = 0.57\tau$, (b) $L^p = 0.14\tau$, where τ and L^p are the current equivalent Kirchhoff stress and hardening modulus, respectively

mode, while in the case (b) of a more pronounced drop of the material stiffness, a shear band mode is clearly visible. The actual tangent stiffness matrix is indefinite for the fundamental homogeneous solution and positive definite for the secondary solutions. Of course, due to inherent nonlinearity of the problem, the magnitude of the bifurcation mode cannot be freely scaled down, contrary to the linearized bifurcation problem discussed in Sect. 1.2.4.

(ii) $\mathbb{E}$ **as the primary eigenstate**

Suppose that at $\mathbb{E}$ the condition (1.126) is satisfied but (1.68) fails. Since (1.126) excludes (1.87), we must have (1.38) so that the primary eigenstate is met. Typically, the fundamental post-critical path becomes then unstable in the dynamic sense associated with fulfillment of (1.127). Depending on the problem, there may or may not be another post-critical branch along which (1.126) would still be satisfied. For instance, if the point $\mathbb{E}$ is met as a limit point then its appearance can be related to an observable "dynamic" instability (snap-through). The limit point can be understood in a generalized sense of an analytic extremum of a controlled loading parameter λ, and not necessarily of a load multiplier (Fig. 1.10). For instance, for a tensile specimen tested in a loading device of finite elastic stiffness, the limit point is met when the load-displacement curve becomes tangent to the (possibly nonlinear) characteristic of the loading device.

(iii) $\mathbb{E}$ **as the onset of material instability**

"Local" or "material" instabilities can be investigated following the same general approach as above, without the need to formulate any auxiliary problem (as is usually done in the "local" bifurcation approach by discussing a rate-problem for an infinite

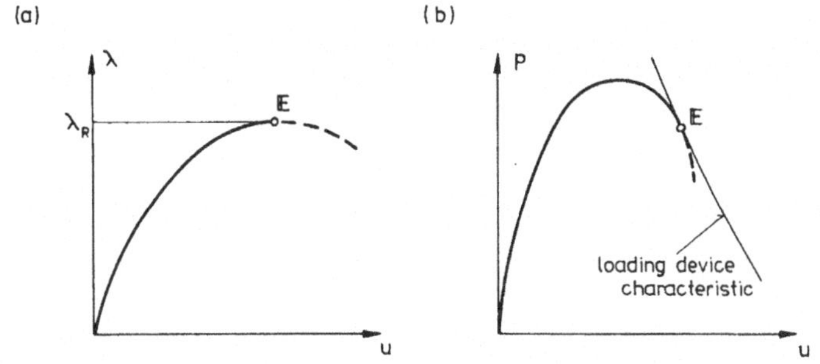

Fig. 1.10 (a) Generalized limit point and (b) its interpretation for an elastic loading device

homogeneous continuum [50]). For, it suffices to consider local conditions whose fulfillment implies (1.125). We do not discuss that problem in much detail here, and restrict ourselves to certain introductory remarks. From the identity (1.115) and the well-known Legendre-Hadamard necessary condition for a minimum of an integral functional, it follows that a *necessary* condition for (1.126) is that at every regular point in the body (i.e. at which the velocity gradient $v^0_{j,i}$ and the respective tangent moduli C^0_{ijkl} are well defined and continuous with respect to ξ_i) the tangent moduli are at least *semi-strongly elliptic,* viz.

$$C^0_{ijkl} n_i n_k g_j g_l \geq 0 \quad \text{for all } n_i, g_j\,. \tag{1.139}$$

Therefore, if the tangent moduli in a deformation process satisfy somewhere in the body the inequality

$$C^0_{ijkl} n_i n_k g_j g_l < 0 \quad \text{for some } n_i, g_j \tag{1.140}$$

then the process is unstable in the energy sense (1.125). It can be seen that the condition of strong ellipticity loss results in a natural way from the energy criterion of instability of a deformation *process*. (Clearly, the condition analogous to (1.140) can be applied to two-dimensional continua intended, for instance, to model a thin metal sheet.) The associated critical condition is (cf. (1.82))

$$\det(C^0_{ijkl}\, n_i n_k) = 0 \tag{1.141}$$

which is commonly interpreted as a critical condition for shear band bifurcation [50]. Observe that here (1.139) results in a natural way as a particular condition for the critical instant $\mathbb{E}$.

In turn, by appealing to a theorem due to Graves [54] we obtain that a necessary condition for the functional $I(\tilde{\mathbf{w}})$ to reach an absolute minimum at $\tilde{\mathbf{w}} = \tilde{\mathbf{0}}$, or equivalently, to take only nonnegative values, is that

$$U(\dot{\mathbf{F}}) \geq 0 \quad \text{for all } \dot{F}_{ji} = g_j n_i \tag{1.142}$$

at every material point. It follows that if somewhere in the body we have

$$U(\dot{\mathbf{F}}) < 0 \quad \text{for some } \dot{F}_{ji} = g_j n_i \tag{1.143}$$

then the equilibrium state is unstable in the energy sense (1.127). Since we have assumed that the constitutive inequality (1.76) holds, (1.143) cannot take place without (1.140), on account of (1.78) which is a consequence of (1.76). The critical condition associated with (1.142) is (cf. (1.71))

$$\dot{N}_{ij}(\dot{\mathbf{F}})\, n_i = 0 \quad \text{for some } \dot{F}_{ji} = g_j n_i \tag{1.144}$$

which corresponds to the possibility of shear band bifurcation when the material outside the band is non-deforming [55].

The conditions (1.140) and (1.143) may be regarded as conditions sufficient for local (or material) instability [34]; it is pointed out that these conditions have been derived for a possibly inhomogeneous, finite body under arbitrary boundary conditions. The distinction between the two conditions is that (1.140) implies an instability of the *process* of uniform straining of a material element while (1.143) implies an instability of *equilibrium* of the element. Transition from the range (1.139) (with strict inequality for nonzero n_i, g_j) to (1.140) without violating (1.142) is related to initiation of quasi-static localization of deformation into shear bands. In turn, in the range (1.143) the shear band formation may take place as a *dynamic* process representing a kind of "internal snap-through".

Remark If (1.76) does not hold then bifurcations can take place before (1.125) is met. This indicates that while the energy approach and the bifurcation approach are closely connected to each other, they are not equivalent. It can be mentioned that if (1.76) does not hold then (1.139) and (1.142) are merely consequences of a more general condition necessary for material stability [34] which is not discussed here.

1.6.4.2 Upper Bound Technique for Determining the Onset of Path Instability

The condition (1.125) provides a basis for a simple upper-bound technique for determining the critical point beyond which the deformation path corresponding to the velocity field $\tilde{\mathbf{v}}^0$ is expected to loose its practical significance. Namely, it suffices to find just one kinematically admissible velocity field satisfying (1.125) to demonstrate instability of the path. This task is further simplified if the implication (1.87) ⇒(1.88) ⇔ (1.125) is taken into account. It follows that we can consider a certain *restricted* class of possible instability modes of required type from $\mathcal{W}$ and then by direct substitution determine the range where (1.87) is satisfied for some field from that class, implying (1.125). To do this, we do not need to solve any equations; of course, the better the choice of the examined class of modes the better is the estimation of the critical point. In particular, *all* admissible modes can be taken into account; then the "most dangerous" instability modes are found as eigenfields for the linearized homogeneous problem for the comparison material. The critical point is then found when the lowest eigenvalue reaches zero, which is nothing else than the usual procedure for determining the primary bifurcation point from (1.60) or an equivalent condition. The upper bound technique with a restricted class of modes is much simpler while the predictions for critical stresses or strains may be quite close to exact bifurcation values. It can be mentioned that the technique works also for stress-strain curves with a dicontinuous tangent, while the bifurcation approach based on the incrementally linear comparison material in general breaks then down.

Example: Upper bound to the onset of necking in 3D-specimens subject to uniaxial tension

Consider a specimen whose cross-section orthogonal to the tensile x_1-axis in the current configuration is independent of x_1 but is otherwise arbitrary (Fig. 1.11a). The usual assumptions of shear-free end conditions and the prescribed elongation rate are adopted. The material is assumed to be homogeneous and orthotropic relative to x_i-axes, but detailed specification of its incremental constitutive law (possibly highly nonlinear) is unknown. Only the uniaxial stress-strain curve and the cross-section evolution in the *fundamental, uniform* deformation process are treated as known, so that the current uniaxial tangent modulus $L^{\mathrm{p}} = \dot{\tau}_1^0/\dot{\varepsilon}_1^0$ and the ratios $\nu_2 = -\dot{\varepsilon}_2^0/\dot{\varepsilon}_1^0$, $\nu_3 = -\dot{\varepsilon}_3^0/\dot{\varepsilon}_1^0$ are given.

In the upper bound technique, a kinematically admissible mode of instability can be assumed arbitrarily. Let the velocity field $\tilde{\mathbf{v}}$, to be substituted in (1.125) as the necking mode, be defined as $\tilde{\mathbf{v}} = \tilde{\mathbf{v}}^0 + \tilde{\mathbf{w}}$, where

$$\begin{aligned} w_1 &= \phi_1\phi_2'\phi_3', \\ w_2 &= -\nu_2\phi_1'\phi_2\phi_3', \\ w_3 &= -\nu_3\phi_1'\phi_2'\phi_3. \end{aligned} \qquad \phi_r = \phi_r(x_r), \quad \phi_r' = d\phi_r/dx_r, \tag{1.145}$$

As explained above, the task of determining the range where (1.125) holds can be simplified by substituting the field $\tilde{\mathbf{w}}$ into (1.87) (with the current configuration as reference). The mode (1.145) has been chosen such that

$$w_{3,3} : w_{2,2} : w_{1,1} = \dot{\varepsilon}_3^0 : \dot{\varepsilon}_2^0 : \dot{\varepsilon}_1^0, \tag{1.146}$$

which eliminates from the integrand in (1.87) all tangent moduli of the type L^0_{rrss} by replacing them by the single uniaxial modulus L^{p}. A further natural assumption

$$\phi_1 = \sin\frac{K\pi x_1}{l}, \quad \phi_2 = \sin\frac{K\pi x_2\sqrt{\nu_2}}{l}, \quad \phi_3 = \sin\frac{K\pi x_3\sqrt{\nu_3}}{l}, \quad K = 1, 2, \ldots, \tag{1.147}$$

compatible with the kinematical constraints $w_1 = 0$ at $x_1 = 0$ and $x_1 = l$, yields $\dot{\varepsilon}_{12} = \dot{\varepsilon}_{13} = 0$. The only remaining unknown modulus L^0_{2323} is assumed to be not greater than the elastic shear modulus G.

On using the above specifications and (1.15), after straightforward transformation we obtain that (1.87) is satisfied if

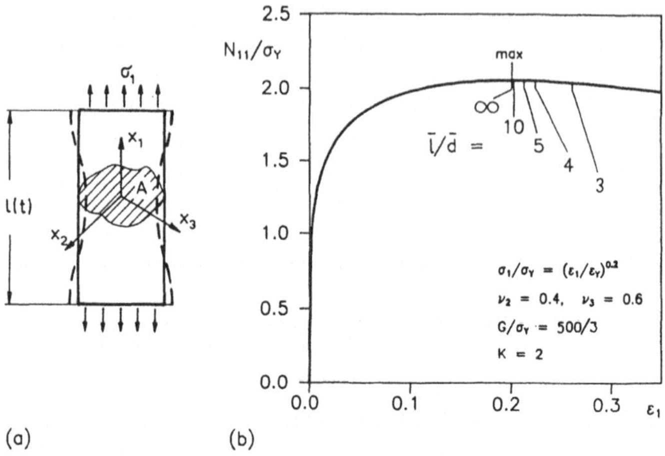

Fig. 1.11 Upper bound to the onset of necking in uniaxial tension; (a) the specimen, (b) results for an orthotropic, initially cylindrical specimen of different initial length/diameter ($\bar{l}/\bar{d}$) ratios on a plot of normalized nominal stress vs logarithmic strain

$$L^{\mathrm{p}} < \sigma_1 - \frac{\nu_2\Phi_2 + \nu_3\Phi_3}{\Psi}\sigma_1 - \frac{4\nu_2\nu_3\Xi}{\Psi}G, \tag{1.148}$$

where

$$\begin{aligned}
\Phi_2 &= \int\int_A \sin^2\frac{K\pi x_2\sqrt{\nu_2}}{l}\cos^2\frac{K\pi x_3\sqrt{\nu_3}}{l}\,dx_2\,dx_3,\\
\Phi_3 &= \int\int_A \sin^2\frac{K\pi x_3\sqrt{\nu_3}}{l}\cos^2\frac{K\pi x_2\sqrt{\nu_2}}{l}\,dx_2\,dx_3,\\
\Psi &= \int\int_A \cos^2\frac{K\pi x_2\sqrt{\nu_2}}{l}\cos^2\frac{K\pi x_3\sqrt{\nu_3}}{l}\,dx_2\,dx_3,\\
\Xi &= \int\int_A \sin^2\frac{K\pi x_2\sqrt{\nu_2}}{l}\sin^2\frac{K\pi x_3\sqrt{\nu_3}}{l}\,dx_2\,dx_3.
\end{aligned} \tag{1.149}$$

In the range (1.148) the uniform deformation *process* is thus necessarily unstable, in the sense discussed in Sect. 1.6.3; in view of the assumed form of the instability mode, necking is the expected symptom of instability. Note that equilibrium can still be stable if the loading device is sufficiently stiff.

In Fig. 1.11b the critical points beyond which (1.148) holds are illustrated for a specimen of an initially circular but currently elliptic cross-section (on account of assumed orthotropy) and for a power hardening uniaxial stress-strain curve. It can be seen that instability is predicted somewhat beyond the maximum load point at which $L^{\mathrm{p}} = \sigma_1$. The last two correction terms in (1.148) tend to zero for infinitely slender specimens, in agreement with the classical Considère criterion. In comparison with the similar results obtained earlier by using the bifurcation approach (cf. [47] for a review), the present conclusion has been obtained for a wider class of materials and for arbitrary cross-sections.

1.6.4.3 A Computational Algorithm for Automatic Choice of the Post-Bifurcation Branch

On the basis of (1.126) as a necessary condition of stability of a deformation process, a computational method has been proposed in [41] for crossing bifurcation points with automatic branch switching. Post-bifurcation branches which are unstable according to (1.125), the fundamental branch in particular, are automatically rejected in the algorithm. The algorithm has been implemented into a finite element code developed by K.Thermann; the results in Fig. 1.9 illustrate applicability of the method also in the case when the stress-strain curve is not smooth. Further details can be found in [41].

References

1. Hill, R.: On the problem of uniqueness in the theory of a rigid-plastic solid, *J. Mech. Phys. Solids,* **5** (1957), 153-161, 302-307.
2. Hill, R.: On uniqueness and stability in the theory of finite elastic strain, *J. Mech. Phys. Solids,* **5** (1957) 229-241.
3. Hill, R.: Stability of rigid-plastic solids, *J. Mech. Phys. Solids,* **6** (1957), 1-8.
4. Hill, R.: A general theory of uniqueness and stability in elastic-plastic solids, *J. Mech. Phys. Solids,* **6** (1958), 236-249.
5. Hill, R.: Some basic principles in the mechanics of solids without a natural time. *J. Mech. Phys. Solids,* **7** (1959), 209-225.
6. Hill, R.: Aspects of invariance in solids mechanics, *Advances in Applied Mechanics, Vol.* **18**, Acad. Press, New York 1978, 1-75.
7. Hill, R.: On the classical constitutive laws for elastic/plastic solids, *Recent Progress in Applied Mechanics, The Folke Odkvist Volume* (Ed B. Broberg *at. al.*), Almqvist & Wiksell, Stockholm 1967, 241-249.
8. Hill, R. and Rice, J.R.: Constitutive analysis of elastic-plastic crystals at arbitrary strain, *J. Mech. Phys. Solids,* **20** (1972), 401-413.
9. Sewell, M.J.: A survey of plastic buckling, in: *Stability* (Ed. H.H.E. Leipholz), Univ. of Waterloo Press, Ontario 1972, 85-197.
10. Koiter, W.T.: Stress-strain relations, uniqueness and variational theorems for elastic-plastic materials with a singular yield surface, *Quart. Appl. Math.,* **11** (1953), 350-353.
11. Mandel, J.: Generalisation de la théorie de plasticité de W.T. Koiter, *Int. J. Solids Structures,* **1** (1965), 273-295.
12. Hill, R.: Generalized constitutive relations for incremental deformation of metal crystals by multislip, *J. Mech. Phys. Solids,* **14** (1966), 95-102.
13. Klushnikov, V.D.: *Stability of Elasic-Plastic Systems* (in Russian), Nauka, Moscow 1980.
14. Petryk, H. and Thermann, K.: Second-order bifurcation in elastic-plastic solids, *J. Mech. Phys. Solids,* **33** (1985), 577-593.
15. Petryk, H. and Mróz, Z.: Time derivatives of integrals and functionals defined on varying volume and surface domains, *Arch. Mech.,* **38** (1986), 697-724.
16. Thomas, T.Y.: *Plastic Flow and Fracture in Solids*, Acad. Press, New York 1961.
17. Hill, R.: Uniqueness criteria and extremum principles in self-adjoint problems of continuum mechanics, *J. Mech. Phys. Solids,* **10** (1962), 185-194.
18. Nguyen, Q.S.: Stabilité et bifurcation des systèmes dissipatifs standarts à comportement indépendant du temps physique, *C. R. Acad. Sci. Paris, t.* **310**, *Serie II,* (1990), 1375-1380.
19. Hill, R.: Eigenmodal deformations in elastic/plastic continua, *J. Mech. Phys. Solids,* **15** (1967), 371-386.

20. Raniecki, B.: Uniqueness criteria in solids with non-associated plastic flow laws at finite deformations, *Bull. Acad. Polon. Sci., Sér. sci. techn.,* **27** (1979), 391-399.
21. Raniecki, B. and Bruhns, O.T.: Bounds to bifurcation stresses in solids with non-associated plastic flow law at finite strain, *J. Mech. Phys. Solids,* **29** (1981), 153-172.
22. Triantafyllidis, N.: On the bifurcation and postbifurcation analysis of elastic-plastic solids under general prebifurcation conditions, *J. Mech. Phys. Solids,* **31** (1983), 499-510.
23. Hill, R.: Bifurcation and uniqueness in non-linear mechanics of continua, in: *Problems of Continuum Mechanics, N. I. Muskhelishvili Anniversary Volume*, SIAM, Philadelphia 1961, 155-164.
24. Hill, R. and Sewell, M.J.: A general theory of inelastic column failure - I, *J. Mech. Phys. Solids,* **8** (1960), 105-111.
25. Hutchinson, J.W.: Post-bifurcation behavior in the plastic range, *J. Mech. Phys. Solids,* **21** (1973), 163-190.
26. Hutchinson, J.W.: Plastic buckling, *Advances in Applied Mechanics, Vol.* **14**, Acad. Press, New York 1974, 67-144.
27. Hill, R. and Sewell, M.J.: A general theory of inelastic column failure - II, *J. Mech. Phys. Solids,* **8** (1960), 112-118.
28. Nguyen, Q.S. and Radenkovic, D.: Stability of equilibrium in elastic plastic solids, *Lecture Notes in Mathematics, Vol.* **503**, Springer, Berlin 1975, 403-414.
29. Shanley, F.R.: Inelastic column theory, *J. Aero. Sci.,* **14** (1947), 261-267.
30. Hill, R.: The essential structure of constitutive laws for metal composites and polycrystals, *J. Mech. Phys. Solids* **15** (1967), 79-95.
31. Lin, T.H.: Physical theory of plasticity, *Advances in Applied Mechanics, Vol.* **11**, Acad. Press, New York 1971, 255-311.
32. Hutchinson, J.W.: Elastic-plastic behaviour of polycrystalline metals and composites. *Proc. Roy. Soc. Lond.,* **A 319** (1970), 247-272.
33. Christoffersen, J.; Hutchinson, W: A class of phenomenological corner theories of plasticity. *J. Mech. Phys. Solids,* **27** (1979), 465-487.
34. Petryk, H.: On energy criteria of plastic instability, in: *Plastic Instability, Proc. Considère Memorial,* Ecole Nat. Ponts Chauss., Paris 1985, 215-226.
35. Petryk, H.: On the theory of bifurcation in elastic-plastic solids with a yield-surface vertex, in: *Inelastic Solids and Structures, Antoni Sawczuk Memorial Volume* (Ed. M.Kleiber and J.A.König), Pineridge Press, Swansea 1990, 131-143.
36. Petryk, H.: The energy criteria of instability in time-independent inelastic solids, *Arch. Mech.,* **43** (1991), No 4, 519-545.
37. Hill, R.: On constitutive macro-variables for heterogeneous solids at finite strain, *Proc. Roy. Soc. Lond.,* **A 326** (1972), 131-147.
38. Petryk, H.: On constitutive inequalities and bifurcation in elastic-plastic solids with a yield-surface vertex, *J. Mech. Phys. Solids,* **37** (1989), 265-291.
39. Nguyen, Q.S. and Petryk, H.: A constitutive inequality for time-independent dissipative solids, *C. R. Acad. Sci. Paris, t.* **312**, *Serie II,* (1991), 7-12.
40. Petryk, H.: On the second-order work in plasticity, *Arch. Mech.,* **43** (1991) No 2/3, 377-397.
41. Petryk, H. and Thermann, K.: On discretized plasticity problems with bifurcations, *Int. J. Solids Structures,* **29** (1992) 745-765.
42. Petryk, H.: A consistent energy approach to defining stability of plastic deformation processes, in: *Stability in the Mechanics of Continua, Proc. IUTAM Symp. Nümbrecht* 1981 (Ed. F.H. Schroeder), Springer, Berlin - Heidelberg 1982, 262-272.
43. Sewell, M.J.: On configuration-dependent loading, *Arch. Rat. Mech. Anal.,* **23** (1967), 327-351.
44. Petryk, H.: On the onset of instability in elastic-plastic solids, in: *Plasticity Today: Modelling, Methods and Applications* (Ed. A. Sawczuk and G. Bianchi), Elsevier, London 1985, 429-447.
45. Hutchinson, J.W.: Imperfection sensitivity in the plastic range, *J. Mech. Phys. Solids,* **21** (1973), 191-204.
46. Storakers, B.: On uniqueness and stability of elastic-plastic deformation, *Arch. Mech.,* **27** (1975), 821-839.

47. Miles, J.P.: On necking phenomena and bifurcation solutions, *Arch. Mech.*, **32** (1980), 909-931.
48. Needleman, A. and Tvergaard, V.: Aspects of plastic postbuckling behavior, in: *Mechanics of Solids, The Rodney Hill 60-th Anniversary Volume* (Ed. H.G. Hopkins and M.J. Sewell), Pergamon Press, Oxford 1982, 453-498.
49. Tvergaard, V.: On bifurcation and stability under elastic-plastic deformation, *Plasticity Today: Modelling, Methods and Applications* (Ed. A. Sawczuk and G. Bianchi), Elsevier, London 1985, 377-398.
50. Rice, J.R.: The localization of plastic deformation, in: *Theoretical and Applied Mechanics* (Ed. W.T. Koiter), North- Holland, Amsterdam 1977, 207-220.
51. Hutchinson J.W. and Tvergaard, V.: Surface instabilities on statically strained plastic solids, *Int. J. Mech. Sci.*, **22** (1980), 339-354.
52. Krasovskii, N.N.: *Stability of Motion,* Stanford Univ. Press 1963.
53. Petryk, H.: On stability and symmetry conditions in time-independent plasticity, *Arch. Mech.*, **37** (1985), 503-520.
54. Graves, L.M.: The Weierstrass condition for multiple integral variation problems, *Duke Math. J.*, **5** (1939), 656-660.
55. Kolymbas, D.: Bifurcation analysis for sand samples with a non-linear constitutive equation, *Ing.-Archiv,* **50** (1981), 131-140.

Chapter 2
Stability and Constitutive Inequalities in Plasticity

Abstract A thermodynamic theory of stability in solids with intrinsic dissipation of plastic type is developed, and various related constitutive inequalities are discussed. The thermodynamic formalism for finite strain elastoplasticity is presented, with rate-dependent plastic behaviour and its rate-independent limit described with the help of internal variables. A general condition sufficient for stability of equilibrium in the sense of Lyapunov is formulated and then successively transformed as additional assumptions are introduced, with special attention focused on isothermal rate-independent plasticity. In the latter case the conditions for stability of a quasi-static process at varying loading are also derived and shown to differ from the respective conditions of stability of equilibrium. The stability conditions are formulated for an arbitrary continuous system with specified boundary conditions as well as for a homogeneous material element embedded in a continuum. Relation between intrinsic instability at the level of a material element and propagation of acceleration waves or strain localization in shear bands is discussed.

2.1 Introduction

The principal topic of these lecture notes is the stability in finite strain plasticity considered from the point of view of thermodynamics. Stability of equilibrium is examined parallel to, and distinguished from, stability of a quasi-static process. In both cases, a continuous, generally inhomogeneous material body subject to specified boundary conditions as well as a material element embedded in a continuum are investigated. Conditions for various types of stability, especially at the level of a material element, lead to a number of constitutive inequalities which form the second, and closely related to the first, topic of these lecture notes.

---

[0] Reprinted from: CISM Courses and Lectures No. 336, Udine 1992 (Ed. W.Muschik), pp. 259-329, Springer-Verlag, Wien 1993. H.Petryk, Stability and constitutive inequalities in plasticity. Copyright © 1993 by CISM, Udine.

H. Petryk, *Theory of Plastic Instability*,
https://doi.org/10.1007/978-3-031-98223-1_2

Stability in continua is known to be a mathematically delicate subject, and plasticity effects introduce additional complications. In the present state of theoretical development and in a general case, one has apparently to be satisfied with stability conditions which are either mathematically rigorous but difficult to be further exploited, or being simpler and verifiable but at the cost of loosing their precise connection with a mathematical definition of stability. The conditions formulated below are frequently of the second type, although an attempt has been made to give them an interpretation as either necessary or sufficient conditions for stability of some specified kind.

Fundamentals of thermodynamics of solids are not discussed here since they were presented in other parts of this CISM Course. In Sect. 2.2 of this Chapter the thermodynamic constitutive formalism for finite strain elasto-plasticity is presented. Rate-dependent (visco-plastic) behaviour and its rate-independent limit are described with the help of internal variables. Two particular classes of rate-independent models are introduced as the examples which will be referred to in the next sections when discussing detailed forms of the stability criteria. Sect. 2.3 is devoted to the general question of stability of thermodynamic equilibrium of an inelastic system. After certain mathematical preliminaries, a general condition for stability of a compound system is formulated. The condition is successively transformed as additional assumptions are introduced; in particular, the energy criterion for isothermal perturbations is investigated in the framework of rate-independent plasticity. In Sect. 2.4 the concept of stability is extended to quasi-static isothermal processes induced by slowly varying external loading. It is shown that the appropriately extended energy criterion of stability is applicable to such processes although under more restrictive assumptions. Stability at the level of a material element embedded in a continuum is investigated in Sect. 2.5. Under the assumption of isothermal deformations and in most cases for rate-independent plasticity, a number of conditions for intrinsic stability of the material are formulated. Connections between the thermodynamic conditions of material stability and the propagation of acceleration waves or the formation of shear bands are then investigated.

In view of a number of essential questions left open, the presentation below certainly does not provide a complete theory of thermodynamic stability in plasticity. In particular, non-isothermal perturbations receive only little attention here. However, it must be pointed out that the assumptions adopted in most previous studies on thermodynamic stability excluded, in fact, the plastic behaviour as understood in the mechanics of solids. The general line of presentation adopted here seems to be novel; also several partial results discussed below have been obtained only recently and some other appear to be new.

2.2 Thermodynamic Formalism in Finite Strain Plasticity

2.2.1 Thermodynamic State of a Material Element

We begin with specifying the notation for basic variables. Denote by $\mathbf{x}$ and $\boldsymbol{\xi}$ the position vector of a material point in a continuum in the actual and reference configuration, respectively. A basic measure of the finite strain and rotation is the deformation gradient $\mathbf{F} = \partial\mathbf{x}/\partial\boldsymbol{\xi}$, expressed according to the polar decomposition theorem as

$$\mathbf{F} \equiv \nabla\mathbf{x} = \mathbf{RU}, \quad \mathbf{R}^T\mathbf{R} = 1, \quad \mathbf{U} = \mathbf{U}^T, \quad \mathbf{e} = \mathbf{e}(\mathbf{U})\,. \tag{2.1}$$

$\mathbf{e}$ stands for an *arbitrary* symmetric strain measure of Lagrangian type. Work-conjugate stress $\mathbf{t}$ is defined by the work differential [1]:

$$\mathrm{d}w = \mathbf{S}\cdot d\mathbf{F} = \mathbf{t}\cdot \mathrm{d}\mathbf{e} \quad \text{(per unit reference volume).} \tag{2.2}$$

A dot between two quantities denotes the appropriate inner product (e.g. $\mathbf{S}\cdot \mathrm{d}\mathbf{F} = S_{ij}\mathrm{d}F_{ij}$ in the indical notation). $\mathbf{S}$ is the first (unsymmetric) Piola-Kirchhoff stress (the transpose of the nominal stress). Among infinitely many work-conjugate pairs $(\mathbf{t}, \mathbf{e})$, one of the most popular ones consists of the Green strain as $\mathbf{e}$ and of the second Piola-Kirchhoff stress as $\mathbf{t}$. However, there is no need here to prefer some specific pair over another, and the reference configuration can be chosen arbitrarily. In the reference configuration any $\mathbf{e}$ vanishes and any $\mathbf{t}$ coincides with the Cauchy stress $\boldsymbol{\sigma}$.

If $\mathbf{e}$ and not $\mathbf{F}$ is used as a work variable for an anisotropic material then in transition to another reference configuration it is necessary to introduce orientation variables, e.g. a director triad or an orientation tensor ($\mathbf{Q}$) [2]. We shall assume that in the reference configuration $\mathbf{Q} = \mathbf{1}$.

Plastic behaviour at a macroscopic scale is considered as a consequence of microstructural rearrangements [3]. They are described (in the reference configuration) by internal variables $\alpha_K, K = 1, 2, \ldots,$ collectively denoted by $\alpha = (\alpha_K)$, whose number and character (scalar or tensorial) depend on the physical mechanisms of plastic deformation as well as on the intended degree of approximation of the real material behaviour. No balance equation for α_K is assumed.

A general thermodynamic theory of solids is presented in the other parts of this CISM course, so that only a brief account is given here. The material element is regarded as a representative macroscopic sample of a heterogeneous material, sufficiently large in size to behave as if it were homogeneous and sufficiently small to be treated as a point in the continuum description. The element is regarded here as a closed system: chemical reactions and diffusion are disregarded so that the chemical composition is ruled out from the list of variables. In the spirit of the classical theory of irreversible processes (e.g. [4]), the material element is assumed to be in local (constrained) equilibrium when its internal energy (u), macroscopic strain ($\mathbf{e}$) and internal microstructure (α) are given. More generally, the local equilibrium state can be viewed as a projection of a current non-equilibrium state under the assumption that relaxation times of non-equilibrium variables (different from α_K) are very short

[5, 6]. The absolute temperature T, specific entropy s and other thermodynamic variables are introduced locally as in thermostatics.

It will be convenient to operate with all quantities and fields defined in the reference configuration (being arbitrary but fixed). To simplify the notation by eliminating the fixed reference mass density $\bar{\rho}$ from the formulae, the densities of: entropy (s), internal energy (u), Helmholtz free energy (ϕ), Gibbs function (ψ), etc., all are taken per unit reference volume.

The Gibbs equation is postulated in the form

$$\mathrm{d}u = \mathbf{t} \cdot \mathrm{d}\mathbf{e} + T\mathrm{d}s - \sum_J \boldsymbol{A}_J \cdot \mathrm{d}\alpha_J\,, \tag{2.3}$$

where $\boldsymbol{A}_J$ are the thermodynamic forces (or affinities) associated with α_J, collectively denoted by $\boldsymbol{A} = (\boldsymbol{A}_J)$. It will be more convenient to work with the Helmholtz free energy density

$$\phi \equiv u - Ts = \phi(\mathbf{e}, T, \alpha) \equiv \phi(\mathfrak{s}), \qquad \mathfrak{s} \equiv (\mathbf{e}, T, \alpha) \tag{2.4}$$

in terms of which (2.3) becomes

$$\mathrm{d}\phi = \mathbf{t} \cdot \mathrm{d}\mathbf{e} - s\mathrm{d}T - \sum_J \boldsymbol{A}_J \cdot \mathrm{d}\alpha_J \tag{2.5}$$

with

$$\mathbf{t} = \partial\phi/\partial\mathbf{e}, \quad s = -\partial\phi/\partial T, \quad \boldsymbol{A}_K = -\partial\phi/\partial\alpha_K\,. \tag{2.6}$$

Viscous effects are neglected so that $\mathbf{t}$ in (2.6) is equal to that in (2.2).

The plastic strain $\hat{\mathbf{e}}$ is, by definition, the strain which remains when the current stress $\mathbf{t}$ and temperature T are reduced at frozen α to zero and to some reference value $\hat{T}$, respectively, to produce a relaxed state. In defining the plastic deformation, there is still an indeterminacy in rotation which is removed if the relaxed state is assumed to be isoclinic [2], where $\mathbf{Q} = \mathbf{1}$ by definition. In an isoclinic relaxed state, we have

$$\mathbf{t} = \mathbf{0}, \quad T = \hat{T}, \quad \mathbf{Q} = \mathbf{1} \quad \Rightarrow \quad \hat{e}, \hat{s}, \hat{u}, \hat{\phi}, \hat{\mathbf{F}} = \hat{\mathbf{R}}\hat{\mathbf{U}} \quad \text{as functions of } \alpha, \tag{2.7}$$

where the quantities with (ˆ) denote in particular: plastic strain $\hat{\mathbf{e}}$, structural entropy ($\hat{s}$), stored internal energy ($\hat{u}$), plastic deformation ($\hat{\mathbf{F}}$), plastic rotation ($\hat{\mathbf{R}}$).

The relaxed state is frequently taken as a moving reference state for a material element, with the free energy dependent explicitly on the elastic and not total strain. Since the elastic strain can be expressed in terms of the total strain $\mathbf{e}$ and α, this is contained in the description with a fixed reference state which, for our present purposes, appears to be more convenient. Various aspects of thermoplastic constitutive laws are discussed e.g. in [2, 3, 7, 8, 9].

2.2.2 Thermodynamic Process

A thermodynamic process $\wp : t \to \jmath(t)$ which takes place in the representative material element in some interval of time t at varying $(\mathbf{e}, T)(t)$ is viewed as a sequence of constrained equilibrium states close to each other. Transitory non-equilibrium processes between such states, and also all instabilities at a microscopic level which correspond to the physical nature of plastic deformation [10] are not examined here. Rapid localized structural changes in a small part of the representative volume are imagined to be "smoothed out" in space and time so that at the macroscopic level they correspond to regular variations of appropriately selected internal variables α_K of averaging type. Consequently, the function $\jmath(t)$ is assumed to be continuous and (at least) piecewise differentiable; a dot over a symbol will denote the right-hand time derivative at a fixed material point, e.g. $\dot{\alpha} = (\mathrm{d}^+/\mathrm{d}t)\alpha$.

The entropy production inequality (per unit reference volume) reads:

$$T\sigma \equiv \mathbf{t} \cdot \dot{\mathbf{e}} - \dot{u} + T\dot{s} - \mathbf{q} \cdot \nabla T/T \geq 0, \tag{2.8}$$

where $\mathbf{q}$ is the heat flux transformed to the reference configuration. The rate of entropy production (σ) is assumed to be split into two nonnegative parts, resulting in:

the thermal dissipation inequality:

$$T\sigma^{\mathrm{T}} \equiv -\mathbf{q} \cdot \nabla T/T \geq 0\,, \tag{2.9}$$

the intrinsic dissipation inequality:

$$\dot{\mathcal{D}} \equiv T\sigma^{\mathrm{p}} \equiv \mathbf{t} \cdot \dot{\mathbf{e}} - (\dot{\phi} + s\dot{T}) \geq 0\,. \tag{2.10}$$

On substituting (2.5), the inequality (2.10) takes the form

$$\dot{\mathcal{D}} = T\sigma^{\mathrm{p}} = \sum_J \boldsymbol{A}_J \cdot \dot{\alpha}_J \equiv \boldsymbol{A} \cdot \dot{\alpha} \geq 0\,. \tag{2.11}$$

In the following, the more condensed notation ($\dot{\mathcal{D}} = \boldsymbol{A} \cdot \dot{\alpha}$) will be used in most cases, with the obvious meaning of the dot between $\boldsymbol{A}$ and α.

Introduce the Gibbs function ψ as the partial Legendre transform of ϕ with respect to $\mathbf{e}$:

$$\psi \equiv \phi - \mathbf{e} \cdot \mathbf{t} = \psi(\mathbf{t}, T, \alpha_K), \quad \mathbf{e} = -\partial\psi/\partial\mathbf{t}, \quad \boldsymbol{A}_K = -\partial\psi/\partial\alpha_K\,. \tag{2.12}$$

By equating the cross-derivatives of ϕ or ψ, we obtain the Maxwell relations which involve $\boldsymbol{A}_K$:

$$\partial\mathbf{t}(\mathbf{e}, T, \alpha)/\partial\alpha_K = -\partial\boldsymbol{A}_K(\mathbf{e}, T, \alpha)/\partial\mathbf{e}, \tag{2.13}$$

$$\partial\mathbf{e}(\mathbf{t}, T, \alpha)/\partial\alpha_K = \partial\boldsymbol{A}_K(\mathbf{t}, T, \alpha)/\partial\mathbf{t}. \tag{2.14}$$

Let the plastic part of stress-rate be defined as

$$\dot{\mathbf{t}}^{\mathrm{p}} \equiv (\partial \mathbf{t}(\mathbf{e}, T, \alpha)/\partial \alpha) \cdot \dot{\alpha} = \dot{\alpha} \cdot \partial^2 \phi/\partial \alpha \partial \mathbf{e} = -\sum_J \dot{\alpha}_J \cdot \partial \boldsymbol{A}_J(\mathbf{e}, T, \alpha)/\partial \mathbf{e} \quad (2.15)$$

and the plastic part of strain-rate as

$$\dot{\mathbf{e}}^{\mathrm{p}} \equiv (\partial \mathbf{e}(\mathbf{t}, T, \alpha)/\partial \alpha) \cdot \dot{\alpha} = -\dot{\alpha} \cdot \partial^2 \psi/\partial \alpha \partial \mathbf{t} = \sum_J \dot{\alpha}_J \cdot \partial \boldsymbol{A}_J(\mathbf{t}, T, \alpha)/\partial \mathbf{t}. \quad (2.16)$$

Then we have

$$\dot{\mathbf{e}} = \dot{\mathbf{e}}^{\mathrm{e}} + \dot{\mathbf{e}}^{\mathrm{p}}, \quad \dot{\mathbf{t}} = \dot{\mathbf{t}}^{\mathrm{e}} + \dot{\mathbf{t}}^{\mathrm{p}}, \quad \dot{\mathbf{t}}^{\mathrm{p}} = -\partial^2 \phi/\partial \mathbf{e} \partial \mathbf{e} \cdot \dot{\mathbf{e}}^{\mathrm{p}}, \quad (2.17)$$

where

$$\dot{\mathbf{t}}^{\mathrm{e}} = \phi_{,\mathbf{ee}} \cdot \dot{\mathbf{e}} + \phi_{,T\mathbf{e}} \dot{T}, \quad \dot{\mathbf{e}}^{\mathrm{e}} = -\psi_{,\mathbf{tt}} \cdot \dot{\mathbf{t}} - \psi_{,T\mathbf{t}} \dot{T} \quad (2.18)$$

are the thermoelastic stress-rate and strain-rate, respectively. The usual short-hand notation has been introduced, where a comma as a lower index denotes partial derivation with respect to the arguments which follow the comma.

In general, we have

$$\dot{\mathbf{e}}^{\mathrm{p}} \neq (\hat{\mathbf{e}})^{\bullet} \quad (2.19)$$

but in many practical cases the difference can be neglected. For further discussion on the structure of incremental relations in plasticity, see e.g. [11].

2.2.3 Kinetics for Rate-Dependent Plastic Solid

Evolution equations for internal variables in rate-dependent plasticity are assumed in the form

$$\dot{\alpha}_K = \dot{\alpha}_K(\boldsymbol{A}, T, \alpha) \quad (2.20)$$

upon which the restriction (2.11) is imposed. Since under the assumption of local equilibrium $\boldsymbol{A}_K$ are functions of $\mathfrak{s}$ by (2.6)$_3$, we can also rewrite (2.20) as $\dot{\alpha}_K = \dot{\alpha}_K(\mathfrak{s})$.

Our special attention will be directed to the cases where the equations (2.20) admit a potential with respect to the thermodynamic forces, viz.

$$\dot{\alpha}_K = \partial \omega(\boldsymbol{A}, T, \alpha)/\partial \boldsymbol{A}_K, \quad (2.21)$$

which is (formally) equivalent to assuming the following generalization of the Onsager reciprocal relations to the present nonlinear kinetic law:

$$\partial \dot{\alpha}_K/\partial \boldsymbol{A}_J = \partial \dot{\alpha}_J/\partial \boldsymbol{A}_K. \quad (2.22)$$

This is, of course, an approximation to reality and not a law of nature. The potential form (2.21) can be derived from the assumption [12, 3] that $\boldsymbol{A}, \alpha$ can be represented by a number of scalars A_K, α_K such that each $\dot{\alpha}_K$ is dependent on the associated force A_K and not on other $A_J, J \neq K$, viz.

$$\dot{\alpha}_K = \dot{\alpha}_K(A_K, T, \alpha)\,. \tag{2.23}$$

(2.23) can evidently be recast in the form

$$\dot{\alpha}_K = \partial\omega(A, T, \alpha)/\partial A_K, \quad \omega = \sum_K \int_0^{A_K} \dot{\alpha}_K(B_K, T, \alpha)\, \mathrm{d}B_K\,.$$

If A_K are expressed through (2.12) as functions of $(\mathbf{t}, T, \boldsymbol{\alpha})$ then from (2.16) we obtain [12, 3] that the plastic part of the macroscopic strain rate is also derivable from a potential, viz.

$$\dot{\mathbf{e}}^{\mathrm{p}} = \partial\omega(\mathbf{t}, T, \boldsymbol{\alpha})/\partial\mathbf{t}, \quad \omega(\mathbf{t}, T, \boldsymbol{\alpha}) \equiv \omega(\boldsymbol{A}(\mathbf{t}, T, \boldsymbol{\alpha}), T, \boldsymbol{\alpha})\,. \tag{2.24}$$

Under the assumption that $\dot{\alpha}_K$ is a non-decreasing function of A_K at fixed T and α, the potential ω is a convex function of A_K.

Analogously, the equation (2.24) results also from a more general assumption (2.21). If ω in (2.21) is a strictly convex (and everywhere differentiable) function of $\boldsymbol{A}_K$ at fixed $T, \boldsymbol{\alpha}$ then we can introduce its partial Legendre transform with respect to $\boldsymbol{A}_K$ to obtain an inverse form of (2.21):

$$\boldsymbol{A}_K = \partial d(\dot{\boldsymbol{\alpha}}, T, \boldsymbol{\alpha})/\partial\dot{\boldsymbol{\alpha}}_K, \quad d(\dot{\boldsymbol{\alpha}}, T, \boldsymbol{\alpha}) \equiv \boldsymbol{A} \cdot \dot{\boldsymbol{\alpha}} - \omega\,. \tag{2.25}$$

The intrinsic dissipation inequality then reads

$$D = d + \omega \geq 0, \quad (d + \omega = 0 \Leftrightarrow \text{ all } \dot{\boldsymbol{\alpha}}_K = \mathbf{0}) \tag{2.26}$$

while the compatibility between $\boldsymbol{A}_K$ in (2.25) and (2.6) requires that

$$\partial\phi/\partial\boldsymbol{\alpha}_K + \partial d/\partial\dot{\boldsymbol{\alpha}}_K = \mathbf{0}\,. \tag{2.27}$$

Note that the potential d does not coincide with the rate $\dot{\mathcal{D}}$ of intrinsic dissipation.

In a scale of time such that creep can be neglected, experiments on metals show that there are no macroscopically observable changes in the state of the material if the stress is kept within some (elastic) domain at T = const. Having assumed that changes of $\boldsymbol{\alpha}_K$ are responsible for the plastic strains, it follows from (2.14) that $\partial\boldsymbol{A}_K/\partial\mathbf{t}$ cannot vanish identically; hence, the elastic domain in stress space corresponds to some elastic domain in $\boldsymbol{A}_K$-space. Accordingly, we assume that

$$\dot{\boldsymbol{\alpha}}_K = \mathbf{0} \quad \text{if } \boldsymbol{A}_K \in C_K(T, \boldsymbol{\alpha}), \quad \mathbf{0} \in C_K; \tag{2.28}$$

the fact that $\dot{\boldsymbol{\alpha}}_K = \mathbf{0}$ need not correspond to the zero point in $\boldsymbol{A}_K$-space but can take place in a finite domain C_K is crucial in plasticity. The elastic domain in $\boldsymbol{A}$-space is obtained as a common part of all C_K :

$$C = C(T, \alpha) \equiv \bigcap_K C_K(T, \alpha). \tag{2.29}$$

Note that even if the boundary ∂C_K of each C_K is smooth, the elastic domain C has in general a boundary ∂C which is only piecewise smooth.

If $\boldsymbol{A}(\mathbf{e}, T, \alpha(t)) \notin \overline{C}(T, \alpha(t))$ where $\overline{C} \equiv C \cup \partial C$ then $\dot{\alpha}(t) \neq 0$; we assume that if $(\mathbf{e}, T)$ are kept constant then an asymptotic state $\mathfrak{z}^\infty$ is eventually approached, viz.

$$\begin{aligned} \mathfrak{z}^\infty(\mathbf{e}, T, \alpha(0)) = (\mathbf{e}, T, \alpha^\infty): \quad \alpha_K^\infty = \alpha_K(0) + \int_0^\infty \dot{\alpha}_K(\mathbf{e}, T, \alpha(t))\, \mathrm{d}t\,, \\ \dot{\alpha}_K(\mathfrak{z}^\infty) = 0, \quad \boldsymbol{A}_K(\mathfrak{z}^\infty) \in \overline{C}_K(T, \alpha^\infty). \end{aligned} \tag{2.30}$$

It is essential that α^∞ depends in general on the initial value $\alpha(0)$ and that $A(\mathfrak{z}^\infty)$ need not vanish in the asymptotic state. The latter property is of fundamental importance in formulating the stability conditions in plasticity. A state $\mathfrak{z}^\infty$ satisfying (2.30) is, by definition, an equilibrium state of the material element at the prescribed strain $\mathbf{e}$ and temperature T.

The loading functions $f_K(\boldsymbol{A}_K, T, \alpha)$ are defined such that

$$f_K < 0 \text{ if } \boldsymbol{A}_K \in C_K(T, \alpha), \quad f_K = 0 \text{ if } \boldsymbol{A}_K \in \partial C_K(T, \alpha), \quad f_K > 0 \text{ otherwise.} \tag{2.31}$$

2.2.4 Rate-Independent Plasticity

Experiments performed on metals show that the stresses during plastic flow often only slightly depend on the strain-rate in a certain range of deformation rates and temperatures. In that range, we can consider a rate-independent limit of plastic behaviour as relaxation times for internal structural rearrangements tend either to zero or to infinity. That limit is not uniquely defined and can depend on the time scale of changes of external loading and on the split of internal variables into completely frozen and possibly active. Suppose that the former have been eliminated from consideration. According to (2.30), in the rate independent limit we have

$$f_K \leq 0, \quad \mathfrak{z} \to \mathfrak{z}^\infty \tag{2.32}$$

at every stage of a process. Alternatively, we can take

$$\omega(\boldsymbol{A}, T, \alpha) = 0 \quad \text{if } \boldsymbol{A} \in \overline{C}(T, \alpha), \qquad \omega(\boldsymbol{A}, T, \alpha) = +\infty \quad \text{if } \boldsymbol{A} \notin \overline{C}(T, \alpha) \tag{2.33}$$

and rewrite (2.21) and (2.25) in terms of subgradients [13, 14] rather than partial derivatives.

If the internal variables α_K describe local rearrangements in the representative material element then the resulting limit process $t \to \mathfrak{z}^\infty(t)$ need not be continuous. For, the attractor α^∞ can suffer a discontinuous jump that corresponds to an instability at the micro-level. As indicated above, such instabilities are beyond the scope of this presentation. Rather, the variables α_K are understood as averaging internal variables which vary continuously in time. Consequently, only continuous processes $\not p$ will be considered.

A complementary assumption to (2.32) is that $\dot{\alpha}_K$ can be non-zero only if the respective yield condition in $\boldsymbol{A}$-space is satisfied:

$$\dot{\alpha}_K \neq \mathbf{0} \quad \Rightarrow \quad f_K(\boldsymbol{A}_K, T, \alpha) = 0 \quad \Leftrightarrow \quad \boldsymbol{A}_K \in \partial C_K(T, \alpha). \tag{2.34}$$

Transformed to **e**-space [1], this reads:

$$g_K(\mathfrak{z}) \equiv g_K(\mathbf{e}, T, \alpha) \leq 0, \quad g_K \dot{\alpha}_K = \mathbf{0}, \quad g_K(\mathfrak{z}) \equiv f_K(\boldsymbol{A}_K(\mathfrak{z}), T, \alpha)\,. \tag{2.35}$$

Contrary to (2.20), the actual values of $\dot{\alpha}_K$ depend now not only on the state $\mathfrak{z}$ but also on $\dot{\mathbf{e}}$ and $\dot{T}$.

As examples, two particular cases of the time-independent constitutive framework are considered below. Note that evolution equations for $\dot{\alpha}_K$ are contained in an indirect form.

Example (1)

Consider the dissipation function of the form:

$$\mathcal{D} = \sum_J \boldsymbol{A}_J \cdot \dot{\alpha}_J = D(T, \alpha, \dot{\alpha}) \geq 0, \quad D(T, \alpha, r\dot{\alpha}) = rD(T, \alpha, \dot{\alpha}) \quad \text{for every } r > 0, \tag{2.36}$$

as a convex (and homogeneous of degree one as indicated) function of $\dot{\alpha}_j$. In place of (2.21) and (2.25), the normality law which follows is written as:

$$\dot{\alpha}_K \in \partial\omega(\boldsymbol{A}_K), \quad \boldsymbol{A}_K \in \partial D(\dot{\alpha}_K), \quad T, \alpha \text{ fixed}, \tag{2.37}$$

where, on account of non-differentiability of D at $\dot{\alpha} = 0$ (cf. (2.36)) and of ω at $\boldsymbol{A} \in \partial C$ (cf. (2.33)), the subdifferential notation [14] has been used in place of partial derivatives. Alternatively, the above convexity and normality assumptions can be replaced by the postulated principle of maximum dissipation rate:

$$(\boldsymbol{A} - \boldsymbol{A}^*) \cdot \dot{\alpha} \geq 0 \quad \text{for every} \quad \boldsymbol{A}^* \in \bar{C}(T, \alpha)\,, \tag{2.38}$$

where $\boldsymbol{A}$ is associated with $\dot{\alpha}$. (2.38) implies that the domain C in $\boldsymbol{A}$-space is convex and that $\dot{\alpha}$ lies within the cone generated by outward normals to the boundary ∂C at given $\boldsymbol{A}(\mathfrak{z})$. For our purposes, validity of (2.38) will be required at the actual $\boldsymbol{A}$ for $|\boldsymbol{A} - \boldsymbol{A}^*| < \varepsilon$ only, where ε is an arbitrarily small positive number. It will

[1] It is common to consider the yield condition in stress-space rather than in strain-space. However, the description adopted here is applicable under more general assumptions.

be shown later how that hypothesis is related to material stability with respect to internal structural rearrangements.

The compatibility between (2.36) and $(2.6)_3$ requires that

$$D(T, \alpha, \dot{\alpha}^*) = \boldsymbol{A}(\mathfrak{z}) \cdot \dot{\alpha}^* \quad \Leftrightarrow \quad \sum_K (\partial\phi/\partial\alpha_K + \partial D/\partial\dot{\alpha}^*_K) \cdot \dot{\alpha}^*_K = 0 \tag{2.39}$$

for all virtual rates $\dot{\alpha}^*$ consistent with the actual thermodynamic forces $\boldsymbol{A}(\mathfrak{z})$; in obtaining the right-hand form, the Euler theorem on homogeneous functions has been used. In a process $\mathfrak{p}$, (2.38) holds for $|\boldsymbol{A} - \boldsymbol{A}^*| < \varepsilon$ at every instant, by assumption. Hence, among all virtual $\dot{\alpha}^*$ which satisfy (2.39), the actual rate $\dot{\alpha}$ in $\mathfrak{p}$ must satisfy the following incremental compatibility conditions

$$\begin{aligned} &(\tfrac{\mathrm{d}^+}{\mathrm{d}t}\boldsymbol{A}(\mathfrak{z})) \cdot \dot{\alpha} = D_{,\alpha}(T, \alpha, \dot{\alpha}) \cdot \dot{\alpha} + D_{,T}(T, \alpha, \dot{\alpha})\dot{T}, \qquad \dot{\alpha} = \tfrac{\mathrm{d}^+}{\mathrm{d}t}\alpha \text{ in } \mathfrak{p}, \\ &(\tfrac{\mathrm{d}^+}{\mathrm{d}t}\boldsymbol{A}(\mathfrak{z})) \cdot \dot{\alpha}^* \leq D_{,\alpha}(T, \alpha, \dot{\alpha}^*) \cdot \dot{\alpha} + D_{,T}(T, \alpha, \dot{\alpha}^*)\dot{T} \ \text{ for all } \dot{\alpha}^* \text{ which satisfy (2.39)}, \end{aligned} \tag{2.40}$$

where the dependence of D on α and T has been assumed smooth (Gateaux differentiable).

By using (2.40) and the expression for $\dot{\boldsymbol{A}}$ obtained from (2.6) it can be shown that $\dot{\alpha}$ at given $\mathfrak{z}$ is uniquely defined by $\dot{T}$ and $\dot{\mathbf{e}}$ if

$$(\dot{\alpha} - \dot{\alpha}^*) \cdot \phi_{,\alpha\alpha} \cdot (\dot{\alpha} - \dot{\alpha}^*) + (\dot{\alpha} - \dot{\alpha}^*) \cdot (D_{,\alpha}(T, \alpha, \dot{\alpha}) - D_{,\alpha}(T, \alpha, \dot{\alpha}^*)) > 0 \quad \text{for all } \dot{\alpha} \neq \dot{\alpha}^* \text{ which satisfy (2.39)}. \tag{2.41}$$

We will make use of the additional symmetry assumption

$$D_{,\alpha}(T, \alpha, \dot{\alpha}) \cdot \dot{\alpha}^* = D_{,\alpha}(T, \alpha, \dot{\alpha}^*) \cdot \dot{\alpha} \tag{2.42}$$

for every $\dot{\alpha}, \dot{\alpha}^*$ which satisfy (2.39). The meaning of the mathematical condition (2.42) will be seen later (cf. the property (2.110)).

The incremental framework with D dependent on α has been recently developed in [15, 16] where references to the related earlier work can be found; the above extension to a variable temperature is straightforward.

Example (2)

The direction (but not the magnitude) of each $\dot{\alpha}_K$ is assumed to be restricted by an individual normality rule in the form

$$\dot{\alpha}_K = \dot{\gamma}_K \boldsymbol{P}_K(\boldsymbol{A}_K, T, \alpha), \quad \boldsymbol{P}_K \equiv \partial f_K(\boldsymbol{A}_K, T, \alpha)/\partial \boldsymbol{A}_K, \quad \dot{\gamma}_K \geq 0, \quad f_K \dot{\gamma}_K = 0\,. \tag{2.43}$$

The related dissipation function can be written as

$$D = D(\boldsymbol{A}_J, T, \alpha; \dot{\gamma}_J) = \sum_J \pi^{\mathrm{c}}_J(\boldsymbol{A}_J, T, \alpha)\dot{\gamma}_J \tag{2.44}$$

while f_K can be identified with the difference between the thermodynamic "driving force" π_K and its threshold value π_K^c , viz.

$$f_K \equiv \pi_K - \pi_K^c, \quad \pi_K = \pi_K(\boldsymbol{A}_K, T, \alpha) \equiv \boldsymbol{A}_K \cdot \boldsymbol{P}_K\,. \tag{2.45}$$

Note that this implies

$$\partial\pi_K^c/\partial\boldsymbol{A}_K = \boldsymbol{A}_K \cdot \partial^2 f_K/\partial\boldsymbol{A}_K\partial\boldsymbol{A}_K = \boldsymbol{A}_K \cdot \partial\boldsymbol{P}_K/\partial\boldsymbol{A}_K\,. \tag{2.46}$$

The yield functions g_K in strain space are given by (2.35), with the compatibility condition

$$g_K(\mathfrak{s})\dot{\gamma}_K^* = 0 \tag{2.47}$$

satisfied by all virtual $\dot{\gamma}_K^*$. The actual $\dot{\gamma}_K$ must satisfy the incremental compatibility condition

$$\dot{g}_K = \boldsymbol{\lambda}_K \cdot \dot{\mathbf{e}} + g_{KT}\dot{T} - \sum_J g_{KJ}\dot{\gamma}_J \leq 0 \quad \text{if } g_K = 0, \quad \text{with } \dot{g}_K\dot{\gamma}_K = 0, \tag{2.48}$$

where, on account of (2.46),

$$\begin{aligned}
\boldsymbol{\lambda}_K &\equiv \partial g_K/\partial\mathbf{e} = \boldsymbol{P}_K \cdot \boldsymbol{A}_{K,\mathrm{e}}\,,\\
g_{KT} &\equiv \partial g_K/\partial T = \boldsymbol{P}_K \cdot \boldsymbol{A}_{K,T} + \boldsymbol{A}_K \cdot \boldsymbol{P}_{K,T} - \pi^c_{K,T}\,,\\
g_{KJ} &\equiv -\partial g_K/\partial\alpha_J \cdot \boldsymbol{P}_J = (\boldsymbol{P}_K \cdot \phi_{,\alpha_K\alpha_J} - \boldsymbol{A}_K \cdot \boldsymbol{P}_{K,\alpha_J} + \pi^c_{K,\alpha_J}) \cdot \boldsymbol{P}_J\,.
\end{aligned} \tag{2.49}$$

It can be shown that $\dot{\gamma}_K$ at given $\mathfrak{s}$ is uniquely defined by $\dot{T}$ and $\dot{\mathbf{e}}$ if

$$(g_{KJ}) \text{ is positive definite,} \quad K, J : g_K = g_J = 0\,. \tag{2.50}$$

An analog to the symmetry property (2.42) is

$$g_{KJ} = g_{JK}, \quad K, J : g_K = g_J = 0\,; \tag{2.51}$$

the meaning of (2.51) will be seen later; cf. the property (2.114). Of course, all the above quantities without a dot depend on the state $\mathfrak{s}$.

The above constitutive framework is analogous to that developed in [2, 17, 18].

In both Examples, from the incremental compatibility condition (2.40) or (2.48) the actual values of $\dot{\alpha}_K$ can be determined for given $\dot{\mathbf{e}}, \dot{T}$. Uniqueness of such $\dot{\alpha}_K$ will be assumed as granted, except in Sect. 2.5.2. Note that this does not ensure uniqueness of $\dot{\alpha}_K$ when $\dot{\mathbf{t}}$ is given in place of $\dot{\mathbf{e}}$. If $\dot{\alpha}_K$ have been found then the incremental constitutive law (the constitutive rate equations) follows from (2.17).

For given $\mathfrak{s}$ and $\dot{T}$, one can distinguish in $\dot{\mathbf{e}}$-space three constitutive cones: the total loading cone, the total unloading cone and the transitory range of partial unloading. The total loading cone contains $\dot{\mathbf{e}}$ such that the associated $\dot{\alpha}_K$ are nonzero for all K such that $g_K(\mathfrak{s}) = 0$. The total unloading condition in Example (2) is

$$\lambda_K \cdot \dot{\mathbf{e}} + g_{KT}\dot{T} < 0 \quad \text{if } g_K(\mathfrak{z}) = 0\,. \tag{2.52}$$

We will use later $(\mathbf{S}, \mathbf{F})$ as a pair of stress and deformation measures particularly useful for examining stability at given boundary conditions. In analogy to $(2.17)_2$ and (2.15), we have

$$\dot{\mathbf{S}} = \dot{\mathbf{S}}(\mathfrak{z}, \dot{\mathbf{F}}, \dot{T}) = \phi_{,\mathbf{FF}} \cdot \dot{\mathbf{F}} + \phi_{,T\mathbf{F}}\dot{T} + \dot{\alpha} \cdot \phi_{,\alpha\mathbf{F}}, \quad \phi = \phi(\mathfrak{z}), \quad \mathfrak{z} = (\mathbf{F}, T, \alpha)\,. \tag{2.53}$$

For isothermal processes and at the assumed constitutive uniqueness in determining $\dot{\alpha}$, each of the symmetry conditions: (2.42) or (2.51) implies existence of the velocity-gradient potential introduced by Hill [19]

$$\dot{\mathbf{S}} = \partial U / \partial \dot{\mathbf{F}}, \quad U = U(\mathfrak{z}, \dot{\mathbf{F}}) = \frac{1}{2}\dot{\mathbf{S}} \cdot \dot{\mathbf{F}}, \quad T = \text{ const}\,. \tag{2.54}$$

This property is of fundamental importance in the isothermal theory of bifurcation and stability in rate-independent plasticity.

2.3 Stability of Equilibrium of Elastic-Plastic Systems

2.3.1 General Concept of Stability. Lyapunov Functional

Any system in an equilibrium state is subject in reality to small disturbing influences (≡ disturbances) which induce a non-equilibrium process in the system. If the process deviates only slightly from (fluctuates about) the equilibrium state then the latter is called stable. If, on the contrary, the deviation becomes finite no matter how small the disturbing influences are then the equilibrium state is called unstable and regarded as unattainable in a physical system.

That intuitive concept of stability of equilibrium has to be made precise by defining (i) the object of stability analysis, (ii) the class of disturbances considered, (iii) the measure of "strength" of disturbance, (iv) the measure of distance from the equilibrium state. Since arbitrariness in formulating such definitions cannot be avoided, the concept of stability is not an absolute one. We begin with the following conventions.

(i) We shall examine stability of an *equilibrium state* $\mathcal{S}^0$ of a given *isolated* system $\mathcal{A}$, in general defined as a compound system. (Stability of a process will be discussed in Sect. 2.4).

(ii) Small changes of the *initial* values (at an instant τ, say) of selected parameters of a (non-equilibrium) state $\mathcal{S}$ are taken as initial disturbances; the *equations* governing the subsequent non-equilibrium process $\mathcal{P} : t \to \mathcal{S}(t), t \geq \tau$, are regarded as unperturbed.

(iii) and (iv) The measures: (ρ^0) of an initial disturbance and (ρ) of a distance between $\mathcal{S}$ and the equilibrium state $\mathcal{S}^0$ in the process $\mathcal{P}$, called metrics[2], are not specified universally; they may depend on the system and on the examined type of stability. We merely impose the obvious restrictions:

$$\rho^0 = \rho^0(\mathcal{S}, \mathcal{S}^0) \geq 0, \quad \rho^0(\mathcal{S}^0, \mathcal{S}^0) = 0, \tag{2.55}$$
$$\rho = \rho(\mathcal{S}, \mathcal{S}^0) \geq 0, \quad \rho(\mathcal{S}^0, \mathcal{S}^0) = 0, \tag{2.56}$$

and assume that:

$$\rho(t) \equiv \rho(\mathcal{S}(t), \mathcal{S}^0) \text{ is continuous with respect to } t \text{ in any process } \mathcal{P}, \tag{2.57}$$

$$\rho \text{ is continuous at } (\mathcal{S}^0, \mathcal{S}^0) \text{ with respect to } \rho^0. \tag{2.58}$$

As discussed below, one can use "natural" measures generated by a Lyapunov functional. In particular, ρ^0 can coincide with ρ.

Following the Lyapunov-Movchan theory of stability [20, 29], the definition of stability for infinitesimal initial disturbances with respect to metrics ρ^0 and ρ is now adopted.

Definition 1 *An equilibrium state $\mathcal{S}^0$ is called stable for initial disturbances with respect to two metrics ρ^0, ρ if and only if for every $\varepsilon > 0$ there is $\delta(\varepsilon) > 0$ such that $\rho^0(\tau) < \delta(\varepsilon)$ implies $\rho(t) < \varepsilon$ at every instant $t > \tau$.*

The *existence* of a (perturbed) process $\mathcal{P}$ in a time interval $[\tau, +\infty)$ initiated by any initial disturbance is taken here as granted. While this is a natural assumption from a physical point of view, serious mathematical difficulties can be met when the existence is attempted to be proved for a continuous nonlinear system.

The initial state of the process $\mathcal{P}$ is regarded as attainable from the examined equilibrium state in some transitory process, e.g. due to imposing appropriate time-dependent perturbations of body forces, surface tractions or temperature in some time interval $[t^0, \tau)$, with $\mathcal{S}(t^0) = \mathcal{S}^0$. (In particular, the initial disturbances must be consistent with the assumed continuity restrictions.) Since any theoretical description is approximate with respect to reality, it is natural to examine stability for such time-dependent perturbations of the governing equations (i.e. for so-called *persistent* disturbances) rather than only for initial disturbances. However, as a first step we restrict ourselves here to considering only the latter.

A powerful tool for establishing stability of equilibrium (or, more generally, of a process) is the Lyapunov direct method. Its essence lies in constructing a functional V^{L} (the Lyapunov functional)[3] such that

(a) V^{L} vanishes at the equilibrium state,
(b) V^{L} is positive definite with respect to the distance ρ,

---

[2] They need not possess all metric properties assumed in topology.

[3] In the following, we will use the notation $V^{\mathrm{L}}(\mathcal{S})$ to indicate that the value of V^{L} depends on the current state. However, V^{L} may also depend on the process which leads from $\mathcal{S}^0$ to $\mathcal{S}$.

(c) V^{L} is non-increasing in any process following an initial disturbance.

The properties (b) and (c) need to hold only in some neighborhood of $\mathcal{S}^0$, e.g. for $\rho < R$, say. If, additionally,

(d) the initial value of V^{L} at τ is continuous with respect to ρ^0,

then stability of equilibrium in the sense of Definition 1 can be inferred. In fact, one can prove the following theorem (Movchan [20] gave a rigorous proof under more general assumptions) on stability with respect to two metrics.

Theorem 1 *Let a functional $V^{\mathrm{L}}(\mathcal{S})$ exists such that*

(a) $V^{\mathrm{L}}(\mathcal{S}^0) = 0$;

(b) $V^{\mathrm{L}}(\mathcal{S}) \geq a(\rho(\mathcal{S}, \mathcal{S}^0))$ *if* $\rho(\mathcal{S}, \mathcal{S}^0) < R, \quad R = const > 0,$
$a(\cdot)$ is a continuous and monotonically increasing function, $a(0) = 0$;

(c) $V^{\mathrm{L}}(t) \equiv V^{\mathrm{L}}(\mathcal{S}(t))$ *is non-increasing with respect to t in any process $\mathcal{P}$ so long as $\rho(\mathcal{S}(t), \mathcal{S}^0) < R$,*

(d) $V^{\mathrm{L}}(\mathcal{S})$ *continuous at $\mathcal{S}^0$ with respect to $\rho^0(\mathcal{S})$.*

Then the equilibrium state $\mathcal{S}^0$ is stable with respect to two metrics ρ^0, ρ in the sense of Definition 1.

In *proof*, suppose that the conditions (a)÷(d) are satisfied by some $V^{\mathrm{L}}(\mathcal{S})$, and take any ε such that $R > \varepsilon > 0$. From (d) it follows that $V^{\mathrm{L}}(\tau) < a(\varepsilon)$ provided ρ^0 is sufficiently small with respect to $a(\varepsilon)$, i.e. if $\rho^0(\mathcal{S}(\tau), \mathcal{S}^0) < \hat{\delta}(a(\varepsilon)) \equiv \delta(\varepsilon)$, say. Then from (b) we obtain that $\rho(\tau) \leq a^{-1}(V^{\mathrm{L}}(\tau)) < \varepsilon$. We also find that $\rho(t)$ cannot reach ε at any $t_\varepsilon > \tau$ since by condition (b) this would imply $V^{\mathrm{L}}(t_\varepsilon) \geq a(\varepsilon) > V^{\mathrm{L}}(\tau)$, in contradiction with (c). From the continuity of $\rho(t)$ we now obtain that $\rho(t) < \varepsilon$ for any $t > \tau$, implying stability according to Definition 1.

From the above shortened proof one can see how the assumed properties of V^{L}, ρ and ρ^0 interact leading to stability with respects to two metrics.

An important advantage of the Lyapunov direct method is that the solutions which describe perturbed processes induced by initial disturbances need not be determined. The property (c) can in principle be established from the general form of the governing equations. If a functional $V(\mathcal{S})$ is found such that its value is continuous and non-increasing in time in any process $\mathcal{P}$ and, moreover,

$$V(\mathcal{S}) - V(\mathcal{S}^0) > 0 \quad \text{for every} \quad \mathcal{S} \neq \mathcal{S}^0 \tag{2.59}$$

then we may identify the difference $V(\mathcal{S}) - V(\mathcal{S}^0)$ with the Lyapunov functional V^{L} discussed above. The pseudo-metric generated by V :

$$\rho_V(\mathcal{S}, \mathcal{S}^0) \equiv \left|V(\mathcal{S}) - V(\mathcal{S}^0)\right| \tag{2.60}$$

can serve as a particular measure ρ of the distance from the unperturbed equilibrium state, the so-called Lyapunov (pseudo-)metric. Moreover, we can also take $\rho^0 = \rho_V$

as a particular measure for initial disturbances. Then all the conditions in Theorem 1 are evidently satisfied. Thus, we have:

Corollary 1 *An equilibrium state $\mathcal{S}^0$ is stable for initial disturbances with respect to the Lyapunov metric generated by a functional V if:*
(i) $V(\mathcal{S}) > V(\mathcal{S}^0)$ *for every* $\mathcal{S} \neq \mathcal{S}^0$,
(ii) $V(t) \equiv V(\mathcal{S}(t))$ *is continuous with respect to t in any process* $\mathcal{P}$,
(iii) $\mathrm{d}^+V/\mathrm{d}t \leq 0$ *in any process* $\mathcal{P}$.

If stability in the sense of Corollary 1 has been established for a given system then one can search for other metrics ρ^0, ρ such that stability with respect to them could be concluded from Theorem 1.

2.3.2 Thermodynamic Condition of Stability of Equilibrium

We restrict ourselves to considering a continuous elastic-plastic body $\mathcal{B}$ subject to *conservative* external loads (dependent on a loading parameter λ) at a fixed environmental temperature $\bar{T}$. In a state $\mathcal{S}^0$ of *thermodynamic equilibrium* of the system, all time derivatives of parameters of state vanish, by definition, at every point of the system. According to the assumptions specified in the previous section, in particular to the assumption of a local state, an equilibrium state $\mathcal{S}^0$ in the body is defined by the triple ($\tilde{\mathbf{x}}^0, \bar{T}, \tilde{\alpha}^0$), or alternatively by ($\tilde{\mathbf{x}}^0, \tilde{s}^0, \tilde{\alpha}^0$), where a tilde over a symbol is used to denote a spatial field over a domain G. To define a *non-equilibrium* state $\mathcal{S}$ of the body (but still under the assumption of local equilibrium within each material element separately), we must add the velocity field $\tilde{\mathbf{v}}, \mathbf{v} \equiv \dot{\mathbf{x}}$, so that $\mathcal{S}$ is defined by the quadruple ($\tilde{\mathbf{x}}, \tilde{T}, \tilde{\alpha}, \tilde{\mathbf{v}}$), or alternatively by ($\tilde{\mathbf{x}}, \tilde{s}, \tilde{\alpha}, \tilde{\mathbf{v}}$). A compatible strain field $\tilde{\mathbf{e}}$ is determined from $\tilde{\mathbf{F}} = \nabla\tilde{\mathbf{x}}$; from now on it will be convenient to use $\mathbf{F}$ rather than $\mathbf{e}$ as the external variable for a material element. For simplicity, we shall assume, if not stated otherwise, that not only $\tilde{\mathbf{x}}$ but also $\tilde{\mathbf{v}}$ and $\tilde{T}$ are continuous fields on $\overline{G}$ (including continuity of $\mathbf{v}$ and T across the body surface) and that $\mathbf{F}, T, \alpha$ are continuous and at least right-hand differentiable with respect to t.

The internal energy $\mathbb{U}$, macroscopic kinetic energy $\mathbb{K}$, entropy $\mathbb{S}$ and Helmholtz free energy Φ of the body $\mathcal{B}$ are defined in the usual manner as volume integrals of the respective local quantities u, $\bar{\rho}\mathbf{v}\cdot\mathbf{v}/2$, s and ϕ over a fixed domain $G \subset \mathcal{R}^3$ occupied by the body in the *reference* state. The assumed conservative loads acting on the body $\mathcal{B}$ are imagined to result from mechanical interaction between the body and a *loading device* which, by definition, can exchange work and not entropy with $\mathcal{B}$. The assumed constant temperature $\bar{T}$ over the body surface ∂G is imagined to result from the thermal contact with an ideal heat reservoir of temperature $\bar{T}$, which can exchange entropy and not work with $\mathcal{B}$. It is assumed that the loading device and heat reservoir undergo only reversible processes, with no entropy production also at the interfaces with the body $\mathcal{B}$.

The global quantities for the body $\mathcal{B}$ are complemented with the potential energy $\Omega = \Omega(\tilde{x}, \lambda)$ of the mechanical loading device and with the internal energy $\mathbb{E}_{\mathrm{h}}$ and entropy $\mathbb{S}_{\mathrm{h}}$ of the heat reservoir. Since the entropy of the mechanical loading device is unessential, it is set equal to zero for simplicity and Ω is interpreted as the internal energy of the device. We shall consider the compound system $\mathcal{A}$ which consists of the body $\mathcal{B}$, the heat reservoir and the loading device, whose total (internal and kinetic) energy $\mathbb{E}_{\mathcal{A}}$ and total entropy $\mathbb{S}_{\mathcal{A}}$ read

$$\mathbb{E}_{\mathcal{A}} = \mathbb{U} + \mathbb{K} + \mathbb{E}_{\mathrm{h}} + \Omega, \quad \mathbb{S}_{\mathcal{A}} = \mathbb{S} + \mathbb{S}_{\mathrm{h}} . \tag{2.61}$$

The compound system $\mathcal{A}$ is regarded as adiabatically insulated if $\lambda = \lambda(t)$ and as fully isolated if $\lambda =$ const. These restrictions can only be violated in a transitory perturbed process due to additional disturbing influences acting independently of the assumed external conditions.

An admissible thermodynamic process $\mathscr{P}$ is subject to the restrictions which follow from the basic laws of thermodynamics. From the energy balance (the first law) for the body $\mathcal{B}$ at the absence of additional supply terms, in a process $\mathscr{P}$ we have

$$\dot{\mathbb{E}}_{\mathcal{A}} = \dot{\mathbb{W}} + \dot{\mathbb{K}} + \dot{\Omega} = 0 \quad \text{at } \lambda = \text{ const} , \quad \dot{\mathbb{W}} = \dot{\mathbb{U}} - \dot{\mathbb{Q}} , \tag{2.62}$$

where $\dot{\mathbb{W}} = \int_G \mathbf{S} \cdot \dot{\mathbf{F}} \, \mathrm{d}\boldsymbol{\xi}$ is the rate of deformation work in $\mathcal{B}$ and $\dot{\mathbb{Q}} = -\dot{\mathbb{E}}_{\mathrm{h}} = -\bar{T}\dot{\mathbb{S}}_{\mathrm{h}}$ is the rate of heat supply to $\mathcal{B}$ from the heat reservoir. From the assumed reversibility of the surroundings and from the second law we obtain

$$\dot{\mathbb{S}}_{\mathcal{A}} = \dot{\mathbb{S}}^{\mathrm{i}} \equiv \dot{\mathbb{S}} - \dot{\mathbb{Q}}/\bar{T} \geq 0 \quad \text{in } \mathscr{P} . \tag{2.63}$$

Under the assumptions introduced, the entropy production rate $\dot{\mathbb{S}}^{\mathrm{i}}$ in the body $\mathcal{B}$ consists of the plastic part $\dot{\mathbb{S}}^{\mathrm{p}}$ and the thermal part $\dot{\mathbb{S}}^{\mathrm{T}}$, being the volume integrals over G of the local nonnegative contributions σ^{p} and σ^{T}, respectively. In general, there may be more non-negative contributions to $\dot{\mathbb{S}}^{\mathrm{i}}$, e.g. due to viscosity, strong discontinuities in $\mathbf{F}$ or T, etc. In any case, we can assume that (cf. (2.9) and (2.10))

$$\dot{\mathbb{S}}^{\mathrm{i}} - \dot{\mathbb{S}}^{\mathrm{p}} \geq 0, \quad \dot{\mathbb{S}}^{\mathrm{p}} \equiv \int_G \sigma^{\mathrm{p}} \, \mathrm{d}\boldsymbol{\xi} \geq 0 . \tag{2.64}$$

The former inequality will play a fundamental role in establishing a general condition of thermodynamic stability of equilibrium of systems with plastic dissipation.

Let the prefix Δ denote the difference of corresponding quantities in states $\mathcal{S}$ and $\mathcal{S}^0$, e.g. $\mathbb{S}_{\mathcal{A}}(\mathcal{S}) - \mathbb{S}_{\mathcal{A}}(\mathcal{S}^0) \equiv \Delta\mathbb{S}_{\mathcal{A}}$, $\mathbb{E}_{\mathcal{A}}(\mathcal{S}) - \mathbb{E}_{\mathcal{A}}(\mathcal{S}^0) \equiv \Delta\mathbb{E}_{\mathcal{A}}$. Suppose first that $\tilde{\alpha}$ is frozen, so that $\dot{\mathbb{S}}^{\mathrm{p}} = 0$. The classical Gibbs condition of stability of equilibrium reads

$$\Delta\mathbb{S}_{\mathcal{A}} < 0 \quad \text{whenever} \quad \Delta\mathbb{E}_{\mathcal{A}} = 0, \quad \dot{\mathbb{S}}^{\mathrm{p}} \equiv 0 . \tag{2.65}$$

The equality constraint can be eliminated in the usual way by introducing a Lagrangian multiplier, which allows the (thermo-elastic) stability condition to be written in the equivalent[4] form

$$V^{\mathrm{L}} \equiv \Delta V_{\mathrm{e}} > 0 \quad \text{for every } \mathcal{S} \neq \mathcal{S}^0, \quad \dot{\mathbb{S}}^{\mathrm{p}} \equiv 0 \tag{2.66}$$

where

$$V_{\mathrm{e}} \equiv \mathbb{E}_{\mathcal{A}} - \bar{T}\mathbb{S}_{\mathcal{A}}. \tag{2.67}$$

The value of V_{e} varies continuously in time since the energy and entropy do, and from (2.62) and (2.63) we obtain that

$$\dot{V}_{\mathrm{e}} \equiv \dot{\mathbb{E}}_{\mathcal{A}} - \bar{T}\dot{\mathbb{S}}^{\mathrm{i}} \leq 0 \tag{2.68}$$

in any process $\mathcal{P}$ which follows an initial disturbance and is free of further (persistent) disturbances. From Corollary 1 we arrive at the conclusion that (2.66) ensures stability of $\mathcal{S}^0$ with respect to the Lyapunov metric generated by V_{e}.

The conclusion would be valid also for variable $\tilde{\alpha}$. However, as it will be shown later, the inequality in (2.66) or (2.65) is generally not satisfied for solids with intrinsic plastic dissipation. As a consequence, if $\dot{\mathbb{S}}^{\mathrm{p}} \neq 0$ then the Gibbs condition ((2.65) or its equivalent) does not provide a satisfactory basis for stability investigations within the present thermodynamic formalism of plasticity.[5] The hypothesis that equilibrium is unstable unless all modes of departure from it contradict the second law must be abandoned. The attempts to derive a stability criterion in plasticity on the basis of that hypothesis [21] refer, in effect, to a hypothetical "equivalent elastic structure" rather than to the actual inelastic solid.

The main idea developed below is to extend Gibbs' stability condition to the systems which exhibit plastic dissipation, by taking as a basis the inequality (2.64) rather than that in (2.63). To construct the corresponding Lyapunov functional for an isolated system $\mathcal{A}$, in place of (2.67) we define

$$V \equiv \mathbb{E}_{\mathcal{A}} - \bar{T}(\mathbb{S}_{\mathcal{A}} - \Delta\mathbb{S}^{\mathrm{p}}) \tag{2.69}$$

where

$$\Delta\mathbb{S}^{\mathrm{p}} \equiv \int_{t^0}^{t} \dot{\mathbb{S}}^{\mathrm{p}}\, \mathrm{d}t \tag{2.70}$$

is the intrinsic (plastic) dissipation in the body $\mathcal{B}$ in a process starting from the equilibrium state $\mathcal{S}^0 = \mathcal{S}(t^0)$ and leading to a current state $\mathcal{S} = \mathcal{S}(t)$.

By using (2.62), (2.63)$_1$ and (2.64)$_1$ we obtain that

$$\dot{V} = \dot{\mathbb{E}}_{\mathcal{A}} - \bar{T}(\dot{\mathbb{S}}^{\mathrm{i}} - \dot{\mathbb{S}}^{\mathrm{p}}) \leq 0 \tag{2.71}$$

---

[4] Implication (2.66) $\Rightarrow$ (2.65) is obvious. The converse implication follows from the possibility of varying $\Delta\mathbb{E}_{\mathcal{A}}$ at constant V_{e} by supplying heat from external sources to the heat reservoir.

[5] This conclusion is not surprising if the macroscopic plastic behaviour is thought of as being associated with a large number of non-equilibrium jumps at a microscopic level between (metastable) equilibrium states in an elastic body. In view of entropy production during any such jump, the relation between the occurrence of macroscopic plastic dissipation and the violation of (2.65) becomes comprehensible.

in any process $\mathscr{P}$ free of persistent disturbances. A condition of stability of the equilibrium state $\mathscr{S}^0$ can thus be formulated as

$$\Delta V > 0 \quad \text{for every } \mathscr{S} \neq \mathscr{S}^0 \tag{2.72}$$

and for every transitory process leading from $\mathscr{S}^0$ to $\mathscr{S}$. From Corollary 1 we arrive at the following conclusion.

Corollary 2 *If the condition (2.72) is satisfied then the equilibrium state $\mathscr{S}^0$ is stable for initial disturbances with respect to the Lyapunov metric generated by the functional V.*

Note that the pseudo-metric (2.60) depends now not only on $\mathscr{S}$ but also on the process leading to $\mathscr{S}$ from $\mathscr{S}^0$.

If $\tilde{\alpha}$ is fixed then V reduces to V_e, so that the conditions for thermoelastic stability do not change. If $\tilde{\alpha}$ is varying then $\Delta\mathbb{S}^p > 0$ and $V > V_e$ so that the condition (2.72) is less restrictive than (2.66) and, as it will be shown below, is applicable to plastic stability problems. The integral in (2.70) is in general path-dependent, i.e. the actual value of V depends not only on the current state but also on the whole process from $\mathscr{S}^0$ to $\mathscr{S}$, no matter how small the final variations of the state fields are. In order not to complicate the notation, this dependence is not indicated in a formal way.

The condition (2.72) is in general not necessary for stability of equilibrium; note that in Theorem 1 only a proper local minimum of a Lyapunov functional with respect to a metric ρ is required and not the absolute minimum property (2.72) in the whole state space. In the absence of plastic dissipation, the condition (2.65) or (2.66) but weakened to define a local minimum can be regarded as a necessary condition for asymptotic stability. For, if V_e vanishes asymptotically and is decreasing in any real (dissipative) process then its initial value must always be positive. In plasticity, the property of asymptotic stability cannot in general be expected.

2.3.3 Thermoelastic Stability

Before proceeding to our main topic – stability in plasticity – it is instructive to discuss first the *relatively* simpler case when the internal variables α are frozen. Substitution of (2.61) into (2.67) yields

$$\Delta V_e = \Delta(\,\mathbb{K} + \mathbb{U} - \bar{T}\mathbb{S} + \Omega\,)\,, \tag{2.73}$$

where the energy and entropy of the heat reservoir have been eliminated.

To begin with, suppose that the body is artificially constrained to undergo no displacements ($\tilde{\mathbf{x}}$ fixed, so that $\tilde{\mathbf{v}} \equiv \tilde{\mathbf{0}}$), and that the entropy field $\tilde{s}$ is treated as the independent variable. The condition (2.66) reduces to

$$\Delta V_e|_{\tilde{\mathbf{x}},\tilde{\alpha}} = \Delta V_e(\tilde{s}) = \int_G (\Delta u - \bar{T}\Delta s)\,\mathrm{d}\boldsymbol{\xi} > 0 \quad \text{for } \tilde{s} \neq \tilde{s}^0\,. \tag{2.74}$$

Assume that $u = u(\mathbf{F}, s, \alpha)$ is continuously twice differentiable with respect to s. For (2.74) to hold, we must have the stationarity property

$$\delta V_e(\tilde{s}) = \int_G (u_{,s} - \bar{T})\,\delta s\,\mathrm{d}\boldsymbol{\xi} = 0 \quad \text{in } \mathscr{S}^0 \tag{2.75}$$

which means that in the equilibrium state $\mathscr{S}^0$ we have

$$u_{,s} \equiv T = \bar{T} \quad \text{in } G \tag{2.76}$$

as the expected condition of *thermal equilibrium*. Moreover, (2.74) implies non-negativeness of the second variation

$$\delta^2 V_e(\tilde{s}) = \int_G \mathrm{u}_{,ss}(\delta s)^2\,\mathrm{d}\boldsymbol{\xi} \geq 0 \quad \text{in } \mathscr{S}^0 \tag{2.77}$$

equivalent to

$$\mathrm{u}_{,ss} \equiv \bar{\rho}\bar{T}/c \geq 0 \quad \text{in } G \text{ in } \mathscr{S}^0. \tag{2.78}$$

In turn, fulfilment of (2.74) in any equilibrium state is ensured if c, the specific heat at constant strain, is always positive

$$c > 0 \tag{2.79}$$

which we assume to be the case. For, the functional in (2.74) is then strictly convex with respect to $\tilde{s}$, so that its stationary point is automatically an absolute minimum point. The inequality (2.79) is the (classical) condition of *thermal stability* of all constrained equilibrium states (where displacements and internal variables are treated as fixed).

Suppose now that the fields $\tilde{\mathbf{x}}$, $\tilde{\mathbf{v}}$ and $\tilde{s}$ can undergo simultaneous perturbations. On substituting (2.73) the condition (2.66) becomes

$$\Delta V_e(\tilde{\mathbf{x}}, \tilde{s}, \tilde{\mathbf{v}}) = \Delta\mathbb{K} + \int_G (\Delta u - \bar{T}\Delta s)\,\mathrm{d}\boldsymbol{\xi} + \Delta\Omega > 0 \quad \text{for } \mathscr{S} \neq \mathscr{S}^0\,. \tag{2.80}$$

We make now use of the identity (with (2.76) accounted for)

$$u(\mathbf{F}, s, \alpha) - u(\mathbf{F}^0, s^0, \alpha) - \bar{T}(s - s^0) = \phi(\mathbf{F}, \bar{T}, \alpha) - \phi(\mathbf{F}^0, \bar{T}, \alpha) + \mathscr{E}_u|_{\mathbf{F},\alpha}\,, \tag{2.81}$$

where

$$\mathscr{E}_u = \mathscr{E}_u(s, \bar{s}) \equiv u(s) - u(\bar{s}) - u_{,s}(\bar{s})(s - \bar{s}), \quad \bar{s} = -\phi_{,T}(\bar{T}), \quad \mathbf{F}, \alpha \text{ fixed} \tag{2.82}$$

is the Weierstrass function associated with $u(s)$ at fixed $\mathbf{F}$ and α, which represents the excess of $u(s)$ over its linear approximation at $\bar{s}$ (with the tangent $u_{,s}(\bar{s}) = \bar{T}$). Substitution of the identity into (2.80) yields

$$\Delta V_{\mathrm{e}}(\tilde{\mathbf{x}}, \tilde{s}, \tilde{\mathbf{v}}) = \Delta \mathbb{K} + \int_G \mathscr{E}_u|_{\mathbf{F},\alpha} \, \mathrm{d}\boldsymbol{\xi} + \Delta\Phi|_{\bar{T}} + \Delta\Omega \tag{2.83}$$

where in the term $\Delta\Phi|_{\bar{T}}$ the free energy density is evaluated at the environmental (equilibrium) temperature $\bar{T}$ and not at the current local temperature T. From the assumed thermal stability condition (2.79) we obtain that $u(s)$ is strictly convex so that $\mathscr{E}_u(s, \bar{s}) > 0$ if $s \neq \bar{s}$. The first two terms on the right-hand side of (2.83) are thus non-negative and vanish only if $\tilde{\mathbf{v}} = \tilde{\mathbf{0}}$ and $\tilde{T} = \bar{T}$, respectively. The thermodynamic stability condition (2.80) reduces, in effect, to the quasi-static isothermal condition [22]

$$\Delta V_{\mathrm{e}}|_{\bar{T},\tilde{\mathbf{v}}=\tilde{\mathbf{0}}} = \Delta V_{\mathrm{e}}(\tilde{\mathbf{x}}) = \Delta\Phi|_{\bar{T}} + \Delta\Omega > 0 \quad \text{for } \tilde{\mathbf{x}} \neq \tilde{\mathbf{x}}^0 \,. \tag{2.84}$$

The inequality (2.84) means that the potential energy of the *system* which consists of the elastic body and the loading device attains under isothermal conditions an absolute minimum in the equilibrium state. As already mentioned, this is not necessary for a physically meaningful stability of equilibrium. If (2.79) is granted then (2.84) *is sufficient for thermodynamic stability of the equilibrium state $\mathscr{S}^0$ with respect to the Lyapunov metric generated by* V_{e} (at frozen $\tilde{\alpha}$ and variable $\tilde{T}$). For further discussion and possible extensions, see [23].

Examine now prerequisites for (2.84). Assuming that there are no unilateral constraints, $V_{\mathrm{e}}(\tilde{\mathbf{x}})$ must be stationary (in the sense of the usual weak variations, i.e. of the identically vanishing Gateaux differential) in the equilibrium state:

$$\delta(\Phi + \Omega) = 0 \quad \text{in } \mathscr{S}^0, \quad T = \text{const} \tag{2.85}$$

which gives the conditions of *mechanical equilibrium*. For instance, suppose that the displacements vanish on a part $S_u \subset \partial G$ of the body surface, the *nominal* surface tractions $\mathbf{T}$ (per unit surface area in the *reference* configuration) are prescribed on the complementary part S_{T} of ∂G, and the *nominal* body forces $\mathbf{b}$ (per unit reference volume) are prescribed in G. Then

$$\Delta\Omega = -\int_G \mathbf{b} \cdot \mathbf{u} \, \mathrm{d}\boldsymbol{\xi} - \int_{S_{\mathrm{T}}} \mathbf{T} \cdot \mathbf{u} \, \mathrm{d}S \,, \tag{2.86}$$

where $\mathbf{u} = \Delta\mathbf{x}$ are displacements from the equilibrium configuration $\tilde{\mathbf{x}}^0$. The condition (2.85) has the form of the virtual work principle written in the reference configuration :

$$\delta(\Phi + \Omega) = \int_G (\mathbf{S} \cdot \nabla(\delta\mathbf{x}) - \mathbf{b} \cdot \delta\mathbf{x}) \, \mathrm{d}\boldsymbol{\xi} - \int_{S_{\mathrm{T}}} \mathbf{T} \cdot \delta\mathbf{x} \, \mathrm{d}S = 0 \quad \text{if } \delta\mathbf{x} = \mathbf{0} \text{ on } S_{\mathrm{u}}. \tag{2.87}$$

Under appropriate regularity conditions, this can be reduced in the standard manner by employing the Green theorem to the local conditions of mechanical equilibrium in the Lagrangian formulation:

$$\operatorname{Div} \mathbf{S} + \mathbf{b} = \mathbf{0} \quad \text{in } G \backslash S_{\mathrm{D}}, \quad [\![\mathbf{S}]\!] \cdot \mathbf{n} = \mathbf{0} \quad \text{on } S_{\mathrm{D}}, \quad \mathbf{S} \cdot \mathbf{n} = \mathbf{T} \quad \text{on } S_{\mathrm{T}}, \tag{2.88}$$

where S_D is a possible surface of the stress discontinuity $[\![\mathbf{S}]\!]$ and $\mathbf{n}$ is the unit normal to a surface (all in the reference configuration).

As another implication of (2.84), we obtain

$$\delta^2(\Phi + \Omega) \geq 0 \quad \text{in } \mathscr{S}^0, \quad T = \text{const}\,, \tag{2.89}$$

where, under the assumed linear dependence of Ω on the displacement field,

$$\delta^2(\Phi + \Omega) = \delta^2\Phi = \int_G \nabla(\delta\mathbf{x}) \cdot \phi_{,\mathbf{FF}} \cdot \nabla(\delta\mathbf{x})\, \mathrm{d}\boldsymbol{\xi}, \quad \delta\mathbf{x} = \mathbf{0} \text{ on } \mathrm{S_u}\,, \tag{2.90}$$

is the second weak variation of the isothermal Helmholtz free energy (whose density is assumed to be continuously twice differentiable with respect to its arguments). The condition (2.89) may be regarded as a *necessary* condition for *mechanical stability of equilibrium*, although more in a physical than mathematically rigorous sense. Evidently, (2.89) is necessary for *asymptotic* stability.

In general, positiveness of the second weak variation of the total potential energy ($\Phi + \Omega$) (i.e. when only trivial equality in (2.89) is allowed) need not imply (2.84) and thus need not ensure elastic stability. This question will be discussed later when examining material stability aspects.

2.3.4 Stability of Equilibrium of Systems With Plastic Dissipation

The above constraint $\tilde{\alpha} = \text{const}$ is relaxed from now on. To justify the statement about inapplicability of the condition (2.65) or (2.66) to systems with plastic dissipation, consider the first weak variation of V_e in an equilibrium state. On account of (2.75) and (2.85), it reads

$$\delta V_e = \delta V_e|_{\mathbf{x},\bar{T}} = \delta\Phi(\tilde{\alpha}) = -\int_G \boldsymbol{A} \cdot \delta\boldsymbol{\alpha}\, \mathrm{d}\boldsymbol{\xi}\,. \tag{2.91}$$

Since the thermodynamic forces need not vanish in an equilibrium state of the elastic-plastic solid as discussed in Sect. 2.2, this expression is generally non-zero, in contrast to the stationarity properties (2.75) and (2.85). If the actual stresses satisfy the yield condition ($g_K = 0$) within a finite part of the body $\mathscr{B}$ then any disturbance $\delta\alpha_K$ compatible with the actual thermodynamic forces makes the expression (2.91) negative. This means that the condition (2.66) or (2.65) can never be satisfied if $\tilde{\alpha}$ is variable, with the exception of the particular equilibrium states where within the body either the current stress point lies everywhere inside the elastic domain (so that $\delta\alpha_K \equiv 0$) or all thermodynamic forces A_K happen to vanish.

As stated in Sect. 2.3.4, the condition (2.72) and not (2.66) is taken as a basis for the stability investigations in solids which exhibit plastic dissipation. Substitution of (2.61) into (2.69) yields

$$\Delta V = \Delta\mathbb{K} + \Delta\mathbb{U} - \bar{T}\,(\Delta\mathbb{S} - \Delta\mathbb{S}^{\mathrm{p}}) + \Delta\Omega\,. \tag{2.92}$$

In the following we shall assume (except in Sect. 2.5.2) that the disturbing agency does not penetrate "inside" a material element so that (cf. (2.11))

$$\dot{\mathcal{D}}(\jmath,\dot{\alpha}) = \boldsymbol{A}(\jmath)\cdot\dot{\alpha} \tag{2.93}$$

also in a transitory perturbed process. From (2.91) and (2.93) it follows that the first one-sided weak variation of V vanishes in any equilibrium state $\mathcal{S}^0$ provided the variation of $\tilde{\alpha}$ is compatible, in the sense of (2.93), with the actual thermodynamic forces in state $\mathcal{S}^0$.

Besides Corollary 2, other interpretations of (2.72) as a sufficient condition for stability can be obtained with the help of Theorem 1. For instance, let the class of admissible disturbances be restricted such that during a transitory process in $[t^0,\tau)$ there is no entropy supply (of neither sign) to the compound system $\mathcal{A}$ from its environment (e.g. transitory processes induced by mechanical excitation with $\mathcal{A}$ thermally insulated), which is denoted by $\Delta^{\mathrm{ext}}\mathbb{S}_{\mathcal{A}} = 0$. Then $\Delta S_{\mathcal{A}} = \Delta S^{\mathrm{i}}$ also for $t<\tau$ so that an increment of the term in parentheses in (2.69) is non-negative by $(2.64)_1$. Suppose that (2.72) holds, and define the metric ρ^0 for *initial* disturbances at τ as the *maximal* amount of energy supplied to $\mathcal{A}$ by a disturbing agency within the time interval $[t^0,\tau]$ in order to reach the state $\mathcal{S}(\tau)$, viz.

$$\rho^0_{\mathrm{E}} = \sup_{t^0\le t\le\tau} \Delta\mathbb{E}_{\mathcal{A}}(t), \quad \Delta^{\mathrm{ext}}\mathbb{S}_{\mathcal{A}} \equiv 0\,. \tag{2.94}$$

Then

$$\rho^0_{\mathrm{E}} \ge \Delta\mathbb{E}_{\mathcal{A}}(\tau) \ge \Delta V(\tau) > 0 \quad \text{in } \mathcal{S}_\tau \ne \mathcal{S}^0 \tag{2.95}$$

so that V and ρ^0_{E} satisfy the condition (d) in Theorem 1. Hence, (2.72) is sufficient for stability of equilibrium with respect to ρ^0_E and the metric ρ^{L} generated by the functional (2.69). If the restriction $\Delta^{\mathrm{ext}}S_{\mathcal{A}} \equiv 0$ is relaxed then $\Delta\mathbb{E}_{\mathcal{A}}$ in the above formulae should be replaced by $(\Delta\mathbb{E}_{\mathcal{A}} - \bar{T}\Delta^{\mathrm{ext}}S_{\mathcal{A}})$; this quantity is not smaller that ΔV, on account of $(2.64)_1$. As discussed below, the problem of determining a physically acceptable metric ρ for a continuum, which could satisfy the condition (b) in Theorem 1, is more complicated.

In analogy to (2.83), we can write

$$\Delta V = \Delta\mathbb{K} + \int_G \mathscr{E}_u|_{\mathbf{F},\alpha}\,\mathrm{d}\xi + \Delta\Phi|_{\bar{T}} + \Delta\Omega + \bar{T}\,\Delta\mathbb{S}^{\mathrm{p}}\,. \tag{2.96}$$

Since the first two terms on the right-hand side in (2.96) are non-negative, the remaining quantity

$$\Delta\mathbb{E} \equiv \Delta\Phi|_{\bar{T}} + \Delta\Omega + \bar{T}\,\Delta\mathbb{S}^{\mathrm{p}} \tag{2.97}$$

is not greater than ΔV. On account of $(2.64)_1$, $\Delta\mathbb{E}$ represents also a lower bound to the amount of energy which has to be supplied to $\mathcal{A}$ by a disturbing agency such that $\Delta^{\mathrm{ext}}\mathbb{S}_{\mathcal{A}} = 0$ in order to carry the system from the equilibrium state $\mathcal{S}^0$ to another (quasi-equilibrium) state $\mathcal{S}^{\mathrm{qe}}$ of uniform temperature $\bar{T}$ and zero velocities. It follows that (2.72) is ensured if

$$\Delta\mathbb{E} > 0 \quad \text{in every } \mathscr{S}^{\text{qe}} \neq \mathscr{S}^0 \tag{2.98}$$

and for every process leading from $\mathscr{S}^0$ to $\mathscr{S}^{\text{qe}}$. The first two terms in (2.97) are defined as functions of $(\tilde{\mathbf{x}}, \tilde{\alpha})$ under isothermal conditions. However, contrary to the case of thermoelastic stability, we cannot conclude that the isothermal stability implies stability also in a non-isothermal case since the last term in (2.96), previously absent, depends in general on the temperature history in the process from $\mathscr{S}^0$ to $\mathscr{S}^{\text{qe}}$.

Isothermal Case

From now on we restrict ourselves to examining isothermal processes only.[6] Then $\sigma^{\text{T}} \equiv 0$ (cf. (2.9)) and we have $\dot{\mathbb{S}}^{\text{i}} = \dot{\mathbb{S}}^{\text{p}}$ since other sources of dissipation have been disregarded.[7] Thus (cf. (2.62))

$$\dot{V} = \dot{\mathbb{K}} + \dot{\mathbb{W}} + \dot{\Omega} = 0 \quad \text{at } T = \text{ const }, \ \lambda = \text{ const }, \tag{2.99}$$

in a process $\mathscr{P}$ free of persisting disturbances; it is recalled that $\dot{\mathbb{W}} = \int_G \mathbf{S} \cdot \dot{\mathbf{F}} \, \mathrm{d}\boldsymbol{\xi}$. The stability condition (2.98) reduces to

$$\Delta\mathbb{E} = \Delta V - \Delta\mathbb{K} = \Delta\mathbb{W} + \Delta\Omega > 0 \quad \text{in every } \mathscr{S}^{\text{qe}} \neq \mathscr{S}^0 \tag{2.100}$$

and for every isothermal process from $\mathscr{S}^0$ to $\mathscr{S}^{\text{qe}}$. The functional $\mathbb{E}$ can be identified with the energy functional introduced in [24] where an increment of $\mathbb{E}$ in an isothermal quasi-static deformation process was interpreted as the amount of energy supplied from external sources to the mechanical system consisting of the deformed body and the loading device. Here, for isothermal processes such that $\Delta^{\text{ext}}\mathbb{S}_{\mathscr{A}} = 0$, $\mathbb{E}$ can be identified with the internal energy of the compound thermodynamic system $\mathscr{A}$ which includes additionally the heat reservoir.

The discussion from now on will be restricted to rate-independent plasticity. Consider a kinematically admissible process of departure from the equilibrium state $\mathscr{S}^0$ with a *continuous* velocity field $\tilde{\mathbf{v}}$ such that $\mathbf{v} = \mathbf{v}^0$ on $\mathrm{S_u}$ and $\|\nabla\mathbf{v}\| < \infty$, where $\| \cdot \| \equiv (\int_G | \cdot |^2 \, \mathrm{d}\boldsymbol{\xi})^{1/2}$. As discussed above, the first time derivative $\dot{\mathbb{E}}$ of $\mathbb{E}$ (representing the first one-sided variation for a *given* direction of departure from $\mathscr{S}^0$) vanishes identically in $\mathscr{S}^0$. For (2.100) to hold it is necessary that the following *second-order energy condition* is satisfied:

$$\ddot{\mathbb{E}} \geq 0 \quad \text{in } \mathscr{S}^0 \tag{2.101}$$

for every kinematically admissible mode of departure from $\mathscr{S}^0$, with the restriction on $\tilde{\mathbf{v}}$ mentioned above. If $\tilde{\alpha}$ is frozen then (2.101) reduces to (2.89).

For dead loading associated with (2.86), the condition (2.101) takes the form

---

[6] An analogous discussion is possible for adiabatic processes but it is omitted here.

[7] If moving strong discontinuities of $\mathbf{F}$ were allowed then we could have $\dot{\mathbb{S}}^{\text{i}} > \dot{\mathbb{S}}^{\text{p}}$ and $\dot{V} < 0$ even in an isothermal process (cf. the discussion in Sect. 2.5.4).

$$\int_G \dot{\mathbf{S}}(\mathscr{s}^0, \dot{\mathbf{F}}) \cdot \dot{\mathbf{F}} \, d\xi \geq 0 \quad \text{for every } \tilde{\mathbf{v}}, \quad \mathbf{v} = \mathbf{0} \text{ on } \mathrm{S_u} \,. \tag{2.102}$$

This inequality (but strict for $\tilde{\mathbf{v}} \neq \tilde{\mathbf{0}}$) was proposed in [25] as a condition sufficient for stability of equilibrium, however, under certain additional assumptions; cf. the discussion below. Strict inequality in (2.102) for $\tilde{\mathbf{v}} \neq \tilde{\mathbf{0}}$ excludes departure from equilibrium on a *straight* deformation path $\Delta\mathbf{x}(\boldsymbol{\xi}, t) = (t - t^0)\mathbf{v}(\boldsymbol{\xi})$ unless the disturbing influences are not arbitrary small. If this is so for any kinematically admissible direction $\tilde{\mathbf{v}}$ then (2.102) is sufficient for a kind of stability which may be called *directional stability of equilibrium.*

More can be said about stability *in the first approximation*, where an auxiliary (*partially* linearized) problem is considered in place of the "true" fully nonlinear problem. That approach, to be used below, can be regarded as an extension of the so-called linear theory of elastic stability to incrementally nonlinear, rate-independent plasticity.

Stability in the First Approximation

The attention is restricted to isothermal processes and rate-independent plasticity. Let a bar over a symbol denote either a quantity in the state $\mathscr{S}^0$ or a function obtained by approximation at $\mathscr{S}^0$. Consider a small neighborhood of $\mathscr{S}^0 \equiv (\bar{\mathscr{z}})$, $\bar{\mathscr{z}}(\boldsymbol{\xi}) \equiv (\bar{\mathbf{x}}, \bar{T}, \bar{\alpha})(\boldsymbol{\xi})$, defined by *pointwise* inequalities $|\alpha_K - \bar{\alpha}_K| < R$ and $|\mathbf{F} - \bar{\mathbf{F}}| < R$ in G, where R is a small positive number. In that neighborhood, we wish to consider the *second-order* approximation of the left-hand expression in the stability condition (2.100), with path-dependence of the dissipation and deformation work taken to that order into account. For this purpose, we shall consider an auxiliary problem where the (isothermal) Helmholtz free energy density $\bar{\phi}$ is defined by

$$\begin{aligned} \bar{\phi}(\mathbf{F}, \alpha) \equiv \; & \phi(\bar{\mathscr{z}}) + \phi_{,\mathbf{F}}(\bar{\mathscr{z}}) \cdot (\mathbf{F} - \bar{\mathbf{F}}) + \phi_{,\alpha}(\bar{\mathscr{z}}) \cdot (\alpha - \bar{\alpha}) \\ & + \tfrac{1}{2}(\mathbf{F} - \bar{\mathbf{F}}) \cdot \phi_{,\mathbf{FF}}(\bar{\mathscr{z}}) \cdot (\mathbf{F} - \bar{\mathbf{F}}) + (\mathbf{F} - \bar{\mathbf{F}}) \cdot \phi_{,\mathbf{F}\alpha}(\bar{\mathscr{z}}) \cdot (\alpha - \bar{\alpha}) \\ & + \tfrac{1}{2}(\alpha - \bar{\alpha}) \cdot \phi_{,\alpha\alpha}(\bar{\mathscr{z}}) \cdot (\alpha - \bar{\alpha}). \end{aligned} \tag{2.103}$$

The two Examples of the constitutive framework from Sect. 2.2 will be discussed in more detail. In Example (1) the dissipation function D is now linearized but only with respect to α, viz.

$$\bar{D}(\alpha, \dot{\alpha}) \equiv D(\bar{\alpha}, \bar{T}, \dot{\alpha}) + (\alpha - \bar{\alpha}) \cdot D_{,\alpha}(\bar{\alpha}, \bar{T}, \dot{\alpha}) \geq -\bar{\phi}_{,\alpha}(\mathbf{F}, \alpha) \cdot \dot{\alpha} \,; \tag{2.104}$$

nonlinearity of D with respect to $\dot{\alpha}$ has been retained. In Example (2), as an analog to (2.104) we assume (cf. (2.35) and (2.48)):

$$\bar{g}_K(\mathbf{F}, \gamma_J) \equiv g_K(\bar{\mathscr{z}}) + \bar{\mathbf{\Lambda}}_K \cdot (\mathbf{F} - \bar{\mathbf{F}}) - \sum_J g_{KJ}(\bar{\mathscr{z}})(\gamma_J - \bar{\gamma}_J) \leq 0 \,, \tag{2.105}$$

$$\alpha_K = \bar{\alpha}_K + (\gamma_K - \bar{\gamma}_K)\bar{\boldsymbol{P}}_K, \;\; \bar{\mathbf{\Lambda}}_K \equiv (\partial g_K / \partial \mathbf{F})(\bar{\mathscr{z}}), \;\; \bar{\boldsymbol{P}}_K \equiv (\partial f_K / \partial \mathbf{A}_K)(\bar{\mathscr{z}}). \tag{2.106}$$

The internal mechanisms of plastic deformation which are not potentially active in $\mathcal{S}^0$ are regarded as frozen in the neighborhood of $\mathcal{S}^0$, that is

$$g_K(\bar{\jmath})\,\dot{\alpha}_K(t) = 0 \tag{2.107}$$

at any t in every perturbed process under consideration. This implies

$$\bar{D}(\bar{\alpha}, \dot{\alpha}(t)) = \boldsymbol{A}(\bar{\jmath}) \cdot \dot{\alpha}(t) \tag{2.108}$$

$$g_K(\bar{\jmath})\dot{\gamma}_K(t) = 0 \tag{2.109}$$

in Example (1) and (2), respectively.

For simplicity, the configuration-insensitive loading corresponding to the expression (2.86) for $\Delta\Omega$ is assumed.

Under the assumptions introduced above, $\Delta\mathbb{E}$ is represented by a quantity $\Delta\bar{\mathbb{E}}$ which is of second-order with respect to $\Delta\mathbf{F}$ and $\Delta\alpha$. For, the assumption (2.107) eliminates the first-order contribution to $\Delta\bar{\mathbb{E}}$, and the adopted linearizations ensure that a local contribution to $\Delta\bar{\mathbb{E}}$ varies proportionally to the square of path length in $(\mathbf{F}, \alpha)$-space when such a path is proportionally scaled down so that its length varies while its complexity is preserved (cf. the discussion of path-dependent second-order work in [26]).

Since various internal mechanisms of plastic deformation may subsequently be unloaded and activated in a perturbed process, the dissipation $\bar{T}\Delta S^{\mathrm{p}}$ and the deformation work $\Delta\mathbb{W}$ in (2.100) are in general dependent on the transitory process leading to some given state $\mathcal{S}^{\mathrm{qe}}$ and not only on $\mathcal{S}^{\mathrm{qe}}$ itself, also if second-order expressions are only considered. This difficulty can be overcome if a *lower bound* to $\Delta\mathbb{W}$ or $\Delta\mathbb{S}^{\mathrm{p}}$ is determined for given $\mathcal{S}^{\mathrm{qe}}$. That approach has been proposed in [25], and used in [27] for particular classes of solids with D independent of α. We present below a generalization of previous results for the two Examples of constitutive framework from Sect. 2.2.

Example (1)

Suppose that the symmetry property (2.42) holds. Then, by appealing to the assumed linearity of $\bar{D}$ with respect to α and convexity with respect to $\dot{\alpha}$ it can be shown [28] [see Chapter 3] that among all paths in α-space satisfying (2.108) at each instant, the dissipation integral is minimized on the straight path, viz.

$$\int_{t^0}^{\tau} \bar{D}(\alpha(t), \dot{\alpha}(t))\mathrm{d}t \geq \bar{D}(\bar{\alpha}, \bar{\bar{\alpha}} - \bar{\alpha}) + \frac{1}{2}(\bar{\bar{\alpha}} - \bar{\alpha}) \cdot \bar{D},\alpha(\bar{\alpha}, \bar{\bar{\alpha}} - \bar{\alpha}), \qquad \alpha(t^0) = \bar{\alpha}, \quad \alpha(\tau) = \bar{\bar{\alpha}}. \tag{2.110}$$

This minimum property and the resulting stability condition given below provide an extension of Nguyen's [16] results.

By using (2.110) and (2.86), we obtain the (second-order) inequality

$$\Delta\bar{V} \geq \Delta\bar{\mathbb{E}} \geq \Delta^2\Phi + \bar{T}\,\Delta^2\mathbb{S}^{\mathrm{p}} \equiv \rho_{\mathrm{E}} \tag{2.111}$$

where

$$\Delta^2\Phi = \int_G \left(\frac{1}{2}\Delta\mathbf{F}\cdot\phi_{,\mathbf{FF}}(\bar{\mathfrak{z}})\cdot\Delta\mathbf{F} + \Delta\mathbf{F}\cdot\phi_{,\mathbf{F}\alpha}(\bar{\mathfrak{z}})\cdot\Delta\alpha + \frac{1}{2}\Delta\alpha\cdot\phi_{,\alpha\alpha}(\bar{\mathfrak{z}})\cdot\Delta\alpha\right)\mathrm{d}\xi,$$
$$\bar{T}\,\Delta^2\mathbb{S}^{\mathrm{p}} = \int_G \tfrac{1}{2}\Delta\alpha\cdot\bar{D}_{,\alpha}(\bar{\alpha},\Delta\alpha)\,\mathrm{d}\xi\,. \tag{2.112}$$

From (2.94), (2.111) and Theorem 1 we arrive at the conclusion that if (2.42) holds then

$$\Delta^2\Phi + \bar{T}\,\Delta^2\mathbb{S}^{\mathrm{p}} > 0 \quad \text{for every } \Delta\tilde{\mathbf{x}},\ \Delta\tilde{\alpha} \neq \tilde{\mathbf{0}}, \quad \Delta\mathbf{x} = \mathbf{0} \text{ on } S_u \tag{2.113}$$

is sufficient for stability of equilibrium *in the first approximation* with respect to metrics ρ_{E}^0 and ρ_{E}.

Example (2)

Suppose that the conditions of uniqueness (2.50) and of symmetry (2.51) are satisfied. On using this and (2.106) it can be proved [26] that among all deformation paths satisfying (2.109) at each instant and leading to a given small *deformation* increment, the work integral is minimized, to second order, on the straight (direct) path, viz.

$$\begin{aligned}
\Delta\bar{w} &\equiv \int_{t^0}^{\tau} \bar{\phi}_{,\mathbf{F}}(t)\cdot\dot{\mathbf{F}}(t)\,\mathrm{d}t \\
&\geq \bar{\phi}_{,\mathbf{F}}(\bar{\mathfrak{z}})\cdot\Delta\mathbf{F} + \frac{1}{2}\Delta\mathbf{F}\cdot\bar{\phi}_{,\mathbf{FF}}\cdot\Delta\mathbf{F} - \frac{1}{2}\sum_{K,J} g_{KJ}(\bar{\mathfrak{z}})\Delta^{\mathrm{d}}\gamma_K\Delta^{\mathrm{d}}\gamma_J \\
&= \mathbf{S}(\bar{\mathfrak{z}})\cdot\Delta\mathbf{F} + \tfrac{1}{2}\Delta^{\mathrm{d}}\mathbf{S}\cdot\Delta\mathbf{F}; \quad \Delta() = ()|_{t^0}^{\tau}, \quad \mathfrak{z}(t^0) = \bar{\mathfrak{z}}\,.
\end{aligned} \tag{2.114}$$

The superscript "d" denotes the increments obtained on a straight (direct) deformation path from $\bar{\mathbf{F}} = \mathbf{F}(t^0)$ to $\bar{\mathbf{F}} + \Delta\mathbf{F}$.

From (2.114), (2.86) and (2.54) we obtain the inequality

$$\Delta\bar{V} \geq \Delta\bar{\mathbb{E}} \geq \Delta^2\mathbb{W} \tag{2.115}$$

where

$$\Delta^2\mathbb{W} = \int_G \frac{1}{2}\Delta\mathbf{S}^{\mathrm{d}}\cdot\Delta\mathbf{F}\,\mathrm{d}\xi = \int_G U(\bar{\mathfrak{z}},\Delta\mathbf{F})\,\mathrm{d}\xi \tag{2.116}$$

is the global expression for the second-order work on straight (direct) deformation paths. Note that if (2.86) holds and $\tilde{\mathbf{v}}$ is continuous then

$$\Delta^2\mathbb{W}(\Delta\tilde{\mathbf{x}}) = \frac{1}{2}\ddot{\mathbb{E}}(\tilde{\mathbf{v}})(\Delta t)^2, \quad \text{for } \Delta\mathbf{x} = \mathbf{v}\Delta t, \quad \mathbf{v} = 0 \text{ on } \mathrm{S_u}. \tag{2.117}$$

Under the assumption that a finite part S_{u} of the body surface is rigidly constrained, define a distance from $\mathcal{S}^0$ by the metric

$$\rho_{\mathrm{F}} = \|\Delta\mathbf{F}\| \equiv \left(\int_G |\Delta\mathbf{F}|^2 \, \mathrm{d}\xi\right)^{1/2} . \tag{2.118}$$

From Theorem 1, the definition (2.94) and the above relationships we arrive at the conclusion that

$$\ddot{\mathbb{E}}(\tilde{\mathbf{v}}) \geq k\|\dot{\mathbf{F}}\|^2 \quad \text{for every } \tilde{\mathbf{v}}, \quad \mathbf{v} = 0 \text{ on } S_u , \quad k > 0 \tag{2.119}$$

is sufficient for stability, *in the first approximation*, of equilibrium state $\mathcal{S}^0$ with respect to metrics $\rho_{\mathrm{E}}^0, \rho_{\mathrm{F}}$.

It must be noted that fulfilment of the pointwise inequality $|\Delta\mathbf{F}| < R$ is not automatically implied by ρ_{F} or ρ_{E} tending to zero. In turn, if that inequality is violated then validity of the second-order approximation can be questioned. We are faced with a serious difficulty which is met when a rigorous justification for the second-order sufficiency condition for stability of a *continuum* is searched for. The difficulty persists also in the case of purely elastic continua (cf. [29, 30]).

Status of (2.101) as a *necessary* condition of stability of equilibrium of a continuous body also is not mathematically clear in general. The fact that (2.101) is necessary for *asymptotic* stability of equilibrium is not of particular importance here since such a property cannot be required in plasticity. Nevertheless, for a *discretized* system and in the first approximation, (2.101) is necessary for isothermal stability in a dynamic sense provided the incremental constitutive relationship admits a velocity-gradient potential (2.54) [31] ; to prove this, none of the sets of constitutive assumptions from Examples (1) and (2) is needed. Extension of this result to more general circumstances requires further investigations.

2.4 Stability of a Quasi-Static Process

2.4.1 The Concept of Stability of a Process

In the preceding section, the loading (i.e. body forces $\mathbf{b}$ in G, surface tractions $\mathbf{T}$ on S_{T}, and displacements $\mathbf{u}$ on S_{u}) did not depend *explicitly* on time t, although in a general case $\mathbf{b}$ and $\mathbf{T}$ could be configuration-dependent and vary thus in a process induced by disturbances. Now we will consider processes which take place when the loading variables $\mathbf{b}, \mathbf{T}$ and $\mathbf{u}$ depend explicitly on a loading parameter $\lambda = \lambda(t)$. We shall assume that the variation of λ in a real time is vanishingly slow so that there exists an isothermal and quasi-static response of the system, called *the fundamental process* $\mathcal{P}^0 : \lambda \to \mathcal{S}^0(\lambda)$, such that $\mathcal{S}^0(\lambda)$ at each λ is in equilibrium.[8] In a (sufficiently short) period of $\mathcal{P}^0$, we chose λ to be strictly increasing and adopt λ as a time-like parameter playing throughout this section the role of time. Accordingly,

[8] This is obviously an approximation, as discussed in Sect. 2.2 in the context of a material element and extended now to a system.

a dot over a symbol denotes now the right-hand material time derivative (forward rate) with respect to λ. The "quasi-static" velocities $\mathbf{v} = \dot{\mathbf{x}} = \partial \mathbf{x}(\boldsymbol{\xi}, \lambda)/\partial \lambda$ have now conventional meaning and are no longer regarded as non-equilibrium state variables.

The question now arises: does stability of each equilibrium state $\mathscr{S}^0(\lambda)$ at *fixed* λ, as examined in the preceding section, guarantee stability of the *process* $\mathscr{P}^0$ when λ is varying? The answer is negative: even if the distance between $\mathscr{S}^0$ and a state $\mathscr{S}$ reached in a perturbed process is arbitrarily small if λ = const, the possibility that the distance grows with a *finite rate with respect to* λ has not been excluded. Hence, it can happen that after a small increment of λ the distance between $\mathscr{S}^0(\lambda)$ and $\mathscr{S}(\lambda)$ exceeds some given positive value (ε, say) no matter how small the initial disturbance is, irrespectively of the preserved stability of equilibrium. This possibility is a peculiar property of rate-independent plastic deformations, closely related to the possibility of *bifurcation*, i.e. of a non-unique incremental response to the loading program in the absence of any disturbances.

The next question is whether an approach similar to that used to examine stability of equilibrium, i.e. via an appropriate Lyapunov functional based on thermodynamics, is possible also when λ is not kept fixed. A general answer is lacking at present; we will restrict ourselves to examining this question under certain additional assumptions. In particular, rate-independent plasticity and isothermal perturbed processes will only be considered.

Moreover, the attention will be confined to *stability in the first approximation*, in the sense analogous to that adopted for stability of equilibrium in Sect. 2.3.4. Take $\mathscr{S}^0(\bar{\lambda}) = (\bar{\mathfrak{z}})$, $\bar{\mathfrak{z}} = (\bar{\mathbf{F}}, \bar{T}, \bar{\alpha}) \equiv (\mathbf{F}^0, \bar{T}, \alpha^0)|_{\bar{\lambda}}$, at some fixed $\bar{\lambda}$ as the initial state of all processes under consideration which take place at non-decreasing[9] λ in a small neighborhood of $\mathscr{S}^0(\bar{\lambda})$, defined by the pointwise inequalities $|\alpha_K - \bar{\alpha}_K| < R, |\mathbf{F} - \bar{\mathbf{F}}| < R$ in G as well as $\lambda - \bar{\lambda} < R$. A bar over a symbol denotes now either a quantity in the state $\mathscr{S}^0(\bar{\lambda})$ or a function obtained by approximation at $\mathscr{S}^0(\bar{\lambda})$. Consider an auxiliary problem where the Helmholtz free energy density $\bar{\phi}$ is defined by the quadratic approximation of ϕ at $\bar{\mathfrak{z}}$, with the related constitutive approximations either (2.104) or (2.106) in Example (1) or (2), respectively. We also assume that disturbances are sufficiently small in order not to violate (2.107) so that either (2.108) or (2.109) holds in every perturbed process under consideration. In the present auxiliary problem, the assumed configuration-independent loading functions are additionally linearized with respect to λ at $\bar{\lambda}$, viz.

$$\begin{aligned} \bar{\mathbf{T}}(\boldsymbol{\xi}, \lambda) &= \mathbf{T}(\boldsymbol{\xi}, \bar{\lambda}) + (\lambda - \bar{\lambda})\, \dot{\mathbf{T}}(\boldsymbol{\xi}, \bar{\lambda}) \quad \text{on } S_{\mathrm{T}}, \\ \bar{\mathbf{b}}(\boldsymbol{\xi}, \lambda) &= \mathbf{b}(\boldsymbol{\xi}, \bar{\lambda}) + (\lambda - \bar{\lambda})\, \dot{\mathbf{b}}(\boldsymbol{\xi}, \bar{\lambda}) \quad \text{in } G, \\ \bar{\mathbf{x}}(\boldsymbol{\xi}, \lambda) &= \mathbf{x}(\boldsymbol{\xi}, \bar{\lambda}) + (\lambda - \bar{\lambda})\, \mathbf{v}(\boldsymbol{\xi}, \bar{\lambda}) \quad \text{on } S_{\mathrm{u}}. \end{aligned} \tag{2.120}$$

Evidently, the fundamental solution $\overline{\mathscr{P}^0} : \lambda \to \bar{\mathscr{S}}^0(\lambda)$ to the auxiliary problem is

$$\mathbf{F}^0(\boldsymbol{\xi}, \lambda) = (\lambda - \bar{\lambda})\dot{\mathbf{F}}^0(\boldsymbol{\xi}, \bar{\lambda}), \quad \alpha^0(\boldsymbol{\xi}, \lambda) = (\lambda - \bar{\lambda})\, \dot{\alpha}^0(\boldsymbol{\xi}, \bar{\lambda}) \quad \text{for } \lambda > \bar{\lambda}, \tag{2.121}$$

[9] The possibility of a dynamic transitory process at fixed λ is not excluded.

where the rates at $\bar{\lambda}$ are taken from the "true" fundamental process $\mathscr{P}^0$. $\overline{\mathscr{P}^0}$ may be regarded as a linearization (with respect to λ) of $\mathscr{P}^0$ in a small interval $[\bar{\lambda}, \bar{\lambda} + \delta\lambda)$. Stability of the solution (2.121) is now examined as stability in the first approximation of the "true" fundamental process $\mathscr{P}^0$.

The mathematical concept of stability of equilibrium with respect to two metrics, discussed in Sect. 2.3.1, can be extended to processes [20]. For our purposes, the following specifications are made. The object of stability analysis is the isothermal process $\overline{\mathscr{P}^0}$ at varying λ in the thermally insulated, compound system $\mathscr{A}$ defined in Sect. 2.3.2. Small changes of the parameters of an *equilibrium* state $\mathscr{S}(\bar{\bar{\lambda}})$ at some $\bar{\bar{\lambda}} \geq \bar{\lambda}$ with respect to those of $\mathscr{S}^0(\bar{\lambda})$ are taken as *initial* disturbances for the subsequent *quasi-static* process $\mathscr{P} : \lambda \to \mathscr{S}(\lambda),\ \lambda \geq \bar{\bar{\lambda}}$, assumed to be free of any further disturbing influences. The restriction is introduced that $\mathscr{S}(\bar{\bar{\lambda}})$ is attainable from $\mathscr{S}^0(\bar{\lambda})$ in some transitory process corresponding to some time-dependent disturbances which are small enough to preserve (2.107), in which the continuity assumptions from Sect. 2.3.2 and the constitutive relations are valid. The metric ρ^0 for the initial disturbances satisfies (2.55), where $\mathscr{S}$ and $\mathscr{S}^0$ should be understood as $\mathscr{S}(\bar{\bar{\lambda}})$ and $\bar{\mathscr{S}}^0(\bar{\bar{\lambda}})$, respectively. The metric ρ in (2.56) refers to the distance between $\mathscr{S}(\lambda)$ and $\bar{\mathscr{S}}^0(\lambda)$ at some $\lambda \geq \bar{\bar{\lambda}}$, such that $\rho(\lambda) \equiv \rho(\mathscr{S}(\lambda), \bar{\mathscr{S}}^0(\lambda))$ is continuous with respect to λ (cf. (2.57)). In analogy to (2.58), it is assumed that $\rho(\bar{\bar{\lambda}})$ is continuous at $(\bar{\mathscr{S}}^0(\bar{\bar{\lambda}}), \bar{\mathscr{S}}^0(\bar{\bar{\lambda}}))$ with respect to ρ^0.

Definition 2 *The process $\overline{\mathscr{P}^0}$ (or $\mathscr{P}^0$) is called stable (or stable in the first approximation at $\bar{\lambda}$) for initial disturbances with respect to metrics ρ^0, ρ if and only if for every $\varepsilon > 0$ there is $\delta(\varepsilon) > 0$ such that $\rho^0(\bar{\bar{\lambda}}) < \delta(\varepsilon)$ implies $\rho(\lambda) < \varepsilon$ at every $\lambda > \bar{\bar{\lambda}}$.*

A theorem analogous to Theorem 1 can be proved [20] which motivates the approach below.

Following the concept of stability in the energy sense proposed in [24], consider the possibility of taking the *difference* of the energy functional in the processes $\mathscr{P}$ and $\overline{\mathscr{P}^0}$, viz.

$$\Delta\mathbb{E}|_{\lambda} = (\Delta\mathbb{W} + \Delta\Omega)|_{\lambda}, \quad \lambda \geq \bar{\lambda} \tag{2.122}$$

as the Lyapunov functional for *quasi-static* processes at varying λ.[10] Here and in the following the prefix Δ denotes the difference of any corresponding quantities *at the same λ* in the perturbed and (linearized) fundamental processes, $\mathscr{P}$ and $\overline{\mathscr{P}^0}$; the quantities in $\overline{\mathscr{P}^0}$ are distinguished by the superscript "0". To simplify the notation, bars over quantities dependent on λ will be omitted if this causes no confusion. Of course, the actual value of $\Delta\mathbb{E}$ at λ depends on how the perturbed state $\mathscr{S}(\lambda)$ has been reached from $\mathscr{S}^0(\bar{\lambda})$.

From the assumption of configuration-independent loading, the formula (2.86) retains its validity with the changed meaning of Δ. By using $(2.120)_3$ and the conditions of mechanical equilibrium satisfied in $\overline{\mathscr{P}^0}$ (cf. (2.87)), after straightforward transformations we obtain

[10] One can admit also dynamic perturbed processes, and then $\mathbb{E}$ has to be replaced by $\mathbb{E} + \mathbb{K}$.

$$\Delta\dot{\Omega} = -\int_G (\mathbf{b}\cdot\Delta\mathbf{x})^\bullet \, d\xi - \int_{S_T} (\mathbf{T}\cdot\Delta\mathbf{x})^\bullet \, dS = -\int_G \dot{\mathbf{S}}^0\cdot\Delta\mathbf{F}\, d\xi - \int_G \mathbf{S}^0\cdot\Delta\dot{\mathbf{F}}\, d\xi\,.$$

Now, one easily shows that

$$\Delta\dot{\mathbb{E}} = \int_G (\Delta\mathbf{S}\cdot\dot{\mathbf{F}}^0 - \dot{\mathbf{S}}^0\cdot\Delta\mathbf{F})\, d\xi + \int_G \Delta\mathbf{S}\cdot\Delta\dot{\mathbf{F}}\, d\xi\,. \tag{2.123}$$

Below we will examine the sign of the former integrand; the second integral vanishes for a quasi-static process $\mathscr{P}$ free of persistent disturbances.

2.4.2 A Constitutive Inequality

Suppose that no abrupt unloading takes place in the fundamental process $\mathscr{P}^0$ at $\lambda = \bar{\lambda} + 0$, which is equivalent to

$$g_K^0 \equiv g_K(\mathfrak{z}^0(\lambda)) = 0 \quad \text{in } \overline{\mathscr{P}}^0 \quad \text{if } g_K(\bar{\mathfrak{z}}) = 0\,. \tag{2.124}$$

In combination with the assumption (2.107), we obtain that

$$\bar{D}(\alpha^0(\lambda), \dot{\alpha}(\lambda)) = \mathbf{A}(\mathfrak{z}^0(\lambda))\cdot\dot{\alpha}(\lambda) \tag{2.125}$$

$$g_K^0\dot{\gamma}_K(\lambda) = 0 \tag{2.126}$$

at any λ and in any process $\mathscr{P}$, in Example (1) or (2), respectively. This implies

$$\bar{D}_{,\alpha}(\bar{\alpha}, \dot{\alpha}(\lambda))\cdot\dot{\alpha}^0 = \dot{\mathbf{A}}^0\cdot\dot{\alpha}(\lambda), \tag{2.127}$$

$$\dot{g}_K^0(\bar{\mathfrak{z}})\dot{\gamma}_K(\lambda) = 0\,. \tag{2.128}$$

These equalities are now assumed to hold as the necessary conditions for non-unloading in the fundamental process $\mathscr{P}^0$ at $\lambda = \bar{\lambda} + 0$. Alternatively, (2.127) and (2.128) result from the (stronger) assumption that the fundamental process $\mathscr{P}^0$ is *smooth* at $\bar{\lambda}$.

It has been shown [32, 33] that the above non-unloading condition in both Examples (1) and (2) results in the following constitutive inequality:

$$\dot{\mathbf{S}}(\bar{\mathfrak{z}}, \dot{\mathbf{F}}^0)\cdot\dot{\mathbf{F}} - \dot{\mathbf{S}}(\bar{\mathfrak{z}}, \dot{\mathbf{F}})\cdot\dot{\mathbf{F}}^0 \geq 0 \quad \text{for every } \dot{\mathbf{F}} \tag{2.129}$$

provided the symmetry property (2.42) or (2.51) has been assumed. A certain consequence of (2.129) will be used later, at the moment (2.129) is quoted to indicate the possibility that the similar inequality

$$\dot{\mathbf{S}}^0\cdot\Delta\mathbf{F} - \Delta\mathbf{S}\cdot\dot{\mathbf{F}}^0 \geq 0 \quad \text{for every } \Delta\mathbf{F} = \mathbf{F} - \mathbf{F}^0,\ \Delta\mathbf{S} = \bar{\phi}_{,\mathbf{F}}(\mathbf{F}, \alpha) - \bar{\phi}_{,\mathbf{F}}(\mathbf{F}^0, \alpha^0) \tag{2.130}$$

may be valid under the same assumptions; note that $\Delta\mathbf{F}$ and $\Delta\mathbf{S}$ in the auxiliary problem under consideration should be regarded as first-order quantities. This possibility is examined below.

Example (1)

The left-hand side of the inequality (2.130) is transformed as follows

$$\begin{aligned}\dot{\mathbf{S}}^0 \cdot \Delta\mathbf{F} - \Delta\mathbf{S} \cdot \dot{\mathbf{F}}^0 &= \Delta\mathbf{F} \cdot \bar{\phi}_{,\mathbf{F}\alpha} \cdot \dot{\alpha}^0 - \dot{\mathbf{F}}^0 \cdot \bar{\phi}_{,\mathbf{F}\alpha} \cdot \Delta\alpha = -\Delta\boldsymbol{A} \cdot \dot{\alpha}^0 + \dot{\boldsymbol{A}}^0 \cdot \Delta\alpha \\ &= -\boldsymbol{A} \cdot \dot{\alpha}^0 + \bar{D}(\alpha^0, \dot{\alpha}^0) + \dot{\boldsymbol{A}}^0 \cdot \Delta\alpha \\ &= -\Delta\alpha \cdot (\bar{D}_{,\alpha}(\bar{\alpha}, \dot{\alpha}^0) - \dot{\boldsymbol{A}}^0) + (\bar{D}(\alpha, \dot{\alpha}^0) - \boldsymbol{A} \cdot \dot{\alpha}^0)\,.\end{aligned} \tag{2.131}$$

By writing $\Delta\alpha = \int(\dot{\alpha} - \dot{\alpha}^0)\,\mathrm{d}\lambda$, using the symmetry property (2.42) and substituting (2.127) we obtain that the former term after the last equality sign in (2.131) vanishes. Since the last term in (2.131) is nonnegative on account of (2.104), we arrive at the conclusion that (2.130) is valid in Example (1) provided (2.127) and (2.42) have been assumed.

Example (2)

In place of (2.131) we obtain

$$\begin{aligned}\dot{\mathbf{S}}^0 \cdot \Delta\mathbf{F} - \Delta\mathbf{S} \cdot \dot{\mathbf{F}}^0 &= \sum_K \bar{\mathbf{\Lambda}}_K \cdot (\dot{\mathbf{F}}^0 \Delta\gamma_K - \Delta\mathbf{F}\dot{\gamma}_K^0) \\ &= \sum_K (\dot{g}_K^0 \cdot \Delta\gamma_K + \sum_J g_{KJ}(\bar{\jmath})(\dot{\gamma}_J^0 \Delta\gamma_K - \dot{\gamma}_K^0 \Delta\gamma_J)) - \sum_K g_K \dot{\gamma}_K^0.\end{aligned} \tag{2.132}$$

The former sum over K after the last equality sign vanishes on account of (2.128) and (2.51). The last sum over K in (2.132) is evidently nonpositive, and validity of (2.130) has been proved also in Example (2) provided (2.128) and (2.51) are valid.

The essential point in the proof of (2.130), besides the assumed symmetry and linearizations, is that in a perturbed process only those internal mechanisms of plastic deformation are activated which are active or at least non-unloaded in the linearized fundamental process $\overline{\mathscr{P}}^0$. In other words, a perturbed process can involve unloading with respect to $\overline{\mathscr{P}}^0$ but not conversely; this is given a mathematical formulation in (2.127) or (2.128), alternatively. As already mentioned, this can be regarded as a consequence of smoothness of the fundamental process at the stage $\bar{\lambda}$ under consideration.

2.4.3 Conditions of Stability in the First Approximation

We are now ready to formulate conditions of stability (in the first approximation) of our (linearized) fundamental process. Referring to Definition 2, suppose that after some transitory, possibly dynamic process which takes place in a sufficiently small R-neighborhood of $\mathcal{S}^0(\bar{\lambda})$ as discussed above, the system is in an equilibrium state $\mathcal{S}(\bar{\bar{\lambda}}) \neq \bar{\mathcal{S}}^0(\bar{\bar{\lambda}})$, $\bar{\bar{\lambda}} \geq \bar{\lambda}$. In a subsequent quasi-static process $\mathcal{P}$ starting from $\mathcal{S}(\bar{\bar{\lambda}})$ and induced by quasi-statically varying loading in the absence of further disturbing influences, the latter integral in (2.123) vanishes by the standard transformation with the use of Green's theorem, while the former is non-positive under the assumptions leading to (2.130). Thus, if (2.130) holds then we arrive at the desired result that

$$\Delta\dot{\mathbb{E}} \leq 0 \quad \text{for } \lambda \geq \bar{\bar{\lambda}} \tag{2.133}$$

in any quasi-static process $\mathcal{P}$ (free of persistent disturbances). However, it must be noted that, contrary to a similar property for the functional V when λ is fixed, (2.133) is not universally valid: it has been derived only for the specified two classes of rate-independent elastic-plastic solids, under the additional assumptions: either (2.127), (2.42) or (2.128), (2.51), and only for the auxiliary problem.

Under the conditions leading to (2.133), a natural condition of stability of the fundamental process is

$$\Delta\mathbb{E}|_{\lambda} > 0. \quad \text{for every } \mathcal{S}(\lambda) \neq \mathcal{S}^0(\lambda), \quad \lambda \geq \bar{\lambda}, \tag{2.134}$$

and for every transitory process from $\mathcal{S}^0(\bar{\lambda})$ to $\mathcal{S}(\lambda)$. Evidently, this is a stronger condition than the similar condition for stability of equilibrium discussed in Sect. 2.3.4 at $\lambda = \text{const}$. Note that $\Delta\mathbb{E}$ in (2.134) is understood as a *second-order quantity* with respect to the increments of $\mathbf{F}$, $\boldsymbol{\alpha}$ and λ from their values in $\mathcal{S}^0(\bar{\lambda})$. For, the first-order contribution to $\Delta\mathbb{E}$ vanishes on account of the equilibrium conditions and (2.107), while the higher-order terms have been eliminated by the linearizations involved in the definition of the auxiliary problem. It is possible not to discuss the auxiliary problem but rather to perform all calculations for the "true" process in a R-neighborhood of $\mathcal{S}^0(\bar{\lambda})$ with accuracy to the second-order terms only.

The quantity $\Delta\mathbb{E}|_{\bar{\bar{\lambda}}}$ can be interpreted as the energy supplied to the compound system $\mathcal{A}$ due to disturbing influences[11] in a transitory (isothermal) process such that $\Delta^{\text{ext}}S_{\mathcal{A}} = 0$. If (2.134) holds then the supremum of $\Delta\mathbb{E}|_{\lambda}$ in $[\bar{\lambda}, \bar{\bar{\lambda}}]$ may be identified with a metric ρ^0_{E} generated by the energy functional $\mathbb{E}$ for the initial disturbance at $\bar{\bar{\lambda}}$ (cf. 2.94). Analogously, $\Delta\mathbb{E}|_{\lambda}$ may be taken as the Lyapunov pseudo-metric ρ_{E} generated by $\mathbb{E}$ for the distance between $\mathcal{S}(\lambda)$ and $\bar{\mathcal{S}}^0(\lambda)$. From the theorem analogous to Theorem 1 it follows that *the condition (2.134) is sufficient for stability of the process $\overline{\mathcal{P}}^0$ with respect to these metrics* in the sense of Definition 2. Another interpretation is discussed below.

[11] Note that $\Delta\mathbb{E}|_{\lambda}$ is not equal in general to the work done directly by some perturbing forces since the energy flux from external sources to the body $\mathcal{B}$ through S_{u}, or to the loading device, is in general also influenced by a disturbance when λ is varying.

For fulfilment of the condition (2.134) it is necessary that

$$\Delta\ddot{\mathbb{E}}(\tilde{\mathbf{v}}) > 0 \quad \text{in } \mathcal{S}^0(\bar{\lambda}) \quad \text{for every } \tilde{\mathbf{v}} \neq \tilde{\mathbf{v}}^0,\ \mathbf{v} = \mathbf{v}^0 \text{ on } S_{\mathrm{u}}\,, \tag{2.135}$$

where, for the configuration-independent loading,

$$\Delta\ddot{\mathbb{E}}(\tilde{\mathbf{v}}) = \Delta\left(\int_G \dot{\mathbf{S}}(\dot{\mathbf{F}}, \mathfrak{z}) \cdot \dot{\mathbf{F}}\, \mathrm{d}\xi - 2\int_G \dot{\mathbf{b}} \cdot \mathbf{v}\, \mathrm{d}\xi - 2\int_{S_T} \dot{\mathbf{T}} \cdot \mathbf{v}\, \mathrm{d}S\right) \quad \text{in } \mathcal{S}^0(\bar{\lambda})\,. \tag{2.136}$$

When multiplied by $\frac{1}{2}(\lambda - \bar{\lambda})^2$, the expression (2.136) represents the value of $\Delta\mathbb{E}|_\lambda$ reached on a straight path in $\tilde{\mathbf{F}}$-space from $\mathcal{S}^0(\bar{\lambda})$ to $\mathcal{S}(\lambda)$. As discussed in Sect. 2.3.4 in the context of stability of equilibrium, if in Example (2) the conditions of symmetry (2.51) and of constitutive uniqueness (2.50) are satisfied then among all transitory processes which satisfy (2.109) at each instant and lead to the same increment of $\tilde{\mathbf{x}}$, the second-order work integral, and consequently $\Delta\mathbb{E}$ as well, are minimized on such direct paths. Hence, under these assumptions, the second-order condition (2.134) is equivalent to (2.135). The equivalence can be demonstrated also in Example (1) provided the constitutive uniqueness condition (2.41) and the symmetry condition (2.42) are satisfied.

Now, we make one more step which was not possible in the case of examining stability of equilibrium. Namely, we may assume that $\dot{\mathbf{S}}$ as a function of $\dot{\mathbf{F}}$ in $\bar{\mathfrak{z}}$ is differentiable at $\dot{\mathbf{F}}^0$ almost everywhere in G. This is a natural requirement provided, of course, that $\dot{\mathbf{F}}^0 \neq \mathbf{0}$. Then the so-called tangent moduli $\mathbf{C}^0(\bar{\mathfrak{z}}) \equiv (\partial\dot{\mathbf{S}}/\partial\dot{\mathbf{F}})(\dot{\mathbf{F}}^0, \bar{\mathfrak{z}})$ are well defined almost everywhere in G. For both the sets of constitutive assumptions used to derive (2.133), we have the potential constitutive relationship (2.54) so that the tangent moduli exhibit the diagonal symmetry, $C^0_{ijkl} = C^0_{klij}$. This symmetry property is of primary importance for the possibility of (2.135) to hold and thus for validity of the whole approach discussed in this section.

The step mentioned above follows from the fact that (2.135) is, as a consequence of (2.129), implied by [31, 32]

$$\bar{I}^0(\tilde{\mathbf{w}}) > 0 \quad \text{in } \mathcal{S}^0(\bar{\lambda}) \quad \text{for every } \tilde{\mathbf{w}} \neq \tilde{\mathbf{0}}, \quad \mathbf{w} = \mathbf{0} \text{ on } S_{\mathrm{u}}, \tag{2.137}$$

where

$$\bar{I}^0(\tilde{\mathbf{w}}) \equiv \frac{1}{2}\int_G \nabla\mathbf{w} \cdot \mathbf{C}^0(\bar{\mathfrak{z}}) \cdot \nabla\mathbf{w}\, \mathrm{d}\xi \tag{2.138}$$

is a *quadratic* functional of the single field $\tilde{\mathbf{w}} = \Delta\tilde{\mathbf{v}}$. Conversely, (2.135) implies nonnegativeness of (2.138) as of the second variation of $\Delta\ddot{\mathbb{E}}$. The inequality (2.137) ensures also uniqueness of the quasi-static solution in velocities at $\bar{\lambda}$, as established originally by Hill [25] for classical elastoplasticity, extended in [17] to Example (2) and extended further in [32] to any case where (2.54) is complemented with (2.129).

To interpret (2.137) as a condition sufficient for *stability* in the first approximation of the fundamental *process*, let us rewrite it in a slightly stronger form

$$\bar{I}^0(\tilde{\mathbf{w}}) \geq k\|\nabla\mathbf{w}\|^2 \quad \text{in } \mathcal{S}^0(\bar{\lambda}) \text{ for every } \tilde{\mathbf{w}}, \quad \mathbf{w} = \mathbf{0} \text{ on } S_{\mathrm{u}}, \quad k > 0, \tag{2.139}$$

where k may be arbitrarily small. This is equivalent to the fulfilment of (2.137) for the moduli C^0_{1jk1} slightly modified by $(-k\delta_{ik}\delta_{jl})$, where δ_{ij} is the Kronecker symbol. Under the assumptions of the above mentioned theorem on the minimum of the second-order work on a straight deformation path, from (2.139) we obtain

$$\Delta\mathbb{E}|_\lambda \geq \frac{1}{2}\Delta\ddot{\mathbb{E}}(\tilde{\mathbf{x}}(\lambda) - \tilde{\mathbf{x}}(\bar{\lambda})) \geq \bar{I}^0(\Delta\tilde{\mathbf{x}}(\lambda)) \geq k\|\Delta\mathbf{F}(\lambda)\| \equiv k\rho_{\mathrm{F}}(\lambda); \tag{2.140}$$

the second inequality in (2.140) can be shown to follow from (2.129) and (2.54).

For any set of the constitutive assumptions used above to derive (2.133) and under the assumed condition of constitutive uniqueness, we arrive thus at the following conclusion: *The condition (2.139) is sufficient for stability (in the first approximation at $\bar{\lambda}$) of the fundamental process $\overline{\mathscr{P}}^0$ ($\mathscr{P}^0$) for initial disturbances with respect to metrics ρ^0_{E}, ρ_{F}.* Uniqueness of the incremental solution follows as a corollary.

For the non-linearized fundamental process $\mathscr{P}^0$, the stability has a meaning in a R-neighborhood of $\mathcal{S}^0(\bar{\lambda})$ only. As already mentioned in the context of stability of equilibrium, ρ_{F} vanishingly small does not guarantee that $|\Delta\mathbf{F}| < R$ at every material point; this is the source of difficulties in obtaining a rigorous proof of a second-order sufficiency condition for stability in a continuum.

The condition

$$\Delta\ddot{\mathbb{E}} \geq 0 \quad \text{in } \mathcal{S}^0(\bar{\lambda}) \quad \text{for every } \tilde{\mathbf{v}}, \quad \mathbf{v} = \mathbf{v}^0 \text{ on } S_{\mathrm{u}}\,, \tag{2.141}$$

slightly weaker than (2.135), can be interpreted as *necessary* for stability of the linearized fundamental process $\overline{\mathscr{P}}^0$, or for stability in the first approximation of $\mathscr{P}^0$, under weaker constitutive assumptions than those used to justify the sufficiency condition above. Namely, one can prove the following statement [31] :

If

(i) the constitutive velocity-gradient potential (2.54) exists,
(ii) an auxiliary problem defined by (2.103) and (2.120) is considered,
(iii) the condition (2.119) of stability of equilibrium holds in $\mathcal{S}^0(\bar{\lambda})$,
(iv) $\Delta\ddot{\mathbb{E}}(\tilde{\mathbf{v}}) < 0$ in $\mathcal{S}^0(\bar{\lambda})$ for some $\tilde{\mathbf{v}}$, $\mathbf{v} = \mathbf{v}^0$ on $S_{\mathbf{u}}$,

then a given distance ρ_{F} from the fundamental process $\overline{\mathscr{P}}^0$ can be reached on a straight deformation path from $\mathcal{S}^0(\bar{\lambda})$ to some $\mathcal{S}(\lambda)$ under action of arbitrarily small perturbing forces.

The condition (iv) can be replaced by

$$\bar{I}^0(\tilde{\mathbf{w}}) < 0 \quad \text{in } \mathcal{S}^0(\bar{\lambda}) \quad \text{for some } \tilde{\mathbf{w}}, \quad \mathbf{w} = \mathbf{0} \text{ on } S_{\mathrm{u}} \tag{2.142}$$

which excludes (2.141) so that (iv) is implied. By combining this instability condition with the stability condition (2.139), we arrive at the conclusion that the sign of the expression (2.138) has fundamental significance for stability of the *process* $\mathscr{P}^0$, *provided* that the symmetry of the tangent moduli (along with conservativeness of the loading) is assumed.

For a *discretized* problem, the perturbing forces mentioned in the theorem above are not required and existence of bifurcation under the conditions (i)÷(iv) can be proved [31]. A process in which (i)+(iv) are satisfied for $\bar{\lambda}$ from some interval represents then *a continuous spectrum of bifurcation points*.

2.5 Material Stability

2.5.1 Intrinsic Stability Within a Material Element

Consider a material element as a particular continuous system whose unperturbed local state is independent of spatial coordinates. Volume of the element is taken sufficiently small so that body forces can be neglected; to indicate a refined length scale, the material coordinates within the element (in the reference configuration) are no longer denoted by $\boldsymbol{\xi}$ but by another symbol $\mathbf{X}$. A constant temperature $\bar{T}$ is prescribed over the element boundary. As particular cases of stability (of equilibrium or of a quasi-static isothermal process) in a general system, we define two kinds of *intrinsic* stability within the material element:

- Stability of equilibrium within the element if displacements are fixed over the element boundary,
- Stability of the process of uniform straining $\mathbf{F}^0(t)$ of the element if placements $\mathbf{x}(\mathbf{X},t) = \mathbf{x}^0(t) + \mathbf{F}^0(t) \cdot \mathbf{X}$ are given over the element boundary.

These properties evidently do not depend on possible configuration-sensitivity of the external loads acting from outside on the element boundary. They are thus intrinsic properties of an infinitesimal material element embedded in a quasi-statically deforming (possibly inhomogeneous) continuum. Conditions for intrinsic stability will be obtained below from stability conditions for a general system as their specifications for the particular boundary value problem and also as their prerequisites. Moreover, we will additionally investigate stability with respect to internal rearrangements described by internal variables $\boldsymbol{\alpha}$ at prescribed $\mathbf{F}$ or $\mathbf{F}(t)$, the question not discussed so far here. All quantities in this section correspond to some fixed regular point $\boldsymbol{\xi}$ of G.

We will distinguish two kinds of disturbances according to whether they are associated with uniform of non-uniform deformations of the element. In the former case, besides the intrinsic stability also a certain other type of material stability is briefly discussed.

2.5.2 Stability at Uniform Deformations

We have already derived the (classical) condition (2.79) for thermal stability. i.e. for the intrinsic stability of equilibrium with respect to (non-uniform) temperature fluctuations within the material element.

In this section, we will examine stability with respect to internal rearrangements within the element, described by variations of α_K at a constant temperature $\bar{T}$ for prescribed $\mathbf{F}$ or $\mathbf{F}(t)$. Contrary to other sections, the compatibility condition (2.93) is no longer assumed in advance and is allowed to be violated by disturbing influences which perturb intrinsic equilibrium within the element. Since α_K are not subject to any boundary conditions, there is no loss in generality in treating here the material element as a point.

Stability of Equilibrium

Contribution $\mathrm{d}\upsilon$ (per unit reference volume) to $\mathrm{d}\mathbb{E}$ from the element in a given *equilibrium state* $\mathfrak{s} = (\mathbf{F}, \bar{T}, \boldsymbol{\alpha})$ at fixed $\mathbf{F}$ and $\bar{T}$ reads (cf. (2.97))

$$\mathrm{d}\upsilon = \phi_{,\boldsymbol{\alpha}}(\mathfrak{s}) \cdot \mathrm{d}\boldsymbol{\alpha} + \bar{T}\sigma^{\mathrm{p}}\, \mathrm{d}t = -\boldsymbol{A}(\mathfrak{s}) \cdot \mathrm{d}\boldsymbol{\alpha} + D(\mathfrak{s}, \mathrm{d}\boldsymbol{\alpha}). \tag{2.143}$$

This expression vanishes if $\mathrm{d}\boldsymbol{\alpha}$ is consistent with the actual thermodynamic forces $\boldsymbol{A}(\mathfrak{s})$. However, we may consider disturbances at a microscopic level which induce a change of a thermodynamic force from the actual value $\boldsymbol{A}_K$ to some $\boldsymbol{A}_K^*$ such that the associated flux $\dot{\boldsymbol{\alpha}}_K^*$ is nonzero even if $g_K(s) < 0$. In order that $\mathrm{d}\upsilon \geq 0$ also for such disturbances, it is necessary that

$$D(\mathfrak{s}, \dot{\boldsymbol{\alpha}}_K^*) \equiv \boldsymbol{A}_K^* \cdot \dot{\boldsymbol{\alpha}}_K^* \geq \boldsymbol{A}_K \cdot \dot{\boldsymbol{\alpha}}_K^* \quad \text{for any admissible } \dot{\boldsymbol{\alpha}}_K^* . \tag{2.144}$$

The problem what $\dot{\boldsymbol{\alpha}}_K^*$ may be regarded as admissible from a physical point of view is not discussed here since this would require a specification of the physical meaning of $\boldsymbol{\alpha}_K$.

Suppose, as a hypothesis, that (2.144) holds in every stable equilibrium state $\mathfrak{s}$ and for any $\dot{\boldsymbol{\alpha}}_K^*$ corresponding to a small perturbation of the associated thermodynamic force $\boldsymbol{A}_K^*$, i.e. for $|\boldsymbol{A}_K^* - \boldsymbol{A}_K| < \varepsilon$, where ε is an arbitrarily small positive number. Suppose also that points on the boundary $\partial C(\bar{T}, \boldsymbol{\alpha})$ of the current elastic domain correspond to stable equilibrium states, at least in vicinity of the actual $\boldsymbol{A}_K$. In rate-independent plasticity this readily implies the local convexity of $C(\bar{T}, \boldsymbol{\alpha})$ in $\boldsymbol{A}$-space at the actual $\boldsymbol{A}_K$ as well as the generalized normality rule for $\dot{\boldsymbol{\alpha}}$, i.e. (2.38) restricted to $|\boldsymbol{A}_K^* - \boldsymbol{A}_K| < \varepsilon$. If all points on $\partial C(\bar{T}, \boldsymbol{\alpha})$ correspond to stable equilibrium then we obtain also the global convexity of the (connected) domain $C(\bar{T}, \boldsymbol{\alpha})$ in $\mathcal{A}$-space and the principle of maximum dissipation rate (2.38). If $(2.43)_1$ is assumed then we additionally obtain $(2.43)_2$, i.e. the normality of $\dot{\boldsymbol{\alpha}}_K$ to the hypersurface $f_K = 0$ at any $\boldsymbol{A}_K \in \partial C(\bar{T}, \boldsymbol{\alpha})$, for each K separately. Hence, essential constitutive assumptions used in previous sections result now from a study of material stability. It must be pointed out, however, that the necessity of the underlying condition $\mathrm{d}\upsilon \geq 0$ for material stability is not clear.

In passing, we note that convexity of the elastic domain in $\boldsymbol{A}$-space does not necessarily imply convexity of the elastic domain in *stress*-space (nor strain-space) if the dependence of $\boldsymbol{A}$ on the stress (or strain) is nonlinear.

For the rate-dependent plasticity discussed in Sect. 2.2.3, if (2.144) is assumed to hold whenever $\boldsymbol{A}_K$ and $\boldsymbol{A}_K^*$ are in the ε-neighborhood of the examined point on $\partial C(\bar{T}, \alpha)$ then we obtain an analogous normality property but in the limit sense, viz.

$$\delta\boldsymbol{A}_K \cdot \dot{\alpha}_K(\boldsymbol{A} + \delta\boldsymbol{A}, \bar{T}, \alpha) \geq 0 \quad \text{at } \boldsymbol{A} \in \partial C(\bar{T}, \alpha), \tag{2.145}$$

where $\delta\boldsymbol{A}_K$ is an infinitesimal variation of $\boldsymbol{A}_K$. Local or global convexity of the elastic domain $C(\bar{T}, \alpha)$ in $\boldsymbol{A}$-space can analogously be concluded as a consequence of $\mathrm{d}\upsilon \geq 0$. It is worth pointing out that $\dot{\alpha}_K$ as a continuous function of $\boldsymbol{A}_K$ cannot be analytic across ∂C_K, so that care is needed when (2.145) is further exploited to obtain explicit conditions in terms of partial derivatives. For instance, (2.145) can be satisfied if

$$\partial_{\mathrm{p}}\dot{\alpha}_K / \partial\boldsymbol{A}_K = r\boldsymbol{P}_K \otimes \boldsymbol{P}_K \neq \boldsymbol{0} \quad \text{at } \boldsymbol{A}_K \in \partial C(\bar{T}, \alpha), \quad \boldsymbol{P}_K = \partial f_K / \partial\boldsymbol{A}_K, \quad r > 0, \tag{2.146}$$

where $\partial_p / \partial\boldsymbol{A}_K$ denotes one-sided partial derivative at the "plastic" side. For a scalar variable α_K, (2.145) means that the conjugate thermodynamic force A_K increases (at least initially) when $\dot{\alpha}_K$ becomes positive (i.e. positive rate-sensitivity at ∂C).

Consider now a departure from the equilibrium state $\mathfrak{z}$ with $\dot{\alpha} \neq 0$ being initially consistent with the thermodynamic forces $\boldsymbol{A}(\mathfrak{z})$. Then (2.93) is satisfied (and $\mathrm{d}^+\upsilon/\mathrm{d}t = 0$) at an *initial instant* of a non-equilibrium process of internal rearrangements. This is possible for initial $\dot{\alpha} \neq \boldsymbol{0}$ only in rate-independent plasticity to which our attention will now be confined. The contribution $\ddot{\upsilon}$ from the material element in state $\mathfrak{z}$ to $\ddot{\mathbb{E}}$ for fixed $\mathbf{F}$ and $\bar{T}$ reads

$$\ddot{\upsilon} = \ddot{\phi} + \dot{D} = \sum_{K:\dot{\alpha}_K \neq 0} (-\boldsymbol{A}_K + D_{,\dot{\alpha}_K}(\alpha, \dot{\alpha})) \cdot \ddot{\alpha}_K + (-\dot{\boldsymbol{A}} + D_{,\alpha}(\alpha, \dot{\alpha})) \cdot \dot{\alpha} \tag{2.147}$$

or

$$\ddot{\upsilon} = \ddot{\phi} + \dot{D} = -\sum_{K:\,\dot{\gamma}_K \neq 0} g_K(\mathfrak{z})\ddot{\gamma}_K + \sum_{K,J} g_{KJ}(\bar{\mathfrak{z}})\dot{\gamma}_K\dot{\gamma}_J \tag{2.148}$$

for the dissipation function D of the form (2.36) or (2.44), respectively. To have $\ddot{\upsilon} > 0$ for nonzero $\dot{\alpha}$ it is necessary that these expressions are positive. In the former case with $\ddot{\alpha}$ unrestricted, this implies the potentiality condition

$$\boldsymbol{A}_K = \partial D / \partial\dot{\alpha}_K \quad \text{if } \dot{\alpha}_K \neq 0 \tag{2.149}$$

and (cf. [16])

$$\dot{\alpha} \cdot \phi_{,\alpha\alpha} \cdot \dot{\alpha} + D_{,\alpha}(\alpha, \dot{\alpha}) \cdot \dot{\alpha} > 0 \quad \text{for every } \dot{\alpha} \neq 0 \text{ such that } D(\alpha, \dot{\alpha}) = \boldsymbol{A}(\mathfrak{z}) \cdot \dot{\alpha}. \tag{2.150}$$

In the latter case, we obtain $g_K(\mathscr{J})\dot{\gamma}_K = 0$ and

$$\sum g_{KJ}\dot{\gamma}_K\dot{\gamma}_J > 0 \quad \text{for every } \dot{\gamma}_K > 0 \text{ such that } \dot{\gamma}_K g_K(\mathscr{J}) = 0. \tag{2.151}$$

The conditions (2.150) or (2.151) ensure that no spontaneous internal rearrangements are possible at fixed T and $\mathbf{F}$ if disturbances are absent. Their sufficiency for the stability with respect to internal rearrangements at fixed T and $\mathbf{F}$ can be demonstrated under additional assumptions, analogous to those at the end of Sect. 2.3.4.

Suppose now that $\dot{\alpha}$ is uniquely defined by $\dot{\mathbf{e}}$ in any $\mathscr{J}$, and consider stability of equilibrium of a material element under *uniform* variations of $\mathbf{F}$ within the element at fixed $\bar{T}$ and at a passive loading on the element boundary. This leads to the stability condition of the type frequently met in the literature:

$$\dot{\mathbf{t}}(\dot{\mathbf{e}}) \cdot \dot{\mathbf{e}} > 0 \quad \text{for every } \dot{\mathbf{e}} \text{ in } \mathscr{J} \tag{2.152}$$

obtained as a local version of $\Delta\ddot{\mathbb{E}} > 0$ (at $\lambda = \text{const}$) when the potential energy of the loading environment of the element is taken as $(-\bar{\mathbf{t}}\cdot\mathbf{e})$ with $\bar{\mathbf{t}} = \mathbf{t}(\mathscr{J})$ fixed. However, that inequality is not invariant with respect to the choice of stress and strain measures [1]. This reflects the fact that (2.152) involves implicitly a definition of the loading environment and can thus hardly be regarded as a condition for intrinsic stability of the material itself.

There is one exception, namely, if the material element is in a relaxed state ($\mathbf{t} = \mathbf{0}$). Then the element can be imagined to be cut off from a continuum and subject to uniform perturbations of $\mathbf{F}$ at zero tractions on its surface. The resulting condition of stability of equilibrium

$$\dot{\mathbf{t}}(\dot{\mathbf{e}}) \cdot \dot{\mathbf{e}} > 0 \quad \text{for every } \dot{\mathbf{e}} \quad \text{at } \mathbf{t} = \mathbf{0} \tag{2.153}$$

is invariant under transformation to another pair $(\mathbf{t}, \mathbf{e})$ of work-conjugate measures of stress and strain. Of course, the stability is neutral with respect to rigid-body rotations of the element. If the relaxed state corresponds to an interior point of the elastic domain then (2.153) reduces to the classical condition that the fourth-order tensor of the (isothermal) elastic moduli at a stress-free state is positive definite in the space of symmetric second-order tensors.

Stability of a Deformation Process

Consider a process of macroscopically uniform deformation, $\mathscr{P}^0_{\mathrm{F}} : \lambda \to \mathbf{F}^0(\lambda)$, in a homogeneous material element. In the fundamental thermodynamic process $\mathscr{P}^0$, the deformation process $\mathscr{P}^0_{\mathrm{F}}$ is accompanied by a process $\mathscr{P}^0_{\alpha} : \lambda \to \alpha^0(\lambda)$ of internal structural rearrangements whose stability for given $\mathscr{P}^0_{\mathrm{F}}$ will be now examined. We shall combine the procedure adopted in Sect. 2.3.4 with the approach used above to study intrinsic stability of equilibrium; the attention is restricted to rate-independent isothermal processes.

An auxiliary problem being a specification of that defined in Sect. 2.4.1 is formulated at some $\bar{\lambda}$ by adopting the assumptions (2.103), (2.107) and either (2.104) or (2.106). The stability of $\mathscr{p}_\alpha^0$ in the first approximation is examined under the restriction that the macroscopic deformation is unperturbed, $\mathscr{p}_\mathrm{F} = \mathscr{p}_\mathrm{F}^0$. The additional assumptions of symmetry and non-unloading in $\bar{\mathscr{p}}^0$, (2.42), (2.127) or (2.51), (2.128), are also introduced. As an appropriate Lyapunov functional, we then take (with Δ indicating the difference of corresponding quantities in a perturbed process and in $\bar{\mathscr{p}}_\alpha^0$):

$$\Delta v|_\lambda \equiv \Delta\bar{\phi}\big|_\lambda + \Delta\int_{\bar{\lambda}}^{\lambda} \bar{D}\,\mathrm{d}\lambda \tag{2.154}$$

so that

$$\Delta\dot{v} = \Delta\mathbf{S}\cdot\dot{\mathbf{F}}^0 + (\bar{D}(\boldsymbol{\alpha},\dot{\boldsymbol{\alpha}}) - \boldsymbol{A}\cdot\dot{\boldsymbol{\alpha}}) \tag{2.155}$$

or

$$\Delta\dot{v} = \Delta\mathbf{S}\cdot\dot{\mathbf{F}}^0 - \sum_K g_K\dot{\gamma}_K \tag{2.156}$$

in Example (1) or (2), respectively. In a process $\mathscr{p}_\alpha$ free of persistent disturbances in which the internal state is always in equilibrium, the second term vanishes while $\Delta\mathbf{S}\cdot\dot{\mathbf{F}}^0$ will be non-positive as a particular case of (2.130). Consequently, $\Delta\dot{v} \le 0$ in such $\mathscr{p}_\alpha$ as required. Note the essential role of the assumed symmetry and non-unloading in $\mathscr{p}^0$ at $\bar{\lambda}$.

The (second-order) condition of stability of $\mathscr{p}_\alpha^0$ with respect to internal rearrangements becomes

$$\Delta v|_\lambda > 0 \quad \text{at every } \boldsymbol{\alpha}(\lambda), \quad \lambda > \bar{\lambda}, \quad |\alpha_K(\lambda) - \alpha_K(\bar{\lambda})| < R \tag{2.157}$$

and for every transitory process leading from $\boldsymbol{\alpha}^0(\bar{\lambda})$ to $\boldsymbol{\alpha}(\lambda)$ (at given $\mathscr{p}_\mathrm{F}^0$) and satisfying (2.107).

In the first instance, the condition (2.157) reduces for straight paths in $\boldsymbol{\alpha}$-space to $\Delta\ddot{v} > 0$ in $\bar{\mathfrak{z}} = \mathfrak{z}^0(\bar{\lambda})$, viz.

$$\begin{aligned} \Delta\ddot{v} = 2\dot{\mathbf{F}}^0\cdot\phi_{,\mathbf{F}\alpha}\cdot\Delta\dot{\boldsymbol{\alpha}} + \Delta\left(\dot{\boldsymbol{\alpha}}\cdot\phi_{,\alpha\alpha}\cdot\dot{\boldsymbol{\alpha}} + \dot{\boldsymbol{\alpha}}\cdot D_{,\alpha}(\bar{\boldsymbol{\alpha}},\dot{\boldsymbol{\alpha}})\right) > 0 \\ \text{for every } \dot{\boldsymbol{\alpha}} \ne \dot{\boldsymbol{\alpha}}^0 \text{ satisfying } D(\bar{\boldsymbol{\alpha}},\dot{\boldsymbol{\alpha}}) = \bar{A}\cdot\dot{\boldsymbol{\alpha}}, \end{aligned} \tag{2.158}$$

or (on using (2.13) and (2.106)):

$$\begin{aligned} \Delta\ddot{v} = -2\dot{\mathbf{F}}^0\cdot\sum_K \boldsymbol{\Lambda}_K(\bar{\mathfrak{z}})\Delta\dot{\gamma}_K + \sum_{K,J} g_{KJ}(\bar{\mathfrak{z}})\,\Delta(\dot{\gamma}_K\dot{\gamma}_J) > 0 \\ \text{for every } \dot{\gamma}_K > 0,\ \dot{\gamma}_K \not\equiv \dot{\gamma}_K^0,\ g_K(\bar{\mathfrak{z}})\dot{\gamma}_K = 0, \end{aligned} \tag{2.159}$$

in Example (1) or (2), respectively.

Example (1)

Using (2.42), (2.127) and the expression for $\dot{\boldsymbol{A}}$ resulting from (2.103), the stability condition (2.158) is transformed to the form

$$\Delta\ddot{v} = \Delta\dot{\alpha}\cdot\phi_{,\alpha\alpha}\cdot\Delta\dot{\alpha} + \Delta\dot{\alpha}\cdot(D_{,\alpha}(\bar{\alpha},\dot{\alpha}) - D_{,\alpha}(\bar{\alpha},\dot{\alpha}^0)) > 0 \quad \text{for every } \dot{\alpha}\neq\dot{\alpha}^0 \text{ satisfying } D(\bar{\alpha},\dot{\alpha}) = \bar{\boldsymbol{A}}\cdot\dot{\alpha}\,. \tag{2.160}$$

As already discussed, under the assumed symmetry property (2.42) and linearity of $\bar{D}$ with respect to α, among all transitory paths in α-space satisfying (2.108) at each instant, the dissipation integral in (2.154) is minimized on the straight paths (cf. (2.110)). In that case (2.160) is equivalent to (2.157) and is thus sufficient for stability (in the first approximation) of the process $\mathscr{P}_\alpha^0$ for given $\mathscr{P}_\mathrm{F}^0$, with Δv as the measure of perturbations. In particular, (2.160) ensures that $\Delta\alpha_K(\lambda) = \mathbf{0}$ (to first order) at the absence of disturbances; cf. the uniqueness condition (2.41).

Example (2)

Using (2.106), (2.128) and (2.51), the condition (2.159) is rearranged as follows

$$\Delta\ddot{v} = \sum_{KJ} g_{KJ}(\bar{\mathfrak{z}})\Delta\dot{\gamma}_K\Delta\dot{\gamma}_J > 0 \quad \text{for every } \Delta\dot{\gamma}_K\neq 0,\; g_K(\bar{\mathfrak{z}})\Delta\dot{\gamma}_K = 0. \tag{2.161}$$

One can show (cf. [26]) that under the assumptions introduced, (2.161) ensures that Δv corresponding to given $\Delta\gamma$ and $\Delta\mathbf{F}\equiv\mathbf{0}$ is minimized on direct paths (to second order), so that (2.161) is equivalent to (2.157). It follows that in Example (2) under the assumptions (2.128) and (2.51) the condition for stability in the first approximation of the process $\mathscr{P}_\alpha^0$ at given $\mathscr{P}_\mathrm{F}^0$ in a neighborhood of $s^0(\bar{\lambda})$ reduces to the requirement of positive definiteness of the matrix (g_{kJ}) for the indices running over the set of potentially active mechanisms in $\mathfrak{z}^0(\bar{\lambda})$. In this case, the stability means that

$$k\,|\Delta\gamma_K(\lambda)|^2 \leq \sum_{K,J} g_{KJ}(\bar{\mathfrak{z}})\,\Delta\gamma_K(\lambda)\,\Delta\gamma_J(\lambda) \leq \Delta v|_{\bar{\bar{\lambda}}} \quad \text{for } \lambda\geq\bar{\bar{\lambda}},\quad k>0, \tag{2.162}$$

so long as $\lambda-\bar{\lambda}$ remains sufficiently small to justify the linearizations made. In particular, it follows that $\Delta\gamma_K(\lambda) = 0$ (to first order) for given $\mathscr{P}_\mathrm{F}^0$ if disturbances are absent; cf. the uniqueness condition (2.50).

It can be concluded that various constitutive inequalities, adopted in Sect. 2.2 simply as convenient working assumptions, can also be derived as conditions of material stability with respect to internal structural rearrangements.

2.5.3 Material Stability at Non-Uniform Deformations

Stability of Equilibrium

Examine now intrinsic stability of an equilibrium state $\mathfrak{z}^0$ of a (continuous and homogeneous) material element with respect to *non-uniform* isothermal perturbations which preserve continuity of displacements and are compatible with zero displacements over the element boundary. Material stability with respect to internal rearrangements at prescribed deformations is taken as granted; in particular, (unlike in the preceding subsection) we again assume that $\dot{\alpha}$ in a given state is uniquely defined by $\dot{\mathbf{F}}$ (at $\dot{T} = 0$) and that (2.93) holds universally.

Conditions of the intrinsic stability of equilibrium within the element are obtained by specification of, and also as necessary conditions for, the conditions formulated in Sect. 2.3 for a general inhomogeneous body. To begin with, let us discuss first *elastic* stability when α is frozen. Assuming that (2.79) always holds, the restriction to considering only isothermal perturbations is in that case fully justified. As a specification of (and a necessary condition for, cf. [34, 35]) (2.84), at an arbitrary regular point $\boldsymbol{\xi}$ of G we obtain

$$\int_{\mathcal{M}} \phi(\mathbf{F}^0 + \nabla\mathbf{u}(\mathbf{X}), \bar{T}, \alpha)\, \mathrm{d}\mathbf{X} \geq |\mathcal{M}|\, \phi(\mathbf{F}^0, \bar{T}, \alpha) \quad \text{for every } \tilde{\mathbf{u}}, \mathbf{u} = \mathbf{0} \text{ over } \partial\mathcal{M}. \tag{2.163}$$

$\tilde{\mathbf{u}}$ is a (continuous and piecewise smooth) variation of the actual placement field $\mathbf{x}^0(\mathbf{X}) = \bar{\mathbf{x}}^0 + \mathbf{F}^0 \cdot \mathbf{X}$ within the element. $\mathbf{X}$ denotes a position vector, $\nabla(\,) = \partial(\,)/\partial\mathbf{X}$, and $\mathcal{M}$ is a domain of volume $|\mathcal{M}|$ occupied by the element, all considered in the reference configuration. Note that the $\mathbf{X}$-variable refers to a spatial scale of observation "finer" than that for a finite body: in writing (2.163), the $\boldsymbol{\xi}$-variable has been fixed, so that $\mathbf{F}^0 = \mathbf{F}^0(\mathfrak{z}^0, \boldsymbol{\xi})$ and $\alpha = \alpha(\mathfrak{z}^0, \boldsymbol{\xi})$ are independent of $\mathbf{X}$. If (2.163) holds then ϕ as a function of $\mathbf{F}$ at fixed $\bar{T}$ and α is called *quasi-convex* at $\mathbf{F}^0$. One easily shows that the shape and size of $\mathcal{M}$ in (2.163) is unessential.

To derive a simpler condition of material stability, we can consider the possibility that the perturbed deformation mode is effectively concentrated in a planar band of the reference orientation $\mathbf{n}$ and of the thickness h negligible in comparison with dimensions of the material element (Fig. 2.1). The band thickness must somehow tend to zero as the element boundary is approached, but the boundary effects can be neglected. The deformation can be assumed constant within the band and of the form $\mathbf{F} = \mathbf{F}^0 + \mathbf{g} \otimes \mathbf{n}$ (in components $F_{ji} = F^0_{ji} + g_j n_i$), where $\mathbf{n}$ is the unit normal to the band in the reference configuration and $\mathbf{g}$ is some *nonzero* vector.

From the Green theorem and the boundary condition $\mathbf{u} = \mathbf{0}$ over $\partial\mathcal{M}$, we readily obtain

$$\int_{\mathcal{M}} \left(\phi(\mathbf{F}^0 + \nabla\mathbf{u}(\mathbf{X})) - \phi(\mathbf{F}^0)\right) \mathrm{d}\mathbf{X} = \int_{\mathcal{M}} \mathscr{E}_\phi(\mathbf{F}^0, \mathbf{F}^0 + \nabla\mathbf{u}(\mathbf{X}))\, \mathrm{d}\mathbf{X}, \tag{2.164}$$

where

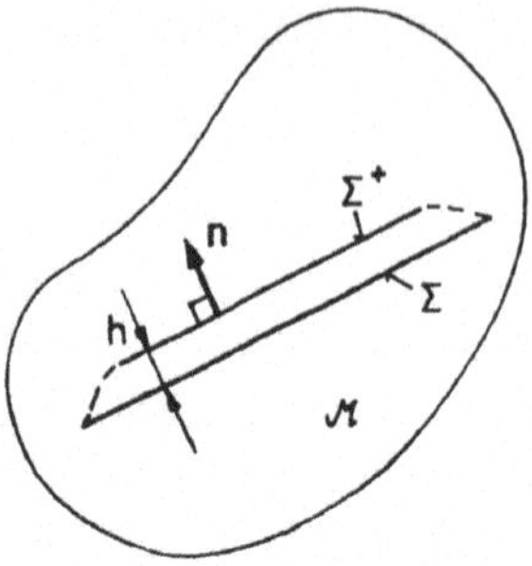

Fig. 2.1 A material element with a planar band of localized deformation.

$$\mathscr{E}_{\phi}(\mathbf{F}^0, \mathbf{F}) \equiv \phi(\mathbf{F}) - \phi(\mathbf{F}^0) - \phi_{,\mathbf{F}}(\mathbf{F}^0) \cdot (\mathbf{F} - \mathbf{F}^0), \quad T, \alpha \text{ fixed} \tag{2.165}$$

is the Weierstrass function associated with ϕ (cf. (2.82)); the frozen arguments $\bar{T}$ and α are omitted.

The displacements at the band boundaries, $\mathbf{u} = \mathbf{0}$ at the plane Σ and $\mathbf{u} = h\mathbf{g}$ at the surface Σ^+, say (Fig. 2.1), can easily be accommodated outside the band by a smooth field $\tilde{\mathbf{u}}$ such that $\nabla\mathbf{u}$ is everywhere of order h provided the band profile has a "cusp" at its perimeter. Then the right-hand integral in (2.164) is of order h^2 everywhere except within the band. For $h \to 0$ we thus obtain

$$\int_M \phi(\mathbf{F}^0 + \nabla\mathbf{u}(\mathbf{X}))\, \mathrm{d}\mathbf{X} - |\mathcal{M}|\phi(\mathbf{F}^0) = \mathscr{E}_{\phi}(\mathbf{F}^0, \mathbf{F}^0 + \mathbf{g}\otimes\mathbf{n}) \int_{\Sigma} h\, \mathrm{d}\Sigma + O(h^2). \tag{2.166}$$

Hence, a necessary condition for (2.163), and thus also for (2.84), is

$$\mathscr{E}_{\phi}(\mathbf{F}^0, \mathbf{F}^0 + \mathbf{g} \otimes \mathbf{n}) \geq 0 \quad \text{for all vectors } \mathbf{g}, \mathbf{n} \tag{2.167}$$

at every regular point in G. A rigorous proof of that result in the general context of the calculus of variations was given first by Graves [36]. In the usually used terminology [35], (2.167) is the condition of *rank 1 convexity* of $\phi(\mathbf{F})$ at $\mathbf{F}^0$, on account of the rank of the matrix $(g_j n_i)$ equal to 1. Further discussion of (2.167) and of its extension to elastic-plastic solids is deferred to the next section.

In the above considerations, *finite* perturbations of $\mathbf{F}$ (which can, however, be associated with arbitrarily small variations of $\mathbf{u}$) are allowed for; in the calculus of variations such perturbations are called *strong* variations. If only infinitesimal perturbations of $\mathbf{F}$ are considered then, as a specification of the condition (2.89) involving the second *weak* variation, in place of the condition (2.163) we obtain

$$\int_{\mathcal{M}} \nabla\mathbf{u} \cdot \phi_{,\mathbf{FF}}(\mathbf{F}^0) \cdot \nabla\mathbf{u}\, \mathrm{d}\mathbf{X} \geq 0 \quad \text{for every } \tilde{\mathbf{u}},\ \mathbf{u} = \mathbf{0} \text{ over } \partial\mathcal{M}\,. \tag{2.168}$$

Perhaps surprisingly at a first glance, (2.168) is equivalent [37] to the Legendre-Hadamard condition

$$(\mathbf{g}\otimes\mathbf{n})\cdot\phi_{,\mathbf{FF}}(\mathbf{F}^0)\cdot(\mathbf{g}\otimes\mathbf{n})\equiv C^{\mathrm{e}}_{jilk}(\mathbf{F}^0)g_jn_ig_ln_k\geq 0 \quad \text{for every } \mathbf{g},\mathbf{n}; \tag{2.169}$$

this provides another counterargument against treatment of positive definiteness of elastic moduli (at non-zero stress) in full space of strain tensors as a thermodynamic condition of stability under *arbitrary* boundary conditions. The condition (2.169) is also obtained as a second-order version of (2.167). In any case, the condition (2.169) can be regarded as necessary for material stability in a thermodynamic sense. If the inequality in (2.169) is strict for nonzero **g**, **n** then it is called the *strong ellipticity condition*, which clearly excludes a non-trivial equality in (2.168).

To find analogous conditions of intrinsic stability of equilibrium for an elastic-plastic material, we begin with the second-order energy condition (2.102). We restrict ourselves to examining the constitutive rate equations in the potential form (2.54). As a specification of and the necessary condition for (2.102), we then obtain

$$\int_{\mathcal{M}} U(\mathfrak{z}^0,\nabla\mathbf{v}(\mathbf{X}))\,\mathrm{d}\mathbf{X}\geq 0 \quad \text{for every } \tilde{\mathbf{v}}, \quad \mathbf{v}=\mathbf{0} \text{ over } \partial\mathcal{M}. \tag{2.170}$$

By using the above construction of a deformation mode effectively concentrated in a band but with velocities in place of displacements, or directly from the Graves theorem, we obtain the following condition necessary for (2.170), and thus also for (2.101):

$$U(\mathfrak{z}^0,\mathbf{g}\otimes\mathbf{n})\geq 0 \quad \text{for all vectors } \mathbf{g},\mathbf{n}; \tag{2.171}$$

note that the velocity-gradient potential U vanishes with its first derivative $\partial U/\partial(\nabla\mathbf{v})$ at $\nabla\mathbf{v}=\mathbf{0}$. Magnitude of the vectors $\mathbf{g},\mathbf{n}$ is unessential in (2.171) since U is a homogeneous function of degree two of $\dot{\mathbf{F}}$. The condition (2.171) can be regarded as a thermodynamic condition necessary for intrinsic stability of equilibrium of the homogeneous material element (under any boundary conditions). It represents an extension of the Legendre-Hadamard condition obtained for *linear* constitutive rate equations to the case when the constitutive rate equations are nonlinear. A mechanical interpretation of (2.171) will be discussed in the next section.

Stability of Deformation Process

The second-order energy condition (2.135) of stability of a quasi-static deformation process obeying (2.54) has a mathematical structure similar to (2.84), with the difference that a velocity field appears in place of a displacement field and the velocity-gradient potential replaces the free energy density. In analogy to (2.163), the quasi-convexity condition for $U(\dot{\mathbf{F}})$ at $\dot{\mathbf{F}}^0$ can be formulated as a local specification and consequence of the stability condition (2.135) for the homogeneous element deforming with uniform velocity gradient $\dot{\mathbf{F}}^0$, as defined in Sect. 2.5.1. However, of more interest will be the consequence of (2.141) in the form of the condition of rank 1 convexity of $U(\dot{\mathbf{F}})$ at $\dot{\mathbf{F}}^0$, viz.

$$\mathscr{E}_U(\mathfrak{z};\dot{\mathbf{F}}^0,\dot{\mathbf{F}}^0+\mathbf{g}\otimes\mathbf{n})\geq 0 \quad \text{for all vectors } \mathbf{g},\mathbf{n} \tag{2.172}$$

at every regular material point in G at any $\bar{\lambda}$, where

$$\mathscr{E}_U(\mathfrak{z};\dot{\mathbf{F}}^0,\dot{\mathbf{F}}) \equiv U(\mathfrak{z};\dot{\mathbf{F}}) - U(\mathfrak{z},\dot{\mathbf{F}}^0) - \dot{\mathbf{S}}(\mathfrak{z},\dot{\mathbf{F}}^0)\cdot(\dot{\mathbf{F}}-\dot{\mathbf{F}}^0) \tag{2.173}$$

is the Weierstrass function associated with the velocity-gradient potential U. The condition (2.172) is obtained in an analogous way as (2.167). Since (2.141) can be interpreted as necessary for stability of the fundamental process for an arbitrary constitutive potential U, (2.172) may be interpreted as necessary for intrinsic stability of a process of uniform deformation with velocity gradient $\dot{\mathbf{F}}^0$ within the material element embedded in a deforming continuum.

If the global stability condition is taken in the form (2.137) then (2.172) reduces to Legendre-Hadamard condition formulated for *the tangent moduli*,

$$(\mathbf{g}\otimes\mathbf{n})\cdot\mathbf{C}^0(\bar{\mathfrak{z}})\cdot(\mathbf{g}\otimes\mathbf{n}) \geq 0 \quad \text{for every } \mathbf{g},\mathbf{n}\,. \tag{2.174}$$

If the constitutive inequality (2.129) holds then (2.174) is equivalent to (2.172). However, (2.171) can still hold even if (2.174) is violated; this reflects at the level of a material element the basic fact that in plasticity the concepts and thus also the conditions of stability of equilibrium and of stability of a quasi-static deformation process are distinct.

2.5.4 Acceleration Waves and Shear Bands

To give another interpretation to the conditions of material stability obtained in the preceding section, consider the problem of propagation of *acceleration* waves in a rate-independent material. An acceleration wave is, by definition, a geometric surface Σ which moves relatively to the material and across which the acceleration $\dot{\mathbf{v}}$, velocity gradient $\dot{\mathbf{F}} = \nabla\mathbf{v}$ and stress-rate $\dot{\mathbf{S}}$ are discontinuous while the displacement, velocity, strain and stress are continuous. Only isothermal deformations are considered and, for simplicity, the reference configuration is chosen to coincide momentarily with the actual configuration ($\mathbf{F} = \mathbf{I}$). The well-known kinematic compatibility and wave propagation conditions read

$$[\![\dot{\mathbf{v}}]\!] = -v\mathbf{g}\,, \quad [\![\dot{\mathbf{F}}]\!] = \mathbf{g}\otimes\mathbf{n}\,, \quad [\![\dot{\mathbf{S}}]\!]\cdot\mathbf{n} = \bar{\rho}v^2\mathbf{g}\,, \tag{2.175}$$

where $[\![\cdot]\!]$ denotes a jump across Σ, $\mathbf{n}$ is the unit normal to Σ, $\mathbf{g}$ is a nonzero vector, $\bar{\rho}$ is the mass density and v is the normal speed of propagation of Σ relative to the material.

It is well known that (2.169) fulfilled at every $\mathbf{F}^0$ is necessary and sufficient for the speed of propagation of all acceleration waves in *elastic* materials to be real. The speed is always real and nonzero if and only if the inequality in (2.169) is strict for nonzero $\mathbf{g}$, $\mathbf{n}$, i.e. if the elastic moduli are strongly elliptic.

Consider now rate-independent inelastic materials obeying the constitutive rate equations which admit an arbitrarily nonlinear potential (2.54). The results given below are taken from the recent paper [38]. It has been proved that for such inelastic materials the condition (2.171) is necessary and sufficient for the speed v of propagation of all acceleration waves into *non-deforming* material to be real. If

$$U(\mathbf{g} \otimes \mathbf{n}) > 0 \quad \text{for all } \mathbf{g}, \mathbf{n} \neq \mathbf{0} \tag{2.176}$$

then the speed is real and nonzero. If (2.171) holds with equality for some nonzero $\mathbf{g}^*, \mathbf{n}^*$ then it can be shown that such a pair $(\mathbf{g}^*, \mathbf{n}^*)$ constitutes a solution to the equation (2.175) with $v = 0$, that is, the condition for a *stationary discontinuity* is met. This corresponds to a nonzero quasi-static incremental solution $\dot{\mathbf{F}}^* = \mathbf{g}^* \otimes \mathbf{n}^*$ in a band of the normal $\mathbf{n}^*$, compatible with constant nominal tractions on the band boundaries moving as rigid planes. If (2.171) is violated then it has been argued that a vanishingly small disturbance can induce a dynamic process of localized deformation in a sufficiently thin band of the orientation $\mathbf{n}$. If $\mathbf{g}^*$ is orthogonal to $\mathbf{n}^*$ then the localization has the form of a *shear band*.

Suppose now that at one side of an acceleration wave the material is *deforming* with a velocity gradient $\dot{\mathbf{F}}^0$. Such a distinction was immaterial for purely elastic deformations, but is important for a non-quadratic potential U. It has been proved that the condition (2.172) is necessary for the speed of propagation of all such acceleration waves to be real. In particular, (2.174) is also necessary for this, and with the interpretation of the left-hand expression in (2.174) as a second-order approximation of that in (2.172), it becomes clear that (2.174) is relevant to the waves of an amplitude sufficiently small for the tangent moduli $\mathbf{C}^0$ to apply on both sides of Σ.

If (2.172) holds with equality for some non-zero $\mathbf{g}^*, \mathbf{n}^*$ then a quasi-static *bifurcation* in a band of orientation $\mathbf{n}^*$ becomes possible, in the sense that the quasi-static incremental solution $\dot{\mathbf{F}}^* = \dot{\mathbf{F}}^0 + \mathbf{g} \otimes \mathbf{n}$ in the band is kinematically and statically compatible with the uniform deformation rate $\dot{\mathbf{F}}^0$ outside the band. Moreover, existence of such a bifurcation in a band of the orientation $\mathbf{n}$ has also been proved when the condition (2.172) of intrinsic stability of the process is violated while the condition (2.176) of stability of equilibrium still holds. This is again a demonstration of the distinction between stability of equilibrium and stability of a quasi-static process in plastic solids.

Let us proceed to the condition (2.163). If the elastic moduli $\mathbf{C}^{\mathrm{e}}(\mathfrak{z}^0)$ are strongly elliptic then it might seem, in view of the resulting strict inequality in (2.168) for $\tilde{\mathbf{u}} \neq \tilde{\mathbf{0}}$, that an energy barrier exists in a vicinity of the uniform equilibrium state $\mathfrak{z}^0$ which excludes a departure from equilibrium unless a positive amount of work is supplied to the element. However, this is not so if (2.167) is violated and *moving strong discontinuities in strain* are allowed for.

Namely, suppose that the band illustrated schematically in Fig. 2.1 is (artificially) created by the movement of one (Σ^+) of its planar boundaries with an initial normal speed (in the reference configuration) $\dot{h} > 0$, *starting from zero initial thickness* at $t = 0$, say. This is kinematically admissible if the jump in $\mathbf{F}$ has the form $[\![\mathbf{F}]\!] = \mathbf{g} \otimes \mathbf{n}$ and is associated with initial velocity at the moving discontinuity surface Σ^+ being equal to zero from inside and to $\dot{h}\mathbf{g}$ from outside of the formed band. The terms "speed" and "velocity" are conventional since t need not be here a natural time. If α is frozen then from (2.166) one obtains the formula

$$\frac{\mathrm{d}}{\mathrm{d}t}\int_{\mathcal{M}}\phi(\mathbf{F}^0+\nabla\mathbf{u}(\mathbf{X}))\,\mathrm{d}\mathbf{X}\bigg|_{\mathfrak{z}^0}=\mathscr{E}_\phi(\mathbf{F}^0,\mathbf{F}^0+\mathbf{g}\otimes\mathbf{n})\int_\Sigma \dot{h}\,\mathrm{d}\Sigma\,. \tag{2.177}$$

It follows that fulfilment of (2.167) at every regular point of the body $\mathcal{B}$ being in equilibrium is necessary for V_e to be nondecreasing for any *kinematically* admissible mode of quasi-static departure from the equilibrium state if moving strong discontinuities in strain are not excluded. We do not discuss here the question of dynamic admissibility of the moving discontinuities (cf. e.g. [39]).

Equilibrium states of an elastic solid body which are stable with respect to small variations of $\mathbf{F}$ but such that (2.167) is violated in some part of G can be called metastable. Metastable states can be treated as stable or not depending on the strength of actual local disturbances or inhomogeneities which are neglected or smoothed out in the macroscopic description but are always present in reality; it should not be forgotten that we are discussing here only *models* of real materials.

Finally, we formulate a condition analogous to (2.167) but for a rate-independent elastic-plastic material where α varies with $\mathbf{F}$ and when the energy condition of isothermal stability takes the form (2.100). Specification of (2.100) for the material element, when transformed in analogy to (2.164), reads

$$\int_{\mathcal{M}}\Delta w(\nabla\mathbf{u})\,\mathrm{d}\mathbf{X}=\int_{\mathcal{M}}\Big(\Delta w(\nabla\mathbf{u})-\mathbf{S}(\mathfrak{z}^0)\cdot\nabla\mathbf{u}\Big)\,\mathrm{d}\mathbf{X}\geq 0 \text{ for every } \tilde{\mathbf{u}}, \mathbf{u}=\mathbf{0} \text{ over } \partial\mathcal{M}. \tag{2.178}$$

The value of the deformation work Δw in a transition from $\mathbf{F}^0$ to $\mathbf{F}^0+\nabla\mathbf{u}$ depends now on the deformation path. We restrict ourselves to examining *straight* deformation paths. Consider a deformation mode concentrated in a thin planar band as in the derivation of (2.166). Proceeding exactly as before, we arrive at

$$\int_M \Delta w(\nabla\mathbf{u})\,\mathrm{d}\mathbf{X}=\mathscr{E}_W(\mathfrak{z}^0,\mathbf{g}\otimes\mathbf{n})\int_\Sigma h\,\mathrm{d}\Sigma+O(h^2), \tag{2.179}$$

where

$$\begin{aligned}\mathscr{E}_W(\mathfrak{z}^0,\mathbf{g}\otimes\mathbf{n})&=\int_0^1\Big(\mathbf{S}(\mathfrak{z}(z))-\mathbf{S}(\mathfrak{z}^0)\Big)\cdot\dot{\mathbf{F}}\,\mathrm{d}z\\&=\phi(\mathfrak{z}(1))-\phi(\mathfrak{z}^0)-\mathbf{S}(\mathfrak{z}^0)\cdot(\mathbf{g}\otimes\mathbf{n})+\bar{T}\int_0^1\sigma^{\mathrm{p}}(\mathfrak{z}(z),\dot{\alpha}(z))\,\mathrm{d}z\,,\\ \mathfrak{z}(z)=(\mathbf{F}(s^0)+z\mathbf{g}\otimes\mathbf{n},\bar{T},\alpha(z)),&\quad \dot{\alpha}(z)=\dot{\alpha}(\mathfrak{z}(z),\dot{\mathbf{F}}),\quad \dot{\mathbf{F}}=\mathbf{g}\otimes\mathbf{n}\,.\end{aligned} \tag{2.180}$$

$\mathscr{E}_W$ represents the difference between the actual work of deformation on the straight path from $\mathbf{F}(s^0)$ to $\mathbf{F}(s^0)+\mathbf{g}\otimes\mathbf{n}$ and the corresponding work done by equilibrium stresses $\mathbf{S}(\mathfrak{z}^0)$. For elastic solids $\mathscr{E}_W$ reduces to $\mathscr{E}_\phi$.

From (2.179) as a necessary condition for (2.178), and thus also necessary for the global stability condition (2.100) to hold, we obtain

$$\mathscr{E}_W(\mathscr{S}^0, \mathbf{g} \otimes \mathbf{n}) \geq 0 \quad \text{for all vectors } \mathbf{g}, \mathbf{n}\,. \tag{2.181}$$

Note that (2.181) must be satisfied if (2.100) is to hold in a neighborhood of $\mathscr{S}^0$ relative to the distance $\|\Delta\mathbf{F}\|$. Equilibrium states which are stable with respect to small variations of $\mathbf{F}$ and α but such that (2.181) is violated in some part of G can be called *metastable*; cf. the discussion above on metastability in elastic solids.

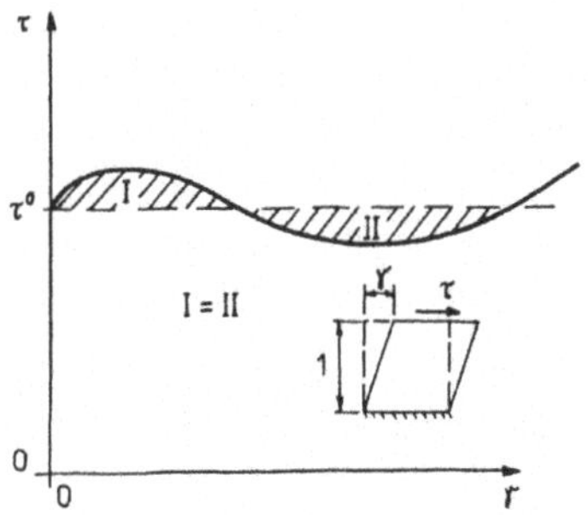

Fig. 2.2 Schematic stress-strain diagram for simple shear superimposed on the critical state defined by condition (2.181).

It is an open question in which cases the metastability associated with violation of (2.181) can be treated as sufficient for formation of shear bands, for instance, in plastically deformed metals. Suppose that existing strong nonuniformities in polycrystals justify that hypothesis, introduced in [40]. The critical stage of the deformation process beyond which the onset of formation of shear bands might be expected is then found by seeking the first Maxwell line for the shear stress - shear strain curves drawn for simple shear (γ) superimposed in various directions on the traversed states (Fig. 2.2). Preliminary calculations performed with the use of simple models of plasticity have given quite realistic results [40], although full justification of that procedure is yet to be found. This question deserves further study.

Acknowledgement

I am indebted to Prof. B. Raniecki for his helpful comments on an earlier version of the manuscript.

References

1. Hill, R.: Aspects of invariance in solids mechanics, Advances in Applied Mechanics, vol. 18, Acad. Press, New York 1978, 1-75.

2. Mandel, J.: Plasticité Classique et Viscoplasticité, CISM Course, Udine, Springer 1971.
3. Rice, J.R.: Inelastic constitutive relations for solids: an internal- variable theory and its application to metal plasticity, J. Mech. Phys. Solids, **19** (1971) 433-455.
4. de Groot, S.R. and Mazur, P.: Non-equilibrium Thermodynamics, NorthHolland, Amsterdam 1962.
5. Muschik, W.: Internal variables in non-equilibrium thermodynamics, J. Non-Equilib. Thermodyn., **15** (1990) 127-137.
6. Kestin, J.: A note on the relation between the hypothesis of local equilibrium and the Clausius-Duhem inequality, J. Non-Equilib. Thermodyn., 15 (1990) 193-212.
7. Rice, J.R.: Continuum mechanics and thermodynamics of plasticity in relation to microscale deformation mechanisms, in: Constitutive Equations in Plasticity (Ed. A.S. Argon), MIT Press, Cambridge Mass. 1975, 23-79.
8. Lehmann, T. (Ed.): The Constitutive Law in Thermoplasticity, CISM Course, Udine, Springer 1984.
9. Kleiber, M. and Raniecki, B.: Elastic-plastic materials at finite strains, in: Plasticity Today: Modelling, Methods and Applications (Ed. A. Sawczuk and G. Bianchi), Elsevier, London 1985, 3-46.
10. Neuhäuser, H.: Physical manifestation of instabilities in plastic flow, in: Mechanical Properties of Solids: Plastic Instabilities (Ed. V. Balakrishnan and E.C. Bottani), World Scientific, Singapore 1986, 209-252.
11. Hill, R. and Rice, J.R.: Elastic potentials and the structure of inelastic constitutive laws, SIAM J. Appl. Math., **25** (1973) 448-461.
12. Kestin, J. and Rice, J.R.: Paradoxes in the application of thermodynamics to strained solids, in: A Critical Review of Thermodynamics (Ed. E.B. Stuart et. al.), Mono Book Corp., Baltimore 1970, 275-298.
13. Morreau, J.J.: Sur les lois de frottement, de plasticité et de viscosité, C. R. Acad. Sci. Paris, t. 271, Serie A (1970) 608-611.
14. Morreau, J.J.: On Unilateral Constraints, Friction and Plasticity, CIME Course, Bressansone, in: New Variational Techniques in Mathematical Physics, Edizioni Cremonese, 1974, 175-322.
15. Nguyen, Q.S.: Stabilité et bifurcation des systèmes dissipatifs standards à comportement indépendant du temps physique, C. R. Acad. Sci. Paris, t. 310, Serie II (1990) 1375-1380.
16. Nguyen, Q.S.: Bifurcation and stability of time-independent standard dissipative systems, CISM Course, Udine 1991.
17. M.J. Sewell: A survey of plastic buckling, in: Stability (H.H.E. Leipholz, ed.), Univ. of Waterloo Press, Ontario 1972, 85-197.
18. Hill, R. and Rice, J.R.: Constitutive analysis of elastic-plastic crystals at arbitrary strain, J. Mech. Phys. Solids, **20** (1972) 401-413.
19. Hill, R.: Some basic principles in the mechanics of solids without a natural time, J. Mech. Phys. Solids, **7** (1959) 209-225.
20. Movchan, A.A.: Stability of processes with respect to two metrics (in Russian), Prikl. Math. Mekh., **24** (1960) 988-1001.
21. Bazant, Z.P.: Stable states and stable paths of propagation of damage zones and interactive fractures, in: Cracking and Damage (Ed. J. Mazars and Z.P. Bazant), Elsevier, London 1989, 183-206.
22. Ericksen, J.L.: A thermo-kinetic view of elastic stability theory, Int. J. Solids Structures, **2** (1966) 573-580.
23. Gurtin, M.E.: Thermodynamics and Stability, Arch. Rational Mech. Anal., **59** (1975) 63-96.
24. Petryk, H.: A consistent energy approach to defining stability of plastic deformation processes, in: Stability in the Mechanics of Cont inua, Proc. IUTAM Symp. (Ed. F.H. Schroeder), Springer, Berlin 1982, 262-272.
25. Hill, R.: A general theory of uniqueness and stability in elastic-plastic solids, J. Mech. Phys. Solids, **6** (1958) 236-249.
26. Petryk, H.: On the second-order work in plasticity, Arch. Mech., **43** (1991) 377-397.
27. Nguyen, Q.S. and Radenkovic, D.: Stability of equilibrium in elastic plastic solids, Springer Lecture Notes in Mathematics, Vol. 503 (1976) 403-414.

28. Petryk, H.: (to be published).
29. Knops, R.J. and Wilkes, E.W.: Theory of Elastic Stability, Handbuch der Physik VI a/3, Springer, Berlin 1973.
30. Koiter, W.T.: A basic open problem in the theory of elastic stability, Springer Lecture Notes in Mathematics, Vol. 503 (1976) 366-373.
31. Petryk, H.: The energy criteria of instability in time-independent inelastic solids, Arch. Mech. **43** (1991) 519-545.
32. Petryk, H.: On constitutive inequalities and bifurcation in elasticplastic solids with a yield-surface vertex, J. Mech. Phys. Solids, **37** (1989) 265-291.
33. Nguyen, Q.S. and Petryk, H.: A constitutive inequality for timeindependent dissipative solids, C. R. Acad. Sci. Paris, t. 312, Serie II, (1991) 7-12.
34. Morrey, C.B.: Quasi-convexity and the lower semicontinuity of multiple integrals, Pacific J. Math., **2** (1952) 25-53.
35. Ball, J.M.: Convexity conditions and existence theorems in nonlinear elasticity, Arch. Rational Mech. Anal., **63** (1977) 337-403.
36. Graves, L.M.: The Weierstrass condition for multiple integral variation problems, Duke Math. J. **5** (1939) 656-660.
37. van Hove, L.: Sur l'extension de la condition de Legendre du calcul des variations aux intégrales multiples à plusieurs fonctions inconnues, Proc. Kon. Ned. Acad. Wet., **50** (1947) 18-23.
38. Petryk, H.: Material instability and strain-rate discontinuities in incrementally nonlinear solids, J. Mech. Phys. Solids, **40** (1992) 1227-1250.
39. Abeyratne, R. and Knowles, J.K.: On the driving traction acting on a surface of strain discontinuity in a continuum, J. Mech. Phys. Solids, **38** (1990) 345-360.
40. Petryk, H.: The energy criteria of instability of plastic flow, XVII IUTAM Congress, Grenoble (1988).

Chapter 3
Theory of Material Instability in Incrementally Nonlinear Plasticity

Abstract Material instability in time-independent elastic-plastic solids is studied as a phenomenon strictly related to qualitative properties of an incremental constitutive law. A broad class of incrementally nonlinear material models is considered which encompasses classical elastoplasticity with a single smooth yield surface, as well as multi-mode plasticity with many internal mechanisms of inelastic deformation which give the yield-surface vertex effect. Three kinds of instability are investigated: with respect to internal microstructural rearrangements, for deviations from uniform deformation under displacement control, and under flexible constraints corresponding to deformation-sensitive loading. It is shown that the respective instability criteria are different and, moreover, dependent on whether the instability concerns a single equilibrium state or a process of quasi-static deformation. Instability of equilibrium is of dynamic type, while instability of a process going through stable equilibrium states is related to a continuous spectrum of quasi-static bifurcation points along the deformation path. Basic concepts are outlined by the example of a one-dimensional discretized tensile bar and extended to incipient localization of deformation in a three-dimensional continuum. Under symmetry restrictions imposed on an incrementally nonlinear constitutive law, a unified approach to material instabilities of various type is presented which is based on the single energy criterion. By specifying the incremental energy consumption to second-order terms and determining the circumstances in which it fails to be minimized along a fundamental deformation path, the onset of material instability of a selected type can be estimated.

[0] Reprinted from: CISM Courses and Lectures No. 414, Udine 1999 (Ed. H.Petryk), pp.261-331, Springer-Verlag, Wien 2000. H.Petryk, Theory of material instability in incrementally nonlinear plasticity. Copyright © 2000 by CISM, Udine.

H. Petryk, *Theory of Plastic Instability*,
https://doi.org/10.1007/978-3-031-98223-1_3

3.1 Introduction

3.1.1 Preliminary Remarks

An observation based on long-standing experience is that an elastic-plastic body deformed under conditions which could result in a quasi-static process of macroscopically uniform deformation, in certain circumstances starts to deform in a significantly different manner. The term 'instability' is commonly used in such cases, although frequently in a more intuitive than precise sense. Typical examples are provided by buckling or necking phenomena understood as a 'geometric' instability, snap-through phenomena as a 'dynamic' instability, or localization of deformation into shear bands as a 'material' instability.

It is well known that the concept of instability of *equilibrium* is too narrow to account for such a variety of instability phenomena in elastic-plastic solids. A suitable framework becomes available if the instability concept is extended to plastic deformation *processes*. For this purpose, a fundamental process is considered which represents a model response of an elastic-plastic body to a given program of slowly varying loading. In reality, there are always certain disturbing influences which are neglected in a theoretical description of the fundamental process; such influences may be represented by additional perturbing forces. If vanishingly small perturbing forces can influence the body response and induce, within a small loading increment, a finite distance from the fundamental process then the latter may be regarded as unstable and, consequently, practically unrealizable in a physical system. This introductory remark indicates how the term instability will be understood below; a more detailed discussion will follow. In plasticity it is frequently more convenient (and physically not less natural) to work with 'persistent disturbances' represented by the perturbing forces mentioned above rather than with 'initial disturbances' common in mathematics.

This Chapter is concerned with *material instability* of various types which may appear independently of the body shape. The emphasis is placed on the influence of the inherent *incremental nonlinearity* of elastic-plastic materials on the onset of material instability and incipient post-critical behaviour. The principal aim is to present in a systematic way the basic theoretical aspects of material instability in slowly loaded incrementally nonlinear solids; thermal and viscous effects are not examined here. No attempt is made to discuss the effect of an internal length scale in the material. This current topic is being examined extensively in the literature [16, 7] and at present it is not easy to summarize the obtained results in the context of material instability.

The considerations below are limited to continuum mechanics. Instabilities occurring in a more refined scale where discrete nature of a solid becomes relevant (e.g. jumps in position of a single dislocation line in a metal crystal) are treated here as physical sources of macroscopic plastic deformation and not as instability in the elastic-plastic material.

This Chapter is to much extent based on the present author's own work in the last years. Among a number of important contributions to the literature which cannot all be mentioned here, the concepts developed by Hadamard [19], Drucker [10, 11], Hill [21, 22, 23, 25, 27, 28, 29], Biot [6], Mandel [38], Maier [37], Rice [66, 67] and Nguyen [40] were particularly helpful in this work. More references can be found in the survey papers [39, 74, 58] and in the previous CISM Lecture Notes by the author [54, 55]. Although concerned with related topics, the author's present and previous Lecture Notes were intended to be more complementary to than overlapping each other.

3.1.2 On the Concept of Material Instability

Consider a solid body of an elastic-plastic material subjected to a given loading program parameterized by a scalar parameter λ which varies in time t in a prescribed manner; the distinction between λ and t will be useful below. Suppose that in the macroscopic scale of observation the material may be regarded as homogeneous and the loading conditions admit a fundamental process of macroscopically uniform deformation. In reality, there are disturbing influences which are neglected when defining the fundamental behaviour of the material under a loading program. Possible instability against physically unavoidable disturbances is to be examined. One may try to give a characterization of material instability as follows: *Material instability is met when finite deviations from a fundamental behaviour of the material in a small interval of λ can be induced by imposing arbitrarily small disturbances,*[1] *irrespectively of the body shape.* This formulation is not free of ambiguities and has to be made precise by defining (cf. [55])

(i) the fundamental behaviour whose stability is examined,
(ii) the class of disturbances considered,
(iii) the measure of a disturbance 'strength',
(iv) the measure of a 'distance' from the fundamental behaviour.

When making the specifications (i)-(iv), some arbitrariness cannot be avoided. We come thus to the conclusion that the concept of material instability is not an absolute one and always involves certain conventions. Moreover, depending on the conventions adopted, the *criteria* of material instability may differ substantially from each other.

At (i), the basic decision to be made is whether we wish to investigate

- instability of *equilibrium* at a fixed loading parameter λ, or
- instability of a *deformation process* at varying λ.

In incrementally nonlinear materials, the corresponding points of stability loss may be quite distinct. It is worth mentioning that instability concerns a *theoretical* behaviour; processes that are unstable in a physically meaningful sense should not

[1] Disturbances should not be confused with *imperfections* which are not considered here. The distinction is that in the stability analysis a perturbed motion of a given material system is investigated while the initial imperfections change the system itself.

be observable in reality. The proposed use of a small interval of λ is motivated by the wish to avoid the possibility that a predicted instability would concern only processes of unrealizable duration.

At (ii), there are various possibilities in perturbing only initial conditions (which does not seem realistic) or also boundary conditions or field equations by applying perturbing forces that may vary in time. In this article, the following three classes of disturbances are distinguished :

• perturbing forces at a micro-level, corresponding to disturbances of the internal state of a material point in the macroscopic description,
• perturbing body forces at a macro-level, corresponding to local disturbances of strain uniformity inside a homogeneized material element,
• perturbing forces at a material element boundary, corresponding to disturbances of overall stress and strain under assumed flexible constraints.

Again, the respective stability conditions can take substantially different forms. Since each class of disturbances can be used to examine stability of an equilibrium state or of a deformation process, the combinations generate already six possible types of material instability.

At (iii), decision is to be made which disturbances are unavoidable from a physical point of view by attributing to them a vanishingly small measure of their 'strength'. For instance, the magnitude of perturbing forces or the energy supplied by a disturbance may be used to generate such measures. As a special case, the possibility of deviations from the fundamental behaviour of the material in the absence of *any* disturbances can be examined; this characterizes the *bifurcation* approach.

At (iv), the choice may depend on the examined type of instability. There is the possibility of neglecting deviations which are trivially possible and unessential for material stability, e.g. rigid-body translations or rotations about a symmetry axis. A careful choice of the measures (iii) and (iv) may be of fundamental importance when precise mathematical statements about stability in a continuum are aimed at.

It must be stressed that the above classification does not cover all aspects of material instability, for instance dynamic propagation of weak or strong disturbances.

It can be concluded that in research on material instability, the adopted meaning of instability should always be clearly specified to avoid misunderstandings. On the other hand, it would be desirable to develop a physically sound and sufficiently wide concept of material instability to provide a common basis for differently oriented investigations of particular aspects.

3.1.3 Intrinsic Instability

The conditions for instability in a finite solid body can be influenced by the body shape or depend on the assumed boundary conditions. In contrast, material instability should rather be viewed as an intrinsic property of the deforming material. Some arbitrariness is again unavoidable here; for instance, a surface instability that leads to waviness of an initially smooth or planar free surface might be regarded as a material instability. That kind of instability is *not* examined in this article.

The fundamental behaviour of the material (cf. point 3.1.2 (i)) may be defined as a *process of macroscopically uniform straining*[2] of a representative volume element *embedded in a continuum* formed by other elements of the material. Uniform perturbations of the deformation gradient within a *bounded* element are generally incompatible with an unperturbed strain in the surrounding material, except in one-dimensional case. Therefore, the loss of material stability in the sense of Drucker [10, 11] need not have consequences in a physical system under sufficiently 'stiff' boundary constraints.

An intrinsic instability in the material is associated below with the possibility of deviation from the fundamental uniform process *within some interior domain* (a localized band in particular) without affecting the behaviour of the surrounding material, cf. [23, 67, 52].

According to the discussion on 3.1.2 (ii) above, we may distinguish three types of so-understood intrinsic instability in the material, either deforming or being at rest.

First, instability against *internal structural rearrangements* in the material can be examined, while the macroscopic deformation gradient is not perturbed at all (cf. Sect. 3.4.2, 3.5.2). The internal rearrangements can be described by changes in *internal state variables*, or by a *history* of activity of internal mechanisms of inelastic deformations. The perturbing forces are then imagined to act at a micro-level, and in the macroscopic description they are represented by perturbations of thermodynamic driving forces. Instability for such perturbations may physically correspond to spontaneous movement, generation or annihilation of dislocations in metal crystals, to rearrangements of the packing of grains in granular materials, to martensitic phase transformations in austenitic steels or shape memory alloys, to uncontrolled propagation of microcracks in a variety of materials, etc.

Second, instability of *uniform straining* within a finite domain can be investigated, while the displacements over the domain boundary remain unaffected by disturbances (cf. Sect. 3.4.3, 3.5.3). In particular, strain localization in a disk-like subdomain may be examined. If a material element is treated as an infinite homogeneous continuum then localization of deformation within a planar band or bands corresponds to that type of intrinsic instability. A special case arises when the uniform straining is of zero rate, i.e. instability concerns a given equilibrium state and is associated with the possibility of dynamic localization of deformation.

Third, the material in a perturbed process may still deform uniformly in an interior *part* of the domain mentioned above, while the remaining outer part (possibly extended up to a specimen surface) compensates that deformation and constitutes flexible constraints for the interior part (cf. Sect. 3.4.4, 3.5.4). This is associated with the concept of stability of the deformation process or equilibrium within a material element under *deformation-sensitive* loading exerted by an outer part. The problem of determining a constitutive law for such loading is not addressed here; rather, we will discuss stability conditions for a whole class of *linearized* constitutive equations

[2] In particular, instability of a single equilibrium state may be regarded as instability of a degenerate deformation process in which nothing changes.

for deformation-sensitive loading. Consequently, the associated instability in general need not be 'intrinsic' in the sense adopted above.

There are other types of material instability in elastic-plastic solids which are not discussed here. This concerns, for instance, flutter instability (e.g. [36, 4]) and shock waves (e.g. [13, 12]).

In material instability studies it is of importance to predict not only the onset of instability but also the post-critical behaviour of the material. Attention is restricted below to the *incipient* behaviour after stability loss, whose character is determined by the encountered type of instability. Analysis of post-critical deformation pattern significantly beyond the first instant of instability is presented in parallel lecture notes (Thermann, 2000).

3.1.4 Stability or Instability in the First Approximation

To simplify the task of establishing conditions or criteria of instability, the classical approach is to use the linear stability analysis. However, in elastic-plastic materials obeying always an *incrementally nonlinear* constitutive law due to the existence of at least two (loading and unloading) constitutive branches, the concept of fully linear (in)stability is too simplified.

The linear (in)stability concept can be generalized to the (in)stability *in the first approximation* by linearizing the material response only along *direct* paths of deformation starting from a given state. More specifically, in the first approximation the increments in forces are assumed proportional to the associated increments in displacements if the latter follow a *straight* segment in the displacement space. In a three-dimensional continuum, this corresponds to the replacement of *rates* of the nominal stress and deformation gradient in the incremental constitutive law by small but finite *increments*, provided the increments are reached on a direct path of constant direction of the velocity gradient. In many cases only such segments need be considered to prove instability. In doing that, the material nonlinearity in transition from one such segment to another differently oriented, as well as the important effect of *incremental* changes in geometry, are still taken into account.

A more detailed characterization of the stability in the first approximation will be given later. In particular, a stability property will be distinguished, called 'directional' stability, restricted to possible departures from the examined equilibrium state or fundamental process along direct paths only. Direct paths may be defined as paths of bounded curvature; if the path length tends to zero then the influence of the curvature on certain quantities essential in stability considerations such as the second-order work becomes negligible. An extension of stability conditions to arbitrarily circuitous paths will also be discussed.

It must be expected that certain instability phenomena (e.g. involving shock waves or cavitation) will not be covered by the concept of instability in the first approximation.

3.2 Illustrative Example: Outline of Basic Concepts

Before proceeding to a three dimensional elastic-plastic continuum, it seems preferable to discuss first the basic concepts by referring to a possibly simple example. The tensile bar with two degrees of freedom, analysed in this section, can hardly be further simplified. Nevertheless, many (perhaps unexpectedly many) aspects and different approaches to material instability can be illustrated by this elementary example. It may be remarked that most of the equations apply, by changing only the meaning of the quantities involved, to the problem of simple shear of a layer in a three-dimensional elastic-plastic continuum discretized (drastically, in fact) across the layer thickness. Of course, the one-dimensional example, in addition to restrictions due to the discretization, has its own peculiarities which do not all transmit to a genuine three-dimensional case.

3.2.1 Example

3.2.1.1 Bar with two degrees of freedom

As a substitute of an elastic-plastic material element, let us consider a one-dimensional *representative bar* of current length $\bar{l}$ under a tensile load P. In an *arbitrarily chosen* but fixed reference configuration, the bar length is denoted by l. According to the remarks in Sect. 3.1.3, suppose that the relative displacement $\bar{l}(t) - l = \hat{u}_o(\lambda(t)) \equiv u_o(t)$ of the end points of the bar is a given function of time t (Fig. 1A); analysis of the case of a given or deformation-sensitive load P is deferred until Sect. 3.2.4.4. To simplify the analysis as much as possible, we shall assume that the bar consists of two parts (a) and (b) of given reference lengths l_a and l_b, respectively, $l_a + l_b = l$, $l_b \leq l_a$, such that each part is enforced to deform uniformly. The part (b) will represent the strain localization zone. For the part (b) of a current length $\bar{l}_b$, the relative displacement of its end points is denoted by $u_b = \bar{l}_b(t) - l_b$, the engineering longitudinal strain by $\epsilon_b = u_b/l_b$, and the volume fraction occupied in the reference configuration by $\eta_b = l_b/l$. The notation for the part (a) is analogous, with an index 'b' replaced by 'a'.

The bar has two degrees of freedom, ϵ_b and ϵ_a, say. They are constrained by the kinematic compatibility condition

$$\eta_b\epsilon_b + \eta_a\epsilon_a = \epsilon_o\,, \tag{3.1}$$

where $\epsilon_o = u_o/l$ is the overall engineering strain of the representative bar, being a given function of time. The remaining degree of freedom is to be determined from the equation of either dynamic or quasi-static motion.

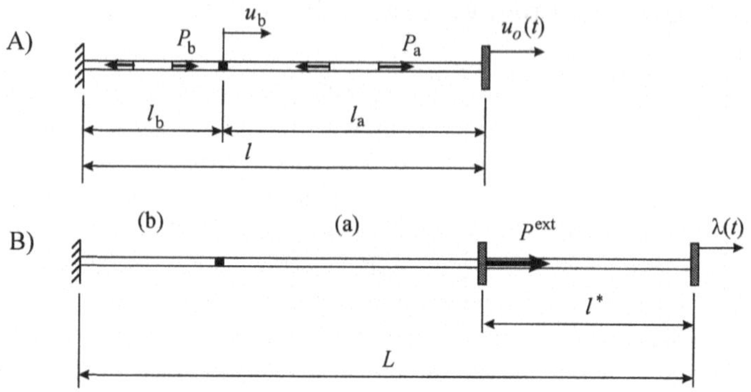

Fig. 3.1 An illustratory example of a representative bar composed of two uniform-strain segments: (A) under end-displacement control, (B) as a part of a longer bar. All dimensions are in the reference configuration.

3.2.1.2 One-dimensional constitutive equations

The incremental constitutive framework to be used involves infinitesimal increments of basic variables. An infinitesimal increment will be replaced for convenience by the *forward* rate, i.e. the one-sided *right-hand* derivative with respect to time t, denoted by a dot superimposed over the symbol of a differentiated quantity.

The material of the bar may deform elastically or plastically. Attention is restricted to extensional plastic straining. In the simplest case discussed, the rate of plastic straining is measured by a scalar plastic multiplier $\dot{\gamma} \geq 0$. In the time-independent framework, $\dot{\gamma}$ can be non-zero only if a loading variable f has reached from below a threshold value, without loss of generality set equal to zero. The basic rule for plastic activity reads

$$\dot{\gamma} \geq 0, \qquad f \leq 0, \qquad f\,\dot{\gamma} = 0\,. \tag{3.2}$$

As the simplest identification for the elastic-plastic bar, we take

$$f = P - P^{\mathrm{y}}, \qquad \dot{\gamma} = \dot{\epsilon}^{\mathrm{p}}\,, \tag{3.3}$$

where $P^{\mathrm{y}} = P^{\mathrm{y}}(\gamma)$ is the actual yield limit, and $\dot{\epsilon}^{\mathrm{p}}$ is the plastic part of strain-rate $\dot{\epsilon}$, which in general is *different* from the rate of plastic strain. The actual value of $\dot{\gamma}$ at $P = P^{\mathrm{y}}$ is to be found from the consistency condition

$$\dot{f} \leq 0, \quad \dot{f}\,\dot{\gamma} = 0 \qquad \text{if } f = 0 \tag{3.4}$$

which follows from (3.2) by differentiation. The constitutive rate equations read

$$\dot{f} = \dot{P} - h\dot{\gamma}\,, \qquad \dot{\epsilon} = (C^{\mathrm{e}})^{-1}\dot{P} + \dot{\gamma}\,, \tag{3.5}$$

where $C^e = \partial P(\epsilon, \gamma)/\partial\epsilon > 0$ is the elastic modulus, and $h = dP^y/d\gamma$, $-C^e < h < C^e$, is the hardening modulus. On eliminating $\dot{\gamma}$ between the equations in (3.5) with the help of (3.4), the rate (or incremental) constitutive relationship between a tensile load $P \geq 0$ and strain ϵ for the elastic-plastic bar takes the form

$$\dot{P} = \begin{cases} C^p\,\dot{\epsilon} & \text{if } P = P^y \text{ and } \dot{\epsilon} > 0, \\ C^e\,\dot{\epsilon} & \text{otherwise,} \end{cases} \tag{3.6}$$

where $C^p = hC^e/(h + C^e) < C^e$ is the tangent (elastoplastic) modulus. In general, all quantities in the above constitutive equations may be different in the parts (b) and (a) of the bar, and will be distinguished by the respective indices. However, the bar is assumed to be initially homogeneous, so that the load-strain relationship that results from (3.6) for a given history $\epsilon(t)$ is the same for the parts (b) and (a).

3.2.1.3 Fundamental deformation process

Suppose now that the controlled relative end displacement u_o, which may be identified with λ, varies linearly in time with a constant speed $v_o \geq 0$, $u_o = v_o t$, $t \geq 0$, and that the external end point of the part (b) is fixed. Then the standard discretization procedure yields the discretized equation of motion of the bar in the form

$$\begin{aligned} P_a(t) - P_b(t) &= \int_0^l \rho_l \ddot{u}(x,t)\,\varphi(x)\,dx \\ &= \frac{1}{3} m l_b \ddot{\epsilon}_b(t)\,, \end{aligned} \tag{3.7}$$

where P_a (or P_b) is a uniform load carried by the part (a) (or (b)) of the bar, ρ_l is the mass density per unit length of the bar in the reference configuration, and φ is a triangular 'shape function' which takes the value 1 at $x = l_b$ and vanishes at the bar ends. The total mass of the representative bar is denoted by m, and $\frac{1}{3}m$ may be regarded as a reduced mass placed between the parts (b) and (a).

Assuming that the bar is in a uniform state at $t = 0$, the fundamental solution to the equation of motion is given by

$$\epsilon_b(t) = \epsilon_a(t) = v_o t/l\,, \qquad t \geq 0\,, \tag{3.8}$$

irrespectively whether the motion is regarded as quasi-static or dynamic; note that (3.8) implies $P_a = P_b$ and $\ddot{u}_b = \frac{d}{dt}(\eta_b v_o) = 0$. In a dynamic analysis it is essential that the velocity in the fundamental motion is constant in time at any material point.

We are primarily interested in instability of the process of slow uniform deformation of the bar. Consequently, v_o in (3.8) is understood as a small parameter, and the increments in time t and in strain ϵ_o are of different order of magnitude in the limit as $v_o \to 0$.

3.2.1.4 The first approximation

We will examine only such perturbations of the fundamental process that satisfy $\Delta_{\max} > \Delta\epsilon_b(t) \geq \Delta\epsilon_o \geq 0$ in the examined interval $\mathcal{T}$ of time $t \geq t_0$, where

$$\Delta\psi(t) \equiv \psi(t) - \psi(t_0) \tag{3.9}$$

denotes an increment of a quantity ψ in time, and t_0 defines the deformation stage at which stability is examined. Moreover, we will restrict ourselves to examine stability or instability in the first approximation at t_0. That is, the number $\Delta_{\max}$ is fixed and taken sufficiently small to allow any changes in C^e, h and C^p to be neglected within the examined time interval $\mathcal{T}$ starting from a fixed t_0. However, C^e, h and C^p may depend on t_0.

Time integration of $(3.5)_2$ now yields

$$\Delta f = \Delta P - h(t_0)\Delta\gamma\,, \qquad \Delta P = C^e(t_0)(\Delta\epsilon - \Delta\gamma) \tag{3.10}$$

(generally, to first order in $\Delta_{\max}$) no matter how the strain at a point of the bar varies in the time interval $\mathcal{T}$. Equations (3.10) provide a definition of the first approximation in the analyzed example. Unless stated otherwise, the first approximation will be tacitly assumed in the following. In particular, $\Delta\epsilon_o(t) < \Delta_{\max}$ for $t \in \mathcal{T}$, which means that path stability will be examined within a small interval of the loading parameter λ.

In two special cases of the deviation on a *direct* path from the fundamental process, the following *partial* linearization follows from (3.10):

$$\Delta P_a - \Delta P_b = \begin{cases} C^p\,(\Delta\epsilon_a - \Delta\epsilon_b) & \text{if } \dot{\gamma}_a > 0 \text{ and } \dot{\gamma}_b > 0 \text{ at every } t \geq t_0\,, \\ C^e\Delta\epsilon_a - C^p\Delta\epsilon_b & \text{if } \dot{\gamma}_a = 0 \text{ and } \dot{\gamma}_b > 0 \text{ at every } t \geq t_0\,. \end{cases} \tag{3.11}$$

3.2.2 Dynamic Instability

3.2.2.1 Linear stability analysis

The classical approach to examining stability of motion is the *linear stability analysis*. For that purpose, the governing equations of motion are fully linearized about the fundamental solution, and a class of perturbed solutions close to the fundamental solution is investigated for those linear equations. The conclusion about (Lyapunov) stability or instability is conventionally drawn according to whether the deviation of perturbed solutions from the fundamental one over the *entire* time domain (usually infinite) tends or does not tend, respectively, to zero with the vanishing strength of an *initial* disturbance.

Let us apply that approach to the fundamental process (3.8). By neglecting the possibility of unloading, the linearized version of the equation of motion (3.7), with the strain difference[3] $\delta\epsilon_\mathrm{b}(t) = \epsilon_\mathrm{b}(t) - \epsilon_\mathrm{o}(t)$ with respect to the fundamental solution taken as the unknown, on substituting $(3.11)_1$ and (3.1) is obtained in the form

$$\frac{1}{3} m l_\mathrm{b} \delta\ddot{\epsilon}_\mathrm{b} + \frac{1}{\eta_\mathrm{a}} C^\mathrm{p} \delta\epsilon_\mathrm{b} = 0\,, \tag{3.12}$$

which has an elementary general solution

$$\delta\epsilon_\mathrm{b}(t) = \begin{cases} c_1 \sin ct + c_2 \cos ct & \text{if } C^\mathrm{p} > 0 \\ c_1 t + c_2 & \text{if } C^\mathrm{p} = 0 \\ c_1 e^{ct} + c_2 e^{-ct} & \text{if } C^\mathrm{p} < 0 \end{cases} \qquad c = \sqrt{\frac{3|C^\mathrm{p}|}{m l_\mathrm{b} \eta_\mathrm{a}}} \tag{3.13}$$

with integration constants c_1, c_2 adjusted to a small initial perturbation. As an initial disturbance consistent with the assumed absence of unloading, we may take $\delta\dot{\epsilon}_\mathrm{b}(t_0) \neq 0$ and $\delta\epsilon_\mathrm{b}(t_0) = 0$ at an initial instant t_0 of a perturbed motion. According to (3.13), the fundamental process of uniform deformation of the bar might appear stable if $C^\mathrm{p} > 0$ and unstable if $C^\mathrm{p} \leq 0$.

That conclusion is, however, not so obvious. It is true that for $C^\mathrm{p} > 0$ a perturbed solution following a small initial disturbance will oscillate in time about the fundamental solution, with the amplitude of oscillations as small as we please if the initial disturbance measured by $|\delta\dot{\epsilon}_\mathrm{b}(t_0)|$ tends to zero. The loading condition is clearly satisfied everywhere for disturbances sufficiently small. Moreover, if viscous effects are included then the oscillations are expected to diminish asymptotically in time. Nevertheless, stability for perturbations that involve temporary unloading remains yet to be proved.

Further, the solution for $C^\mathrm{p} = 0$ may change qualitatively if higher-order effects are included; this is why the usual term 'neutral stability' is more appropriate in this case. To show that the conclusion about instability need not be valid, one may consider a special case when C^p is positive along the fundamental straining path, with the exception for $C^\mathrm{p}(t_0) = 0$.

Next, the perturbed solutions at fixed c depend continuously on the initial data over any *bounded* interval of time. It follows that the conclusion about instability for $C^\mathrm{p} < 0$ requires formally an unbounded interval of time. This can be made consistent with the assumption of small strain increments bounded by Δ_max (and with the linearization assumption) if and only if $\dot{\epsilon}_\mathrm{o}$ is vanishingly small. Moreover, $\delta\epsilon_\mathrm{b}(t_0) = 0$ in the last solution in (3.13) implies $c_1 e^{ct_0} = -c_2 e^{-ct_0}$ so that

$$\begin{aligned} |\delta\dot{\epsilon}_\mathrm{b}(t)| &= c\,|(c_1 e^{ct} - c_2 e^{-ct})| \\ &= c\,|c_1\,(e^{ct} + e^{c(2t_0 - t)})| \\ &> c\,|c_1\,(e^{ct} - e^{c(2t_0 - t)})| \\ &= c\,|\delta\epsilon_\mathrm{b}(t)|\,. \end{aligned} \tag{3.14}$$

---

[3] Note the essential difference between the meaning of the prefixes Δ and δ adopted here: $\Delta\psi$ denotes an increment in ψ between two time instants, and $\delta\psi$ denotes a variation of ψ at a single instant of time.

From (3.14) and (3.1) it follows that either $\delta\dot{\epsilon}_{\rm b}(t) < -c\,|\delta\epsilon_{\rm b}(t)|$ or $\delta\dot{\epsilon}_{\rm a}(t) < -c\,\frac{l_{\rm b}}{l_{\rm a}}|\delta\epsilon_{\rm b}(t)|$. Consequently, the exponential growth of $|\delta\epsilon_{\rm b}(t)|$ at $\dot{\epsilon}_{\rm o} \to 0$ enforces the solution to cross the limit of validity of the no-unloading assumption, before a finite distance in strain can be reached. The conclusion about instability for $C^{\rm p} < 0$ drawn on the basis of (3.13) is thus not fully convincing.

If c is allowed to be unbounded by taking $\eta_{\rm b} \to 0$ (rather than being constant as above) then the use of an infinite time interval and infinitesimal $\dot{\epsilon}_{\rm o}$ is not required in the instability analysis. However, the estimate (3.14) implies again, now on account of $c \to \infty$, that the limit of validity of the no-unloading assumption must be crossed before a finite distance in strain can be reached following a vanishingly small disturbance.

One may conclude that the nonlinearity of the incremental elastic-plastic constitutive law must be taken into account to provide a more satisfactory picture of stability or instability of the fundamental process.

3.2.2.2 Instability of equilibrium

The necessity of taking into account the incremental nonlinearity of the material is evident when examining instability of a degenerate fundamental process with vanishing velocities, i.e. of an equilibrium state at constant $\epsilon_{\rm o}$. For, if the ends of the representative elastic-plastic bar are fixed then further plastic straining in one part of the bar necessarily requires elastic unloading in the other part.

Consider a motion at constant $\epsilon_{\rm o}$ (i.e. at fixed λ) that follows an initial disturbance in velocities superimposed on a uniform equilibrium state at $P = P^{\rm y}$. Attention is restricted to a *direct* path of departure from the equilibrium state, such that in each part of the bar only one constitutive branch is activated, i.e. here $\dot{\epsilon}_{\rm b}(t) > 0$ and $\dot{\epsilon}_{\rm a}(t) < 0$ at every $t \geq t_0$. Accordingly, on applying $(3.11)_2$ and (3.6), the equation of motion (3.7) reduces to another linear differential equation

$$\frac{1}{3}ml_{\rm b}\Delta\ddot{\epsilon}_{\rm b} + C^{\rm r}\Delta\epsilon_{\rm b} = 0\,, \qquad C^{\rm r} \equiv C^{\rm p} + \frac{l_{\rm b}}{l_{\rm a}}C^{\rm e}\,, \tag{3.15}$$

which has an elementary general solution analogous to (3.13). The oscillatory solution for $C^{\rm r} > 0$ is now inconsistent with the non-unloading assumption $\Delta\epsilon_{\rm b} = \dot{\epsilon}_{\rm b} > 0$, so that stability of equilibrium cannot be inferred; we will see later that the energy argument can be used for that purpose. The transitory case $C^{\rm r} = 0$ is also not conclusive and corresponds to the 'neutral' stability (or instability) of equilibrium, cf. the similar remark above. However, the condition

$$C^{\rm r} < 0 \tag{3.16}$$

is sufficient for dynamic instability of equilibrium (in the first approximation) since an arbitrarily small initial perturbation can grow exponentially in time in a free inertial motion within the assumed constitutive branches. As a possible perturbed solution one can take

$$\Delta\epsilon_{\rm b} = \bar{\delta}\,\sinh\left(\sqrt{\frac{-3C^{\rm r}}{ml_{\rm b}}}\Delta t\right) \tag{3.17}$$

with an arbitrarily small constant $\bar{\delta} > 0$. The requirement of an infinite time interval creates no problem here: the second time scale, related previously to given variations of λ and $\epsilon_{\rm o}$ in time, does not appear at $\epsilon_{\rm o} = \text{const.}$

The linearized analysis of dynamic instability in this simple example, although not complete as discussed above, does indicate that the conditions for stability of the fundamental deformation process and of an equilibrium state must be expected to be different in general. The distinction between the two concepts of stability will be demonstrated below by a nonlinear analysis of stability of a *quasi-static* fundamental process.

Exercise: Extend the instability condition (3.16) to the case of an arbitrary sign of $\dot{\epsilon}_{\rm b}$.

3.2.3 Instability of a Quasi-Static Deformation Process

3.2.3.1 Persistent disturbances

The disturbances examined above influenced the initial conditions while the subsequent motion was assumed free of further disturbances. That assumption, although made frequently, is not realistic. In every mechanical model, there are always certain influences neglected in comparison with the behaviour of the modeled physical system, which are perpetual *during* the examined deformation process. Such influences are called *persistent disturbances* and may be represented by additional, unknown terms in the governing equations.

In the illustratory example, persistent disturbances may be represented by a small additional load $P^{\rm dist}(t)$ varying in time, applied by an external agency at the point between the parts (a) and (b) of the bar and taken positive when pointed towards the part (a). Let us examine stability of the uniform deformation process (3.8) with respect to such a disturbance.

Suppose that the overall strain $\epsilon_{\rm o}$ varies slowly with respect to a natural time. The inertia term on the right-hand side of (3.7) is regarded as a small part of the perturbing load $P^{\rm dist}$. The perturbed equation of a quasi-static motion reads

$$P_{\rm a}(t) - P_{\rm b}(t) + P^{\rm dist}(t) = 0\,, \tag{3.18}$$

while all kinematic constraints remain unperturbed.

A special case is now examined when $P^{\rm dist}$ varies in time proportionally to Δt, and thus to $\Delta\epsilon_{\rm o}$, starting from zero value at an instant t_0. A general rate solution to (3.18) satisfies

$$(C_{\rm b} + \frac{l_{\rm b}}{l_{\rm a}} C_{\rm a})\delta\dot{\epsilon}_{\rm b} = \dot{P}^{\rm dist} + (C_{\rm a} - C_{\rm b})\dot{\epsilon}_0\,, \qquad \delta\dot{\epsilon}_{\rm a} = -\frac{l_{\rm b}}{l_{\rm a}}\delta\dot{\epsilon}_{\rm b}\,. \tag{3.19}$$

The incremental stiffness moduli $C_{\rm a}$ and $C_{\rm b}$ in the part (a) and (b) of the bar, respectively, must be consistent with (3.6). Among possible combinations, consider the following two solutions corresponding to *direct* paths:

(i) (fully active loading) $C_{\rm b} = C_{\rm a} = C^{\rm p}$, $\dot{\epsilon}_{\rm b} > \dot{\epsilon}_{\rm o} > 0$, $\dot{\epsilon}_{\rm a} > 0$, so that, in the first approximation,

$$C^{\rm p}(\Delta\epsilon_{\rm b} - \Delta\epsilon_0) = \eta_{\rm a} P^{\rm dist}\,, \qquad C^{\rm p}(\Delta\epsilon_{\rm a} - \Delta\epsilon_0) = -\eta_{\rm b} P^{\rm dist}\,. \tag{3.20}$$

(ii) (elastic unloading in the part (a)) $C_{\rm b} = C^{\rm p}$, $C_{\rm a} = C^{\rm e}$, $C^{\rm r} > 0$, $\dot{\epsilon}_{\rm b} > \dot{\epsilon}_{\rm o} > 0$, $\dot{\epsilon}_{\rm a} < 0$, so that, in the first approximation,

$$\Delta\epsilon_{\rm b} - \Delta\epsilon_{\rm o} = (C^{\rm r})^{-1}(P^{\rm dist} + (C^{\rm e} - C^{\rm p})\Delta\epsilon_0)\,, \qquad \dot{\epsilon}_{\rm a} = \frac{l_{\rm b}}{l_{\rm a}}(C^{\rm r})^{-1}(-\dot{P}^{\rm dist} + \frac{l}{l_{\rm b}} C^{\rm p}\dot{\epsilon}_0)\,. \tag{3.21}$$

In the case (i), if $C^{\rm p} \neq 0$ then the distance between the perturbed and fundamental solutions for a given $\Delta\epsilon_{\rm o}$ decreases to zero as the final perturbing force $\Delta P^{\rm dist}$ tends to zero. This is so no matter whether $C^{\rm p} > 0$ or $C^{\rm p} < 0$, so that the expected instability range $C^{\rm p} < 0$ is not found in that way. We can only conclude that the rate solution becomes indefinitely sensitive to $\dot{P}^{\rm dist}$ as $C^{\rm p} \to 0$, and is indeterminate when $C^{\rm p} = 0$. The latter case is called 'bifurcation in velocities' since more than one velocity solution exists. This corresponds to bifurcation of the quasi-static deformation process (path) in the first approximation when $C^{\rm p}$ is assumed constant. However, if $C^{\rm p}$ varies along the path then the bifurcation in velocities at $C^{\rm p} = 0$ usually does *not* imply a genuine path branching *without unloading* in the absence of any perturbing force. One can check that the fundamental solution (3.8) remains a unique unperturbed solution *without unloading*, satisfying (3.18) for $P^{\rm dist} \equiv 0$, if the slope of the load-strain curve changes sign and vanishes at a *single* value of strain. The conclusion about path bifurcation can then only be obtained by appealing to the incremental nonlinearity of the material, as discussed below.

In the case (ii) with partial unloading, definite conclusions can be drawn. The consistency with the assumed unloading ($\dot{\epsilon}_{\rm a} < 0$) in the part (a) at $\dot{P}^{\rm dist} \to 0$ is ensured if $C^{\rm p} < 0$, by the second equation in (3.21). In that case the left-hand difference between the perturbed and fundamental solutions in the first equation in (3.21) does not tend to zero as $|P^{\rm dist}| \to 0$ but remains finite for any finite overall strain increment $\Delta\epsilon_{\rm o}$. Also without appealing to the first-order approximation, if $C^{\rm p} < 0$ at t_0 then the solution path obtained by time integration of (3.19) under the assumptions of (ii) does not tend to the fundamental solution (3.8) as $\dot{P}^{\rm dist} \to 0$ but deviates from it with a finite rate, which in the limit reads

$$\dot{\epsilon}_{\rm b} - \dot{\epsilon}_{\rm o} = \frac{C^{\rm e} - C^{\rm p}}{C^{\rm r}}\,\dot{\epsilon}_{\rm o} > 0\,, \qquad \dot{P}_{\rm b} - \dot{P}_{\rm o} = \frac{C^{\rm e} - C^{\rm p}}{C^{\rm r}}\,\dot{P}_{\rm o} < 0\,, \qquad \dot{P}_{\rm o} = C^{\rm p}\dot{\epsilon}_{\rm o}\,. \tag{3.22}$$

We thus conclude that the fundamental process of slow uniform straining of the representative bar in the range

$$C^{\mathrm{p}} < 0 \qquad \text{and} \qquad C^{\mathrm{r}} > 0 \tag{3.23}$$

is unstable for persistent disturbances of infinitesimal strength defined by $|P^{\mathrm{dist}}|$. The use of the incremental nonlinearity of the elastic-plastic constitutive law has been essential in arriving at that conclusion.

Exercise: Show that the above conclusion remains valid if $P^{\mathrm{dist}} \neq 0$ only within a vanishingly short time interval.

3.2.3.2 Quasi-static bifurcations

From (3.22) it follows that the existence of *exact bifurcations* of a quasi-static deformation path can be established by taking the incremental nonlinearity of the material into account. *At every point* on a solution path with C^{p} varying continuously within the range defined by (3.23), there exists a non-trivial, secondary rate solution given by (3.22) to the quasi-static *unperturbed* rate-problem. That solution initiates a quasi-static branching path with elastic unloading in the part (a), which can be constructed by time integration of (3.22) also if C^{p} is varying in time. Consequently, we have a *continuous spectrum* of genuine bifurcation points along the fundamental solution path within the whole range (3.23) (Fig. 3.2). If C^{p} is continuously decreasing at the critical instant when $C^{\mathrm{p}} = 0$ then, by the same argument as above, at this instant the primary bifurcation takes place with neutral loading, followed by *elastic unloading just beyond the critical instant*. The secondary branch emanating at that point is the only post-critical quasi-static deformation path that can be stable with respect to small persistent disturbances.

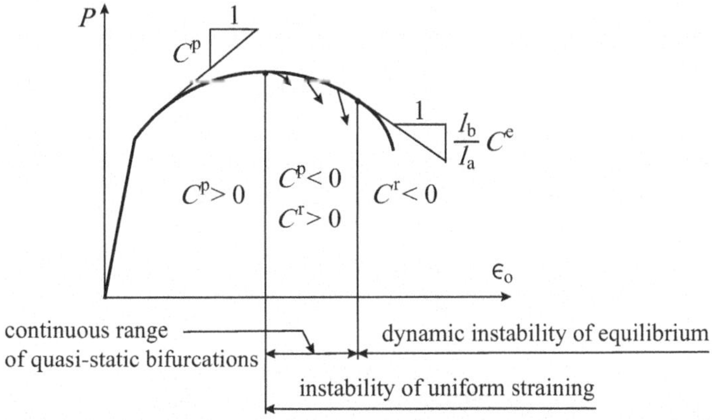

Fig. 3.2 Instability ranges for the discretized bar in quasi-static tension under end displacement control

The concept of instability of a quasi-static deformation process for persistent disturbances is thus closely related to the existence of a continuous range of bifurcation points on the deformation path. The instability concept is, however, more general than the bifurcation concept. If the fundamental process is governed by the dynamic equation of motions as in Sect. 3.2.2.1, and not by its approximate quasi-static version as above, then there is no exact bifurcation. Any deviation from the fundamental solution (3.8) requires some disturbance, irrespectively of the sign of C^{p} and no matter how slow the fundamental motion is. However, if $C^{\mathrm{p}} < 0$ then a finite deviation from the fundamental motion (3.8) can be induced within a given increment $\Delta\epsilon_{\mathrm{o}}$ by a perturbing force $(P^{\mathrm{dist}}(t) - \frac{1}{3}ml_{\mathrm{b}}\ddot{\epsilon}_{\mathrm{b}})$ whose magnitude tends to zero as $\dot{\epsilon}_{\mathrm{o}} \to 0$. This can be shown by 'smoothing out' in time the initial discontinuous change in velocities in the quasi-static bifurcation solution; details are omitted here.

3.2.4 The Energy Criterion

Analysis of particular perturbations may not always be effective, especially for more complex systems. Moreover, to obtain a condition sufficient for stability, all possible disturbances from a specified class should be taken into account. For a class of materials, encompassing that characterized in Sect. 3.2.1.2 which will be generalized later, the approach based on energy considerations can be effectively applied.

3.2.4.1 The second-order work

Consider a uniform equilibrium state $\mathcal{G}_0$ of the bar at an instant t_0 at which

$$P_{\mathrm{a}}(t_0) = P_{\mathrm{b}}(t_0) = P^{\mathrm{y}}(\gamma(t_0))\,. \tag{3.24}$$

On any path starting from $\mathcal{G}_0$, the work of deformation in the representative bar in a time interval $[t_0, t]$ reads

$$\begin{aligned}\Delta W(t) &\equiv l_{\mathrm{b}} \int_{t_0}^{t} P_{\mathrm{b}}(\theta)\dot{\epsilon}_{\mathrm{b}}(\theta)\, d\theta + l_{\mathrm{a}} \int_{t_0}^{t} P_{\mathrm{a}}(\theta)\dot{\epsilon}_{\mathrm{a}}(\theta)\, d\theta \\ &= \Delta_1 W(t) + l_{\mathrm{b}} \int_{t_0}^{t} \Delta P_{\mathrm{b}}(\theta)\dot{\epsilon}_{\mathrm{b}}(\theta)\, d\theta + l_{\mathrm{a}} \int_{t_0}^{t} \Delta P_{\mathrm{a}}(\theta)\dot{\epsilon}_{\mathrm{a}}(\theta)\, d\theta\,.\end{aligned} \tag{3.25}$$

where

$$\begin{aligned}\Delta_1 W &= l_{\mathrm{b}} P_{\mathrm{b}}(t_0)\Delta\epsilon_{\mathrm{b}}(t) + l_{\mathrm{a}} P_{\mathrm{a}}(t_0)\Delta\epsilon_{\mathrm{a}}(t) \\ &= l P^{\mathrm{y}}(t_0)\Delta\epsilon_{\mathrm{o}}\end{aligned} \tag{3.26}$$

is the *first-order* work, with the later expression obtained from (3.1) and (3.24). Further,

$$\int_{t_0}^{t} \Delta P(\theta)\dot{\epsilon}(\theta)\, d\theta = \Delta_2 w(t) + o((\Delta_{\max})^2)\,, \tag{3.27}$$

where $\Delta_2 w$ is the *second-order* work density per unit length of the bar in the reference configuration, and o is the classical order symbol used to indicate higher-order terms; recall that, by definition, $o(r^2)/r^2 \to 0$ as $r \to 0$. We thus have

$$\Delta W = \Delta_1 W + \Delta_2 W \qquad \text{(to second order)}, \tag{3.28}$$

where the second-order work

$$\Delta_2 W = l_b\, \Delta_2 w_b + l_a\, \Delta_2 w_a \tag{3.29}$$

is generally *dependent on the strain path.*[4] By using the formulae (3.10) of the first approximation and rearranging with the help of $(3.2)_3$, (3.5) and (3.24), one can show [50] that

$$\Delta_2 w = \tfrac{1}{2}\Delta P \Delta\epsilon - \tfrac{1}{2}\Delta f \Delta\gamma \tag{3.30}$$

for an *arbitrary* straining path (with or without unloading) starting from $\mathcal{G}_0$. Along *direct* paths to be examined below, we have $\dot\epsilon_b(\theta) \geq 0$ and either $\dot\epsilon_a(\theta) \geq 0$ or $\dot\epsilon_a(\theta) < 0$ at all instants. Along a direct path the last term in (3.30) vanishes, and the associated second-order work density, distinguished by a superscript 'd', reduces to

$$\Delta_2 w^d = \begin{cases} \tfrac{1}{2}C^p(\Delta\epsilon)^2 & \text{if } \dot\epsilon > 0, \\ \tfrac{1}{2}C^e(\Delta\epsilon)^2 & \text{if } \dot\epsilon < 0. \end{cases} \tag{3.31}$$

Accordingly, for $C_b = C^p$ and either $C_a = C^p$ or $C_a = C^e$, the second-order work for the bar deformed along a direct path reads

$$\Delta_2 W^d = \frac{1}{2} l_b C^p (\Delta\epsilon_b)^2 + \frac{1}{2} l_a C_a (\Delta\epsilon_a)^2 \tag{3.32}$$

which is dependent only on the final strain increments.

Consider two strain paths, one direct and another indirect, that lead to the same final strain increment (Fig. 3.3). From (3.2) and (3.24) we obtain $\Delta f \Delta\gamma \leq 0$. One can show that for the assumed constitutive law, and within the class of all (arbitrarily short) paths leading to the same final strain increment $\Delta\epsilon$, the second-order work density $\Delta_2 w$ attains a minimum $\Delta_2 w^d$ on a direct path [50]:

$$\Delta_2 w \geq \Delta_2 w^d\,. \tag{3.33}$$

This obviously implies that $\Delta_2 W \geq \Delta_2 W^d$.

Exercise: Prove validity of (3.30) and (3.33) for the assumed one-dimensional constitutive law.

[4] In this one-dimensional example, strain paths may only differ by involving or not segments of compression. In a three-dimensional solid, paths may be more diverse.

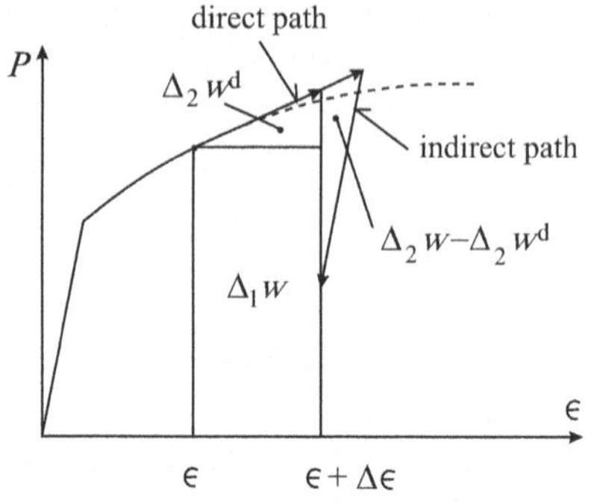

Fig. 3.3 The second-order work on a direct (loading) path and on an indirect path leading to the same final strain increment $\Delta\epsilon$

3.2.4.2 Stability of equilibrium

Without specifying the type of disturbance, the disturbance strength may be defined by determining the amount of energy ΔW^* supplied to the system due to the disturbance. Consider the homogeneous representative bar from Fig. 3.1A with fixed ends, ϵ_o =const, initially at equilibrium under load $P = P^y$ at $\epsilon_a = \epsilon_b$, set in motion at $t = t_0$ by a disturbance. At fixed boundary displacements, the mechanical energy balance implies

$$\Delta W + \Delta K = \Delta W^* \qquad \text{at fixed } \lambda\,, \tag{3.34}$$

where K is the kinetic energy. On account of (3.1), the second-order work (3.32) evaluated on a *direct* path now reads

$$\Delta_2 W^{\mathrm{d}} = \frac{1}{2} l_{\mathrm{b}} (\Delta\epsilon_{\mathrm{b}})^2\, C^{\mathrm{r}}\,. \tag{3.35}$$

According to (3.28), (3.26) and (3.33), we have $\Delta W = \Delta_2 W \geq \Delta_2 W^{\mathrm{d}}$. Moreover, $\Delta K \geq 0$ since the kinetic energy is non-negative and vanishes at equilibrium. It follows that

$$\Delta W^* \geq \Delta_2 W^{\mathrm{d}}\,. \tag{3.36}$$

Consequently, if

$$\Delta_2 W^{\mathrm{d}} > 0 \qquad \text{at fixed } \lambda\,, \tag{3.37}$$

or equivalently, if

$$C^{\mathrm{r}} > 0 \tag{3.38}$$

then any departure from equilibrium at fixed ϵ_0 requires a positive amount of energy ΔW^* to be supplied by a disturbing agency. If ΔW^* tends to zero then the maximum possible distance from the equilibrium configuration, measured by $|\Delta\epsilon_{\mathrm{b}}(t)|$, tends also to zero by (3.36) and (3.35). This means that (3.37) is *sufficient* for stability of equilibrium of the bar with fixed ends, for any disturbances whose strength is defined by the energy measure.

Exercise: Derive a condition sufficient for stability of equilibrium of a bar divided into an arbitrary finite number of uniformly deforming segments.

3.2.4.3 Stability of uniform straining

The object of stability analysis is now the uniform, *quasi-static* plastic straining of the bar. To extend the energy criterion of stability of equilibrium to the uniform straining, imagine our representative bar as a part of a longer bar (Fig. 3.1B) viewed as a substitute of a continuum surrounding a material element. Attention is restricted first to local disturbances (cf. Sect. 3.1.3) that leave the motion outside the representative bar unperturbed, so that $\bar{l}(t) = \bar{l}^0(t)$, where $\bar{l}^0$ is the current length of the representative bar in the fundamental process. Consider a perturbed deformation process in the bar that deviates from the fundamental process of uniform deformation, starting from a common configuration at t_0. The mechanical energy balance implies

$$\Delta W + \Delta K - \Delta W^0 = \Delta W^* \qquad \text{at varying } \lambda\,. \tag{3.39}$$

In comparison with (3.34), the energy balance includes now additionally the work done by the force $P^{\rm ext}$ applied *from the outside* to the movable end point of the representative bar, which under the above assumptions is equal to the deformation work ΔW^0 within the representative bar in the quasi-static fundamental process.

The work difference in the perturbed and fundamental processes involves only the second-order works,

$$\Delta W - \Delta W^0 = \Delta_2 W - \Delta_2 W^0\,, \tag{3.40}$$

since the first-order terms cancel each other and higher order terms have been neglected. Further, $\Delta_2 W$ is minimized on direct paths by (3.33), and $\Delta K \geq 0$ since the kinetic energy is nonnegative and vanishes in the quasi-static motion. We thus obtain

$$\begin{aligned} \Delta_2 W^* &\geq \Delta_2 W^{\rm d} - \Delta_2 W^0 \\ &= \tfrac{1}{2} l_{\rm b} C^{\rm p} (\Delta\epsilon_{\rm b})^2 + \tfrac{1}{2} l_{\rm a} C_{\rm a} (\Delta\epsilon_{\rm a})^2 - \tfrac{1}{2} l C^{\rm p} (\Delta\epsilon_{\rm o})^2 \\ &\geq \tfrac{1}{2} C^{\rm p} (l_{\rm b} (\Delta\epsilon_{\rm b})^2 + l_{\rm a} (\Delta\epsilon_{\rm a})^2 - l(\Delta\epsilon_{\rm o})^2) \\ &= \tfrac{1}{2} C^{\rm p} \big(l_{\rm b} (\delta\epsilon_{\rm b})^2 + l_{\rm a} (\delta\epsilon_{\rm a})^2\big)\,. \end{aligned} \tag{3.41}$$

The latter inequality follows from $C_{\rm a} \geq C^{\rm p}$, and the final equality from $l_{\rm b}(\Delta\epsilon_{\rm b} - \Delta\epsilon_{\rm o}) + l_{\rm a}(\Delta\epsilon_{\rm a} - \Delta\epsilon_{\rm o}) = l_{\rm b}\delta\epsilon_{\rm b} + l_{\rm a}\delta\epsilon_{\rm a} = 0$, cf. (3.1). It follows that the fulfillment of the condition

$$\Delta_2 W^{\rm d} > \Delta_2 W^0 \qquad \text{at varying } \lambda \tag{3.42}$$

for $\Delta\epsilon_{\rm o} \neq \Delta\epsilon_{\rm b} \geq 0$, which is equivalent to

$$C^{\mathrm{p}} > 0\,, \tag{3.43}$$

ensures positiveness of the expression (3.39) for all examined deviations from the fundamental process within the representative bar. Hence, if (3.42) holds (or if $C^{\mathrm{p}} > 0$) then any deviation from the homogeneous plastic flow within the representative bar, embedded in an otherwise unperturbed longer bar, requires a positive amount of energy ΔW^* to be supplied by a disturbing agency. If ΔW^* tends to zero then the maximum possible distance from the fundamental process, measured by $|\delta\epsilon_{\mathrm{b}}(t)|$, tends also to zero by (3.41). This means that (3.42) or (3.43) is *sufficient* for stability of the quasi-static uniform plastic straining of a bar for perturbations restricted to its *interior* representative part. It is clear that the dimensions of the parts (a) and (b) of the representative bar are irrelevant for this conclusion.

3.2.4.4 Stability under deformation-sensitive loading

Consider again the representative bar as a part of a longer bar of reference length $L = l + l^*$, say (cf. Fig. 3.1B), but now with the flexibility of the outer part of the length l^* taken into account. This allows the current length $\bar{l}$ of the representative bar, previously fixed at λ=const, to vary arbitrarily in a perturbed process. The force P^{ext} applied to the representative bar from the side of the outer part depends on $\bar{l}$, constituting thus *deformation-sensitive loading*.

Consider stability of equilibrium under the assumption that there are no perturbations outside the *longer* bar of length L. Assume for simplicity that the part of the length l^* deforms uniformly and only elastically with the elastic modulus C^{e}, and denote by ϵ^* an increment of the respective engineering strain. The previous energy balance (3.34) must now be completed by changes in the potential energy $\Delta\Omega$ in that part, viz.

$$\Delta W + \Delta K + \Delta\Omega = \Delta W^* \qquad \text{at fixed } \lambda\,, \tag{3.44}$$

where

$$\Delta\Omega = l^* P^{\mathrm{ext}}(t_0)\Delta\epsilon^* + \Delta_2\Omega\,, \qquad \Delta_2\Omega = \frac{1}{2} l^* \Delta P^{\mathrm{ext}}\Delta\epsilon^*\,, \qquad \Delta P^{\mathrm{ext}} = C^{\mathrm{e}}\Delta\epsilon^*\,. \tag{3.45}$$

Displacement continuity at λ=const requires $l^*\Delta\epsilon^* = -l\Delta\epsilon_{\mathrm{o}}$, which means that in an equilibrium state at t_0 the first-order terms of ΔW and $\Delta\Omega$ cancel each other. By proceeding similarly as previously, in place of (3.37) as the condition sufficient for stability of equilibrium of the longer bar we obtain

$$\Delta_2 W^{\mathrm{d}} + \Delta_2\Omega > 0 \qquad \text{at fixed } \lambda\,, \tag{3.46}$$

equivalent to

$$C^{\mathrm{p}} + \frac{l_{\mathrm{b}}}{l_{\mathrm{a}} + l^*} C^{\mathrm{e}} > 0 \tag{3.47}$$

or to

$$C^{\mathrm{p}} + \frac{l}{l^*} C^{\mathrm{e}} > 0 \qquad \text{under restriction } \epsilon_{\mathrm{a}} \equiv \epsilon_{\mathrm{b}} . \tag{3.48}$$

The case of a prescribed load P^{ext} =const is met in the limit as $l/l^* \longrightarrow 0$, $\Delta_2\Omega \longrightarrow 0$.

Similarly, in place of (3.42) we now obtain

$$\Delta_2 W^{\mathrm{d}} + \Delta_2\Omega > \Delta_2 W^0 + \Delta_2\Omega^0 \qquad \text{at varying } \lambda \tag{3.49}$$

as a condition of stability of the quasi-static fundamental *process* in the longer bar. It can be checked that (3.49) is still equivalent to (3.43). However, under restriction $\epsilon_{\mathrm{a}} \equiv \epsilon_{\mathrm{b}}$, (3.49) becomes equivalent to (3.48).

Exercise: Extend the analysis from Sect. 3.2.3 to the present case of deformation-sensitive loading.

3.2.4.5 Stability against internal rearrangements

Examine now disturbances at a micro-level in the representative bar, associated with internal rearrangements *not* satisfying the consistency condition (3.4). Suppose that the actual loading level P within the bar at equilibrium is locally increased by $\hat{P}^{\mathrm{dist}}$ by superimposing a *self-equilibriated* system of internal perturbing forces. The rule for plastic activity $(P - P^{\mathrm{y}})\dot{\gamma} = 0$ (cf. $(3.2)_3$) is accordingly extended to

$$(P + \hat{P}^{\mathrm{dist}} - P^{\mathrm{y}})\dot{\gamma} = 0 . \tag{3.50}$$

To simplify the presentation, assume that C^{e} is constant; the corresponding elastic strain energy density ϕ^{e}, per unit reference length of the bar, satisfies

$$\phi^{\mathrm{e}} = \frac{1}{2}(C^{\mathrm{e}})^{-1}P^2 , \qquad \dot{\phi}^{\mathrm{e}} = (C^{\mathrm{e}})^{-1}P\dot{P} = P\dot{\epsilon} - P\dot{\gamma} , \tag{3.51}$$

the last equality being obtained from $(3.5)_2$.

The rate of work to be supplied to the deforming bar, denoted by $\dot{w}'$ and defined per unit reference length, can be expressed as the sum of rates of the elastic and plastic work,

$$\dot{w}' = \dot{\phi}^{\mathrm{e}} + \dot{w}^{\mathrm{p}} , \qquad \dot{w}^{\mathrm{p}} = P^{\mathrm{y}}\dot{\gamma} . \tag{3.52}$$

Due to the additional work w^{dist} done by $\hat{P}^{\mathrm{dist}}$, the supplied work-rate $\dot{w}'$ is no longer equal to $\dot{w} = P\dot{\epsilon}$ but on substituting (3.50), (3.51) and $(3.5)_1$ reads

$$\dot{w}' = \dot{w} + \dot{w}^{\mathrm{dist}} , \qquad \dot{w} = P\dot{\epsilon} , \qquad \dot{w}^{\mathrm{dist}} = \hat{P}^{\mathrm{dist}}\dot{\gamma} = -f\dot{\gamma} . \tag{3.53}$$

Stability of equilibrium with respect to internal rearrangements within the representative bar is now examined under the restriction to deformations which are *uniform* both plastically and elastically. To limit attention to infinitesimal perturbing forces, we shall assume that in the current equilibrium state of the bar in which stability is investigated, $P = P^{\mathrm{y}}$ and $\hat{P}^{\mathrm{dist}} = 0$ at t_0. The first-order work expression remains thus the same as previously. The second-order work is split, according to (3.52), into two parts: $\Delta_2\phi^{\mathrm{e}}$ dependent on ΔP, and $\Delta_2 w^{\mathrm{p}}$ dependent on $\Delta\gamma$. Its final value (in this one-dimensional example) is thus independent of the actual path in (P,γ)-plane leading to the final state. On using (3.53) and (3.5) or (3.10) along a direct path in (P,γ)-plane corresponding to a proportional variation of $\hat{P}^{\mathrm{dist}}$ in time, the second-order work expression reads

$$
\begin{aligned}
\Delta_2 W &= \tfrac{1}{2} l \Delta_2 w' \qquad \text{(at uniform strain)} \\
&= \tfrac{1}{2} l (\Delta P \Delta\epsilon - \Delta f \Delta\gamma) \\
&= \tfrac{1}{2} l (\Delta P (\Delta\epsilon - \Delta\gamma) + h (\Delta\gamma)^2) \\
&= \tfrac{1}{2} l (C^{\mathrm{e}} \Delta\epsilon (\Delta\epsilon - 2\Delta\gamma) + (h + C^{\mathrm{e}}) (\Delta\gamma)^2)\,.
\end{aligned}
\tag{3.54}
$$

Note that an analogous expression for the second-order work was obtained previously in (3.30) for arbirtary strain paths in the absence of internal perturbing forces $\hat{P}^{\mathrm{dist}}$.

Under a fixed load P or strain ϵ, the condition sufficient for stability of equilibrium against internal rearrangements is $\Delta_2 W > 0$, from the energy balance as done previously. On using appropriate expressions in (3.54), this reduces to

$$
\begin{aligned}
h + C^{\mathrm{e}} &> 0 \qquad \text{at fixed } \epsilon_0\,, \\
h &> 0 \qquad \text{at fixed } P\,.
\end{aligned}
\tag{3.55}
$$

Note that $C^{\mathrm{p}} \to -\infty$ as $h + C^{\mathrm{e}} \to 0$. The analysis of stability of a *process* of internal rearrangements is deferred until Sect. 3.5.2.

3.2.4.6 Unified approach based on the energy criterion: a summary

The results obtained for different types of instability in the illustrative example can be put in a unified form as follows. Detailed assumptions given previously are not repeated.

Denote by

$$
\Delta E = \Delta W + \Delta\Omega \qquad \text{(to second order)} \tag{3.56}
$$

the increment of energy to be supplied to the representative bar *and* its outer (elastic) part along a kinematically admissible deformation path. If stability with respect to internal rearrangements is studied then the deformation work ΔW should include the work done by internal perturbing forces, e.g. by integrating (3.52) or (3.53). The part $\Delta\Omega$ expresses a change in the potential energy of the load P^{ext} applied to the representative bar from the outside; if λ is varying in time then $\Delta\Omega$ *cannot* be expressed by the work done by P^{ext} alone.

The energy condition sufficient for stability of *equilibrium* of the bar can be written in the form

$$
\Delta E > 0 \qquad \text{at fixed } \lambda \tag{3.57}
$$

to hold along any path of departure from equilibrium. Along a *direct* path of departure we may write

$$\Delta E^{\mathrm{d}} = \dot{E}\Delta t + \frac{1}{2}\ddot{E}(\Delta t)^2 \qquad \text{(to second order)}, \tag{3.58}$$

where the (right-hand) time derivatives are calculated in the examined equilibrium state. Since $\dot{E}$ vanishes in equilibrium under zero perturbing forces, the condition sufficient for directional stability of equilibrium takes the form

$$\ddot{E} > 0 \qquad \text{for all admissible directions at fixed } \lambda\,. \tag{3.59}$$

On account of (3.33), arbitrarily circuitous paths require not less energy supply (to second order) than direct paths, and therefore (3.59) is sufficient for stability of equilibrium, and not only for directional stability. This condition reduces to (3.37), (3.46) or (3.55) under the assumptions given in the respective sections. 'Directions' mean here non-zero vectors of four components $(\dot{\epsilon}_{\mathrm{a}}, \dot{\epsilon}_{\mathrm{b}}, \dot{\gamma}_{\mathrm{a}}, \dot{\gamma}_{\mathrm{b}})$, always restricted by $\dot{\gamma} \geq 0$ and with further restrictions dependent on the examined type of stability.

On the basis of the result of Sect. 3.2.2.2 and in view of the equality (3.35), under assumptions adopted therein the condition

$$\ddot{E} \geq 0 \qquad \text{for all admissible directions at fixed } \lambda \tag{3.60}$$

is *necessary* for stability of equilibrium. However, in more general circumstances this assertion requires a separate proof.

As shown in Sect. 3.2.4.3, stability (in the first approximation) of the quasi-static *process* of uniform straining of the *interior* representative bar is ensured for perturbations vanishing outside the bar if the second-order work inequality (3.42) holds. In particular, this excludes quasi-static bifurcations of the deformation path. The extended inequality (3.49), rewritten as,

$$\ddot{E} > \ddot{E}^0 \qquad \text{for all (except fundamental) admissible directions at varying } \lambda\,, \tag{3.61}$$

may be regarded as a condition sufficient for directional stability of the quasi-static fundamental process under deformation-sensitive loading. The superscript '0' refers to the fundamental direction of straining.

In turn, the results of Sect. 3.2.3 along with the equivalence (3.42) ⇔ (3.43) indicate that the condition

$$\ddot{E} \geq \ddot{E}^0 \qquad \text{for all admissible directions at varying } \lambda \tag{3.62}$$

is *necessary* for directional stability of the quasi-static fundamental process in the representative bar under the end displacement control. Interpretation of the path instability predicted by failure of (3.62) will be discussed in Sect. 3.5 in more general circumstances.

Four versions of the second-order energy criterion of material stability have been thus distinguished. However, it can be shown (cf. [47]) that (3.62) implies (3.60), which indicates that the energy conditions for stability of a quasi-static process are generally more restrictive than those for stability of equilibrium. In other words, if the former are satisfied then the latter are not needed. On the other hand, if the former fail then the latter can provide the information whether the resulting instability is quasi-static in character, as in Sect. 3.2.3, or is of dynamic type.

Moreover, the quasi-static process can be parameterized by a time-like parameter t in a special way such that an equilibrium state is kept fixed at constant λ in some interval of $t \geq t_0$, so that $\ddot{E}^0 = 0$. The conditions for stability of a quasi-static process within such a particular but arbitrarily long period *reduce* to those for stability of equilibrium. Consequently, the conditions (3.59) and (3.60) may be regarded as particular cases of (3.61) and (3.62), respectively. Finally, since the critical deformation stage associated with failure of (3.61) or (3.62) along a deformation path in usual circumstances will be the same, we arrive in effect at *a single energy criterion of material instability*, associated with the inequality (3.62).

All this concerned the simple example of the bar as a (one-dimensional and discretized) substitute of a material element. Generalization of the energy criterion to material instability in a three-dimensional case is not immediate, and will be discussed in Sect. 3.4 and 3.5, following earlier papers by the author [46, 47, 51, 52, 59]. The above considerations limited to a highly idealized example may, however, be helpful in conceptually similar but mathematically more complex investigations of three-dimensional problems.

3.2.5 Second-Order Geometric Effect

The critical condition

$$C^{\mathrm{p}} = 0 \tag{3.63}$$

which, according to (3.23), marks the onset of instability of uniform straining of the bar, coincides with the classical Considère criterion. It states that necking in a slender tensile bar begins when the *load* (and not the current stress) attains a maximum. At a first glance it could seem that the difference between the maxima of the load and stress may be neglected in the range of sufficiently small strains. That conjecture is incorrect. Denote by σ the true tensile stress, by $\bar{\rho}$ the mass density per unit current volume, and by $\bar{A}$ the current cross-section area of the bar. During uniform straining of the bar, we have

$$\begin{aligned} P &= \sigma \bar{A} \\ &= \frac{\tau}{1+\epsilon} A \,, \end{aligned} \tag{3.64}$$

where A is the (fixed) cross-section area of the bar in the reference configuration, and $\tau = \sigma\rho/\bar{\rho}$ is the uniaxial Kirchhoff stress with ρ as the mass density per unit reference volume. Consequently, for the plastic loading, we have

$$C^{\mathrm{p}} = \frac{dP}{d\epsilon} = \left(\frac{d\tau}{d\epsilon} - \frac{\tau}{1+\epsilon} \right) \frac{A}{1+\epsilon} \,. \tag{3.65}$$

The critical condition takes thus the form

$$C^{\mathrm{p}} = 0 \qquad \Leftrightarrow \qquad \frac{d\tau}{d\varepsilon} = \tau\,, \qquad \varepsilon = \ln(1+\epsilon)\,, \tag{3.66}$$

where ε is the true (logarithmic) strain. If volume changes are neglected then the Kirchhoff stress τ reduces to the true (Cauchy) stress σ, and the condition in (3.66) reduces to $d\sigma/d\varepsilon = \sigma$, the classical condition due to Considère. It is evident from (3.66) that even for vanishingly small strain ϵ (which can always be met by choosing the reference configuration as coinciding momentarily with the current configuration of the bar), instability of the uniform deformation process in the bar occurs when the (Kirchhoff) stress-strain curve is still increasing, with the slope equal to the current stress.

One might argue that the geometric effect discussed above is strictly connected with the necking instability. This is true; however, the effect of changes in geometry is also important for intrinsic instability in the material. The difference between current values of the Cauchy and nominal stresses (per unit current and reference area, respectively) may by attributed to the first-order geometric effect. If the strain and rotation are sufficiently small then the difference can safely be neglected in calculations of the *first-order* work. However, in calculations of the *second-order* work, which is essential in stability analysis, the *second-order* geometric effect is not negligible in general. It is related to the difference between the *rates* (or increments) of the Cauchy and nominal stresses which persists irrespectively whether the current strain is small or not. For instance, if the second-order geometric effect were not taken into account, the right-hand term τ in (3.66) would be erroneously set equal to zero. Similar errors can appear in a three-dimensional analysis. The recommended procedure of calculating the rates or differentials of loads and stresses is thus to perform the needed time differentiation within a geometrically exact framework *before* introducing any small-strain simplifications, and not conversely. As in the simple calculations above, in more general studies of material instabilities this helps in avoiding the over-simplification caused by neglecting the important second-order geometric effect.

Exercise: Carry out a detailed proof of (3.64) to arrive at the equivalence in (3.66).

3.3 Constitutive Framework in Incrementally Nonlinear Plasticity

Notation

From now on we extend investigations to three-dimensional solids. In the tensor notation employed, bold-face letters denote vectors (in $\mathcal{R}^3$) or second-order tensors, and 'sans serif' bold capitals (like A) stand for fourth-order tensors. Direct juxtaposition of two vectors or tensors denotes simple contraction, separation by a dot - double contraction, and the symbol $\otimes$ - the tensor product. In components on a fixed orthonormal basis, which are denoted by lower case Latin subscripts with the summation convention for repeated indices, it means that $s_j n_j = \mathbf{sn}$, $F_{ij} n_j = (\mathbf{Fn})_i$, $\sigma_{ij} W_{jk} = (\boldsymbol{\sigma}\mathbf{W})_{ik}$, $m_{ij} W_{ij} = \mathbf{m}\cdot\mathbf{W}$, $C_{ijkl} F_{kl} = (\mathbf{C}\cdot\mathbf{F})_{ij}$, $C_{ijkl} P_{klpq} = (\mathbf{C}\cdot\mathbf{P})_{ijpq}$, $s_i n_j = (\mathbf{s}\otimes\mathbf{n})_{ij}$, $\lambda_{ij}\mu_{kl} = (\boldsymbol{\lambda}\otimes\boldsymbol{\mu})_{ijkl}$. $\mathbf{I}$ stands for the second-order unit tensor, an upper index T denotes a transpose, $(\mathbf{F}^{\mathrm{T}})_{ij} = F_{ji}$ or $(\mathbf{A}^{\mathrm{T}})_{ijkl} = A_{klij}$, $\operatorname{tr}\mathbf{a} = a_{qq}$ denotes the trace, and subscripts 's' and 'a' denote the symmetric and antisymmetric part, $\frac{1}{2}(\mathbf{a}+\mathbf{a}^{\mathrm{T}})$ and $\frac{1}{2}(\mathbf{a}-\mathbf{a}^{\mathrm{T}})$, respectively, of a second-order tensor $\mathbf{a}$. A superimposed dot over a symbol denotes the *forward* rate, i.e. the one-sided time derivative from the right with respect to a time-like variable t at a fixed material point.

3.3.1 Rates of Strain and Stress

Constitutive rate equations in plasticity involve forward rates of strain and stress as basic quantities. Any objective measures of strain-rate and stress-rate which do not depend on the material spin may be employed. As a basic pair of spin-insensitive, symmetric measures of strain-rate and stress-rate, it is convenient to take $\mathbf{D}$, the Eulerian strain-rate, and $\overset{\nabla}{\boldsymbol{\tau}}$, the Zaremba-Jaumann flux of the Kirchhoff stress based on the *current* configuration. Recall that the Kirchhoff stress $\boldsymbol{\tau}$ is related to the true or Cauchy stress $\boldsymbol{\sigma} = \boldsymbol{\sigma}^{\mathrm{T}}$ by $\boldsymbol{\tau} = J\boldsymbol{\sigma}$, where J is the ratio of reference-to-current material density, and that the Zaremba-Jaumann flux is corotational with the material spin $\mathbf{W}$. If $\boldsymbol{\tau}$ is based on the current configuration then $\boldsymbol{\tau}$ and $\boldsymbol{\sigma}$ momentarily coincide, but their rates are still different if $J\operatorname{tr}\mathbf{D} = \dot{J} \neq 0$. The relevant standard formulae are

$$\mathbf{D} = (\bar{\nabla}\mathbf{v})_{\mathrm{s}}\,, \qquad \mathbf{W} = (\bar{\nabla}\mathbf{v})_{\mathrm{a}}\,, \qquad \overset{\nabla}{\boldsymbol{\tau}} = \overset{\nabla}{\boldsymbol{\sigma}} + \boldsymbol{\sigma}\operatorname{tr}\mathbf{D}\,, \qquad \overset{\nabla}{\boldsymbol{\sigma}} = \dot{\boldsymbol{\sigma}} + \boldsymbol{\sigma}\mathbf{W} - \mathbf{W}\boldsymbol{\sigma}\,, \tag{3.67}$$

where $\bar{\nabla}\mathbf{v}$ is the spatial gradient of a material velocity $\mathbf{v}$ in the current configuration.

Consider a class of of finite strain and stress measures provided by symmetric second-order tensors $\mathbf{e}$ and $\mathbf{t}$, respectively, which are objective in the Lagrangean sense [28, 43]. Denote by $\dot{w}$ the rate of virtual work by current tractions on the material element per unit *reference* volume, assuming that the tractions are in equilibrium with the internal state of stress. The identity

$$\dot{w} \equiv J\boldsymbol{\sigma}\cdot\mathbf{D} = \mathbf{t}\cdot\dot{\mathbf{e}} \tag{3.68}$$

is used to define the stress $\mathbf{t}$ *work-conjugate to* $\mathbf{e}$ [27, 28]. The material (right-hand) time derivatives $(\dot{\mathbf{e}}, \dot{\mathbf{t}})$ provide a class of objective measures of strain-rate and stress-rate, respectively. The basic pair $(\mathbf{D}, \overset{\nabla}{\boldsymbol{\tau}})$ belongs to this class, which motivated the use of $\overset{\nabla}{\boldsymbol{\tau}}$ rather than $\overset{\nabla}{\boldsymbol{\sigma}}$ as the basic stress-rate measure. If the reference configuration is chosen to coincide momentarily with the current configuration then the rate of *any* strain measure $\bar{\mathbf{e}}$ based on that reference configuration is assumed equal to $\mathbf{D}$, while $\bar{\mathbf{e}} = \mathbf{0}$ instantaneously. In turn, $\overset{\nabla}{\boldsymbol{\tau}}$ is then equal to the rate of the *particular* stress measure $\bar{\mathbf{t}}^{(0)}$ that is work-conjugate to the logarithmic strain $\bar{\mathbf{e}}^{(0)}$ based on the current configuration, although every stress $\bar{\mathbf{t}}$ work-conjugate to $\bar{\mathbf{e}}$ coincides momentarily with $\boldsymbol{\sigma}$ [27].

Transformation formulae under the transition from one work-conjugate pair to another were given by Hill [28, 29] cf. also Ogden [44]. In particular, it has been shown that (3.68) implies[5]

$$\mathbf{t} = J\bar{\mathbf{t}}^{(0)} \cdot \frac{\partial \bar{\mathbf{e}}^{(0)}}{\partial \mathbf{e}}, \qquad \dot{\mathbf{t}} = J\overset{\nabla}{\boldsymbol{\tau}} \cdot \frac{\partial \bar{\mathbf{e}}^{(0)}}{\partial \mathbf{e}} + J\bar{\mathbf{t}}^{(0)} \cdot \frac{\partial^2 \bar{\mathbf{e}}^{(0)}}{\partial \mathbf{e} \partial \mathbf{e}} \cdot \dot{\mathbf{e}}. \tag{3.69}$$

Hence,

$$\dot{\mathbf{e}} = \mathsf{A} \cdot \mathbf{D} \quad \text{and} \quad \dot{\mathbf{t}} = \mathsf{B}_{\mathrm{w}} \cdot \overset{\nabla}{\boldsymbol{\tau}} + \mathsf{Z}_{\mathrm{w}} \cdot \mathbf{D}, \tag{3.70}$$

with the following identification

$$\mathsf{A} = \frac{\partial \mathbf{e}}{\partial \bar{\mathbf{e}}^{(0)}}, \qquad \mathsf{B}_{\mathrm{w}}^{\mathrm{T}} = J\mathsf{A}^{-1}, \qquad \mathsf{Z}_{\mathrm{w}} = J\mathsf{Y} \cdot \mathsf{A}, \qquad \mathsf{Y} = \mathsf{Y}^{\mathrm{T}} = \boldsymbol{\sigma} \cdot \frac{\partial^2 \bar{\mathbf{e}}^{(0)}}{\partial \mathbf{e} \partial \mathbf{e}}. \tag{3.71}$$

A, B_{w}, Z_{w} are fourth-order tensors with minor symmetries, independent of $\mathbf{D}$ and $\overset{\nabla}{\boldsymbol{\tau}}$ and dependent in general on the material state; the index 'w' is used to indicate that we are dealing with work-conjugate stress measures. A and B_{w} are assumed invertible when treated as linear operators acting in the space of symmetric second-order tensors. That is, $\mathsf{A} \cdot \mathsf{A}^{-1} = \mathsf{A}^{-1} \cdot \mathsf{A} = \mathsf{I}_{\mathrm{s}}$, where I_{s} is the unit tensor with minor symmetries, of Cartesian components $(\mathsf{I}_{\mathrm{s}})_{ijkl} = \frac{1}{2}(\delta_{ik}\delta_{jl} + \delta_{il}\delta_{jk})$ on a fixed orthonormal basis, and δ_{ij} denotes the Kronecker delta. If $\mathbf{F} = \mathbf{I}$ momentarily (one of the assumptions of the 'small strain' theory) then $\mathsf{A} = \mathsf{B}_{\mathrm{w}} = \mathsf{I}_{\mathrm{s}}$, but Z_{w} does not vanish and is generally of order of the stress $\boldsymbol{\sigma}$. It can be seen that contrary to another assumption of the 'small strain' theory, the resulting differences between various stress-rate measures are generally *not* negligible in plasticity even at small strains.

While the above rate-variables are useful in defining objective incremental constitutive laws, in the analysis of bifurcation and instability problems it is convenient to use the non-symmetric rate-variables $(\dot{\mathbf{F}}, \dot{\mathbf{S}})$ that include the effect of a material spin. They are generated by the material time derivatives of the deformation gradient $\mathbf{F}$ (restricted by $J \equiv \det \mathbf{F} > 0$) and nominal stress $\mathbf{N}$ or its transpose $\mathbf{S} = \mathbf{N}^{\mathrm{T}}$ that satisfy

$$\mathbf{F} = \frac{\partial \mathbf{x}}{\partial \boldsymbol{\xi}}, \qquad \mathbf{S}\mathbf{F}^{\mathrm{T}} = J\boldsymbol{\sigma} = \mathbf{F}\mathbf{N}, \tag{3.72}$$

where $\mathbf{x}$ and $\boldsymbol{\xi}$ denote the position vector of a material point in the current and reference configurations, respectively. Below the tensor $\mathbf{S}$ will be used and for brevity referred to as the Piola stress (rather than the first Piola-Kirchhoff stress tensor). Time differentiation of (3.72) after some rearrangements shows that the rate-variables $(\dot{\mathbf{S}}, \dot{\mathbf{F}})$ are related to these in (3.70) through

$$\dot{\mathbf{F}} = (\mathbf{D} + \mathbf{W})\mathbf{F}, \qquad \dot{\mathbf{S}} = J(\overset{\nabla}{\boldsymbol{\tau}} + \mathbf{W}\boldsymbol{\sigma} - \boldsymbol{\sigma}\mathbf{D})\mathbf{F}^{-1^{\mathrm{T}}}. \tag{3.73}$$

[5] The dot on the left (right) side of a partial derivative denotes the contraction over two first (last) indices of the numerator (denominator), respectively.

3.3.2 The Second-Order Work

The concept of the second-order work plays an important role in material stability investigations. If all changes in geometry of a material element are disregarded then an infinitesimal increment of work per unit volume is usually written as

$$\sigma \cdot d\epsilon + \frac{1}{2} d\sigma \cdot d\epsilon + \ldots, \tag{3.74}$$

where ϵ stands for the small-strain tensor and a prefix d denotes a differential. The former term which is a linear form with respect to an infinitesimal strain increment $d\epsilon$ is called the first-order work. The second term with $d\sigma$ related to $d\epsilon$ is called the second-order work. Positiveness of the second-order work:

$$d\sigma \cdot d\epsilon > 0 \qquad \text{or} \qquad \dot{\sigma} \cdot \dot{\epsilon} > 0 \qquad \text{for all } d\epsilon = \dot{\epsilon}\, dt \tag{3.75}$$

was introduced by Drucker [10] in his postulate of stability of material in relation to the definition of work-hardening.

As shown by Hill [27], the above postulate is not sufficiently precise unless the measure of stress or stress-rate is specified to define the symbols $d\sigma$ or $\dot{\sigma}$, respectively. The correct formula for an increment Δw of the deformation work per unit reference volume evaluated to second order along a *direct* path[6] in strain space is the 'trapezoid rule of quadrature' [28]

$$\Delta w^{\mathrm{d}} = \Delta_1 w + \Delta_2 w^{\mathrm{d}} + o(|\Delta \mathbf{e}|^2)\,, \qquad \Delta_1 w = \mathbf{t}_0 \cdot \Delta \mathbf{e}\,, \qquad \Delta_2 w^{\mathrm{d}} = \frac{1}{2} \Delta \mathbf{t} \cdot \Delta \mathbf{e}\,, \tag{3.76}$$

where $(\mathbf{e}, \mathbf{t})$ stands for any *work-conjugate* pair of strain and stress measures, $\Delta\mathbf{t}$ is the stress increment corresponding to a strain increment $\Delta\mathbf{e}$ along the path, and $\mathbf{t}_0$ is the initial stress. Under transition to another work-conjugate pair $(\mathbf{e}, \mathbf{t})$ the expression for Δw^{d} in (3.76) remains invariant to second order. However, the decomposition (3.76) itself is not invariant since *both* terms $\Delta_1 w$ and $\Delta_2 w$ change by a second-order quantity with opposite signs. As explained by Hill, the lack of invariance is related to the 'curvature' term in the expression for $\Delta\mathbf{e}$ correct to *second* order (e.g. $\Delta\mathbf{e} = \dot{\mathbf{e}}\Delta t + \frac{1}{2}\ddot{\mathbf{e}}(\Delta t)^2 + o(|\Delta\mathbf{e}|^2)$) which should be substituted into $\Delta_1 w$ when used in (3.76). Accordingly, positiveness of the second-order work written as

$$d\mathbf{t} \cdot d\mathbf{e} > 0 \qquad \text{or} \qquad \dot{\mathbf{t}} \cdot \dot{\mathbf{e}} > 0 \qquad \text{for all } d\mathbf{e} = \dot{\mathbf{e}}\, dt \tag{3.77}$$

is *not* a measure-invariant concept [27].

The distinction between (3.75) and (3.77) is generally not negligible even if the current configuration is taken as reference, i.e. when the current strain is exactly zero

[6] A strain path is understood here as a directed, continuous and piecewise smooth line in a strain space. A direct path in strain-space may be identified with a class of paths of uniformly bounded curvature, connecting given initial and final strains arbitrarily close to each other. The concept of *path-dependent* second-order work applicable to indirect, arbitrarily circuitous paths, which will be needed later, was developed in [50].

and $\dot{\mathbf{e}} = \mathbf{D}$. For instance, if $\dot{\sigma}$ in (3.75) is literally understood as the material time derivative of the Cauchy stress then

$$\overset{\nabla}{\tau} \cdot \mathbf{D} - \dot{\sigma} \cdot \dot{\epsilon} = \sigma \cdot (\mathbf{D}\,\mathrm{tr}\,\mathbf{D} + 2\mathbf{W}\mathbf{D})\,. \tag{3.78}$$

More generally, from (3.70) with (3.71) it follows that

$$\frac{1}{J}\dot{\mathbf{t}} \cdot \dot{\mathbf{e}} - \overset{\nabla}{\tau} \cdot \mathbf{D} = \dot{\mathbf{e}} \cdot \mathbf{Y} \cdot \dot{\mathbf{e}}\,. \tag{3.79}$$

In turn, for the work-conjugate pair of non-symmetric measures $(\mathbf{F}, \mathbf{S})$ useful in boundary-value problems, we have

$$\frac{1}{J}\dot{\mathbf{S}} \cdot \dot{\mathbf{F}} - \overset{\nabla}{\tau} \cdot \mathbf{D} = -\sigma \cdot (\mathbf{D}\mathbf{D} + \mathbf{W}\mathbf{W} + 2\mathbf{W}\mathbf{D})\,. \tag{3.80}$$

The differences make explicit the second-order geometric effect. It is generally not negligible for a non-zero current stress being of the order of a tangent stiffness modulus, no matter whether the current strain is 'small' or not.

3.3.3 Constitutive Rate Equations in Plasticity

3.3.3.1 Incremental stiffness moduli

In a given state of the material, denoted here by a symbol $\mathcal{G}$, a stress-rate is assumed to be a continuous function of a strain-rate[7],

$$\dot{\mathbf{t}} = \dot{\mathbf{t}}(\dot{\mathbf{e}}, \mathcal{G}) \qquad \text{such that} \qquad \dot{\mathbf{t}}(r\dot{\mathbf{e}}, \mathcal{G}) = r\dot{\mathbf{t}}(\dot{\mathbf{e}}, \mathcal{G}) \quad \text{for all } r > 0\,. \tag{3.81}$$

The above homogeneity property results from the assumption that the mechanical properties of the material do not depend in any way on a natural time (*time-independent material*). For the sake of simplicity of the notation, the same symbol is used for a function and its value. By Euler's theorem on homogeneous functions applied at every point of differentiability of $\dot{\mathbf{t}}$ it follows that

$$\dot{\mathbf{t}} = \mathsf{L}(\dot{\mathbf{e}}, \mathcal{G}) \cdot \dot{\mathbf{e}}\,, \qquad \mathsf{L} \equiv \frac{\partial \dot{\mathbf{t}}}{\partial \dot{\mathbf{e}}}\,, \qquad \mathsf{L}(r\dot{\mathbf{e}}, \mathcal{G}) = \mathsf{L}(\dot{\mathbf{e}}, \mathcal{G}) \quad \text{for all } r > 0\,. \tag{3.82}$$

Components of the fourth-order tensor L are called *incremental stiffness moduli*. The tensor L depends in general on the *direction* (but not magnitude) of strain-rate $\dot{\mathbf{e}}$; the dependence may be discontinuous and nonlinear. A domain (necessarily cone) in the strain-rate space within which L does not depend on $\dot{\mathbf{e}}$ is called *a constitutive*

[7] Note that the function itself might be dependent on strain *gradients* or on the size of a material element. This way of introducing an internal length scale in the constitutive theory deserves separate analysis, not attempted here.

cone or a constitutive branch. On the interface between two constitutive cones, $\mathbf{L}$ is not uniquely defined. Henceforth, it will be tacitly assumed that the incremental stiffness moduli tensor, whenever appears in the formulae, is well defined.

Irrespectively of the variables in which it is originally formulated to ensure objectivity, the constitutive rate-equation (3.82) can be expressed as

$$\dot{\mathbf{S}} = \dot{\mathbf{S}}(\dot{\mathbf{F}}, \mathcal{G}) = \mathbf{C}(\dot{\mathbf{F}}, \mathcal{G})\cdot\dot{\mathbf{F}}, \qquad \mathbf{C} \equiv \frac{\partial \dot{\mathbf{S}}}{\partial \dot{\mathbf{F}}}, \tag{3.83}$$

e.g., by using the formulae given in Sect. 3.3.1. The moduli tensor $\mathbf{C}$ will play an essential role later on.

The simplest example of a non-linear constitutive rate equation (3.82) or (3.83) is provided by the classical elastoplasticity with two constitutive branches, where

$$\overset{\triangledown}{\tau} = \bar{\mathbf{L}}\cdot\mathbf{D}, \qquad \bar{\mathbf{L}} = \bar{\mathbf{L}}^{\mathrm{e}} - \frac{\kappa}{g}\bar{\lambda}^{\mathrm{p}}\otimes\bar{\lambda}, \qquad \kappa = \begin{cases} 1 \ \text{ if } \bar{\lambda}\cdot\mathbf{D} > 0 & \text{(loading)}, \\ 0 \ \text{ if } \bar{\lambda}\cdot\mathbf{D} < 0 & \text{(unloading)}. \end{cases} \tag{3.84}$$

The notation will be explained in more detail below when discussing the extension to multi-mode plasticity. At the level of generality adopted here, there is no need to appeal to the usual multiplicative decomposition of $\mathbf{F}$. The present incremental description follows Hill and is more general since the plastic deformation gradient need not even be defined.

To make a model complete, the constitutive rate equations must be complemented with evolution rules for the variables dependent on a material state. This means that the dependence of $\mathbf{L}(\dot{\mathbf{e}})$ on $\mathcal{G}$ must be specified. However, in the analysis of material instability in elastic-plastic solids, as already illustrated in Sect. 2, of primary importance are the incremental stiffness moduli, and definite conclusions can be drawn without specifying how $\mathbf{L}$ as a *function* of $\dot{\mathbf{e}}$ varies along a deformation path. To simplify the notation, the dependence of $\mathbf{L}$ on $\mathcal{G}$ will not be indicated explicitly in the following.

3.3.3.2 Multi-mode plasticity and the vertex effect

A basic feature of an incremental constitutive law of the classical bi-linear form (3.84) is that the *direction* of the plastic part of strain-rate, $\mathbf{D}^{\mathrm{p}} \equiv \mathbf{D} - (\bar{\mathbf{L}}^{\mathrm{e}})^{-1}\overset{\triangledown}{\tau}$, in a given state is prescribed (for the *fixed* rate-measures) and hence independent of the stress-rate. According to experimental and micromechanical studies, not reviewed here, that assumption may be an oversimplification for real materials, for instance metal crystals or polycrystals as well as geomaterials. A more accurate approximation of the incremental elastoplastic behaviour can be obtained by considering activity of *multiple internal mechanisms of plastic deformation*. The constitutive relations of classical elastoplasticity are included as the special case $N = 1$.

Let the rate of activity of K-th mechanism, $K = 1, \ldots, N$, be measured by a scalar rate-variable $\mathring{\gamma}_K$, taken nonnegative by convention. It can be nonzero only if the respective loading variable f_K has reached from below a threshold value, without loss of generality set equal to zero. Thus

$$\mathring{\gamma}_K \geq 0, \qquad f_K \leq 0, \qquad f_K\,\mathring{\gamma}_K = 0\,. \tag{3.85}$$

No implicit summation convention is adopted for the capital Latin indices of the mechanisms: the summation over an index K will always be indicated explicitly. If all the $\mathring{\gamma}_K$'s are zero then the material undergoes purely elastic deformation. A circle rather than a dot over γ_K is used to emphasize that $\mathring{\gamma}_K$'s need not represent rates of any internal state variables (cf. Rice [66]).

Current values of the strain, stress and f_K's are obtained by the integration of their rates along a straining path with given initial conditions. The basic constitutive expressions for the rate-variables $\dot{\mathbf{t}}$ and $\dot{f}_K$ in a given state of the material are assumed as linear forms in $(\dot{\mathbf{e}}, \mathring{\gamma}_K)$:

$$\dot{\mathbf{t}} = \mathbf{L}^{\mathrm{e}} \cdot \dot{\mathbf{e}} - \sum_K \lambda_K^{\mathrm{p}} \mathring{\gamma}_K\,, \qquad \dot{f}_K = \lambda_K \cdot \dot{\mathbf{e}} - \sum_L g_{KL} \mathring{\gamma}_L\,. \tag{3.86}$$

A dual form, fully equivalent to (3.86) under the sole additional assumption that $\mathbf{L}^{\mathrm{e}}$ represents an invertible linear operator acting in the space of symmetric second-order tensors, reads

$$\dot{\mathbf{e}} = \mathbf{M}^{\mathrm{e}} \cdot \dot{\mathbf{t}} + \sum_K \mu_K^{\mathrm{p}} \mathring{\gamma}_K\,, \qquad \dot{f}_K = \mu_K \cdot \dot{\mathbf{t}} - \sum_L h_{KL} \mathring{\gamma}_L\,. \tag{3.87}$$

The elastic moduli tensor $\mathbf{L}^{\mathrm{e}}$, direction tensors λ_K^{p} of inelastic stress-rate, normals λ_K to loading hyperplanes in $\dot{\mathbf{e}}$-space, and interaction moduli g_{KL}, depend on the material state and not on values of any rate-variables. In the dual formulation, the elastic compliance tensor $\mathbf{M}^{\mathrm{e}}$, direction tensors μ_K^{p} of inelastic strain-rate, normals μ_K to loading hyperplanes in $\dot{\mathbf{t}}$-space, and *effective* hardening (or softening) moduli h_{KL}, are likewise state-dependent, with the mutual relationships:

$$\mathbf{M}^{\mathrm{e}} = (\mathbf{L}^{\mathrm{e}})^{-1}, \qquad \mu_K^{\mathrm{p}} = \mathbf{M}^{\mathrm{e}} \cdot \lambda_K^{\mathrm{p}}, \qquad \mu_K = \lambda_K \cdot \mathbf{M}^{\mathrm{e}}, \qquad h_{KL} = g_{KL} - \lambda_K \cdot \mathbf{M}^{\mathrm{e}} \cdot \lambda_L^{\mathrm{p}}\,. \tag{3.88}$$

The existence of an elastic strain energy function is assumed, which jointly with the work-conjugacy of $(\mathbf{e}, \mathbf{t})$ implies that $\mathbf{L}^{\mathrm{e}} = \mathbf{L}^{\mathrm{eT}}$, $\mathbf{M}^{\mathrm{e}} = \mathbf{M}^{\mathrm{eT}}$.

In the finite-strain theory of single crystals deformed plastically by multislip, $\mathring{\gamma}_K$ is identified as the *invariant* rate of shearing on K-th slip system, and $f_K = \tau_K - \tau_K^{\mathrm{c}}$, where τ_K is the *generalized* resolved shear stress on that system and τ_K^{c} is its critical value; cf. [30, 1, 20].

The nonnegative $\mathring{\gamma}_K$'s for potentially active mechanisms are to be determined from the consistency conditions that follow from (3.85) by differentiation, viz.

$$\dot{f}_K \leq 0, \quad \dot{f}_K\,\mathring{\gamma}_K = 0 \qquad \text{for } K \in \mathcal{P} \equiv \{K \;:\; f_K = 0\}\,, \tag{3.89}$$

while $\mathring{\gamma}_K = 0$ for $K \notin \mathcal{P}$. A vector of $\mathring{\gamma}_K$'s with $K \in \mathcal{P}$ will be denoted by $[\mathring{\gamma}]$.

If $\dot{\mathbf{e}}$ is prescribed then (3.86) and (3.89) lead to the following rate problem to be solved:

Problem (P1) *In a given material state and for given* $\dot{\mathbf{e}}$*, find* $\overset{\circ}{\gamma}_K$*'s such that*

$$\dot{f}_K = \boldsymbol{\lambda}_K \cdot \dot{\mathbf{e}} - \sum_{L \in \mathcal{P}} g_{KL} \overset{\circ}{\gamma}_L, \quad \overset{\circ}{\gamma}_K \geq 0, \quad \dot{f}_K \leq 0, \quad \dot{f}_K \overset{\circ}{\gamma}_K = 0 \qquad \text{for } K \in \mathcal{P}. \tag{3.90}$$

If (P1) is solved then the stress-rate is determined from $(3.86)_1$.

Eqs. (3.87) and (3.89) lead to the dual problem $(\mathbf{P1})^{\mathrm{dual}}$ formulated analogously to (P1) if $\dot{\mathbf{t}}$ is prescribed, with obvious replacements.

Suppose that Problem (P1) has a unique solution $\overset{\circ}{\gamma}_K(\dot{\mathbf{e}})$; a condition sufficient for that will be given in Sect. 3.6.1. Denote by $\mathcal{A} \subseteq \mathcal{P}$ the set of indices of the active mechanisms ($\overset{\circ}{\gamma}_K > 0$ if and only if $K \in \mathcal{A}$) in the solution for a given $\dot{\mathbf{e}}$, and by $(g_{\mathcal{A}})$ the submatrix g_{KL} with the indices $K, L \in \mathcal{A}$. It is clear that the non-zero $\overset{\circ}{\gamma}_K$'s must satisfy the set of linear equations

$$\dot{f}_K = \boldsymbol{\lambda}_K \cdot \dot{\mathbf{e}} - \sum_{K,L \in \mathcal{A}} (g_{\mathcal{A}})_{KL} \overset{\circ}{\gamma}_L = 0 \qquad \text{for } K \in \mathcal{A}. \tag{3.91}$$

On solving this under the assumption $\det(g_{\mathcal{A}}) \neq 0$ and substituting into $(3.86)_1$, we obtain

$$\dot{\mathbf{t}} = \mathsf{L}(\dot{\mathbf{e}}) \cdot \dot{\mathbf{e}}, \qquad \mathsf{L} = \mathsf{L}^{\mathrm{e}} - \sum_{L \in \mathcal{A}} (g_{\mathcal{A}})^{-1}_{KL} \boldsymbol{\lambda}^{\mathrm{p}}_K \otimes \boldsymbol{\lambda}_L, \tag{3.92}$$

where the dependence of L on $\dot{\mathbf{e}}$ is implicit and solely due to the dependence of $\mathcal{A}$ on $\dot{\mathbf{e}}$.

Analogous conclusions can formally be drawn from the dual formulation (3.87) by regarding $\dot{\mathbf{t}}$ as given. However, in this case there are problems, to be clarified later, connected with the fact that a given $\dot{\mathbf{t}}$ for one choice of the stress-rate measure results in a *deformation-sensitive* $\dot{\mathbf{t}}$ for the same loading condition expressed through another stress-rate measure.

The problem (P1) admits a purely elastic solution with all $\overset{\circ}{\gamma}_K = 0$ if and only if

$$\boldsymbol{\lambda}_K \cdot \dot{\mathbf{e}} \leq 0 \qquad \text{for all } K \in \mathcal{P}. \tag{3.93}$$

Thus, (3.93) with the strict inequality for all K is the condition for elastic unloading of all potentially active mechanisms. The set of all $\dot{\mathbf{e}}$ that satisfy the strict inequalities in (3.93), if non-empty, constitutes the *elastic unloading cone* (Fig. 3.4), which can be identified with a *corner* on the theoretical yield surface representing the boundary of an elastic domain in strain-space. The actual strain represents a *vertex* of this domain. In the dual formulation, the associated corner on the yield surface in stress-space is obtained.

Another constitutive cone in $\dot{\mathbf{e}}$-space in a given state under the assumption $\det(g_{\mathcal{P}}) \neq 0$ is defined by

$$\boldsymbol{\lambda}^{\mathrm{a}}_K \cdot \dot{\mathbf{e}} > 0 \qquad \text{for all } K \in \mathcal{P}, \qquad \boldsymbol{\lambda}^{\mathrm{a}}_K = \sum_{L \in \mathcal{P}} (g_{\mathcal{P}})^{-1}_{KL} \boldsymbol{\lambda}_L. \tag{3.94}$$

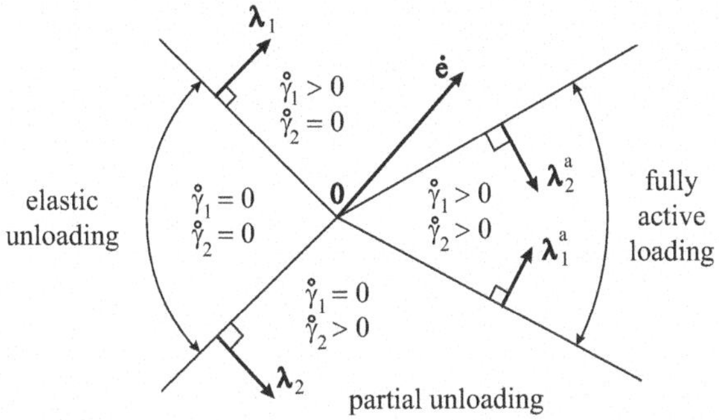

Fig. 3.4 Decomposition of strain-rate space in double-mode plasticity

Every $\dot{\mathbf{e}}$ from this cone corresponds to a solution with $\overset{\circ}{\gamma}_K > 0$ for all $K \in \mathcal{P}$, which can be verified by substituting $\mathcal{A} = \mathcal{P}$ into (3.91) and (3.92). The set of all $\dot{\mathbf{e}}$ that satisfy (3.94) constitutes the *cone of fully active loading* in strain-rate space; cf. [24, 49] for a discussion of this concept.

It can be checked that if more than one mechanisms are simultaneously active then the direction of the plastic part of strain-rate varies generally with the stress-rate direction. This effect is known as the *yield-surface vertex effect*, of fundamental importance for material stability since it can significantly decrease the material stiffness against an abrupt change of the direction of straining, cf. Hutchinson [32].

3.3.3.3 Transformation rules

An important question, investigated thoroughly by Hill [26, 28, 29] and more recently by the author in [61] is the following: how the quantities in constitutive relations for a given material change under transition to another pair of strain-rate and stress-rate measures.

It is assumed here [8] that

$$J^{-1} f_K \text{ and } \overset{\circ}{\gamma}_K \text{ are measure-invariant,} \tag{3.95}$$

Let the quantities appearing in the constitutive rate equations above be distinguished by a superimposed bar when associated with the pair $(\mathbf{D}, \overset{\nabla}{\boldsymbol{\tau}})$. Then, in order for the material properties to be invariant under the transition to any other pair of strain-rate and stress-rate measures from the class (3.70), it can be shown along the lines in [28] and [61] that the following transformation rules must hold:
For the incremental stiffness moduli :

[8] In comparison with [61], the factor J^{-1} is introduced here in order to ensure identical transformation rules for $\boldsymbol{\lambda}_K$ and $\boldsymbol{\lambda}_K^{\mathrm{p}}$ (cf. $(3.71)_2$), at the cost of loosing full invariance of g_{KL}

$$\mathsf{L} = (\mathsf{B}_{\mathrm{w}} \cdot \bar{\mathsf{L}} + \mathsf{Z}_{\mathrm{w}}) \cdot \mathsf{A}^{-1} \qquad \text{if } \dot{\mathbf{t}} = \mathsf{L} \cdot \dot{\mathbf{e}}, \ \overset{\nabla}{\boldsymbol{\tau}} = \bar{\mathsf{L}} \cdot \mathbf{D}. \tag{3.96}$$

For the quantities in (3.86):

$$\mathsf{L}^{\mathrm{e}} = (\mathsf{B}_{\mathrm{w}} \cdot \bar{\mathsf{L}}^{\mathrm{e}} + \mathsf{Z}_{\mathrm{w}}) \cdot \mathsf{A}^{-1}, \quad \lambda_K^{\mathrm{p}} = \mathsf{B}_{\mathrm{w}} \cdot \bar{\lambda}_K^{\mathrm{p}}, \quad \lambda_K = J\bar{\lambda}_K \cdot \mathsf{A}^{-1}, \quad g_{KL} = J\bar{g}_{KL}. \tag{3.97}$$

For the hardening (softening) moduli in (3.87):

$$h_{KL} = J(\bar{h}_{KL} + \mu_K \cdot \mathsf{Z}_{\mathrm{w}} \cdot \bar{\mu}_L^{\mathrm{p}}). \tag{3.98}$$

If the non-symmetric measures (3.73) are used then:
Equations (3.86) transform to

$$\dot{\mathbf{S}} = \mathsf{C}^{\mathrm{e}} \cdot \dot{\mathbf{F}} - \sum_K \boldsymbol{\Lambda}_K^{\mathrm{p}} \mathring{\gamma}_K, \qquad \dot{f}_K = \boldsymbol{\Lambda}_K \cdot \dot{\mathbf{F}} - \sum_L g_{KL} \mathring{\gamma}_L, \tag{3.99}$$

with the relationships

$$\boldsymbol{\Lambda}_K^{\mathrm{p}} = J\bar{\lambda}_K^{\mathrm{p}} \mathbf{F}^{-1^{\mathrm{T}}}, \qquad \boldsymbol{\Lambda}_K = \bar{\lambda}_K \mathbf{F}^{-1^{\mathrm{T}}}. \tag{3.100}$$

Equations (3.92) transforms to

$$\dot{\mathbf{S}} = \mathsf{C}(\dot{\mathbf{F}}) \cdot \dot{\mathbf{F}}, \qquad \mathsf{C} = \mathsf{C}^{\mathrm{e}} - \sum_{K,L \in \mathcal{A}} (g_{\mathcal{A}})^{-1}_{KL} \boldsymbol{\Lambda}_K^{\mathrm{p}} \otimes \boldsymbol{\Lambda}_L. \tag{3.101}$$

Incremental stiffness moduli (elastic or elastoplastic) transform as follows:

$$\begin{aligned} &J^{-1} F_{ip} F_{kq} C_{jplq} = \bar{L}_{ijkl} - \tfrac{1}{2}(\sigma_{jl}\delta_{ik} + \sigma_{jk}\delta_{il} + \sigma_{il}\delta_{jk} - \sigma_{ik}\delta_{jl}) \\ &\text{if } \dot{\mathbf{S}} = \mathsf{C} \cdot \dot{\mathbf{F}}, \ \overset{\nabla}{\boldsymbol{\tau}} = \bar{\mathsf{L}} \cdot \mathbf{D}. \end{aligned} \tag{3.102}$$

Exercise: Show that substitution of the above transformation rules leads to the constitutive rate equations (3.86) and (3.87) expressed in terms of $(\mathbf{D}, \overset{\nabla}{\boldsymbol{\tau}})$ and the respective barred quantities.

3.3.4 Thermodynamic Framework

The material element treated at the macroscopic level as a material point in a continuum, in a more refined scale exhibits an internal structure which may be very complex and strongly heterogeneous. Macroscopic plastic deformation is considered as a consequence of microstructural rearrangements [66]. From a physical point of view this is an inherently nonuniform process, with local instabilities such as jumps of dislocation loops at a micro-level. Consequently, the concept of a macroscopic strain-rate or stress-rate involves averaging over a *representative volume* of the material, as well as over a *representative time increment*.

In a thermodynamic description, changes in the internal structure of the material are frequently described with the help of *internal state variables*. In this article, internal state variables play only an auxiliary role. Their number and character (scalar or tensorial) are left unspecified; they are denoted collectively by α and considered in the *reference* configuration. As in the classical theory of irreversible processes (cf. [34]), a material element is assumed to be in local (constrained) thermodynamic equilibrium when its absolute temperature (T), macroscopic strain ($\mathbf{e}$) and internal microstructure (α) are given. The triple $(T, \mathbf{e}, \alpha)$ may define the material state $\mathcal{G}$.

In the following the temperature T is assumed constant and removed from the list of variables.

To 'smooth out' the influence of possible discontinuous jumps in the material state at a micro-level and to admit the rate-description, the internal variables α are defined as *averaging* variables over *both* a representative volume and a representative time increment [56]. The rate of the isothermal Helmholtz free energy density ϕ per unit reference volume may be written in the usual form as

$$\phi = \phi(\mathbf{e}, \alpha), \qquad \dot{\phi} = \mathbf{t} \cdot \dot{\mathbf{e}} - A \cdot \dot{\alpha}, \qquad \mathbf{t} = \frac{\partial \phi}{\partial \mathbf{e}}, \qquad A = -\frac{\partial \phi}{\partial \alpha}. \tag{3.103}$$

The symbol A (not to be confused with the fourth-order tensor A used above) denotes collectively the set of thermodynamic forces conjugate to α, and the central dot between A and $\dot{\alpha}$ denotes full contraction over all components of internal variables.

Identification of appropriate internal variables α may be not easy. Parameters γ_K obtained by formal time integration of $\mathring{\gamma}_K$ in the multi-mode plasticity theory are generally not internal state variables since the actual pattern of internal rearrangement may also depend on the prior history of the $\mathring{\gamma}_K$'s. Generally, the internal microstructure may be regarded as a *functional* of the prior history $\mathcal{H}$ of non-elastic deformation. Then the material state $\mathcal{G}$ may be identified with the triple $(T, \mathbf{e}, \mathcal{H})$. If the relation between $\mathbf{t}$ and $\mathbf{e}$ at fixed $\mathcal{H}$ is one-to-one then $\mathcal{G}$ may also be identified with $(T, \mathbf{t}, \mathcal{H})$. In the isothermal treatment below, any dependence on T is omitted.

Following Rice [66] and Hill and Rice [31], ϕ in the multi plasticity framework is treated below as a known function of $\mathbf{e}$ at each fixed $\mathcal{H}$, while the dependence of ϕ on $\mathcal{H}$ is prescribed only incrementally, viz.

$$\phi = \phi(\mathbf{e}, \mathcal{H}), \qquad \dot{\phi} = \mathbf{t} \cdot \dot{\mathbf{e}} - \sum_K \pi_K \mathring{\gamma}_K, \qquad \mathbf{t} = \frac{\partial \phi}{\partial \mathbf{e}}, \qquad \pi_K = \pi_K(\mathcal{G}), \tag{3.104}$$

instead of (3.103). Here, a material state is defined by $\mathcal{G} = (\mathbf{e}, \mathcal{H})$, and π_K are state-dependent scalar thermodynamic forces conjugate to $\mathring{\gamma}_K$. When writing (3.104), no explicit reference to internal *state* variables is needed. Links between (3.104) and (3.103) are obtained if every $\mathring{\gamma}_K$ is regarded as a state-dependent, positively homogeneous of degree one function of $\dot{\alpha}$ (or conversely). We assume that that function is independent of elastic strain or stress, that is

$$\mathring{\gamma}_K = \mathring{\gamma}_K(\alpha, \dot{\alpha}), \qquad \mathring{\gamma}_K(\alpha, r\dot{\alpha}) = r\mathring{\gamma}_K(\alpha, \dot{\alpha}) \qquad \text{for all } r > 0. \tag{3.105}$$

The loading variables f_K are given the thermodynamic interpretation:

$$f_K = \pi_K - \pi_K^{\rm c}, \qquad \pi_K^{\rm c} \geq 0, \qquad \pi_K^{\rm c} = \pi_K^{\rm c}(\mathcal{G}), \tag{3.106}$$

where a threshold value $\pi_K^{\rm c}$ may depend here not only on $\mathcal{H}$ (and T) as in Rice's theory [66], but also on *elastic* changes in strain or stress. In that way non-associative plasticity is included here. Current values of π_K and $\pi_K^{\rm c}$ may be determined by the integration of their rates, e.g., assumed in the form analogous to that for $\dot{f}_K$ in (3.86), along a straining path with given initial conditions.

As in the case of ϕ, any other state-dependent variable is regarded as an explicit function of $\mathbf{e}$ (or $\mathbf{t}$, alternatively) while its dependence in $\mathcal{H}$ may be prescribed only incrementally. The *plastic* part of the stress-rate satisfies [66, 31]

$$\dot{\mathbf{t}}^{\rm p} \equiv \dot{\mathbf{t}} - \mathbf{L}^{\rm e} \cdot \dot{\mathbf{e}} = \frac{\partial}{\partial \mathbf{e}}(\dot{\phi} - \mathbf{t} \cdot \dot{\mathbf{e}}) \,. \tag{3.107}$$

By combining this result with (3.86), (3.104) and using (3.106), and proceeding analogously in the dual formulation, one can show [60] that

$$\begin{array}{llll} \lambda_K = f_{K,\mathbf{e}} & \lambda_K^{\rm p} = \pi_{K,\mathbf{e}} & \lambda_K^{\rm p} - \lambda_K = \pi_{K,\mathbf{e}}^{\rm c} & \text{if } \mathcal{G} = (\mathbf{e}, \mathcal{H}) \,, \\ \mu_K = f_{K,\mathbf{t}} & \mu_K^{\rm p} = \pi_{K,\mathbf{t}} & \mu_K^{\rm p} - \mu_K = \pi_{K,\mathbf{t}}^{\rm c} & \text{if } \mathcal{G} = (\mathbf{t}, \mathcal{H}) \,. \end{array} \tag{3.108}$$

Here, the usual short-hand notation is employed, where a comma as a lower index denotes partial differentiation with respect to the arguments that follow the comma. Also, a single symbol is used for the same physical *quantity* dependent on either $\mathbf{e}$ or $\mathbf{t}$, remembering that the respective *functions* are mathematically different. The formulae (3.108) provide thermodynamic interpretation to the quantities involved in (3.86) and (3.87).

In the spirit of the theory of 'standard' materials based on two thermodynamic potentials (cf. [40]) but without the need of using internal *state* variables, another modification to Rice's theory is introduced by defining the *dissipation function* D:

$$D = D(\mathcal{G}, \mathring{\gamma}_K) \equiv \sum_K \pi_K^{\rm c}(\mathcal{G})\, \mathring{\gamma}_K \,. \tag{3.109}$$

The dissipation function D divided by T defines the rate of entropy production due to *virtual* $\mathring{\gamma}_K$'s.

The first condition in (3.85) now ensures that any $\mathring{\gamma}_K$ satisfying it corresponds to a non-negative intrinsic dissipation rate in view of the assumed non-negativeness of the $\pi_K^{\rm c}$'s, the second condition defines an admissible (elastic) domain in the space of thermodynamic forces π_K where $D(\mathcal{G}, \mathring{\gamma}_K) \geq \sum_K \pi_K(\mathcal{G})\, \mathring{\gamma}_K$, and the third condition ensures that the actual plastic dissipation rate in an unperturbed process is equal to D, also when each mechanism is considered separately.

To clarify the meaning of the virtual dissipation rate D, consider (possibly infinitesimal) perturbations of thermodynamic forces π_K by some additional disturbing forces $\pi_K^{\rm dist}$. A perturbed loading variable f'_K should now be determined by using a perturbed value of π_K, that is $f'_K = \pi_K + \pi_K^{\rm dist} - \pi_K^{\rm c}$. The perturbed activity condition in place of $(3.85)_3$ reads

$$f'_K \mathring{\gamma}_K = (\pi_K + \pi_K^{\mathrm{dist}} - \pi_K^{\mathrm{c}})\mathring{\gamma}_K = 0\,. \tag{3.110}$$

The total rate of work to be supplied to a deforming material element, per unit reference volume, includes the rate of work done by the π_K^{dist}'s, viz.

$$\dot{w}' = \dot{w} + \dot{w}^{\mathrm{dist}}, \qquad \dot{w} = \mathbf{t}\cdot\dot{\mathbf{e}}\,, \qquad \dot{w}^{\mathrm{dist}} = \sum_K \pi_K^{\mathrm{dist}}\mathring{\gamma}_K\,. \tag{3.111}$$

From (3.111), (3.110) and (3.106) it follows that

$$\dot{w}' = \mathbf{t}\cdot\dot{\mathbf{e}} - \sum_K f_K\mathring{\gamma}_K\,. \tag{3.112}$$

In turn, by combining (3.109), (3.110), (3.111) and (3.104), we obtain

$$D = \dot{w}' - \dot{\phi} \tag{3.113}$$

in agreement with the standard definition of isothermal dissipation.

3.3.5 Normality and Symmetry

It will be shown in later sections that much more can be said about material stability or instability if a general constitutive law is restricted by the normality and symmetry assumptions. This motivated a more detailed study of these assumptions presented in this section.

3.3.5.1 Basic assumptions

The normality fow rule in classical elastoplasticity correctly extended to finite strain can be written down as follows (cf. [28])

$$\dot{\mathbf{e}}^{\mathrm{p}} = \mathring{\gamma}\,\frac{\partial \hat{f}(\mathbf{t},\mathcal{H})}{\partial \mathbf{t}} \quad\Leftrightarrow\quad \dot{\mathbf{t}}^{\mathrm{p}} = -\mathring{\gamma}\,\frac{\partial f(\mathbf{e},\mathcal{H})}{\partial \mathbf{e}}\,, \qquad \text{where } \dot{\mathbf{e}}^{\mathrm{p}} \equiv \dot{\mathbf{e}} - \mathsf{M}^{\mathrm{e}}\cdot\dot{\mathbf{t}}\,,\ \dot{\mathbf{t}}^{\mathrm{p}} \equiv \dot{\mathbf{t}} - \mathsf{L}^{\mathrm{e}}\cdot\dot{\mathbf{e}} \tag{3.114}$$

and $f(\mathbf{e},\mathcal{H}) = \hat{f}(\mathbf{t},\mathcal{H})$. It is essential that the stress and strain measures are work-conjugate and that the plastic parts $\dot{\mathbf{e}}^{\mathrm{p}}$ of strain-rate and $\dot{\mathbf{t}}^{\mathrm{p}}$ of stress-rate are defined in the way shown above. The equivalence in (3.114) follows from the chain rule of differentiation of f at fixed $\mathcal{H}$ along with the assumed invertibility and diagonal symmetry of L^{e} and M^{e}. The normality rule (3.114) can be deduced from the well-known Ilyushin postulate applied at finite strain [27].

In multi-mode plasticity, the normality rule can be extended to each mechanism separately by adding an index K at both f and $\mathring{\gamma}$ in (3.114). Comparison with (3.87) and (3.86) shows that the extended normality rule is equivalent to

$$\mu_K^{\mathrm{p}} = \mu_K \qquad \Longleftrightarrow \qquad \lambda_K^{\mathrm{p}} = \lambda_K \,. \tag{3.115}$$

In the thermodynamic description, this means that π_K^{c} and the dissipation function D are independent of elastic changes in strain or stress, cf. (3.108). In the case of a single mechanism of plastic deformation as in (3.84), the normality rule (3.114) is equivalent to imposing on (3.83) the symmetry restriction

$$\mathbf{C} = \mathbf{C}^{\mathrm{T}} \qquad \Longleftrightarrow \qquad \mathbf{L} = \mathbf{L}^{\mathrm{T}} \,. \tag{3.116}$$

In turn, (3.116) means that the relationship (3.83) admits a potential, $U(\dot{\mathbf{F}}, \mathcal{G})$ say, such that

$$\dot{\mathbf{S}} = \frac{\partial U}{\partial \dot{\mathbf{F}}}, \qquad \mathbf{C} = \frac{\partial^2 U}{\partial \dot{\mathbf{F}} \partial \dot{\mathbf{F}}}, \qquad U = \frac{1}{2} \dot{\mathbf{S}} \cdot \dot{\mathbf{F}} \,. \tag{3.117}$$

More generally, one may postulate the symmetry property (3.116) for an arbitrary (also infinite) number of internal mechanisms of plastic deformation. This leads to (3.117) where U may be any homogeneous of degree two, continuously differentiable and piecewise continuously twice differentiable function of $\dot{\mathbf{F}}$ [22, 28].

In multi-mode plasticity, normality does not imply the symmetry property (3.116). From (3.92) it follows that (3.116) is ensured if besides normality it is additionally required that

$$g_{KL} = g_{LK} \qquad \Longleftrightarrow \qquad h_{KL} = h_{LK} \qquad \text{for } K, L \in \mathcal{A} \,, \tag{3.118}$$

i.e. for K, L from the set of indices of *active* moduli.

It can be shown, e.g., by applying the transformation rules given in Sect. 3.3.3.3, that the above normality and symmetry properties are invariant under transition to another pair of *work-conjugate* stress and strain measures [28].

3.3.5.2 Minimum of path-dependent dissipation

It will be shown below that the above normality and symmetry assumptions can be deduced from the following single postulate [60].

Postulate of minimum incremental dissipation *The dissipation evaluated to the* second-order *terms along a path connecting any two states arbitrarily close to each other is minimized on a* straight *(or direct) path* •

A straight path for a chosen set of state variables, $(\mathbf{e}, \boldsymbol{\alpha})$ say, generally becomes a curved but *direct* path after transition to other variables. A finite curvature will not influence the value of dissipation up to the second-order with respect to the path length. The dissipation is understood here as *virtual*, i.e. obtained by time integration of the dissipation function D along a path.

If the postulate is not satisfied then the existence of alternative, more complex but energetically more preferable paths *might* lead to an instability of the fundamental smooth path. This is indicated only as a possibility, and the above postulate, even if it seems to be natural, is not obligatory.

The above postulate is analogous to the 'radial path lemma' formulated by Nguyen [40] for generalizing the stability condition to arbitrarily complex paths. Validity of the lemma was stated for a state-independent convex dissipation function (and obviously also for path independent dissipation). Petryk [55] extended validity of the lemma to a convex dissipation function dependent linearly on internal variables under the symmetry assumption (3.127). These previous results are completed in the Appendix, following [60], by establishing necessary and sufficient conditions for the above postulate of minimum incremental dissipation.

The preliminary condition for the fulfillment of the postulate is, of course, that the virtual dissipation evaluated to the *first* order be minimized on a direct path. In the formalism of multi-mode plasticity, the first-order dissipation on a *direct* path starting from $\mathcal{G} = (\mathbf{e}, \alpha)$ reads

$$\Delta_1 \mathcal{D}^{\mathrm{d}} \equiv \sum_K \pi_K^{\mathrm{c}}(\mathcal{G}) \Delta\gamma_K = D(\mathcal{G}, \Delta\gamma_K)\,, \tag{3.119}$$

where $\Delta\gamma_K = \overset{\circ}{\gamma}_K \Delta t$. An arbitrary small increment $\Delta\gamma_K$ can also be imagined to be produced by imposing first a small elastic increment $\delta\mathbf{e}$ in strain (of order $\Delta\gamma_K$), then applying the increment $\Delta\gamma_K$ and next removing $\delta\mathbf{e}$ elastically. The final state will be the same, to first order, on account of (3.105). If some π_K^{c} depends on the elastic strain then the resulting value of the first-order dissipation

$$\Delta_1 \mathcal{D} = D(\mathcal{G}, \Delta\gamma_K) + \delta\mathbf{e} \cdot D_{,\mathbf{e}}(\mathbf{e}, \alpha, \Delta\gamma_K) \tag{3.120}$$

can be made less than (3.119) by selecting $\Delta\gamma_K$ and $\delta\mathbf{e}$ such that the term involving $\delta\mathbf{e}$ be negative. The postulate excludes this, so that the π_K^{c}'s in (3.108) must be independent of elastic changes in strain or stress. This implies the normality rule (3.115) as a consequence of the postulate.

Alternatively, from (3.113), (3.112) and (3.103) it follows that the dissipation function expressed in terms of the rates of internal variables takes the form

$$D(\mathcal{G}, \dot{\alpha}) = \mathbf{A}(\mathcal{G}) \cdot \dot{\alpha} - \sum_K f_K(\mathcal{G}) \overset{\circ}{\gamma}_K(\alpha, \dot{\alpha})\,. \tag{3.121}$$

Partial differentiation of this expression with respect to $\mathbf{e}$, on using (3.103), (3.107) and (3.86), yields

$$D_{,\mathbf{e}} = \mathbf{A}_{,\mathbf{e}} \cdot \dot{\alpha} - \sum_K f_{K,\mathbf{e}} \overset{\circ}{\gamma}_K = -\dot{\mathbf{t}}^{\mathrm{p}} - \sum_K \boldsymbol{\lambda}_K \overset{\circ}{\gamma}_K = \sum_K (\boldsymbol{\lambda}_K^{\mathrm{p}} - \boldsymbol{\lambda}_K) \overset{\circ}{\gamma}_K\,, \tag{3.122}$$

so that the normality rule in the form (3.115) follows again as a consequence of $D_{,\mathbf{e}} = 0$ if $\overset{\circ}{\gamma}_K \geq 0$ is regarded as arbitrary.

The condition $D_{,\mathbf{e}} = \mathbf{0}$ leads to

$$D = D(\alpha, \dot{\alpha}), \qquad D(\alpha, r\dot{\alpha}) = rD(\alpha, \dot{\alpha}) \quad \text{for all } r > 0, \tag{3.123}$$

as it is assumed in the theory of time-independent 'standard' materials (cf. Nguyen [40]). The following lemma holds (cf. Appendix).

Lemma A1 *Under the assumptions (3.123), the first-order dissipation is always minimized on a straight path if and only if*

$$D(\alpha, \dot{\alpha}) \textit{ is convex in } \dot{\alpha} \;\bullet \tag{3.124}$$

All $\dot{\alpha}$ compatible with the current thermodynamic forces $\boldsymbol{A}(\mathcal{G})$ constitute a cone $\mathcal{L}(\mathcal{G})$ in $\dot{\alpha}$-space within which the dissipation function D yields the actual value of the dissipation rate, that is

$$D(\alpha, \dot{\alpha}) = \boldsymbol{A}(\mathcal{G}) \cdot \dot{\alpha} \qquad \textit{for all } \dot{\alpha} \in \mathcal{L}(\mathcal{G}). \tag{3.125}$$

Thus, D is linear in $\dot{\alpha}$ within that cone.

Suppose that at least one $\dot{\alpha} = \dot{\alpha}^{\mathrm{a}}$ in (3.125) is associated with $\dot{\mathbf{e}}^{\mathrm{a}}$ from the cone of *fully active loading* in $\dot{\mathbf{e}}$-space. For such a pair $(\dot{\mathbf{e}}^{\mathrm{a}}, \dot{\alpha}^{\mathrm{a}})$ the cone $\mathcal{L}(\mathcal{G})$ is assumed to vary right-hand continuously in time, so that as a consequence of (3.125) we obtain

$$\dot{\alpha} \cdot D_{,\alpha}(\alpha, \hat{\alpha}) = \dot{\boldsymbol{A}} \cdot \hat{\alpha} \qquad \text{for all } \hat{\alpha} \in \mathcal{L}(\mathcal{G}) \quad \text{if } (\dot{\mathbf{e}}, \dot{\alpha}) = (\dot{\mathbf{e}}^{\mathrm{a}}, \dot{\alpha}^{\mathrm{a}}). \tag{3.126}$$

It is assumed here that the dissipation function $D(\alpha, \dot{\alpha})$ is continuously (Fréchet) differentiable with respect to α at each $\dot{\alpha}$, and the derivative is denoted by $D_{,\alpha}$.

In the Apendix the following result is proven.

Lemma A2 *Under the assumptions (3.123), (3.124), (3.125) and (3.126), the dissipation evaluated to second-order is always minimized on a straight path if and only if*

$$\hat{\alpha} \cdot D_{,\alpha}(\alpha, \dot{\alpha}) = \dot{\alpha} \cdot D_{,\alpha}(\alpha, \hat{\alpha}) \quad \textit{for all } \dot{\alpha}, \hat{\alpha} \in \mathcal{L}(\mathcal{G}) \;\bullet \tag{3.127}$$

The symmetry property (3.127) is thus obtained as a consequence of the postulate of minimum incremental dissipation. The symmetry (3.116) of the incremental stiffness moduli can be deduced from (3.127) by appealing to an integral constitutive inequality [49, 41].

The symmetry (3.118) of the moduli matrix in the multi-mode plasticity can be deduced more directly as follows. A 'plastic' variation of (3.121) in a direction $\hat{\alpha}$ at fixed $\dot{\alpha}$ gives

$$\hat{\alpha}\cdot D_{,\alpha}(\mathcal{G}, \dot{\alpha}) = \hat{\alpha}\cdot \boldsymbol{A}_{,\alpha}(\mathcal{G})\cdot\dot{\alpha} - \hat{\alpha}\cdot\sum_K \big(f_{K,\alpha}(\mathcal{G})\mathring{\gamma}_K(\alpha, \dot{\alpha}) + f_K(\mathcal{G})\mathring{\gamma}_{K,\alpha}(\alpha, \dot{\alpha})\big) \tag{3.128}$$

with $\mathcal{G} = (\mathbf{e}, \alpha)$. It is assumed here that the functions $f_K(\mathbf{e}, \alpha)$ and $\mathring{\gamma}_K(\alpha, \dot{\alpha})$ are continuous and (Gateaux) differentiable with respect to α.

Consider admissible directions of inelastic straining in the multi-mode plasticity framework, subject to (3.85) at every instant. Then $f_K(\mathcal{G}) < 0$ implies $\mathring{\gamma}_K = 0$ in some neighborhood of $\mathcal{G}$, by $(3.85)_3$ and continuity of f_K. Consequently, the last term in (3.128) vanishes. From $(3.86)_2$ the 'plastic' variation of f_K reads

$$\hat{\alpha} \cdot f_{K,\alpha}(\mathbf{e}, \alpha) = -\sum_{KL} g_{KL}(\mathbf{e}, \alpha)\mathring{\gamma}_L(\alpha, \hat{\alpha}) . \tag{3.129}$$

On account of symmetry of $\phi_{,\alpha\alpha} = -\boldsymbol{A}_{,\alpha}$, from (3.128) we thus obtain

$$\hat{\alpha} \cdot D_{,\alpha}(\mathcal{G}, \dot{\alpha}) - \dot{\alpha} \cdot D_{,\alpha}(\mathcal{G}, \hat{\alpha}) = \sum_{KL}(g_{KL}(\mathcal{G}) - g_{LK}(\mathcal{G}))\mathring{\gamma}_K(\alpha, \dot{\alpha})\mathring{\gamma}_L(\alpha, \hat{\alpha}) \tag{3.130}$$

for any two directions $\dot{\alpha}$, $\hat{\alpha}$ admissible in $\mathcal{G}$. On applying $(3.123)_1$ and (3.127), as a consequence of the postulate of minimum incremental dissipation we conclude that the expression (3.130) vanishes. The symmetry property (3.118) follows if $\dot{\alpha}$, $\hat{\alpha}$ vary within the constitutive cone where $\mathcal{A}$ is the set of indices of active mechanisms.

3.3.5.3 Linear comparison solid

A widely used approach to deriving conditions for stability or uniqueness is to linearize the incremental problem formulated in a given state. The question is whether such linearized conditions pertain to the actual elastic-plastic material which is incrementally nonlinear. An affirmative answer can be given for a class of materials for which suitable constitutive inequalities are established.

In the analysis of bifurcation in materials obeying the constitutive law (3.117) with a potential U, the 'relative convexity property' introduced by [21, 22] is very useful. It states that the actual potential U for an elastic-plastic material is 'more convex' than the quadratic potential U^{L} for an incrementally linear *comparison solid*, in the sense that $(U - U^{\mathrm{L}})$ in a given state is a convex function of $\dot{\mathbf{F}}$. This can be written as

$$(\dot{\mathbf{S}} - \dot{\mathbf{S}}^*) \cdot (\dot{\mathbf{F}} - \dot{\mathbf{F}}^*) \geq (\dot{\mathbf{F}} - \dot{\mathbf{F}}^*) \cdot \mathbf{C}^{\mathrm{L}} \cdot (\dot{\mathbf{F}} - \dot{\mathbf{F}}^*) \qquad \text{for every } \dot{\mathbf{F}}, \dot{\mathbf{F}}^* , \tag{3.131}$$

where the quantities in each pair $(\dot{\mathbf{F}}, \dot{\mathbf{S}})$ and $(\dot{\mathbf{F}}^*, \dot{\mathbf{S}}^*)$ are mutually related by (3.83). The moduli tensor $\mathbf{C}^{\mathrm{L}}$ of the comparison solid does not depend on the actual strain-rate and must be suitably defined for a given material model. From (3.73) and (3.70) it follows that (3.131) is equivalent to

$$(\dot{\mathbf{t}} - \dot{\mathbf{t}}^*) \cdot (\dot{\mathbf{e}} - \dot{\mathbf{e}}^*) \geq (\dot{\mathbf{e}} - \dot{\mathbf{e}}^*) \cdot \mathbf{L}^{\mathrm{L}} \cdot (\dot{\mathbf{e}} - \dot{\mathbf{e}}^*) \qquad \text{for every } \dot{\mathbf{e}}, \dot{\mathbf{e}}^* \tag{3.132}$$

for *all* work-conjugate measures $(\mathbf{e}, \mathbf{t})$ [28], with the analogous meaning of the quantities involved. Hill [21] showed that (3.131) holds for the classical plasticity (3.84) with normality (3.114) when $\mathbf{L}^{\mathrm{L}}$ is taken as the moduli tensor from the loading branch. For the classical elastoplasticity *without* normality, Raniecki [64] (cf. also [65]) constructed a comparison solid that verifies (3.131) and whose moduli do not coincide with the actual stiffness moduli from the loading branch.

Validity of (3.131) was extended to multi-mode plasticity by Sewell [71] under the normality and symmetry assumptions given above and under the additional assumption that

$$\text{the matrix } (g_{KL}) \text{ with } K, L \in \mathcal{P} \text{ is positive definite.} \tag{3.133}$$

In that case, the comparison solid is defined by the moduli from the branch of fully active loading, that is, by (3.92) with $\mathcal{A} = \mathcal{P}$.

If a fundamental solution is known, which is usually the case, then the *tangent moduli tensor* $\mathbf{L}^0 \equiv \mathbf{L}(\dot{\mathbf{e}}^0)$ corresponding to the fundamental strain-rate $\dot{\mathbf{e}}^0$ defines the *tangent comparison solid*. Instead of (3.132), for the analogous purpose the following constitutive inequality has been proposed [49]

$$(\dot{\mathbf{t}} - \dot{\mathbf{t}}^0) \cdot (\dot{\mathbf{e}} - \dot{\mathbf{e}}^0) \geq (\dot{\mathbf{e}} - \dot{\mathbf{e}}^0) \cdot \mathbf{L}^0 \cdot (\dot{\mathbf{e}} - \dot{\mathbf{e}}^0) \qquad \text{for every } \dot{\mathbf{e}}\,, \tag{3.134}$$

equivalent to

$$(\dot{\mathbf{S}} - \dot{\mathbf{S}}^0) \cdot (\dot{\mathbf{F}} - \dot{\mathbf{F}}^0) \geq (\dot{\mathbf{F}} - \dot{\mathbf{F}}^0) \cdot \mathbf{C}^0 \cdot (\dot{\mathbf{F}} - \dot{\mathbf{F}}^0) \qquad \text{for every } \dot{\mathbf{F}}\,, \tag{3.135}$$

where the fundamental pair $(\dot{\mathbf{e}}^0, \dot{\mathbf{t}}^0)$ or $(\dot{\mathbf{F}}^0, \dot{\mathbf{S}}^0)$ is known and fixed. This is a less restrictive requirement than (3.131) with $\mathbf{C}^{\mathrm{L}} = \mathbf{C}^0$ since now only one pair $(\dot{\mathbf{t}}, \dot{\mathbf{e}})$ is variable. Validity of (3.135) has been extended beyond the classical and multi-mode plasticity to an unspecified constitutive law (3.117) under the assumption that the framework of multi-mode plasticity with the normality and symmetry conditions (3.115) and (3.118) (but without assuming (3.133)) applies *at a micro-level*. This was done by appealing to the inequality

$$\dot{\mathbf{t}}^0 \cdot \dot{\mathbf{e}} - \dot{\mathbf{t}} \cdot \dot{\mathbf{e}}^0 \geq 0 \qquad \text{for every } \dot{\mathbf{e}} \tag{3.136}$$

transmissible from the micro- to macro-level. The inequality (3.134) applies at a *regular* point of a straining path where $\mathbf{L}^0$ does not correspond to any abrupt unloading at the micro-level; for more details, cf. [49, 54].

3.3.5.4 Path-dependent second-order work

In stability investigations based on the energy criterion, an estimate is needed for the work done on arbitrary circuitous paths of departure from a given state [21]. Generally, the work must be calculated with accuracy to second-order terms. It might seem at a first glance that piecewise-regular strain paths become 'direct' if their length is sufficiently small, so that there are no problems in using the expression (3.76). This is really so when any path is regarded as *fixed* as its length tends to zero. However, when proving stability we should also consider a *sequence* of paths which may substantially differ in shape from a smooth arc of bounded curvature, no matter how short they may be. More specifically, a given path in strain space may be scaled down such that it length $\Delta\theta$ tends to zero while its complexity is preserved. Then, in place of $(3.76)_1$ we can write

$$\Delta w = \Delta_1 w + \Delta_2 w + o(|\Delta\theta|^2)\,, \qquad \Delta\theta = \int (d\mathbf{e}\cdot d\mathbf{e})^{1/2}\,, \tag{3.137}$$

where $\Delta_2 w$ is the *path-dependent* second-order work. The concept is discussed in more detail in [51] where the following result, useful in stability investigations for justifying the convenient restriction to direct strain paths, has been proven.

Theorem (on minimum second-order work) *In the framework of multi-mode plasticity from Sect. 3.3.3.2 under the normality (3.115) and symmetry (3.118) assumptions, the positive-definiteness property (3.133) implies*

$$\Delta_2 w \geq \Delta_2 w^{\mathrm{d}} \tag{3.138}$$

within the class of all paths leading to the same final strain increment $\Delta\mathbf{e}$, *to second-order with respect to* $\Delta\theta$ • [9]

Although the value of the second-order work on each side in (3.138) depends on the chosen pair of work-conjugate measures of strain and stress, the inequality itself is measure-invariant.

3.3.6 Constitutive Equations for External Loading

A material element surrounded by similar elements is generally subjected to tractions dependent on the element configuration, also if a larger sample of the material is tested under the control of external surface tractions or displacements. The actual interaction rule between the configuration of a material element and its loading environment may be difficult to determine, especially if the surrounding material undergoes nonuniform, somewhere plastic and elsewhere elastic deformation. In general investigations of the influence of such flexible internal constraints on material instability, it seems natural to restrict attention in the first instance to a *linearized* interaction rule. The surrounding material is accordingly treated as nonuniformly deformed but incrementally *linear* (not necessarily elastic), while the interior material element is incrementally nonlinear but assumed to deform *uniformly*. A special case of interest arises when the interior element is imagined to be of ellipsoidal shape and embedded in an infinite homogeneous linear continuum, which is consistent with uniform straining of the ellipsoid [14]. In that case the external stiffness moduli (distinguished below by a star) can be effectively determined in a manner which is not discussed here.

[9] In contrast to (3.244), the final *state* (including the final stress) need not be the same for the paths involved in (3.138). In obtaining (3.138), the conditions (3.4) were assumed to hold along every strain path under consideration, and no reference to thermodynamics was needed.

Following [61], the rate of work done by tractions applied to the element boundary from the outside is assumed as a linear form in $\dot{\mathbf{F}}$,

$$\dot{w}^{\mathrm{ext}} = \mathbf{S}^{\mathrm{ext}} \cdot \dot{\mathbf{F}}\,, \tag{3.139}$$

per unit reference volume of the material. In analogy to Hill's concept of a *symmetric* work-conjugate stress (cf. Sect. 3.3.1), the rate of work is understood here as a basic quantity, and (3.139) serves as a definition of the *external* Piola stress $\mathbf{S}^{\mathrm{ext}}$. The product in (3.139) can be decomposed as

$$\begin{gathered} \mathbf{S}^{\mathrm{ext}} \cdot \dot{\mathbf{F}} = J(\sigma^{\mathrm{ext}} \cdot \mathbf{D} + \mathbf{m}^{\mathrm{ext}} \cdot \mathbf{W})\,, \\ \sigma^{\mathrm{ext}} = \sigma^{\mathrm{ext}\mathrm{T}} = J^{-1}(\mathbf{S}^{\mathrm{ext}}\mathbf{F}^{\mathrm{T}})_{\mathrm{s}}\,, \qquad \mathbf{m}^{\mathrm{ext}} = -\mathbf{m}^{\mathrm{ext}\mathrm{T}} = J^{-1}(\mathbf{S}^{\mathrm{ext}}\mathbf{F}^{\mathrm{T}})_{\mathrm{a}}\,. \end{gathered} \tag{3.140}$$

The need for distinguishing between the external ($\mathbf{S}^{\mathrm{ext}}$) and internal ($\mathbf{S}$) Piola stress can be seen from the absence of a counterpart to $\mathbf{m}^{\mathrm{ext}}$ in the constitutive framework for the material without couple stresses. The requirement $\mathbf{S}\mathbf{F}^{\mathrm{T}} = \mathbf{F}\mathbf{S}^{\mathrm{T}}$ (equivalent to $\sigma = \sigma^{\mathrm{T}}$) of rotational equilibrium need not apply *in advance* to $\mathbf{S}^{\mathrm{ext}}$, so that $\mathbf{m}^{\mathrm{ext}} \neq \mathbf{0}$ in general. A simple illustration is provided by 'dead' loading with prescribed $\mathbf{S}^{\mathrm{ext}}$, when $\mathbf{S}^{\mathrm{ext}}\mathbf{F}^{\mathrm{T}}$ need not be equal to $\mathbf{F}\mathbf{S}^{\mathrm{ext}\mathrm{T}}$ for an arbitrary $\mathbf{F}$.

The actual history of the deformation gradient $\mathbf{F}$ in a quasi-static motion must be calculated from the equilibrium condition. Suppose that $\mathbf{F}$ is kinematically unconstrained. Then in a state of mechanical equilibrium we must have

$$\mathbf{S} = \mathbf{S}^{\mathrm{ext}} \qquad \Leftrightarrow \qquad (\,\sigma = \sigma^{\mathrm{ext}} \quad \text{and} \quad \mathbf{m}^{\mathrm{ext}} = \mathbf{0}\,), \tag{3.141}$$

by the principle of virtual power. In view of the incremental character of the constitutive law for the material, the basic condition of continuing mechanical equilibrium of a material element is formulated for the stress rates:

$$\dot{\mathbf{S}} = \dot{\mathbf{S}}^{\mathrm{ext}}\,. \tag{3.142}$$

To specify the loading conditions, we need a constitutive expression for $\dot{\mathbf{S}}^{\mathrm{ext}}$ (which is different from that for $\dot{\mathbf{S}}$). In a given equilibrium state, we shall consider a class of deformation-sensitive incremental loadings of a material element, defined by

$$\dot{\mathbf{S}}^{\mathrm{ext}} = \dot{\mathbf{S}}^{*} - \mathbf{C}^{*} \cdot \dot{\mathbf{F}}\,. \tag{3.143}$$

Here, a fourth-order tensor $\mathbf{C}^{*}$ of external stiffness moduli and an applied stress-rate $\dot{\mathbf{S}}^{*}$ are given quantities in a given state of the material element. $\dot{\mathbf{S}}^{*}$ is a controllable part and $(-\mathbf{C}^{*} \cdot \dot{\mathbf{F}})$ is a deformation-sensitive part of $\dot{\mathbf{S}}^{\mathrm{ext}}$. For example, for the loading by all-around hydrostatic pressure ($\sigma^{\mathrm{ext}} = -p\mathbf{I}$) and in the current configuration taken as reference ($\mathbf{F} = \mathbf{I}$ instantaneously), we have $\dot{\mathbf{S}}^{*} = -\dot{p}\mathbf{I}$ and $C^{*}_{ijkl} = (\delta_{il}\delta_{jk} - \delta_{ij}\delta_{kl})p$.

Eq. (3.142) provides nine equations for nine unknowns: six components of the strain-rate and three components of the material spin. Suppose that the components of the material spin are either known or expressible uniquely in terms of the strain-rate components, cf. [25, 61]. Then the equation (3.143) can be reduced to the following form not involving material spin

$$\dot{\mathbf{t}}^{\mathrm{ext}} = \dot{\mathbf{t}}^{*} - \mathsf{L}^{*} \cdot \dot{\mathbf{e}}\,, \tag{3.144}$$

where the external stress-rate measure $\dot{\mathbf{t}}^{\mathrm{ext}}$ is defined by $\dot{\mathbf{t}}^{\mathrm{ext}} = \mathsf{B}_{\mathrm{w}} \cdot \overset{\nabla}{\tau}{}^{\mathrm{ext}} + \mathsf{Z}_{\mathrm{w}} \cdot \mathbf{D}$ with the same B_{w} and Z_{w} as in (3.70). Here, $\dot{\mathbf{t}}^{*}$ and L^{*} are regarded as known quantities. The following transformation rules analogous to those in (3.97) apply:

$$\dot{\mathbf{t}}^{*} = \mathsf{B}_{\mathrm{w}} \cdot \overset{\nabla}{\tau}{}^{*}, \qquad \mathsf{L}^{*} = (\mathsf{B}_{\mathrm{w}} \cdot \bar{\mathsf{L}}^{*} - \mathsf{Z}_{\mathrm{w}}) \cdot \mathsf{A}^{-1} \qquad \text{if } \overset{\nabla}{\tau}{}^{\mathrm{ext}} = \overset{\nabla}{\tau}{}^{*} - \bar{\mathsf{L}}^{*} \cdot \mathbf{D}\,. \tag{3.145}$$

Note that if $\mathsf{L}^{*} = \mathbf{0}$ for one choice of a pair $(\dot{\mathbf{e}}, \dot{\mathbf{t}}^{\mathrm{ext}})$ then generally $\mathsf{L}^{*} \neq \mathbf{0}$ for another choice. This means that if one stress-rate measure is fully controlled then another stress-rate measure is generally dependent on the strain-rate, cf. [26, 28].

3.4 Stability of Equilibrium

3.4.1 The Energy Condition for Stability of Equilibrium

The concepts of stability of equilibrium in the energy sense, presented in Sect. 2 with reference to the one-dimensional example, are now generalized to a three-dimensional continuum.

There are different ways to derive a criterion of stability of equilibrium based on the energy considerations. Without using thermodynamics, the basic way (cf. [21]) is to begin with the mechanical energy balance for a solid body taken away from equilibrium, viz. (cf. (3.44))

$$\Delta W - \Delta W^{\mathrm{ext}} + K = \Delta W^{*}\,, \tag{3.146}$$

where ΔW is the work to be supplied to the deformed body along a path of departure from the examined equilibrium state, ΔW^{ext} is the work done by external loads along that path, and K is the (non-negative) kinetic energy at the path end. The unbalanced amount of energy ΔW^{*} must be supplied to the deformed body by a disturbing agency. ΔW is to be determined by integration of the work-rate density (3.111) over the reference volume V of a material element and over a time interval, viz.

$$\Delta W = \int \dot{W}(t)\, dt\,, \qquad \dot{W} = \int_V (\dot{w} + \dot{w}^{\mathrm{dist}})\, dV\,. \tag{3.147}$$

Equivalently, in the thermodynamic framework with the dissipation function D, we can use the following decomposition (cf. [40])

$$\Delta W = \int_V (\Delta\phi + \Delta\mathcal{D})\, dV\,, \qquad \Delta\mathcal{D} = \int D(t)\, dt\,. \tag{3.148}$$

The time integrals above are generally path-dependent. Further specifications of the quantities involved in (3.146) may depend on the kind of material stability as discussed below.

If

$$\Delta W - \Delta W^{\text{ext}} > 0 \tag{3.149}$$

along every kinematically admissible deformation path that starts from the examined equilibrium state then any departure from that state requires a positive amount of energy ΔW^* to be supplied by a disturbing agency. The condition (3.149) may be only imposed in a certain neighborhood of the equilibrium state and for a specified class of paths of departure from equilibrium. This is usually regarded as a condition sufficient for dynamic stability of equilibrium (for the specified class of paths). It is justified provided the (positive) left-hand quantity in (3.149) is accepted as the measure of the disturbance strength as well as of the distance from the equilibrium state. If another distance measure $d \geq 0$ is used (e.g., geometric or in state variables) then (3.149) should be strengthened to

$$\Delta W - \Delta W^{\text{ext}} \geq a(d) \qquad \text{if } d \leq R, \tag{3.150}$$

where $a(\cdot)$ is an arbitrary continuous and monotonically increasing function with $a(0) = 0$, and $R = \text{const} > 0$. In turn, for a quasi-static path of departure from equilibrium in a time interval $[t_0, t]$, the quantity

$$d_0(t) = \sup_{t_0 \leq \theta \leq t} (\Delta W(\theta) - \Delta W^{\text{ext}}(\theta)) \tag{3.151}$$

may serve as a natural measure for a disturbance strength. Clearly, for a dynamic path the disturbance strength is not less on acount of a non-negative kinetic energy.

The stability of equilibrium with respect to two measures (metrics) d_0, d in the rigorous sense of the Lyapunov-Movchan theory follows from (3.150), as discussed in more detail in other CISM Lecture Notes [55]; cf. also [42]. An extension to thermodynamic stability of equilibrium in plasticity under possible perturbations of the temperature field was given in [55, 56].

A natural first step in verifying (3.149) is to consider *direct* paths of departure from equilibrium by developing the work expressions into a Taylor series with respect to 'time' t along a path,

$$\Delta W = \dot{W}\,\Delta t + \frac{1}{2}\ddot{W}\,(\Delta t)^2 + \ldots, \qquad \Delta W^{\text{ext}} = \dot{W}^{\text{ext}}\,\Delta t + \frac{1}{2}\ddot{W}^{\text{ext}}\,(\Delta t)^2 + \ldots, \tag{3.152}$$

where the superimposed dots denote the right-hand rates evaluated in the equilibrium state. If (3.149) holds after substituting the series (3.152) truncated to the needed (usually the second) order terms then the resulting stability will be called 'directional'. The second-order condition for directional stability reads

$$\ddot{W} - \ddot{W}^{\text{ext}} > 0, \tag{3.153}$$

to hold for all admissible directions of departure from equilibrium. A strengthened version of (3.153) obtained from (3.150) can also be examined.

It must be noted that, in some analogy to structural stability under follower loads, directional stability does not necessarily imply dynamic stability for arbitrarily circuitous paths.

3.4.2 Stability for Internal Rearrangements

In this section, the material stability is examined at uniform macroscopic strain against possible spontaneous internal structural rearrangements at a micro-level, cf. Sect. 3.1.3. We will study parallely two cases when either the strain $\mathbf{e}$ is fixed (a material element with a clamped boundary), or a stress measure $\mathbf{t} = \mathbf{t}^{\text{ext}}$ is fixed (irrespectively of unrestricted changes in the uniform material strain). Internal rearrangements are described macroscopically in the framework of multi-mode plasticity, cf. Sect. 3.3.3.2 and in the thermodynamic framework, cf. Sect. 3.3.4.

Consider a departure from a given equilibrium state, induced by some disturbances at a micro-level which correspond to perturbations of thermodynamic forces in the macroscopic description. Since $\mathbf{t} \cdot \dot{\mathbf{e}} = \mathbf{t}^{\text{ext}} \cdot \dot{\mathbf{e}}$ by assumption, the rate of extra work $(W - W^{\text{ext}})$ to be supplied to the element of a reference volume V reduces to

$$\frac{1}{V}(\dot{W} - \dot{W}^{\text{ext}}) = \dot{w}' - \dot{w} = -\sum_K f_K \mathring{\gamma}_K \tag{3.154}$$

according to (3.112), or to

$$\frac{1}{V}(\dot{W} - \dot{W}^{\text{ext}}) = \dot{\phi} + D - \dot{w} = D - \boldsymbol{A} \cdot \dot{\boldsymbol{\alpha}} \tag{3.155}$$

from (3.113) and (3.103).

A prerequisite for (3.149) is that in the examined equilibrium state

$$\dot{W} - \dot{W}^{\text{ext}} \geq 0 \tag{3.156}$$

for all possible rearrangements. Since all $\mathring{\gamma}_K \geq 0$ by assumption, in this way from (3.156) and (3.154) we recover the condition $f_K \leq 0$ which has been assumed in advance in (3.85) as the restriction on admissible states in the multi-mode plasticity framework. Similarly, on using (3.155) we obtain the inequality

$$D(\boldsymbol{\alpha}, \dot{\boldsymbol{\alpha}}) - \boldsymbol{A}(\mathcal{G}) \cdot \dot{\boldsymbol{\alpha}} \geq 0 \quad \text{for all } \dot{\boldsymbol{\alpha}} \tag{3.157}$$

as the restriction on admissible states $\mathcal{G}$ in the framework with internal state variables $\boldsymbol{\alpha}$.

However, this is not a 'proof' of these restrictions on admissible states since the status of (3.156) as a condition necessary for material stability is not fully clear.

In an admissible ($f_K \leq 0$) equilibrium state, on account of (3.156) we may restrict attention to the directions of straining that make the expression (3.154) equal to zero. That is, all the conditions in the basic rule (3.85) are now assumed to hold *in the examined equilibrium state*, but the consistency conditions (3.89) may be violated. This means that any vector $[\mathring{\gamma}]$ of non-negative components represents an admissible perturbation (recall that $[\mathring{\gamma}]$ denotes a vector $(\mathring{\gamma}_K)$ with $K \in \mathcal{P}$), while $\mathring{\gamma}_K$ is assumed to vanish if $f_K < 0$. According to the discussion in Sect. 3.3.4, we consider in this way directional stability of equilibrium for *arbitrarily small* perturbing forces $\pi_K^{\rm dist}$ (if $K \in \mathcal{P}$) that have zero values at equilibrium and vary proportionally in time (to first order) with non-zero *rates*.

Since (3.156) holds universally in all admissible equilibrium states, the condition sufficient for directional stability of equilibrium takes the form (3.153) with

$$\frac{1}{V}(\ddot{W} - \ddot{W}^{\rm ext}) = \ddot{w}^{\rm dist} = -\sum_K \dot{f}_K \mathring{\gamma}_K - \sum_K f_K \frac{d}{dt}\mathring{\gamma}_K\,. \tag{3.158}$$

At equilibrium we have $f_K \frac{d}{dt}\mathring{\gamma}_K \leq 0$ since (3.85) holds initially and $\mathring{\gamma}_K \geq 0$ at subsequent instants. The value of $\dot{f}_K$ is determined from $(3.86)_2$ or $(3.87)_2$ for $\dot{\mathbf{e}} = \mathbf{0}$ or $\dot{\mathbf{t}} = \mathbf{0}$, respectively. In that way from (3.153) we arrive at the following condition sufficient for directional stability of equilibrium with respect to spontaneous internal rearrangements :

$$\sum_{K,L\in\mathcal{P}} g_{KL}\mathring{\gamma}_K\mathring{\gamma}_L > 0 \quad \text{for every } [\mathring{\gamma}] \geq [0],\ [\mathring{\gamma}] \neq [0] \quad \text{at fixed } \mathbf{e}\,, \tag{3.159}$$

or

$$\sum_{K,L\in\mathcal{P}} h_{KL}\mathring{\gamma}_K\mathring{\gamma}_L > 0 \quad \text{for every } [\mathring{\gamma}] \geq [0],\ [\mathring{\gamma}] \neq [0] \quad \text{at fixed } \mathbf{t}\,. \tag{3.160}$$

The inequalities between vectors should be read componentwise (for each $K \in \mathcal{P}$). Under the assumptions introduced, the condition (3.153) is equivalent to (3.159) or (3.160).

The inequalities (3.159) and (3.160) as conditions for material stability at finite strain against spontaneous internal rearrangements under fixed strain or stress were derived in [55] and [59], respectively. A precursor to (3.159) in the formalism with internal state variables reads (cf. [40])

$$\dot{\alpha}\cdot D_{,\alpha}(\alpha,\dot{\alpha}) + \dot{\alpha}\cdot\phi_{,\alpha\alpha}\cdot\dot{\alpha} > 0 \qquad \text{for all } \dot{\alpha} \in \mathcal{L}(\mathcal{G})\,, \quad \dot{\alpha} \neq \mathbf{0}\,. \tag{3.161}$$

It is pointed out that in distinction to (3.161), the stability conditions (3.159) and (3.160) can be used within the purely mechanical, phenomenological framework of multi-mode plasticity presented in Sect. 3.3.3.2, and their derivation does not require specific reference to internal state variables.

The inequalities in (3.159) and (3.160) appeared earlier in the literature on crystal plasticity in a different context, cf. [20]. It should be noted that while the inequality in (3.159) holds either for all or for none of work-conjugate pairs $(\mathbf{t}, \mathbf{e})$ of stress and strain measures, this is not so for the inequality in (3.160) involving h_{KL}. This can be seen from the transformation formulae (3.97) and (3.98) given in Sect. 3.3.3.3. The lack of invariance of (3.160) is related to the dependence of stability on *which* stress measure is kept constant. In spite of this, for a reasonable choice of the measures, the condition (3.160) is acceptable for effectively hardening solids, for instance, it holds automatically if all h_{KL} are positive. The stronger condition of positive definiteness of the matrix (h_{KL}) fails in crystals when effective latent hardening of the slip systems is greater than their self-hardening. The invariant condition imposed on g_{KL} in (3.159) is usually still less restrictive than (3.160) and refers to the most rigid boundary constraints. The inequality (3.159) may thus be regarded as the condition for intrinsic (thermodynamic) stability of the material with respect to internal microstructural rearrangements.

It should be added that (3.159) is sufficient for the existence of a solution to the constitutive rate-problem (P1) of multi-mode plasticity, which follows from the mathematical theory of the linear complementarity problem, cf. [9]. Similarly, (3.160) is sufficient for the existence of a solution to the dual rate-problem $(\text{P1})^{\text{dual}}$ at given $\dot{\mathbf{t}}$.

3.4.3 Dynamic Localization of Deformation

Intrinsic stability of equilibrium in a homogeneous material element in a uniform state of deformation and stress is examined below under fixed *overall* deformation gradient. In particular, attention is primarily restricted to the case of fixed displacements over the element boundary. The restriction to strain uniformity within the element, adopted in Sect. 3.4.2, is now removed, and the possibility of dynamic instability of equilibrium due to spontaneous localization of deformation is investigated.

Suppose that instability of equilibrium with respect to internal structural rearrangements has been excluded. We shall assume that constitutive equations for the material have been transformed to the form (3.83). This could have been done by transforming an objective constitutive law (3.81) or (3.82) with the help of the transformation formulae from Sect. 3.3.1 and 3.3.3.3. Starting from multi-mode plasticity or from a thermodynamic formalism, that transformation could be preceded by elimination of the inelastic multipliers or rates of internal variables.

No perturbations of the constitutive equations are now admitted, so that $\dot{w}^{\text{dist}} \equiv 0$. Rather, arbitrarily small perturbing body forces are considered in connection with the possibility of spontaneous localization of deformation. The corresponding virtual deformation modes within a material element are non-uniform and are described by a velocity field $\tilde{\mathbf{v}}$ considered in the *reference* configuration.

3.4.3.1 Exclusion condition in integral form

Consider a material element of a reference volume V and with fixed displacements over its boundary ∂V. Under the assumption that a velocity field $\tilde{\mathbf{v}}$ within the element is continuous and piecewise continuously differentiable, along a non-uniform deformation path of departure from equilibrium we have

$$\dot{W} - \dot{W}^{\mathrm{ext}} = \dot{W}(\tilde{\mathbf{v}}) - 0 = \int_V \mathbf{S} \cdot \dot{\mathbf{F}}\, dV \qquad \text{for } \mathbf{v} = \mathbf{0} \text{ over } \partial V\,, \tag{3.162}$$

with $\dot{\mathbf{F}} = (\bar{\nabla}\mathbf{v})\mathbf{F} = \nabla\mathbf{v}$, the velocity gradient in the reference configuration. By the divergence theorem, $\int_V \mathbf{S}^0 \cdot \ddot{\mathbf{F}}\, dV = 0$ for an equilibrium stress $\mathbf{S}^0$ and under the boundary constraint $\dot{\mathbf{v}} = 0$. On using this after differentiating (3.162), the second-order condition (3.153) of directional stability of equilibrium takes the form (cf. [21])

$$\int_V \dot{\mathbf{S}} \cdot \dot{\mathbf{F}}\, dV > 0 \qquad \text{for all } \tilde{\mathbf{v}} \neq \tilde{\mathbf{0}} \text{ with } \mathbf{v} = \mathbf{0} \text{ over } \partial V\,. \tag{3.163}$$

Here and in the following, $\dot{\mathbf{S}}$ is assumed to be constitutively related to $\dot{\mathbf{F}}$ through (3.83).

If we start from the stronger stability condition (3.150) with

$$d^2 = \int_V \Delta\mathbf{F} \cdot \Delta\mathbf{F}\, dV\,, \qquad a(d) = \frac{1}{2} b d^2\,, \qquad b > 0 \tag{3.164}$$

then we arrive at the following condition sufficient for *directional* stability of equilibrium with respect to the distance d :

$$\int_V \dot{\mathbf{S}} \cdot \dot{\mathbf{F}}\, dV \geq \int_V b\dot{\mathbf{F}} \cdot \dot{\mathbf{F}}\, dV \qquad \text{for all } \tilde{\mathbf{v}} \text{ with } \mathbf{v} = \mathbf{0} \text{ over } \partial V\,. \tag{3.165}$$

It is pointed out again that this condition without additional assumptions does not necessarily ensure stability for arbitrarily circuitous paths. For instance, (3.149) is always violated on certain deformation cycles within a continuum obeying a non-associative plasticity law (3.84) with different directions of $\bar{\boldsymbol{\lambda}}^{\mathrm{p}}$ and $\bar{\boldsymbol{\lambda}}$ [48].

3.4.3.2 Local condition for stability of equilibrium

Suppose that $\dot{\mathbf{F}}$ is effectively concentrated in a disk-like volume B of predominantly uniform thickness h which is small compared with the dimensions of the element (Fig. 3.5). This is possible if $\nabla\mathbf{v} = \mathbf{g} \otimes \mathbf{n}$ within the disk, where $\mathbf{n}$ is the normal to the disk mid-plane in the reference configuration and $\mathbf{g}$ is an arbitrary vector. In this case, the velocity gradient outside the disk can be made arbitrarily small as $h \to 0$ at given $\dot{\mathbf{F}}$ within the disk. (Any other form of $\dot{\mathbf{F}}$ in B gives a finite contribution $\dot{\mathbf{F}}\mathbf{s}$ to the velocity gradient *outside* the disk in some direction $\mathbf{s}$ tangent to the disk boundary, by continuity of $\mathbf{v}$). By proceeding as in [23, 52], from (3.163) in the limit as $h \to 0$ we obtain[10]

$$\dot{\mathbf{S}} \cdot \dot{\mathbf{F}} = (\mathbf{g} \otimes \mathbf{n}) \cdot \mathsf{C}(\mathbf{g} \otimes \mathbf{n}) \cdot (\mathbf{g} \otimes \mathbf{n}) \geq 0 \qquad \text{for all } \dot{\mathbf{F}} = \mathbf{g} \otimes \mathbf{n}\,. \tag{3.166}$$

Similarly, as a consequence of (3.165) we obtain

$$\dot{\mathbf{S}} \cdot \dot{\mathbf{F}} = (\mathbf{g} \otimes \mathbf{n}) \cdot \mathsf{C}(\mathbf{g} \otimes \mathbf{n}) \cdot (\mathbf{g} \otimes \mathbf{n}) > 0 \qquad \text{for all } \dot{\mathbf{F}} = \mathbf{g} \otimes \mathbf{n} \neq \mathbf{0} \tag{3.167}$$

as the exclusion condition for spontaneous dynamic localization of deformation in a *vanishingly thin* band within a finite material element (or in a band of finite thickness in an infinite continuum).

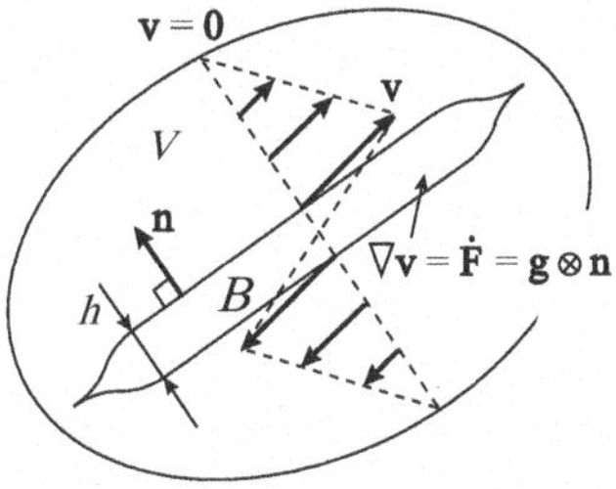

Fig. 3.5 A perturbed velocity field corresponding to incipient localization of deformation

It is clear that in the above conditions refer only to the symmetric part $\frac{1}{2}(\mathsf{C} + \mathsf{C}^{\mathrm{T}})$ of the moduli tensor C. The condition (3.166) is *distinct* from the *semi-strong ellipticity* condition, which has an analogous form but with the difference that a moduli tensor C is independent of $\mathbf{n}, \mathbf{g}$. However, in the special case of classical elastoplasticity with two linear constitutive branches as in (3.84), the condition (3.166) *implies* (and is obviously implied by) semi-strong ellipticity of the moduli from each branch. For, the sense of a velocity gradient $\mathbf{g} \otimes \mathbf{n}$ can always be adjusted to fall within a chosen constitutive half-space. In conclusion, semi-strong ellipticity of the incremental stiffness moduli from both constitutive branches in the classical elastoplasticity follows as a consequence of the restricted form (3.163) of Hill's condition of stability of equilibrium [54, 70].

Similar remarks apply to the relation between (3.167) and the condition of *strong ellipticity*.

From (3.80) it follows that the second-order effect of geometric changes should not be neglected *in advance* when evaluating the above left-hand expression. Nevertheless, with the help of (3.73) or (3.80) it is easy to show that (3.167) is equivalent to

$$\overset{\circ}{\sigma} \cdot \mathbf{D} > 0 \qquad \text{for all } \dot{\mathbf{F}} = \mathbf{g} \otimes \mathbf{n} \neq \mathbf{0}\,. \tag{3.168}$$

[10] The implication (3.163) $\Longrightarrow$ (3.166) follows also from the theorem due to Graves [18], as noted in [47].

In contrast to the Drucker condition (3.75), the constraint on the velocity gradient in (3.168) provides a definite coupling between the material spin and strain-rate, removing thus ambiguities raised when the material spin is unspecified.

3.4.3.3 Condition sufficient for instability

Fulfillment of the condition

$$\dot{\mathbf{S}}\mathbf{n} = \mu \mathbf{g} \qquad \text{for some } \dot{\mathbf{F}} = \mathbf{g} \otimes \mathbf{n} \neq \mathbf{0}\,,\ \mu < 0\,. \tag{3.169}$$

may be regarded as sufficient for the local instability of equilibrium, related to spontaneous dynamic localization of deformation into a planar band in a homogeneous, uniformly stressed infinite continuum. This was demonstrated by Mandel [38] for an incrementally *linear* material in the small-strain framework, with application to classical elastoplasticity without normality. An extension to an incrementally nonlinear material at finite deformations was given in [52][11] in the potential case (3.117) with the left-hand side in (3.169) written as $\partial U(\mathbf{g} \otimes \mathbf{n})/\partial \mathbf{g}$. The present extension to non-symmetric $\mathbf{C}$ is immediate.

To see better both the analogies and differences between the linear and nonlinear cases, (3.169) can be rewritten as[12]

$$\dot{\mathbf{S}}\mathbf{n} \equiv \mathbf{Q}(\mathbf{n}, \dot{\mathbf{F}})\mathbf{g} = \mu \mathbf{g} \qquad \text{for some } \dot{\mathbf{F}} = \mathbf{g} \otimes \mathbf{n} \neq \mathbf{0}\,,\ \mu < 0\,, \tag{3.170}$$

where $\mathbf{Q}$ of components

$$Q_{ik}(\mathbf{n}, \dot{\mathbf{F}}) = C_{ipkq}(\dot{\mathbf{F}}) n_p n_q \tag{3.171}$$

is known as the (elastoplastic) 'acoustic tensor'. In contrast to the linear case, here $\mathbf{Q}$ is generally dependent on $\dot{\mathbf{F}}$, and thus on both $\mathbf{n}$ and $\mathbf{g}$, due to the dependence of $\mathbf{C}$ on the direction of strain-rate. The equation (3.170) represents thus a *nonlinear* eigenvalue problem, while in the classical elastoplasticity (3.170) reduces to a *linear* eigenvalue problem. The distinction is essential; for instance, the critical condition associated with (3.170) for some $\mathbf{n}$ is *not* the same as $\det \mathbf{Q}(\mathbf{n}, \mathbf{g} \otimes \mathbf{n}) = 0$ for a given $\mathbf{g}$, since the vector $\mathbf{g}$ and an eigenvector $\mathbf{g}^*$ corresponding to the zero eigenvalue of $\mathbf{Q}(\mathbf{n}, \mathbf{g} \otimes \mathbf{n})$ need not coincide. The critical condition associated with (3.170) is

$$\dot{\mathbf{S}}\mathbf{n} = \mathbf{Q}(\mathbf{n}, \mathbf{g} \otimes \mathbf{n})\mathbf{g} = \mathbf{0} \qquad \text{for some } \dot{\mathbf{F}} = \mathbf{g} \otimes \mathbf{n} \neq \mathbf{0}\,, \tag{3.172}$$

which corresponds to the quasi-static bifurcation in a band under the restriction to rigid-body motion outside the band [35]. A more general case of bifurcation in a band will be examined in Sect. 3.5.3 and 3.6.3.

[11] As mentioned in the reference, the proof is not fully rigorous, not even under the assumptions of the first approximation, since a rigid body motion outside the band is not a free inertial motion. However, the meaning of instability associated with the fulfillment of (3.169) seems to be clear.

[12] The left-hand expression in (3.166) and (3.167) takes the form $\mathbf{g}\mathbf{Q}(\mathbf{n}, \mathbf{g} \otimes \mathbf{n})\mathbf{g}$.

For a class of incrementally nonlinear models proposed for geomaterials, the condition (3.172) has been found to be the most critical in the sense that it can be met earlier along a deformation path than any other bifurcation in a band (Chambon et al., 2000).

Note that (3.172) *differs* from the critical condition $\mathbf{Q}_s(\mathbf{n}, \mathbf{g} \otimes \mathbf{n})\mathbf{g} = \mathbf{0}$ associated with the failure of (3.167), unless $\mathbf{Q}$ is symmetric, i.e. $\mathbf{Q} = \mathbf{Q}_s$. Failure of (3.167) along a deformation path cannot appear *later* than fulfillment of (3.172) since (3.172) obviously excludes (3.167). However, failure of (3.167) without satisfying (3.170) or (3.172) has not been shown, to the author's knowledge, to imply a dynamic instability of equilibrium for arbitrarily small disturbances.

Material instability associated with fulfillment of (3.170) can also be seen by comparison with the condition of propagation of an *acceleration wave* into *non-deforming* material [23, 52]:

$$\dot{\mathbf{S}}\mathbf{n} = \mathbf{Q}(\mathbf{n}, \dot{\mathbf{F}})\mathbf{g} = \rho c^2 \mathbf{g} \qquad \text{for } \dot{\mathbf{F}} = \mathbf{g} \otimes \mathbf{n} \neq \mathbf{0}, \tag{3.173}$$

where $\dot{\mathbf{F}}$ represents the velocity gradient on the deforming side of the discontinuity surface propagating with normal speed c into the material being at rest, and ρ is the mass density in the reference configuration. It follows that the normal speed of propagation of all acceleration waves into the material being at rest is real (positive or zero) only if (3.170) is excluded.

3.4.4 Stability Under Deformation-Sensitive Loading

Examine now the kind of instability of equilibrium, announced in Sect. 3.1.3, which may appear when the environment of a stressed material element is not sufficiently 'stiff', also when (3.159) and (3.165) are satisfied. Attention is restricted again to uniform straining within the element, and for simplicity the *rotational* equilibrium is assumed to be not disturbed. The constitutive equations for external loading are assumed in the spin-invariant form (3.144), which at constant λ reduces to $\dot{\mathbf{t}}^{\mathrm{ext}} = -\mathbf{L}^* \cdot \dot{\mathbf{e}}$. This includes, of course, the particular case when *some* measure of external stress $\mathbf{t}^{\mathrm{ext}}$ is fixed.

Stability of an equilibrium state at $\mathbf{t} = \mathbf{t}^{\mathrm{ext}}$ is investigated, assuming that $\dot{w}^{\mathrm{dist}} \equiv 0$. The constitutive rate equations for the material are taken in the form (3.82). Along a path of departure from equilibrium we now have

$$\frac{1}{V}(\dot{W} - \dot{W}^{\mathrm{ext}}) = \dot{w} - \dot{w}^{\mathrm{ext}} = (\mathbf{t} - \mathbf{t}^{\mathrm{ext}}) \cdot \dot{\mathbf{e}}\,. \tag{3.174}$$

The second-order condition (3.153) sufficient for directional stability of equilibrium takes the form

$$(\dot{\mathbf{t}} - \dot{\mathbf{t}}^{\mathrm{ext}}) \cdot \dot{\mathbf{e}} \equiv \dot{\mathbf{e}} \cdot (\mathsf{L}(\dot{\mathbf{e}}) + \mathsf{L}^*) \cdot \dot{\mathbf{e}} > 0 \qquad \text{for all } \dot{\mathbf{e}} \neq \mathbf{0}\,. \tag{3.175}$$

If $\mathbf{t}^{\text{ext}}$ is fixed then (3.175) reduces to (3.77). However, as already stated, the inequality (3.77) is not measure-invariant, and under the transition to another work-conjugate measures we must return to the more general form (3.175) [28]. From (3.96) and (3.145) it follows that (3.175) is measure-invariant as required.

If L and L^* are diagonally symmetric and, moreover, the second-order work inequality (3.138) holds then (3.175) becomes sufficient for material stability of the type examined here also for arbitrary circuitous paths.

Exercise: Prove the above statement by showing validity of (3.150) with $d = |\Delta\mathbf{e}|$.

3.5 Instability of a Quasi-Static Process

3.5.1 The Energy Criterion

Henceforth, we shall examine possible instability of a quasi-static deformation *process* (or *path*, in equivalent terminology). It was already shown by the illustrative example in Sect. 2 that the process or path going through stable equilibrium states may itself be unstable. The *fundamental process* (path) whose stability is to be studied is regarded as known, and all corresponding quantities are distinguished by a superscript '0'. The fundamental path is obtained as a quasi-static solution for a continuous solid body subjected to a loading program parameterized by a scalar loading parameter $\lambda = \lambda(t)$ (which need not be a monotonic function).

Throughout this Sect. 3.5, time-independent constitutive equations given in Sect. 3.3.3 are assumed jointly with *the symmetry and normality conditions* from Sect. 3.3.5.1. Since the normality rule may be regarded as a consequence of symmetry of incremental stiffness moduli if a single internal mechanism of plastic deformation is operating, the symmetry conditions are in effect most important. A thermodynamic argument supporting the symmetry assumptions was presented in Sect. 3.3.5.2.

Moreover, the incremental loading is assumed to be *conservative* in an overall sense, implying that a potential energy $\Omega(\tilde{\mathbf{u}}, \lambda)$ of the loading device can be defined (at least to second order) as a functional of the displacement field $\tilde{\mathbf{u}}$ and λ. For the external loading from Sect. 3.3.6, this reduces to another symmetry restriction $\mathsf{C}^* = \mathsf{C}^{*\mathrm{T}}$.

Under these symmetry assumptions, the energy criterion of stability of equilibrium can be extended to deformation processes. [13] For this purpose, *the energy functional* [46]

$$E = E^{\text{suppl}} + W^{\text{dist}} = W + \Omega \tag{3.176}$$

is defined for a class of kinematically admissible processes of isothermal deformation. An increment ΔE in the value of E represents the amount of energy to be supplied from external sources to the mechanical *system*, consisting of the deformed

[13] This concept is described here only in outline; for more details, see [57] and the relevant references therein.

body *and* the loading device, in order to produce quasi-statically a deformation increment, generally with the help of additional disturbing forces. In the former decomposition of E, E^{suppl} is the energy supplied from external sources *to the loading device*, and W^{dist} is the work done directly by all disturbing forces. In turn, $E^{\mathrm{suppl}} = \Omega + W^{\mathrm{ext}}$, where the part Ω is stored in the loading device and the part W^{ext} is transmitted to the deformed body. The mechanical energy balance involved in (3.176) is completed by the equality $W^{\mathrm{dist}} = W - W^{\mathrm{ext}}$, where W is the work to be supplied to the deforming body according to the constitutive law, as defined by (3.147). It is not essential here what part of E is stored in the material, dissipated, or transformed into heat. A further insight into the meaning of E is gained if the mechanical system is imagined to be placed in a heat reservoir. Then an increment of E can be identified with a change in the internal energy of the resulting compound thermodynamic system {deforming body + loading device + heat reservoir}, cf. [55]. The choice of the system is not arbitrary: at a constant loading parameter λ and in the absence of disturbances, the compound thermodynamic system can be treated as isolated.

Under the symmetry assumptions indicated above, the following energy criterion has been proposed by Petryk [46, 47] for concluding whether a quasi-static process of elastoplastic deformation can be stable or not.

The energy criterion of path instability *In a stable fundamental process of quasi-static deformation, the increment of E calculated with accuracy to second-order terms is minimized within the class of all kinematically admissible deformation increments*, viz.

$$\Delta E \geq \Delta E^0 \qquad \text{(to second order)} \bullet \tag{3.177}$$

This may be regarded as a specification of the intuitive engineering hypothesis that a real deformation mode exhibits a tendency to minimize the energy consumption.

Justification of the criterion by relating it to a necessity condition for stability (and a sufficiency condition for instability) in the first approximation against persistent disturbances was given in [51] for the constitutive law (3.117), and extended to perturbations of thermodynamic forces in multi-mode plasticity in [59]. Consistency of the criterion with Hill's bifurcation theory was shown earlier in [47] (cf. also [54], and with the initial imperfection approach in [57]). Thermodynamic aspects of the criterion were discussed in [55] and [15]. Other relevant references can be found in these cited above.

From the virtual work principle it follows that in an equilibrium state where all disturbing forces vanish and are not discontinuous in time, $\dot{E} = \dot{E}^0$ identically. Accordingly, the above criterion can be reformulated as follows: *In a stable fundamental process of quasi-static deformation,*

$$\ddot{E} \geq \ddot{E}^0 \tag{3.178}$$

for every admissible direction of a deformation increment.

Detailed specification of (3.177) and (3.178) for material instability problems will be presented below. Fulfillment of (3.178) has been shown to imply that the fundamental velocity field represents automatically a quasi-static solution. This essential conclusion requires the symmetry assumptions introduced above. In turn, not every quasi-static solution minimizes $\ddot{E}$. It follows that if a solution is not unique then (3.178) can provide a criterion of choice within the solution class, which is lacking in the bifurcation theory alone. The energy criterion (3.178) constitutes a basis for the computational method proposed by Petryk and Thermann [62] where a minimization procedure is used to exclude unstable solution paths in an automatic way.

The above energy criterion can be used to examine stability of a degenerate deformation process in which an equilibrium state does not undergo any changes. This is so if the loading parameter λ is kept constant in an arbitrarily long time interval, which can always be assumed on account of the freedom in defining a time-like parameter t for quasi-static deformation of a time-independent material. Then $\Delta E^0 = 0$, $\Delta E^{\text{suppl}} = 0$, $\Delta \Omega = -\Delta W^{\text{ext}}$ and (3.177) reduces to

$$\Delta E = \Delta W - \Delta W^{\text{ext}} \geq 0 \qquad \text{at fixed } \lambda \tag{3.179}$$

within that interval, which is nothing else than the energy criterion (3.149) of stability of *equilibrium* in a somewhat weakened form. Under the symmetry assumptions and certain additional reservations, the weakened form (3.179) may be treated as a *necessity* condition for stability [51], while (3.149) was rather interpreted as a *sufficiency* condition for directional stability. Nevertheless, the critical state associated with both conditions may be expected to be the same in sufficiently regular processes. In conclusion, the condition (3.177) may be regarded as an extension of the energy criterion of stability of equilibrium to quasi-static deformation processes.

3.5.2 Instability of a Process of Internal Rearrangements

This section represents in a sense a continuation of Sect. 3.4.2. Here, as the object of stability investigations we take the fundamental *process* of internal rearrangements within a material element of reference volume V under the assumption of uniform macroscopic strain. Two cases are examined when the process is induced by prescribed variations in time of either strain $\mathbf{e}(\lambda)$ or a certain measure of stress $\mathbf{t} \equiv \mathbf{t}^{\text{ext}}(\lambda)$, which corresponds to $\Omega = 0$ or $\Omega = -V\mathbf{t}^{\text{ext}} \cdot \mathbf{e}$, respectively. The study in Sect. 3.4.2 corresponded to λ kept fixed; here $\lambda = \lambda(t)$ is a given function. The framework of multi-mode plasticity from Sect. 3.3.3.2 is adopted, with the normality (3.115) and symmetry (3.118) assumptions. The rates $(\dot{\mathbf{e}}^0, \dot{\mathbf{t}}^0, \mathring{\gamma}^0_K)$ in fundamental process are obtained as solutions to (P1) or $(\text{P1})^{\text{dual}}$ at every instant.

For any other admissible mode of deformation within the element at a micro-level in the presence of disturbances of thermodynamic forces, the rate of the energy functional (3.176) is specified as follows

$$
\begin{aligned}
\dot{E}/V &= \dot{w} + \dot{w}^{\text{dist}} + \dot{\Omega}/V \\
&= \mathbf{t}\cdot\dot{\mathbf{e}} - \sum_K f_K \mathring{\gamma}_K \qquad \text{for controlled } \mathbf{e}, \\
&= -\dot{\mathbf{t}}\cdot\mathbf{e} - \sum_K f_K \mathring{\gamma}_K \qquad \text{for controlled } \mathbf{t}.
\end{aligned}
\tag{3.180}
$$

Stability of the fundamental process can be investigated in a neighborhood of any given equilibrium state $\mathcal{G}_0$ reached at some λ. To establish *necessary* conditions for path stability, consider deviations from the fundamental process along direct paths initiated in all possible directions corresponding to infinitesimal disturbing forces. Such directions are described by vectors $[\mathring{\gamma}]$ (of non-negative components $\mathring{\gamma}_K$ with $K \in \mathcal{P}$), while $\mathring{\gamma}_K$ vanishes if $f_K(\mathcal{G}_0) < 0$, also at subsequent instants. Then $\dot{E} = \dot{E}^0$ in an equilibrium state $\mathcal{G}_0$.

By differentiating (3.180) outside equilibrium, substituting the constitutive rate equations (3.86) or (3.87) along with the normality rule (3.115), and comparing the terms in the equilibrium state, we obtain the identity

$$
\frac{1}{2V}(\ddot{E} - \ddot{E}^0) = \mathcal{J}([\mathring{\gamma}]) - \mathcal{J}([\mathring{\gamma}^0]) \qquad \text{for all } [\mathring{\gamma}], [\mathring{\gamma}^0] \quad \text{in } \mathcal{G}_0, \tag{3.181}
$$

where

$$
\mathcal{J}([\mathring{\gamma}]) = \begin{cases} \dfrac{1}{2}\displaystyle\sum_{K,L\in\mathcal{P}} g_{KL}\mathring{\gamma}_K\mathring{\gamma}_L - \dot{\mathbf{e}}\cdot\sum_{K\in\mathcal{P}} \boldsymbol{\lambda}_K\mathring{\gamma}_K & \text{for given } \dot{\mathbf{e}}, \\ \dfrac{1}{2}\displaystyle\sum_{K,L\in\mathcal{P}} h_{KL}\mathring{\gamma}_K\mathring{\gamma}_L - \dot{\mathbf{t}}\cdot\sum_{K\in\mathcal{P}} \boldsymbol{\mu}_K\mathring{\gamma}_K & \text{for given } \dot{\mathbf{t}}. \end{cases} \tag{3.182}
$$

The symmetry condition (3.118) and the fact that $[\mathring{\gamma}^0]$ is a solution to (P1) or $(\text{P1})^{\text{dual}}$ are not needed for validity of (3.181).

From (3.178) and (3.181) we arrive at the energy condition for stability of the fundamental process of internal rearrangements described by $\mathring{\gamma}^0_K(t)$, in the form

$$
\mathcal{J}([\mathring{\gamma}]) \geq \mathcal{J}([\mathring{\gamma}^0]) \qquad \text{for all } [\mathring{\gamma}] \geq [0]. \tag{3.183}
$$

The interpretation of (3.183) as a condition *necessary* for stability of a process is supported by the following result proven in [59].

Theorem (on instability of a process of internal rearrangements) *Suppose that along a segment of a solution path* $(\dot{\mathbf{e}}^0, \mathring{\gamma}^0_K)(t)$ *the condition (3.183) is not satisfied while the condition (3.159) or (3.160) of stability of equilibrium holds.*

Then at every point of that segment there exists another solution to (P1) or $(P1)^{\text{dual}}$*, respectively, that is energetically preferable to* $[\mathring{\gamma}]^0$ *in the sense of assigning to* $\ddot{E}$ *a minimum value* •

For any solution $[\mathring{\gamma}]$ to (P1) or $(\text{P1})^{\text{dual}}$, the expression for $\mathcal{J}([\mathring{\gamma}])$ simplifies on account of $\dot{f}_K\mathring{\gamma}_K = 0$ to :

$$\mathcal{J}([\mathring{\gamma}]) = \begin{cases} \frac{1}{2}\dot{\mathbf{t}}^{\mathrm{p}} \cdot \dot{\mathbf{e}} & \text{if } [\mathring{\gamma}] \text{ is a solution to (P1) for controlled } \mathbf{e}\,, \\ -\frac{1}{2}\dot{\mathbf{t}} \cdot \dot{\mathbf{e}}^{\mathrm{p}} & \text{if } [\mathring{\gamma}] \text{ is a solution to (P1)}^{\mathrm{dual}} \text{ for controlled } \mathbf{t}\,, \end{cases} \tag{3.184}$$

with $\dot{\mathbf{t}}^{\mathrm{p}}$, $\dot{\mathbf{e}}^{\mathrm{p}}$ defined as in (3.114).

Hence, among all solutions to (P1) or (P1)^{dual}, the particular solution $[\mathring{\gamma}^0]$ which satisfies (3.183) also:

minimizes $\dot{\mathbf{t}} \cdot \dot{\mathbf{e}}$ (and $\dot{\mathbf{t}}^{\mathrm{p}} \cdot \dot{\mathbf{e}}$) for controlled $\mathbf{e}$,
maximizes $\dot{\mathbf{t}} \cdot \dot{\mathbf{e}}$ (and $\dot{\mathbf{t}} \cdot \dot{\mathbf{e}}^{\mathrm{p}}$) for controlled $\mathbf{t}$.

This corresponds to the criterion of selection of an 'optimal' solution which was proposed by Franciosi and Zaoui [17] in relation to the energy criterion but without assuming the symmetry property (3.118).

A necessary condition for $[\mathring{\gamma}^0]$ to satisfy (3.183) is that

$$\begin{aligned} &\sum_{K,L\in\mathcal{A}^0} g_{KL} w_K w_L \geq 0 && \text{for controlled } \mathbf{e}, \\ &\qquad\qquad \text{for every } [w],\ \mathcal{A}^0 = \{K : \mathring{\gamma}^0_K > 0\} \\ &\sum_{K,L\in\mathcal{A}^0} h_{KL} w_K w_L \geq 0 && \text{for controlled } \mathbf{t}\,. \end{aligned} \tag{3.185}$$

More details can be found in [59]. We quote a simple illustrative example of double slipping in a single crystal with the effective hardening matrix for a selected pair of stress and strain measures given by

$$\begin{pmatrix} h_{11} & h_{12} \\ h_{21} & h_{22} \end{pmatrix} = \begin{pmatrix} 1 & 1+q \\ 1+q & 1 \end{pmatrix} h\,, \qquad h > 0, \qquad q > 0. \tag{3.186}$$

It is clear that the conditions (3.160) and (3.159) (for $\mathbf{L}^{\mathrm{e}}$ positive definite) of stability of equilibrium under both strain and stress control are satisfied, although the matrix (h_{KL}) is indefinite. The process of uniform double slipping is unstable under stress control according to the condition $(3.185)_2$, and may be stable under strain control. According to the Theorem quoted above, an incremental solution under stress control exists which does satisfy (3.183), and in the loading case this will be single slipping $\mathring{\gamma}^0_1 > 0$, $\mathring{\gamma}^0_2 = 0$ or $\mathring{\gamma}^0_1 = 0$, $\mathring{\gamma}^0_2 > 0$.

3.5.3 Instability of Uniform Straining

In this section, the energy criterion of intrinsic stability of equilibrium, examined in Sect. 3.4.3, is extended to a quasi-static *process* of uniform straining and related to the bifurcation approach. A potential constitutive law (3.117) is assumed. Stability of the fundamental process of uniform deformation within a homogeneous material element is examined under controlled *overall* deformation gradient varying in time, so that $\Delta\Omega \equiv 0$. In particular, the case of a displacement history prescribed over the element boundary is considered. In comparison with Sect. 3.5.2, the restriction to strain uniformity within the material element is now removed, and path instability associated with spontaneous deviation from uniform straining is studied.

3.5.3.1 The energy criterion of instability of uniform straining

The integral conditions of path stability that result from the energy criterion formulated in the Sect. 3.5.1 have been discussed in detail elsewhere [47, 51, 54, 55]. Following [52], as the basic condition necessary for stability of the fundamental, uniform deformation process $\dot{\mathbf{F}}^0(t)$ within a homogeneous material element we take

$$\int_V U(\dot{\mathbf{F}}^0 + \nabla \mathbf{w})\,\mathrm{d}V \geq V\,U(\dot{\mathbf{F}}^0) \qquad \text{for every } \widetilde{\mathbf{w}} : \ \mathbf{w} = \mathbf{0} \text{ over } \partial V \tag{3.187}$$

which is a consequence of (3.178). Here, $\tilde{\mathbf{w}}$ belongs to the space of continuous and piecewise smooth vector fields, and the integration domain may be chosen arbitrarily. (3.187) is called the condition of *quasi-convexity of* $U(\dot{\mathbf{F}})$ *at* $\dot{\mathbf{F}}^0$.

From the Green theorem and the boundary condition $\mathbf{w} = \mathbf{0}$ over ∂V it follows that

$$\int_V U(\dot{\mathbf{F}}^0 + \nabla \mathbf{w})\,\mathrm{d}V = V\,U(\dot{\mathbf{F}}^0) + \int_V \mathcal{E}(\dot{\mathbf{F}}^0, \dot{\mathbf{F}}^0 + \nabla \mathbf{w})\,dV\,, \tag{3.188}$$

where $\mathcal{E}$ is the Weierstrass function associated with the constitutive potential $U(\dot{\mathbf{F}})$:

$$\mathcal{E}(\dot{\mathbf{F}}, \dot{\mathbf{F}}^*) = U(\dot{\mathbf{F}}^*) - U(\dot{\mathbf{F}}) - \dot{\mathbf{S}} \cdot (\dot{\mathbf{F}}^* - \dot{\mathbf{F}})\,. \tag{3.189}$$

Let $\nabla \mathbf{w}$ be effectively concentrated in a disk-like volume of predominantly uniform thickness h, taken sufficiently small. As already discussed, such perturbation has a form $\nabla \mathbf{w} = \mathbf{g} \otimes \mathbf{n}$ which previously represented the velocity gradient itself (Fig. 3.5) and now it represents the *difference* between the perturbed and fundamental velocity gradients (Fig. 3.6). By using (3.188), in the limit as $h \longrightarrow 0$ as a consequence of (3.187) we obtain that

$$\mathcal{E}(\dot{\mathbf{F}}^0, \dot{\mathbf{F}}^0 + \mathbf{g} \otimes \mathbf{n}) \geq 0 \qquad \text{for all vectors } \mathbf{g}, \mathbf{n}\,, \tag{3.190}$$

that is, U must be *rank* 1 *convex at* $\dot{\mathbf{F}}^0$.[14] Hence, if (3.190) does not hold then the concentration of deformation within a band (or bands) of orientation $\mathbf{n}$ is shown to be *energetically preferable* to uniform straining with the velocity gradient $\dot{\mathbf{F}}^0$, in the sence of requiring a smaller value of ΔE (to second-order).

In the limit as $|\mathbf{g}||\mathbf{n}| \longrightarrow 0$, (3.190) implies semi-strong ellipticity of the current tangent moduli:

$$(\mathbf{g} \otimes \mathbf{n}) \cdot \mathbf{C}(\dot{\mathbf{F}}^0) \cdot (\mathbf{g} \otimes \mathbf{n}) \geq 0 \qquad \text{for every } \mathbf{g}, \mathbf{n}\,. \tag{3.191}$$

[14] The implication (3.187) $\Rightarrow$ (3.190) follows also as a corollary from the theorem due to Graves [18].

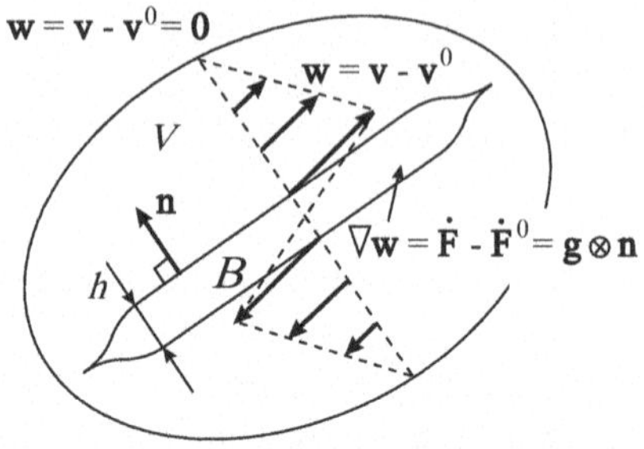

Fig. 3.6 A perturbation superimposed on the fundamental velocity field

If $|\mathbf{g}||\mathbf{n}| \longrightarrow \infty$ then as a consequence of (3.190) we obtain another condition [47]

$$U(\mathbf{g} \otimes \mathbf{n}) \geq 0 \qquad \text{for every } \mathbf{g}, \mathbf{n}, \tag{3.192}$$

which already appeared above as the condition (3.166) of stability of equilibrium.

The conditions (3.190), (3.191) and (3.192) are interpreted as *necessary* for stability of the fundamental process since they are derived from the global energy condition (3.178). This also makes explicit the connection between material instability and the instability in a finite solid body with specified boundary conditions.

The conditon (3.190) is necessary for the speed of dynamic propagation of all acceleration waves into the material *deforming with the rate* $\dot{\mathbf{F}}^0$ to be real [52]. In particular, fulfillment of (3.191) is necessary for this (recall that $\mathbf{C}$ is symmetric here). The distinction between the two conditions is that (3.190) concerns arbitrary acceleration waves, while (3.191) concerns only those acceleration waves which carry a jump in $\dot{\mathbf{F}}$ sufficiently small to remain on *both* sides of the discontinuity surface within the fundamental constitutive cone where $\mathbf{C}^0$ apply.

In turn, if (3.192) fails then the condition (3.169) for dynamic localization of deformation is satisfied. According to the discussion in Sect. 3.4.3.3, the condition (3.192) may thus be regarded as necessary for local stability of equilibrium under arbitrary boundary data. Moreover, the normal speed of propagation of all acceleration waves into the material being at rest is real (positive or zero) if and only if (3.192) holds.

Examine now the critical instants on a smooth deformation path beyond which the conditions (3.190), (3.191) or (3.192) fail.

The first critical instant is met when (3.190) still holds but with equality for some nonzero vectors $\mathbf{g}, \mathbf{n}$. Then the partial derivative of $\mathcal{E}(\dot{\mathbf{F}}^0, \dot{\mathbf{F}}^0 + \mathbf{g} \otimes \mathbf{n})$ with respect to $\mathbf{g}$ vanishes, which reads

$$\mathcal{E}_{,\mathbf{g}}(\dot{\mathbf{F}}^0, \dot{\mathbf{F}}^0 + \mathbf{g} \otimes \mathbf{n}) = (\dot{\mathbf{S}} - \dot{\mathbf{S}}^0)\mathbf{n} = \mathbf{0} \qquad \text{for some } \dot{\mathbf{F}} = \dot{\mathbf{F}}^0 + \mathbf{g} \otimes \mathbf{n} \neq \dot{\mathbf{F}}^0 \,. \tag{3.193}$$

This coincides with the condition for bifurcation in a band, to be discussed below.

Analogously, by appealing to the assumed symmetry of $\mathbf{C}$ and using the definition (3.171) of the acoustic tensor, the critical condition associated with (3.191) reads

$$\mathbf{Q}(\mathbf{n}, \dot{\mathbf{F}}^0)\mathbf{g} = \mathbf{0} \qquad \text{for some } \mathbf{n}, \mathbf{g} \neq \mathbf{0}\,, \tag{3.194}$$

which is the *linearized* condition for bifurcation in a band.

The critical condition associated with (3.192) coincides, for symmetric C, with (3.172) discussed in Sect. 3.4.3.

If the constitutive inequality (3.135) holds then (3.187) ⇔ (3.190) ⇔ (3.191) [52]. However, even then the condition (3.192) represents a different requirement, less restrictive on account of the implication (3.190) ⇒ (3.192). This illustrates once again the fundamental distinction between the instability conditions for a deformation process and for an equilibrium state.

In the special case of classical elastoplasticity, the distinction between (3.191) for all $\dot{\mathbf{F}}^0$ and (3.192) disappears, as already discussed in Sect. 3.4.3, and all the conditions necessary for local (intrinsic) stability of uniform straining reduce to the single requirement of semi-strong ellipticity of the tangent moduli.

3.5.3.2 Bifurcation in a band

To eliminate the influence of boundary conditions, a material element may also be treated, in another scale but under the current uniform stress, as an infinite homogeneous continuum. The loading conditions at infinity are imagined to be such that uniform and constitutively compatible quasi-static rate fields $\dot{\mathbf{F}}^0, \dot{\mathbf{S}}^0$ represent the fundamental solution. Then, non-trivial solutions that correspond to a material instability in the form of bifurcation within a planar band can be investigated. The gradient $\nabla\mathbf{w}$ of admissible variations in velocities (the meaning of $\mathbf{w}$ is analogous to that in Fig. 3.6) is assumed to vanish outside the band whose unit normal in the reference configuration is denoted by $\mathbf{n}$. Within the band, $\mathbf{w}$ is assumed constant on planes $\boldsymbol{\xi} \cdot \mathbf{n} = \text{const}$ parallel to the band boundaries. The traction rates applied along the band boundaries result from the uniform fundamental solution $(\dot{\mathbf{F}}^0, \dot{\mathbf{S}}^0)$. From the spatial continuity of $\widetilde{\mathbf{w}}$ and from the conditions of continuing equilibrium, we obtain the well-known kinematic and static conditions to be satisfied by a pair $(\mathbf{F}, \mathbf{S})$ within the bifurcation band (cf. [67]), viz.

$$(\dot{\mathbf{S}} - \dot{\mathbf{S}}^0)\mathbf{n} = \mathbf{0} \qquad \text{and} \qquad \dot{\mathbf{F}} - \dot{\mathbf{F}}^0 = \mathbf{g} \otimes \mathbf{n}\,, \tag{3.195}$$

respectively, where $\mathbf{g} \in \mathcal{R}^3$ is some function of $(\boldsymbol{\xi} \cdot \mathbf{n})$. Existence of a nontrivial solution to (3.195) is referred to as *bifurcation in a band*. The common term 'shear band bifurcation' is also suitable but it concerns, strictly speaking, the special case when $\mathbf{g} \otimes \mathbf{n}$ represents simple shearing.

The problem of bifurcation in a band in the framework of multi-mode plasticity will be examined in Sect. 3.6.3 without any symmetry assumptions. The discussion below concerns an incrementally nonlinear constitutive law obeying a potential (3.117).

In the case when the fundamental tangent stiffness moduli tensor $\mathsf{C}^0 = \mathsf{C}(\dot{\mathbf{F}}^0)$ applies also inside the band (which is possible for a piecewise-linear law) then the bifurcation condition takes the commonly assumed form [23, 67]

$$\det \mathbf{Q}^0(\mathbf{n}) = 0, \qquad \mathbf{Q}^0(\mathbf{n}) = \mathbf{Q}(\mathbf{n}, \dot{\mathbf{F}}^0), \tag{3.196}$$

equivalent to (3.194). However, it should be noted that (3.194) represents merely a *linearization* of the general condition (3.195). Generally, bifurcation in a band can take place along a deformation path before (3.196) is met, cf. e.g. examples calculated in [52, 53]. An additional constitutive restriction is needed to justify the use of (3.196). For instance, the use of the classical elastoplasticity (3.84) represent such a restriction also if normality does not hold [68]. In incrementally nonlinear plasticity, the justification for using (3.196) is provided by fulfillment of the constitutive inequality (3.135). Then positive definiteness of $\mathbf{Q}^0$ excludes bifurcation in a band [49]; to see this, it suffices to observe that any two pairs $(\dot{\mathbf{F}}, \dot{\mathbf{S}})$ and $(\dot{\mathbf{F}}^0, \dot{\mathbf{S}}^0)$ satisfying (3.195) make the left-hand expression in (3.135) equal to zero.

While the fulfillment of (3.196) may be treated as being *sufficient* for bifurcation in a band,[15] in incrementally nonlinear materials it is *not* necessary. The following statement has been proven in [52].[16]

Theorem (on continuous spectrum of bifurcation points) *Along a given deformation path* $\mathbf{F}^0(t)$ *there is (energetically preferable) bifurcation in a band of a normal* $\mathbf{n}$ *at every instant at which*

$$\mathcal{E}(\dot{\mathbf{F}}^0, \dot{\mathbf{F}}^0 + \mathbf{g}\otimes\mathbf{n}) < 0 \quad \text{for some } \mathbf{g} \quad \textit{and} \quad U(\mathbf{g}\otimes\mathbf{n}) > 0 \quad \text{for every } \mathbf{g}\,. \tag{3.197}$$

Of course, the conclusion holds true if the former condition in (3.197) is replaced by negation of the condition (3.191). The critical instants (3.196) (e.g., when (3.135) holds) and (3.172) may constitute the limiting points of the bifurcation range (3.197).

The above Theorem jointly with the previous interpretation of (3.192) provide the following justification for treating (3.190) and (3.191) as conditions *necessary* for material stability :

Interpretation of instability of uniform straining *If (3.190) (or (3.191)) does not hold then formation of localization bands can be initiated either as energetically preferable bifurcation of a quasi-static path if (3.192) holds, or as a dynamic process if (3.192) fails*.

Exercise: Show that within the range (3.197) there exists a continuous range of band orientations for which shear band bifurcation occurs in a given state.

[15] For a thoroughly nonlinear relation between $\dot{\mathbf{S}}$ and $\dot{\mathbf{F}}$, fulfillment of (3.196) implies bifurcation in a band in *accelerations* rather than in velocities [49].

[16] The existence of a continuous spectrum of bifurcations points along a deformation path has been established for a general incrementally-nonlinear discretized system by Petryk [51]; another version of the proof was given by Petryk and Thermann [62]. Examples of a continuous range of shear band bifurcations were calculated in [52, 53].

3.5.3.3 Incipient volume fraction of localization zone

Consider a smooth fundamental path $\mathbf{F}^0(t)$ of uniform deformation of an infinite homogeneous continuum obeying an incrementally nonlinear constitutive law in the potential form (3.117). Suppose that the range of validity of (3.190) and (3.191) along the fundamental path terminates at the critical instant t_0 at which (3.196) is met. Fully localized deformations need not occur immediately beyond that instant, unless the critical instant defined by (3.196) happens to coincide with that defined by (3.172). The coincidence is enforced in the classical elastoplasticity, which can be seen by adjusting appropriately the sense of $\mathbf{g}$. On the contrary, in an incrementally nonlinear material the critical instants may be distinct when $\mathbf{g} \otimes \mathbf{n}$ in (3.194) is directed outside the constitutive domain of C^0.

Suppose that this is so and that (3.167) still holds in a certain range beyond t_0. Then, concentration of post-critical deformation can develop gradually, and can also saturate as in the post-critical solution with a single band analysed by Hutchinson and Tvergaard [33]. On the other hand, a quasi-static solution with a single bifurcation band in an infinite continuum violates the stability requirement (3.191) outside the band if the uniform solution does, and may analogously be regarded as unrealizable in practice.

Another difficulty, more widely recognized, is associated with an indeterminate band width. It might seem that this difficulty cannot be overcome without enhancing the material model to incorporate an internal length scale, as it is recently being developed in several different ways in the literature in the context of classical elasto-plasticity. Nevertheless, a definite (although not complete) conclusion concerning the incipient post-critical deformation pattern can be drawn from the energy criterion (3.178), in particular from its consequence (3.191), still using the constitutive framework for time-independent incrementally nonlinear materials not possessing any internal length scale. Namely, the incipient *volume fraction* of the whole localization zone and the velocity gradient therein can be determined, although the width, number and spatial arrangement of individual localization bands are indeterminate. The analysis below follows closely the author's paper [57].

A class of velocity solutions in the infinite continuum at the critical instant t_0 for a pair of unit vectors $(\mathbf{n}, \mathbf{g})$ that satisfy (3.194) can be constructed as follows

$$\dot{\mathbf{F}}(\boldsymbol{\xi}) = \dot{\mathbf{F}}^0 + s(\xi_{\mathbf{n}})\, \mathbf{g} \otimes \mathbf{n}\,, \qquad \xi_{\mathbf{n}} = \boldsymbol{\xi} \cdot \mathbf{n} \in (-\infty, +\infty) \tag{3.198}$$

with the restrictions

$$s^- \leq s \leq s^+\,, \qquad \left| \int_{-\infty}^{+\infty} s\, \mathrm{d}\xi_{\mathbf{n}} \right| < \infty\,. \tag{3.199}$$

The limiting values s^- and s^+ are chosen such that $\mathsf{C}(\dot{\mathbf{F}}) = \mathsf{C}(\dot{\mathbf{F}}^0)$ within the corresponding range of $\dot{\mathbf{F}}$. The restrictions (3.199) ensure that the *mean* velocity gradients in the fundamental and bifurcation solutions are the same, and that the pointwise difference in velocities is infinitesimal relative to fundamental velocities at the re-scaled element boundary if $\dot{\mathbf{F}}^0 \neq \mathbf{0}$. The bifurcation solutions represent thus the branching of the material response under *macroscopically* uniform kinematic boundary data. The sense of $\mathbf{g}$ is chosen below such that $\mathbf{D}^* \cdot \mathbf{D}^0 \geq 0$, where $\mathbf{D}^0$ is the fundamental Eulerian strain-rate and $\mathbf{D}^* = \frac{1}{2}(\mathbf{g} \otimes \mathbf{n} + \mathbf{n} \otimes \mathbf{g})$. The current configuration is taken for simplicity as reference so that $\mathbf{F}^0(t_0) = \mathbf{I}$, and $\mathbf{n}$ may be interpreted as the current normal to a band.

In the particular case when $s \neq 0$ only in a finite interval of $\xi_\mathbf{n}$, the solution is obtained which was analyzed in many papers after Rice [67]. That solution violates the path stability conditions (3.190) and (3.191) outside the localization band just beyond the critical instant, assuming that (3.191) then fails along the fundamental path. It is a natural assumption that the *rate* of the left-hand expression in (3.191), starting from $\mathbf{g}, \mathbf{n}$ as above, can be made negative at t_0 not only for $\dot{\mathbf{F}} = \dot{\mathbf{F}}^0$ but also for all $\dot{\mathbf{F}}$ defined by (3.198) with (3.199). By the continuity argument, this is so if the rate of this expression is negative along the fundamental path and if the current constitutive cone is sufficiently sharp. If this is so then (3.191) can be satisfied just beyond the critical instant only due to the transition to a more 'stiff' constitutive branch, which is possible if the strain-rate $\mathbf{D} = \mathbf{D}^0 + s\mathbf{D}^*$, wherever defined, lies at t_0 precisely *on* the boundary of the constitutive cone containing $\mathbf{D}^0$. This means that $s(\xi_\mathbf{n})$ in (3.198) can only take two values, s^- or s^+, which define the intersection points, $\mathbf{D}^-$ and $\mathbf{D}^+$, respectively, of a straight line $\mathbf{D}(s) = \mathbf{D}^0 + s\mathbf{D}^*$ in $\mathbf{D}$-space with the cone boundary (Fig. 3.7a). The bifurcation solution takes the form of a mixture of layers of a constant velocity gradient, $.../\dot{\mathbf{F}}^-/\dot{\mathbf{F}}^+/\dot{\mathbf{F}}^-/\dot{\mathbf{F}}^+/...$ (Fig. 3.7b), in some analogy to fine phase mixtures discussed by Ball and James [3]. The initial localization zone, which occupies a volume fraction η^+, say, is defined as the collection of all layers where $\mathbf{D} = \mathbf{D}^+$.

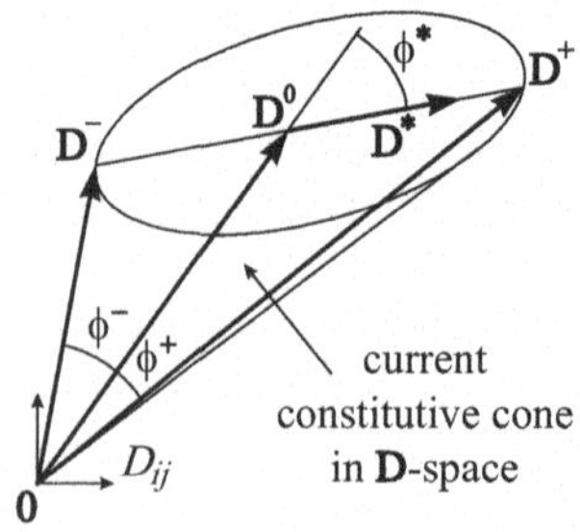

(a)

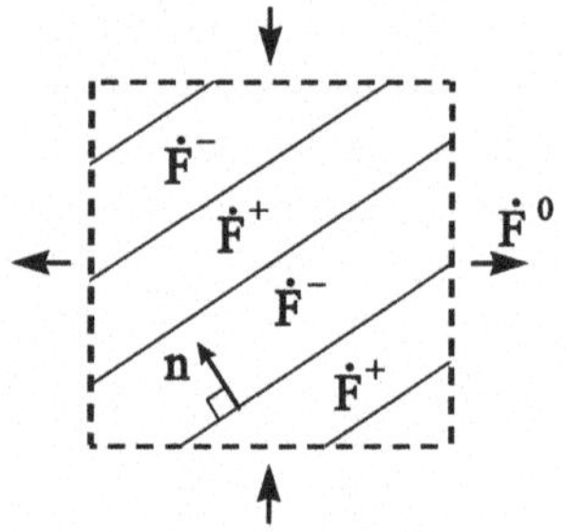

(b)

Fig. 3.7 Schematic sketch (a) of the strain rates and (b) of spatial distribution of the velocity gradient, in a bifurcation solution within a material element

From the integral condition in (3.199) we obtain the equation

$$\eta^+ s^+ + (1 - \eta^+) s^- = 0 \qquad \text{or} \qquad \eta^+ \mathbf{D}^+ + (1 - \eta^+)\mathbf{D}^- = \mathbf{D}^0 . \tag{3.200}$$

On substituting trigonometric relationships within the hyperplane spanned by $(\mathbf{D}^0, \mathbf{D}^*)$ in $\mathbf{D}$-space , the *incipient* volume fraction occupied by the localization zone is determined [57] as

$$\eta^+ = \frac{\cot(\phi^+) - \cot(\phi^*)}{\cot(\phi^+) + \cot(\phi^-)} \qquad \text{for } \phi^+ < \phi^* . \tag{3.201}$$

Here, ϕ^+, ϕ^- and ϕ^* stand for positive angles, according to a definition based on a scalar product, between $\mathbf{D}^0$ and $\mathbf{D}^+, \mathbf{D}^-$ and $\mathbf{D}^*$, respectively (Fig. 3.7a). In the particular case when $\phi^+ = \phi^-$, (3.201) reduces to the formula given by Petryk and Thermann [63] and verified numerically in the context of strain localization in biaxially stretched sheets.

Exercise: Derive the formula (3.201) from (3.200) by using trigonometric relationships which result from Fig. 3.7a.

Exercise: Show that the value of η^+ is independent of the choice of an angular measure in $\mathbf{D}$-space.

3.5.3.4 Instability in heterogeneous materials

The energy criterion in its general form (3.177) can also be applied to heterogeneous materials (as metal polycrystals or composites) to investigate not only local but also overall instabilities. The remarks below (cf. also [57]) are intended only to illustrate the possibilities.

Let curly brackets $\{\cdot\}$ denote an unweighted volume average over a representative sample of a heterogeneous material in a reference configuration. Let a continuous field $\widetilde{\mathbf{w}}$ have the same meaning as above and vanish over the sample boundary. The path stability condition (3.178) applied to the quasi-static velocity field $\widetilde{\mathbf{v}}^0$ within the sample gives

$$\{U(\nabla\mathbf{v}^0 + \nabla\mathbf{w})\} \geq \{U(\nabla\mathbf{v}^0)\} \qquad \text{for every } \widetilde{\mathbf{w}} . \tag{3.202}$$

Accordingly, a minimization procedure: $\{U(\nabla\mathbf{v})\} \to \min$ can be used to determine numerically the material response in the non-uniqueness range, as shown in parallel lecture notes (Thermann, 2000). Fulfillment of (3.202) implies (cf. [51])

$$\{\nabla\mathbf{w} \cdot \mathsf{C}(\nabla\mathbf{v}^0) \cdot \nabla\mathbf{w}\} \geq 0 \qquad \text{and} \qquad \{U(\nabla\mathbf{w})\} \geq 0 \qquad \text{for every } \widetilde{\mathbf{w}} . \tag{3.203}$$

If $\nabla\mathbf{w}$ is effectively concentrated in a disk-like volume of predominantly uniform thickness, taken sufficiently small as in Sect. 3.5.3.1 but now within a heterogeneous material element, (3.203) implies

$$(\mathbf{g}\otimes\mathbf{n}) \cdot \{\mathsf{C}(\nabla\mathbf{v}^0)\}_{\mathrm{n}} \cdot (\mathbf{g}\otimes\mathbf{n}) \geq 0 \quad \text{and} \quad \{U(\mathbf{g}\otimes\mathbf{n})\}_{\mathrm{n}} \geq 0 \quad \text{for every } \mathbf{n}, \mathbf{g} , \tag{3.204}$$

where $\{\cdot\}_{\mathrm{n}}$ denotes averaging over a plane section of reference normal $\mathbf{n}$ through the element.

The conditions (3.204) can be used, for instance, to detect instability in a solution of a layered structure as in Fig. 3.7b in a scale larger than that set by a width of individual bands.

3.5.4 Instability Under Deformation-Sensitive Loading

Stability of a process of uniform straining under controlled overall strain does not imply stability when the environment of a stressed material element is not sufficiently 'stiff'. As in Sect. 3.4.4, attention is restricted in this section to uniform straining within a material element at unperturbed rotational equilibrium and in the absence of disturbances of internal thermodynamic forces. The constitutive equations for the material and external loading are assumed in the spin-insensitive form (3.82) and (3.144), respectively, with the symmetry restrictions $\mathsf{L} = \mathsf{L}^{\mathrm{T}}$ and $\mathsf{L}^* = \mathsf{L}^{*\mathrm{T}}$. Stability of the fundamental process $\mathbf{e}^0(t)$ for infinitesimal perturbations of the external surface tractions associated with $\mathbf{t}^{\mathrm{ext}}$ is investigated by using the energy criterion (3.177).

The potential energy $\Omega(\mathbf{e}, \lambda)$ of the (unperturbed) external loading of the element satisfies, by definition, $V\mathbf{t}^{\mathrm{ext}} = -\partial\Omega(\mathbf{e}, \lambda)/\partial\mathbf{e}$. Specification of (3.176) yields

$$\dot{E} = V\,\dot{w} + \dot{\Omega} = V\,(\mathbf{t} - \mathbf{t}^{\mathrm{ext}}) \cdot \dot{\mathbf{e}} + \Omega_{,\lambda}\dot{\lambda} \tag{3.205}$$

and

$$\ddot{E} = V\,(\dot{\mathbf{t}} - \dot{\mathbf{t}}^{\mathrm{ext}}) \cdot \dot{\mathbf{e}} - V\dot{\lambda}\mathbf{t}^{\mathrm{ext}}_{,\lambda} \cdot \dot{\mathbf{e}} + V\,(\mathbf{t} - \mathbf{t}^{\mathrm{ext}}) \cdot \ddot{\mathbf{e}} + \Omega_{,\lambda\lambda}(\dot{\lambda})^2 + \Omega_{,\lambda}\ddot{\lambda}\,. \tag{3.206}$$

In any given equilibrium state $\mathcal{G}_0$, the term with $(\mathbf{t} - \mathbf{t}^{\mathrm{ext}})$ vanishes and the last two terms are given. On substituting (3.82) and (3.144) with $\dot{\mathbf{t}}^* = \mathbf{t}^{\mathrm{ext}}_{,\lambda}\dot{\lambda}$, we obtain the identity

$$\frac{1}{2V}(\ddot{E} - \ddot{E}^0) = \mathcal{J}(\dot{\mathbf{e}}) - \mathcal{J}(\dot{\mathbf{e}}^0) \qquad \text{for all } \dot{\mathbf{e}}, \dot{\mathbf{e}}^0 \quad \text{in } \mathcal{G}_0\,, \tag{3.207}$$

where

$$\mathcal{J}(\dot{\mathbf{e}}) = \frac{1}{2}\dot{\mathbf{e}} \cdot (\mathsf{L}(\dot{\mathbf{e}}) + \mathsf{L}^*) \cdot \dot{\mathbf{e}} - \dot{\mathbf{t}}^* \cdot \dot{\mathbf{e}}\,. \tag{3.208}$$

The condition (3.178) necessary for stability of the fundamental path reduces to

$$\mathcal{J}(\dot{\mathbf{e}}) \geq \mathcal{J}(\dot{\mathbf{e}}^0) \qquad \text{for all } \dot{\mathbf{e}}\,. \tag{3.209}$$

The interpretation of the instability associated with failure of (3.209) *under the symmetry assumptions* follows from the following statement, which is a particular case of a more general conclusion obtained in [51].

Interpretation of path instability under deformation-sensitive loading *If (3.209) fails then:*

either the inequality sign in (3.175) is reversed for some $\dot{\mathbf{e}}$*, which is interpreted as dynamic instability of equilibrium (in the first approximation),*

or there exists another rate-solution $\dot{\mathbf{e}}$ *consistent with* $\dot{\mathbf{t}} = \dot{\mathbf{t}}^{\rm ext}$ *which represents constitutive branching energetically preferable to* $\dot{\mathbf{e}}^0$ *in the sense of a smaller value of* $\ddot{E}$ •

Note that without the present symmetry restrictions, failure of (3.175) was not necessarily connected with an instability, and (3.175) was treated only as a condition *sufficient* for directional stability of equilibrium. Under the symmetry assumptions, the critical state associated with (3.175) is an *eigenstate* (cf. [29]) where a non-trivial solution exists to the rate-problem $\dot{\mathbf{t}} = -\mathbf{L}^* \cdot \dot{\mathbf{e}}$. Beyond the primary eigenstate on the fundamental path, the inequality sign in (3.175) is usually reversed for some $\dot{\mathbf{e}}$. Then it can be shown (cf. [51]) that there exists a direct path of departure from equilibrium that satisfies (in the first approximation and discretized) equations of motion and follows an arbitrarily small initial disturbance.

The range of validity of (3.209) along a smooth fundamental path usually terminates either at an eigenstate or at the point of primary constitutive branching. At the latter critical point, the minimum of $\mathcal{J}(\dot{\mathbf{e}})$ is still attained at $\dot{\mathbf{e}}^0$ but also at some other $\dot{\mathbf{e}}^* \neq \dot{\mathbf{e}}^0$, say. It follows that $\dot{\mathbf{e}}^*$ is a stationarity point of $\mathcal{J}(\dot{\mathbf{e}})$, and hence another solution to the rate-problem $\dot{\mathbf{t}} = \dot{\mathbf{t}}^{\rm ext}$ under the deformation-sensitive loading (3.144). A typical point of primary constitutive branching can be determiend from the following *linearized* condition with $\mathbf{L}^0 = \mathbf{L}(\dot{\mathbf{e}}^0)$:

$$(\mathbf{L}^0 + \mathbf{L}^*) \cdot \delta\dot{\mathbf{e}} = \mathbf{0} \qquad \text{for some } \delta\dot{\mathbf{e}} \neq 0 \tag{3.210}$$

related to the quadratic condition necessary for (3.209):

$$\delta\dot{\mathbf{e}} \cdot (\mathbf{L}^0 + \mathbf{L}^*) \cdot \delta\dot{\mathbf{e}} \geq 0 \qquad \text{for all } \delta\dot{\mathbf{e}}\,. \tag{3.211}$$

Constitutive branching may precede an eigenstate, as expected from Hill's [22, 28] general bifurcation theory. In turn, the spectrum of points of constitutive branching in incrementally nonlinear materials in the range where (3.209) fails at stable equilibrium is *continuous* along a strain path, as follows from the theorem on discretized bifurcation problems in incrementally nonlinear solids [51, 62].

3.5.5 Energy Conditions for Material Instability: a Summary

The consequences of the energy condition (3.177) and of its reduced form (3.179) presented above may be interpreted as *necessity* conditions for path stability and for stability of equilibrium, respectively. By specifying the incremental energy consumption to second-order terms and determinig the circumstances in which it fails to be minimized along the fundamental deformation path, the onset of material instability of a selected type can be estimated.

Six selected conditions, to be interpreted as *sufficient for material instability*, are reformulated below for the three distinguished kinds of instability. Each of these conditions implies

$$\ddot{E} < \ddot{E}^0 \qquad \text{for some admissible direction of a deformation increment}, \qquad (3.212)$$

which represents, as the negation of (3.178), a unified form of sufficiency conditions for instability in the energy sense. The left column below refers to instability of equilibrium ($\dot{\mathbf{F}}^0 = \mathbf{0}$), the right column refers to instability in a deforming material ($\dot{\mathbf{F}}^0 \neq \mathbf{0}$). The interpretation of instability in the right column refers to the cases when equilibrium is stable.

Instability for internal rearrangements

$$\sum_{K,L\in\mathcal{P}} g_{KL}\mathring{\gamma}_K\mathring{\gamma}_L < 0 \quad \text{for some } \mathring{\gamma}_K \geq 0 \quad \textit{or} \quad \sum_{K,L\in\mathcal{A}^0} g_{KL}\delta\mathring{\gamma}_K\delta\mathring{\gamma}_L < 0 \quad \text{for some } \delta\mathring{\gamma}_K$$

$$\downarrow \qquad\qquad \downarrow$$

snap-through at micro-level — non-uniqueness of active mechanisms

under assumptions $\lambda_K^{\mathrm{p}} = \lambda_K$ and $g_{KL} = g_{LK}$ (cf. Sect. 3.4.2 and 3.5.2).

Instability of uniform straining

$$(\mathbf{g}\otimes\mathbf{n})\cdot\mathsf{C}(\mathbf{g}\otimes\mathbf{n})\cdot(\mathbf{g}\otimes\mathbf{n}) < 0 \text{ for some } \mathbf{g},\, \mathbf{n} \quad \textit{or} \quad (\mathbf{g}\otimes\mathbf{n})\cdot\mathsf{C}^0\cdot(\mathbf{g}\otimes\mathbf{n}) < 0 \text{ for some } \mathbf{g},\, \mathbf{n}$$

$$\downarrow \qquad\qquad \downarrow$$

dynamic localization of deformation — quasi-static bifurcation in a band

under assumption $\mathsf{C} = \mathsf{C}^{\mathrm{T}}$ (cf. Sect. 3.4.3 and 3.5.3).

Instability under deformation-sensitive loading

$$\dot{\mathbf{e}}\cdot(\mathsf{L}(\dot{\mathbf{e}}) + \mathsf{L}^*)\cdot\dot{\mathbf{e}} < 0 \quad \text{for some } \dot{\mathbf{e}} \quad \textit{or} \quad \delta\dot{\mathbf{e}}\cdot(\mathsf{L}^0 + \mathsf{L}^*)\cdot\delta\dot{\mathbf{e}} < 0 \quad \text{for some } \delta\dot{\mathbf{e}}$$

$$\downarrow \qquad\qquad \downarrow$$

snap-through or collapse — constitutive branching

under assumptions $\mathsf{L} = \mathsf{L}^{\mathrm{T}}$ and $\mathsf{L}^* = \mathsf{L}^{*\mathrm{T}}$ (cf. Sect. 3.4.4 and 3.5.4).

3.6 Uniqueness in Multi-Mode Plasticity Without Symmetry

The framework of multi-mode plasticity from Sect. 3.3.3.2 is now adopted without any additional assumptions concerning symmetry or normality. Attention is restricted to the problem of incremental uniqueness of the material response under incremental loading of three different types. In Sect. 3.6.1, the internal response under fully controlled deformation gradient is represented by plastic multipliers only. In Sect. 3.6.2, the material response has more freedom under the assumed deformation-sensitive loading, but the macroscopic strain-rate $\dot{\mathbf{e}}$, to be determined, is still restricted to remain uniform. In Sect. 3.6.3, the problem of bifurcation in a band is addressed in a novel way, without using an incremental stiffness moduli tensor other than purely elastic one.

The presentation below follows closely the author's recent work [61], where an extension to partial kinematic constraints can also be found.

3.6.1 Constitutive Uniqueness

In this section, the problem of uniqueness of plastic multipliers is addressed under the assumption of fully controlled, uniform deformation gradient. Accordingly, the rate-problem (P1) is considered.

Denote by M the number of the elements of $\mathcal{P}$, i.e. the number of potentially active mechanisms ($f_K = 0$) in a given material state. Introduce the matrix notation, where vectors from the real M-dimensional space R^M and square $M \times M$ real matrices, both of the indices from $\mathcal{P}$, are denoted by square brackets. In that notation, (3.90) takes the following form

$$[-\dot{f}] = [g^*][\mathring{\gamma}] + [-b^*], \quad [\mathring{\gamma}] \geq [0], \quad [-\dot{f}] \geq [0], \quad [-\dot{f}]^{\mathrm{T}}[\mathring{\gamma}] = 0, \quad (3.213)$$

where inequalities between vectors should be read componentwise. (3.90) is recovered after the identification $[g^*] = (g_{KL})$ and $[b^*] = (\boldsymbol{\lambda}_K \cdot \dot{\mathbf{e}})$ with $K, L \in \mathcal{P}$.

The mathematical structure of the system (3.213) is known as *the linear complementarity problem* [9]. The fundamental correspondence between the incremental constitutive law of multi-mode plasticity in the small-strain framework and the linear complementarity problem was emphasized by Maier [37].

The fundamental theorem concerning the existence and uniqueness of a solution to the linear complementarity problem (cf. [9]) in the present notation reads

Theorem 1 *The problem (3.213) has a unique solution* $[\mathring{\gamma}]$ *for all vectors* $[b^*] \in R^M$ *if and only if*

$$\textit{all principal minors of the matrix } [g^*] \textit{ are positive} \;\bullet \quad (3.214)$$

Recall that for an index set $\mathcal{A} \subseteq \mathcal{P}$, the determinant of a principal submatrix (a_{KL}), $K, L \in \mathcal{A}$, of a matrix $[a]$ is called a principal minor of $[a]$.

Generally, (3.214) is less restrictive than the requirement (3.133) of positive definiteness of $[g^*]$. It is well known that if the matrix $[g^*]$ is symmetric then these conditions are equivalent to each other.

From the above Theorem we directly obtain the following

Corollary 1.1 *The problem (P1) has a unique solution for all* $\dot{\mathbf{e}}$ *if all principal minors of the matrix* (g_{KL}), $K, L \in \mathcal{P}$, *are positive* •

This result is sharper than that given by Hill and Rice [30], in view of the implication (3.214) $\Leftarrow$ (3.133) for $[g^*] = (g_{KL})$ with $K, L \in \mathcal{P}$.

The 'only if' part of Theorem 6.1 requires some care since it concerns *all* vectors $[b^*] \in R^M$. Vectors $[b^*] = (\boldsymbol{\lambda}_K \cdot \dot{\mathbf{e}})$ span the whole space R^M if and only if $\boldsymbol{\lambda}_K$ are linearly independent for $K \in \mathcal{P}$. From Theorem 6.1 we thus obtain:

Corollary 1.2 *Suppose that* $\boldsymbol{\lambda}_K$, $K \in \mathcal{P}$, *are linearly independent symmetric tensors. Then the problem (P1) has a unique solution for all* $\dot{\mathbf{e}}$ *if and only if all principal minors of the matrix* (g_{KL}), $K, L \in \mathcal{P}$, *are positive* •

Both requirements in Corollary 1.2 are measure-invariant conditions, as follows from (3.97).

The above statements may be regarded as extensions of those given by Maier [37] to finite strain. The extension is formally immediate, however, the question of measure-invariance of the conditions did not appear in the small-strain framework.

3.6.2 Uniqueness Under Deformation-Sensitive Loading

In this section, the spin-insensitive form (3.144) of the loading condition is taken with $\mathbf{L}^*$ and $\dot{\mathbf{t}}^*$ regarded as known. In comparison with the problem discussed in the preceding section, $\dot{\mathbf{e}}$ represents an additional unknown. The quasi-static rate-problem is defined as follows:

Problem (P2) *In a given material state, find* $(\dot{\mathbf{e}}, \mathring{\gamma}_K)$ *such that*

$$\dot{\mathbf{t}} = \dot{\mathbf{t}}^{\mathrm{ext}} \qquad \text{at unconstrained strain-rate}\,, \tag{3.215}$$

where $\dot{\mathbf{t}}$ *and* $\dot{\mathbf{t}}^{\mathrm{ext}}$ *are defined by* $(3.86)_1$ *and (3.144), respectively, and the* $\mathring{\gamma}_K$*'s are subjected to the constraints (3.85) and (3.89)* •

On eliminating $\dot{\mathbf{t}}$ between (3.215) and $(3.86)_1$ and substituting (3.144), we obtain

$$(\mathbf{L}^{\mathrm{e}} + \mathbf{L}^*) \cdot \dot{\mathbf{e}} = \dot{\mathbf{t}}^* + \sum_K \boldsymbol{\lambda}^{\mathrm{p}}_K \mathring{\gamma}_K \,. \tag{3.216}$$

It is worth noting that an identical equation is obtained when $(3.87)_1$ (multiplied by $\mathbf{L}^e$) is used instead of $(3.86)_1$. The tensor $(\mathbf{L}^e + \mathbf{L}^*)$ is related to the moduli tensors $\bar{\mathbf{L}}^e$ and $\bar{\mathbf{L}}^*$ by the formula

$$\mathbf{L}^e + \mathbf{L}^* = \mathbf{B}_w \cdot (\bar{\mathbf{L}}^e + \bar{\mathbf{L}}^*) \cdot \mathbf{A}^{-1} \tag{3.217}$$

which follows from (3.97) and (3.145).

Although $\mathbf{L}^e$ or $\mathbf{L}^*$ taken separately may be singular, we shall assume that their sum $(\mathbf{L}^e + \mathbf{L}^*)$ possesses a unique inverse $(\mathbf{L}^e + \mathbf{L}^*)^{-1}$ when regarded as a linear operator acting in $\mathcal{G}$, cf. the definition of $\mathbf{A}^{-1}$. It is evident that this condition is necessary and sufficient for the uniqueness of a solution to the purely elastic rate-problem (3.216) with all $\mathring{\gamma}_K$ given in advance (e.g., set equal to zero), the consistency conditions (3.89) being then disregarded. An essential conclusion drawn from (3.217) is that the invertibility of $(\mathbf{L}^e + \mathbf{L}^*)$ is a measure-invariant property.

The elimination of $\dot{\mathbf{e}}$ between (3.216) and $(3.86)_2$ now yields

$$\dot{f}_K = \boldsymbol{\lambda}_K \cdot (\mathbf{L}^e + \mathbf{L}^*)^{-1} \cdot \dot{\mathbf{t}}^* - \sum_{L \in \mathcal{P}} (g_{KL} - \boldsymbol{\lambda}_K \cdot (\mathbf{L}^e + \mathbf{L}^*)^{-1} \cdot \boldsymbol{\lambda}_L^p) \mathring{\gamma}_L \,. \tag{3.218}$$

Introduce the notation

$$\mathbf{G}^* = (\mathbf{L}^e + \mathbf{L}^*)^{-1}, \qquad \mathbf{H}^* = \mathbf{L}^* \cdot (\mathbf{L}^e + \mathbf{L}^*)^{-1} \cdot \mathbf{L}^e \tag{3.219}$$

and denote the expressions in (3.218) by

$$\begin{aligned} b_K^* &= \boldsymbol{\lambda}_K \cdot \mathbf{G}^* \cdot \dot{\mathbf{t}}^* & g_{KL}^* &= g_{KL} - \boldsymbol{\lambda}_K \cdot \mathbf{G}^* \cdot \boldsymbol{\lambda}_L^p \\ &= \boldsymbol{\mu}_K \cdot \mathbf{L}^e \cdot \mathbf{G}^* \cdot \dot{\mathbf{t}}^* \,, & &= h_{KL} + \boldsymbol{\mu}_K \cdot \mathbf{H}^* \cdot \boldsymbol{\mu}_L^p \,. \end{aligned} \tag{3.220}$$

The dual expressions in (3.220) involving $\boldsymbol{\mu}_K$ follow from (3.88) and require the assumption of the invertibility of $\mathbf{L}^e$. The equivalence of the two expressions for b_K^* is obvious, and the equivalence of the second pair can be verified as follows:

$$\begin{aligned} g_{KL}^* &= h_{KL} + \boldsymbol{\mu}_K \cdot \mathbf{L}^e \cdot \boldsymbol{\mu}_L^p - \boldsymbol{\mu}_K \cdot \mathbf{L}^e \cdot \mathbf{G}^* \cdot \mathbf{L}^e \cdot \boldsymbol{\mu}_L^p \\ &= h_{KL} + \boldsymbol{\mu}_K \cdot ((\mathbf{L}^e + \mathbf{L}^*) \cdot \mathbf{G}^* - \mathbf{L}^e \cdot \mathbf{G}^*) \cdot \mathbf{L}^e \cdot \boldsymbol{\mu}_L^p \\ &= h_{KL} + \boldsymbol{\mu}_K \cdot \mathbf{H}^* \cdot \boldsymbol{\mu}_L^p \,. \end{aligned} \tag{3.221}$$

The same expressions appear in the formula for $\dot{f}_K$ obtained by starting from $(3.87)_2$ instead of $(3.86)_2$.

Further transformations yield

$$\begin{aligned} \mathbf{H}^* &= \mathbf{L}^* \cdot (\mathbf{I}_s + \mathbf{M}^e \cdot \mathbf{L}^*)^{-1} & &\text{if } \mathbf{M}^e = (\mathbf{L}^e)^{-1} \text{ exists,} \\ &= (\mathbf{M}^e + \mathbf{M}^*)^{-1} & &\text{if } \mathbf{M}^* = (\mathbf{L}^*)^{-1} \text{ exists.} \end{aligned} \tag{3.222}$$

From (3.89), (3.218) and (3.220) we arrive at the following rate problem to be solved with respect to $\mathring{\gamma}_K$'s in a given material state :

$$\dot{f}_K = b^*_K - \sum_{L\in\mathcal{P}} g^*_{KL}\overset{\circ}{\gamma}_L\,, \quad \overset{\circ}{\gamma}_K \geq 0\,, \quad \dot{f}_K \leq 0, \quad \dot{f}_K\,\overset{\circ}{\gamma}_K = 0 \qquad \text{for } K \in \mathcal{P}. \quad (3.223)$$

If this is done then a full solution to (3.215) is completed by using the expression for $\dot{\mathbf{e}}$:

$$\dot{\mathbf{e}} = \mathbf{G}^* \cdot \dot{\mathbf{t}}^* + \mathbf{G}^* \cdot \sum_K \boldsymbol{\lambda}^{\mathrm{p}}_K \overset{\circ}{\gamma}_K \qquad (3.224)$$

obtained from (3.216).

In a given state of the material element and for given incremental constitutive laws (3.86) or (3.87) for the material and (3.144) for the loading, the quantities b^*_K and g^*_{KL} are known. The substitution of the transformation formulae (3.97) and (3.145) into (3.219) and (3.220) shows that both $J^{-1}b^*_K$ and $J^{-1}g^*_{KL}$ are measure-invariant, as expected from (3.95).

On applying Theorem 1 with the identification of $[g^*]$ and $[b^*]$ as above, we obtain [61]:

Theorem 2 *Suppose that the problem (P2) is uniquely solvable for purely elastic rates (so that the inverse tensor* $\mathbf{G}^*$ *defined by (3.219) exists).*
(i) If (3.214) holds then the problem (P2) has a unique solution for every $\dot{\mathbf{t}}^*$.
(ii) Suppose that $\boldsymbol{\lambda}_K$, $K \in \mathcal{P}$, *are linearly independent; then the problem (P2) has a unique solution for every* $\dot{\mathbf{t}}^*$ *if and only if (3.214) holds* •

3.6.3 Bounds to Bifurcation in a Band

3.6.3.1 Exclusion conditions

We return to the problem (3.195) of bifurcation in a band of a unit normal **n**, which is now addressed in a different way. To simplify some formulae, the current configuration is taken as reference, so that **F** coincides instantaneously with the unit tensor **I** (and $\mathbf{A} = \mathbf{B}_{\mathrm{w}} = \mathbf{I}_{\mathrm{s}}$). The fundamental incremental solution is defined by $(\dot{\mathbf{F}}^*, \overset{\circ}{\gamma}{}^*_K)$ uniform in space; note that a superscript '*' is used in place of previous '0'. The problem of bifurcation in a band is formulated as follows.

Problem (P3) *In a given material state and for given* $(\dot{\mathbf{F}}^*, \overset{\circ}{\gamma}{}^*_K)$ *and* **n**, *find* $(\mathbf{g}, \overset{\circ}{\gamma}_K)$ *such that*

$$(\dot{\mathbf{S}} - \dot{\mathbf{S}}^*)\mathbf{n} = \mathbf{0} \qquad \text{and} \qquad \dot{\mathbf{F}} - \dot{\mathbf{F}}^* = \mathbf{g} \otimes \mathbf{n}\,, \qquad (3.225)$$

where $\dot{\mathbf{S}}$ *and* $\dot{\mathbf{S}}^*$ *are constitutively related through (3.99) to* $(\dot{\mathbf{F}}, \overset{\circ}{\gamma}_K)$ *and* $(\dot{\mathbf{F}}^*, \overset{\circ}{\gamma}{}^*_K)$, *respectively, and the* $\overset{\circ}{\gamma}_K$*'s and* $\overset{\circ}{\gamma}{}^*_K$*'s are subjected to the constraints (3.85) and (3.89)*•

The problem (P3) admits a trivial solution $\mathbf{g} = \mathbf{0}$, $\overset{\circ}{\gamma}_K = \overset{\circ}{\gamma}{}^*_K$. Existence of a non-trivial solution to (P3) is identified with bifurcation in a band.

By combining (3.225) and (3.99) and substituting $\mathbf{F} = \mathbf{I}$, we obtain

$$\mathbf{Q}^{e}\mathbf{g} = \sum_{K} (\lambda_{K}^{p}\mathbf{n})(\mathring{\gamma}_{K} - \mathring{\gamma}_{K}^{*}), \tag{3.226}$$

where the second-order tensor $\mathbf{Q}^{e}$ is the acoustic tensor (dependent on $\mathbf{n}$) for the purely elastic constitutive branch,

$$\mathbf{Q}^{e} = \mathbf{n}\bar{\mathbf{L}}^{e}\mathbf{n} - \frac{1}{2}(\mathbf{n}\sigma \otimes \mathbf{n} + \mathbf{n} \otimes \sigma\mathbf{n} + \sigma - (\mathbf{n}\sigma\mathbf{n})\mathbf{I}), \qquad Q_{ik}^{e} = C_{ipkq}^{e} n_{p} n_{q}. \tag{3.227}$$

Suppose that

$$\det \mathbf{Q}^{e} \neq 0, \tag{3.228}$$

which is the classical ellipticity condition for elastic moduli which excludes a non-trivial solution to the problem (P3) posed for purely elastic rates. Then $\mathbf{g}$ can be expressed in terms of unknown $\mathring{\gamma}_{K}$ from (3.226). On substituting that $\mathbf{g}$ into (3.225) and next into $(3.99)_2$ with $\mathbf{F} = \mathbf{I}$, we obtain

$$\dot{f}_{K} = \lambda_{K} \cdot \dot{\mathbf{F}}^{*} - (\lambda_{K}\mathbf{n})(\mathbf{Q}^{e})^{-1} \sum_{L} (\lambda_{L}^{p}\mathbf{n})\mathring{\gamma}_{L}^{*} - \sum_{L} g_{KL}^{\text{band}} \mathring{\gamma}_{L}, \tag{3.229}$$

where

$$\begin{aligned} g_{KL}^{\text{band}} &= g_{KL} - \mathbf{n}\lambda_{K}(\mathbf{Q}^{e})^{-1}\lambda_{L}^{p}\mathbf{n} \\ &= h_{KL} + \mu_{K} \cdot \mathbf{L}^{e} \cdot \mu_{L}^{p} - (\mu_{K} \cdot \mathbf{L}^{e}\mathbf{n})(\mathbf{Q}^{e})^{-1}(\mathbf{n}\mathbf{L}^{e} \cdot \mu_{L}^{p}). \end{aligned} \tag{3.230}$$

All the quantities that appear in (3.230), and also the set $\mathcal{P}$ of the indices of potentially active mechanisms, are known in the material state in which the problem (P3) is posed. Hence, the matrix $[g^{\text{band}}]$ is known and independent on $(\dot{\mathbf{F}}^{*}, \mathring{\gamma}_{K}^{*})$.

By applying Theorem 1, we arrive at the following conclusion [61].

Exclusion conditions for bifurcation in a band *Problem (P3) admits only a trivial solution if simultaneously :*
(a) *the elastic moduli satisfy the ellipticity condition* (3.228),
(b) *all principal minors of the matrix* $[g^{\text{band}}]$ *defined by* (3.230) *with* $K, L \in \mathcal{P}$ *are positive* •

It is pointed out that (a) and (b) jointly exclude the bifurcation in a band for any possible unloading within or outside the band at the bifurcation instant.

For convenience, the condition (b) is rewritten as

$$\det(g_{\mathcal{A}}^{\text{band}}) > 0 \qquad \text{for all } \mathcal{A} \subseteq \mathcal{P}, \tag{3.231}$$

where $(g_{\mathcal{A}}^{\text{band}})$ stands for a principal submatrix of $[g^{\text{band}}]$ with indices from $\mathcal{A} \subseteq \mathcal{P}$.

3.6.3.2 Lower and upper bounds for the primary bifurcation

Consider now a (rather typical) homogeneous deformation path $(\dot{\mathbf{F}}^*, \mathring{\gamma}^*_K)(t)$, parameterized by a time-like scalar $t \geq t_0$, along which the following conditions are satisfied:

(i) the condition (a) above holds true at all t,
(ii) the set $\mathcal{P}$ of the indices of potentially active mechanisms is constant,
(iii) $[\mathring{\gamma}^*] > [0]$,
(iv) the matrix $[g^{\text{band}}]$ for any $\mathbf{n}$ varies continuously with t,
(v) the condition (b) above is satisfied for all $\mathbf{n}$ at an initial instant $t = t_0$.

The following bounds to the instant of primary bifurcation in a band are obtained [61].

Bounds for the primary bifurcation *Along a deformation path that satisfies the conditions (i)-(v) above, the instant t^{c} of the primary bifurcation (in the first-order rates) in a band is bounded by*

$$t_{\mathcal{A}} \leq t^{\text{c}} \leq t_{\mathcal{P}}, \tag{3.232}$$

where the lower bound is determined from

$$t_{\mathcal{A}} = \min t \ : \quad \det(g^{\text{band}}_{\mathcal{A}})(t) = 0 \quad \textit{for some } \mathcal{A} \subseteq \mathcal{P} \textit{ and some } \mathbf{n}, \tag{3.233}$$

and the upper bound from

$$t_{\mathcal{P}} = \min t \ : \quad \det[g^{\text{band}}](t) = 0 \quad \textit{for some } \mathbf{n}\,\text{.} \tag{3.234}$$

The proof of the lower bound follows readily from the exclusion condition (3.231) by the continuity argument applied to $\det(g^{\text{band}}_{\mathcal{A}})$. The upper bound is obtained as a particular bifurcation point at which none of the potentially active mechanisms undergoes unloading; an explicit proof is given in the reference.

There is a variety of possible applications of the above results to specific material models.

If $M = 1$ (only one mechanism of a given index K is considered) then the lower and upper bounds coincide. If $\mathsf{M}^{\text{e}} = (\mathsf{L}^{\text{e}})^{-1}$ exists for the measures chosen then the uniqueness range $t < t^{\text{c}}$ can be expressed from (3.230) as $h_{KK} > h^{\text{c}}_K$. Here, h_{KK} is the effective hardening modulus for the mechanism, and its critical value h^{c}_K reached at $t = t^{\text{c}}$ reads

$$\begin{aligned} h^{\text{c}}_K &= -\lambda_K \cdot \mathsf{M}^{\text{e}} \cdot \lambda^{\text{p}}_K + \max_{\mathbf{n}} \left(\mathbf{n}\lambda_K (\mathbf{Q}^{\text{e}})^{-1} \lambda^{\text{p}}_K \mathbf{n}\right) \\ &= -\mu_K \cdot \mathsf{L}^{\text{e}} \cdot \mu^{\text{p}}_K + \max_{\mathbf{n}} \left((\mu_K \cdot \mathsf{L}^{\text{e}}\mathbf{n})(\mathbf{Q}^{\text{e}})^{-1}(\mathbf{n}\mathsf{L}^{\text{e}} \cdot \mu^{\text{p}}_K)\right). \end{aligned} \tag{3.235}$$

The formula (3.235) is equivalent to that derived by Rice [67] and Asaro and Rice [2] for the linearized problem without unloading. The validity of the formula (3.235) for the critical hardening modulus was extended to the case of possible unloading by Rice and Rudnicki [68]; cf. also [65, 5].

If $M \geq 2$ then the upper bound (3.234) can be reduced to the formula given in [45] for finite strain elastoplasticity of single crystals with the normality flow rule, and without the restriction to normality in [69] and [72] in the small strain and finite strain framework, respectively. In those investigations, the *linearized* condition (3.196) for the bifurcation in a band has been taken as a starting point, which was not used above. The resulting coincidence of the instants of $\det[g^{\mathrm{band}}] = 0$ and $\det \mathbf{Q} = 0$ is not unexpected since both conditions, although obtained on different routes, correspond to the same bifurcation in a band *without unloading*.

It is pointed out that the exclusion condition (3.231) and the related lower bound (3.233), for the bifurcation in a band in the incrementally piecewise-linear material with multiple ($M \geq 2$) mechanisms of time-independent inelastic deformation, are valid for any possible unloading at the bifurcation instant.

If $M = 2$ and $h_{KK} > h^{\mathrm{c}}_{K}$ for each mechanism at every instant then (3.231) reduces to $\det[g^{\mathrm{band}}] > 0$, and the bounds (3.233) and (3.234) coincide. It follows that in that case and along a deformation path that satisfies (i)-(v) above, the condition (3.234) defines the exact instant of the primary bifurcation without unloading, any earlier bifurcation with unloading being excluded. This particular result is of interest for certain material models being in use (double slip, classical plasticity coupled with damage or phase transformation, etc.).

Finally, consider the case of symmetry $g^{\mathrm{band}}_{KL} = g^{\mathrm{band}}_{LK}$. Then the condition (3.231) coincides with the requirement (3.133) of positive definiteness of $[g^{\mathrm{band}}]$. Under the assumptions (v) and (iv) above, this requirement is satisfied up to the critical instant at which $[g^{\mathrm{band}}]$ becomes positive *semi*-definite, i.e. singular. But this marks the upper bound. We have thus the corollary that if $g^{\mathrm{band}}_{KL} = g^{\mathrm{band}}_{LK}$ then the above lower and upper bounds coincide for arbitrary $M \geq 1$, and the condition (3.234) defines the exact instant of the primary bifurcation in a band. This corollary follows also from the relative convexity property (3.131) with the comparison moduli C^{L} defined by (3.101) with $\mathcal{A} = \mathcal{P}$.

Appendix

We will use below the formalism with internal state variables α defined in a fixed reference configuration and assume that a small parameter is provided by a norm $||\Delta\alpha||$ in a Hilbert space containing all α. We shall consider a class of short piecewise-smooth paths in the α-space, leading from $\alpha_0 = \alpha(t_0)$ to a given $\alpha_0 + \Delta\alpha = \alpha(t)$, whose length is of order $||\Delta\alpha||$. The *virtual* dissipation $\Delta\mathcal{D}$ along such a path, defined by time integration of the dissipation function (3.123), can be decomposed as follows:

$$\Delta\mathcal{D} \equiv \int_{t_0}^{t} D(\alpha(\theta), \dot{\alpha}(\theta))\, d\theta = \Delta_1\mathcal{D} + \Delta_2\mathcal{D} + o((||\Delta\alpha||)^2)\,, \tag{3.236}$$

where

$$\Delta_1\mathcal{D} = \int_{t_0}^{t} D(\alpha_0, \dot{\alpha}(\theta))\, d\theta \tag{3.237}$$

is the *path-dependent* first-order dissipation, while $\Delta_2\mathcal{D}$ defined by

$$\Delta_2\mathcal{D} = \int_{t_0}^{t} D_1(\hat{\alpha}(\theta), \dot{\alpha}(\theta))\, d\theta\,, \quad \hat{\alpha} = \alpha - \alpha_0\,, \quad D_1(\hat{\alpha}, \dot{\alpha}) \equiv \hat{\alpha} \cdot D_{,\alpha}(\alpha_0, \dot{\alpha})\,, \tag{3.238}$$

is the *path-dependent* second-order dissipation, an analog of the path-dependent second-order work, cf. [50]. Note that $D_1(\hat{\alpha}, r\dot{\alpha}) = rD_1(\hat{\alpha}, \dot{\alpha})$ for every $r > 0$ and that $D_1(\hat{\alpha}, \dot{\alpha})$ is linear with respect to $\hat{\alpha}$. Recall that $D_{,\alpha}$ is a Fréchet derivative of $D(\alpha, \dot{\alpha}$ with respect to α.

Evaluation of the first- and second-order dissipation for a class of *direct* paths in α-space reduces to the integration in (3.237) and (3.238) along a straight path, which on account of (3.123) yields

$$\Delta_1\mathcal{D}^{\mathrm{d}} = D(\alpha_0, \Delta\alpha)\,, \qquad \Delta_2\mathcal{D}^{\mathrm{d}} = \frac{1}{2} D_1(\Delta\alpha, \Delta\alpha)\,. \tag{3.239}$$

The following simple lemma holds, cf. Nguyen [40].

Lemma A1 *Under the assumptions (3.123), the first-order dissipation (3.237) is always minimized on a straight path, that is,*

$$\Delta_1\mathcal{D} \geq \Delta_1\mathcal{D}^{\mathrm{d}} \tag{3.240}$$

if and only if

$$D(\alpha, \dot{\alpha}) \textit{ is convex in } \dot{\alpha}\ \bullet \tag{3.241}$$

Proof (*Sufficiency*). Suppose that (3.241) holds. Then the convexity and homogeneity of $D(\alpha, \dot{\alpha})$ with respect to $\dot{\alpha}$ imply

$$D(\alpha, \dot{\alpha}) + D(\alpha, \dot{\alpha}^*) \geq 2D(\alpha, \frac{1}{2}(\dot{\alpha} + \dot{\alpha}^*)) = D(\alpha, \dot{\alpha} + \dot{\alpha}^*) \text{ for all } \dot{\alpha}, \dot{\alpha}^*, \alpha\,. \tag{3.242}$$

This can be extended to any finite or infinite sum of any $\dot{\alpha}$'s, which implies

$$\int_{t_0}^{t} D(\alpha_0, \dot{\alpha}(\theta))\, d\theta \geq D(\alpha_0, \int_{t_0}^{t} \dot{\alpha}(\theta)\, d\theta) = D(\alpha_0, \Delta\alpha)\,. \tag{3.243}$$

This proves validity of (3.240).

(*Necessity*). Suppose that (3.240) always holds. This means that (3.243) holds for any $(\alpha_0, \Delta\alpha)$, which implies (3.242) as a particular case of (3.243) for a path consisting of two arbitrary straight segments. This implies (3.241), which completes the proof of Lemma A1. □

For the dissipation evaluated to *second* order to be minimized on a straight path, the convexity property (3.241) alone is not sufficient. Denote by $\mathcal{G}_0$ a material state corresponding to α_0. If the path in (3.236) is chosen such that $\dot{\alpha}(\theta) \in \mathcal{L}(\mathcal{G}_0)$ at each θ (cf. (3.125)) then the difference in the first-order dissipation vanishes, while the difference in the second-order dissipation may be of either sign in general. However, we have the following lemma, cf. [55].

Lemma A2 *Under the assumptions (3.123), (3.241), (3.125) and (3.126), the dissipation (3.236) evaluated to second-order is always minimized on a straight path, that is,*

$$\Delta\mathcal{D} \geq \Delta_1\mathcal{D}^{\mathrm{d}} + \Delta_2\mathcal{D}^{\mathrm{d}} + o((\|\Delta\alpha\|)^2) \tag{3.244}$$

if and only if

$$D_1(\hat{\alpha}, \dot{\alpha}) = D_1(\dot{\alpha}, \hat{\alpha}) \qquad \textit{for all } \dot{\alpha}, \hat{\alpha} \in \mathcal{L}(\mathcal{G}_0)\,\bullet \tag{3.245}$$

Proof. The integrand in (3.238) is transformed as follows

$$\begin{aligned} 2D_1(\hat{\alpha}(\theta), \dot{\alpha}(\theta)) &= D_1(\hat{\alpha}(\theta), \dot{\alpha}(\theta)) + \textstyle\int_{t_0}^{\theta} D_1(\dot{\alpha}(\zeta), \dot{\alpha}(\theta))\, d\zeta \\ &= D_1(\hat{\alpha}(\theta), \dot{\alpha}(\theta)) + \textstyle\int_{t_0}^{\theta} D_1(\dot{\alpha}(\theta), \dot{\alpha}(\zeta))\, d\zeta + 2D_1^{\mathrm{a}}(\theta) \\ &= \tfrac{d}{d\theta} \textstyle\int_{t_0}^{\theta} D_1(\hat{\alpha}(\theta), \dot{\alpha}(\zeta))\, d\zeta + 2D_1^{\mathrm{a}}(\theta)\,, \end{aligned} \tag{3.246}$$

where

$$D_1^{\mathrm{a}}(\theta) = \int_{t_0}^{\theta} \frac{1}{2}\big(D_1(\dot{\alpha}(\zeta), \dot{\alpha}(\theta)) - D_1(\dot{\alpha}(\theta), \dot{\alpha}(\zeta))\big)\, d\zeta\,. \tag{3.247}$$

On substituting (3.246) into (3.238) we obtain

$$\Delta_2\mathcal{D} = \int_{t_0}^{t} \big(D_1(\frac{1}{2}\Delta\alpha, \dot{\alpha}(\theta)) + D_1^{\mathrm{a}}(\theta)\big)\, d\theta\,. \tag{3.248}$$

By combining this with the expression for $\Delta_1\mathcal{D}$, from (3.236) we arrive at

$$\Delta\mathcal{D} = \int_{t_0}^{t} D(\alpha_0 + \frac{1}{2}\Delta\alpha, \dot{\alpha}(\theta))\, d\theta + \int_{t_0}^{t} D_1^{\mathrm{a}}(\theta)\, d\theta + o((\|\Delta\alpha\|)^2)\,. \tag{3.249}$$

(*Sufficiency*). Suppose that (3.241) and (3.245) hold. If $\dot{\alpha}(\theta) \notin \mathcal{L}(\mathcal{G}_0)$ for some θ along a piecewise smooth path then (3.240) holds with strict inequality, implying (3.244). It remains thus to consider the case $\dot{\alpha}(\theta) \in \mathcal{L}(\mathcal{G}_0)$ at every θ. On using (3.243) with $\alpha_0 + \frac{1}{2}\Delta\alpha$ in place of α_0 and (3.249) with $D_1^{\mathrm{a}} = 0$, it follows that

$$\begin{aligned} \Delta\mathcal{D} &= \textstyle\int_{t_0}^{t} D(\alpha_0 + \frac{1}{2}\Delta\alpha, \dot{\alpha}(\theta))\, d\theta + o((\|\Delta\alpha\|)^2) \\ &\geq D(\alpha_0 + \tfrac{1}{2}\Delta\alpha, \Delta\alpha) + o((\|\Delta\alpha\|)^2) \\ &= D(\alpha_0, \Delta\alpha) + \tfrac{1}{2}D_1(\Delta\alpha, \Delta\alpha) + o((\|\Delta\alpha\|)^2)\,, \end{aligned} \tag{3.250}$$

which shows validity of (3.244).

(*Necessity*). Suppose that (3.244) holds universally. Take $\Delta\alpha$ proportional to $\dot{\alpha}^{\mathrm{a}}$ in (3.126) corresponding to fully active loading in $\mathcal{G}_0$, and consider the class of paths in (3.236) such that $\dot{\alpha}(\theta) \in \mathcal{L}(\mathcal{G}_0)$ at every θ. Then (3.240) holds with equality sign by (3.125), and D_1 in (3.248) becomes linear in $\dot{\alpha}(\theta) \in \mathcal{L}(\mathcal{G}_0)$ as follows from (3.126). Hence, from (3.236),

$$\begin{aligned}\Delta\mathcal{D} = D(\alpha_0, \Delta\alpha) + \tfrac{1}{2}D_1(\Delta\alpha, \Delta\alpha) + o((\|\Delta\alpha\|)^2) \\ + \int_{t_0}^{t} D_1^{\mathrm{a}}(\theta)\, d\theta \qquad \text{if } \dot{\alpha}(\theta) \in \mathcal{L}(\mathcal{G}_0)\,.\end{aligned} \tag{3.251}$$

The integral of D_1^{a} can be of second order and of either sign unless (3.245) holds, e.g., along a path consisting of two arbitrary straight segments directed within $\mathcal{L}(\mathcal{G}_0)$. It follows that (3.245) is necessary for (3.244), which completes the proof of Lemma A2.

□

Note that the assumption (3.126) is not used in the sufficiency part of the proof.

References

1. Asaro, R.J. (1983) Micromechanics of crystals and polycrystals. In *Advances in Applied Mechanics, Vol.* 23. New York: Acad. Press. 1-115.
2. Asaro, R.J. and Rice, J.R. (1977) Strain localization in ductile single crystals. *J. Mech. Phys. Solids* **25**:309-338.
3. Ball, J.M. and James, R.D. (1987) Fine phase mixtures as minimizars of energy. *Arch. Rat. Mech. Anal.* **100**:13-52.
4. Bigoni, D. and Loret, B. (1999) Effects of elastic anisotropy on strain localization and flutetr instability in plastic solids. *J. Mech. Phys. Solids* **47**:1409-1436.
5. Bigoni, D. and Zaccaria, D. (1992) Loss of strong ellipticity in non-associative elastoplasticity. *J. Mech. Phys. Solids* 40:1313-1331.
6. Biot, M.A. (1965) *Mechanics of Incremental Deformations*. New York: John Wiley & Sons.
7. de Borst, R. and van der Giessen, E., eds. (1998) *Material Instabilities in Solids*. Chichester: John Wiley & Sons.
8. Chambon, R., Crochepeyre, S. and J. Desrues (2000) Localization criteria for non-linear constitutive equations of geomaterials. *Mech. Cohesive Frictional Mat.* **5**:61-82.
9. Cottle, R.W., Pang, J.-S. and Stone, R.E. (1992) *The Linear Complementarity Problem*. San Diego: Acad. Press.
10. Drucker, D.C. (1950) Some implications of work hardening and ideal plasticity. *Q. Appl. Math.* **7**:411-418.
11. Drucker, D.C. (1964) On the postulate of stability of material in the mechanics of continua. *J. Mécanique* **3**:235-249.
12. Drugan, W.J. (1998) Thermodynamic equivalence of steady-state shocks and smooth waves in general media; applications to elastic-plastic shocks and dynamic fracture. *J. Mech. Phys. Solids* **46**:313-336.
13. Drugan, W.J. and Shen, Y. (1990) Finite deformation analysis of restrictions on moving strong discontinuity surfaces in elastic-plastic materials: quasi-static and dynamic deformations. *J. Mech. Phys. Solids* **38**:553-574.
14. Eshelby, J.D. (1957) The determination of the elastic field of an ellipsoidal inclusion, and related problems. *Proc. Roy. Soc. London A* **241**:376-396.

15. Fedelich, B. and Ehrlaher, A. (1997) An analysis of stability of equilibrium and of quasi-static transformations on the basis of the dissipation function. *Eur. J. Mech. A/Solids* **16**:833-855.
16. Fleck, N.A. and Hutchinson, J.W. (1997) Strain gradient plasticity. In *Advances in Applied Mechanics, Vol.* 33. New York: Acad. Press. 295-361.
17. Franciosi, P. and Zaoui, A. (1991) Crystal hardening and the issue of uniqueness, *Int. J. Plasticity* **7**:295-311.
18. Graves, L. (1939) The Weierstrass condition for multiple integral variation problem. *Duke Math. J.* **5**:656-660.
19. Hadamard, J. (1903) *Leçons sur la Propagation des Ondes et les Equations de l'Hydrodynamique.* Paris: Hermann.
20. Havner, K.S. (1992) *Finite Plastic Deformation of Crystalline Solids.* Cambridge: University Press.
21. Hill, R. (1958) A general theory of uniqueness and stability in elastic-plastic solids. *J. Mech. Phys. Solids* **6**:236-249.
22. Hill, R. (1959) Some basic principles in the mechanics of solids without a natural time. *J. Mech. Phys. Solids* **7**:209-225.
23. Hill, R. (1962) Acceleration waves in solids. *J. Mech. Phys. Solids* **10**:1-16.
24. Hill, R. (1967a) The essential structure of constitutive laws for metal composites and polycrystals. *J. Mech. Phys. Solids* **15**:79-95.
25. Hill, R. (1967b) Eigenmodal deformations in elastic/plastic continua. *J. Mech. Phys. Solids* **15**:371-386.
26. Hill, R. (1967c) On the classical constitutive relations for elastic/plastic solids. In Broberg, B. at al., eds., *Recent Progress in Applied Mechanics, The Folke Odkvist Volume.* Stockholm: Almqvist & Wiksell. 241-249.
27. Hill, R. (1968) On constitutive inequalities for simple materials. *J. Mech. Phys. Solids* **16**:229-242, 315-322.
28. Hill, R. (1978) Aspects of invariance in solids mechanics. In *Advances in Applied Mechanics, Vol.* 18. New York: Acad. Press. 1-75.
29. Hill, R. (1983) On intrinsic eigenstates in plasticity with generalized variables. *Math. Proc. Camb. Phil. Soc.* **93**:177-189.
30. Hill, R. and Rice, J.R. (1972) Constitutive analysis of elastic- plastic crystals at arbitrary strain. *J. Mech. Phys. Solids* **20**:401-413.
31. Hill, R. and Rice, J.R. (1973) Elastic potentials and the structure of inelastic constitutive laws. *SIAM J. Appl. Math.* **25**:448-461.
32. Hutchinson, J.W. (1970) Elastic-plastic behaviour of polycrystalline metals and composites. *Proc. Roy. Soc. Lond. A* **319**, 247-272.
33. Hutchinson, J.W., and Tvergaard, V. (1981) Shear band formation in plane strain. Int. J. Solids Structures **17**:451-470.
34. de Groot, S.R. and Mazur, P. (1962) *Non-equilibrium Thermodynamics.* Amsterdam: North Holland.
35. Kolymbas, D. (1981) Bifurcation analysis for sand samples with a non-linear constitutive equation. *Ing.-Archiv* **50**:131-140.
36. Loret, B. (1992) Does deviation from deviatoric associativity lead to the onset of flutter instability? *J. Mech. Phys. Solids* **40**:1363-1375.
37. Maier, G. (1969) "Linear" flow-laws of elastoplasticity: a unified general approach. *Rend. Acc. Naz. Lincei, Ser. VIII* Vol. XLVII:266-276.
38. Mandel, J. (1966) Conditions de stabilité et postulat de Drucker. In Kravtchenko, J., ed., *Rheology and Soil Mechanics.* Berlin: Springer. 58-67.
39. Needleman, A. and Tvergaard, V. (1992) Analyses of plastic flow localization in metals. *Appl. Mech. Rev.* **45**: no 3, part 2, S3-S18.
40. Nguyen, Q.S. (1993) Bifurcation and stability of time-independent standard dissipative systems. In Nguyen, Q.S., ed., *CISM Lecture Notes No.* 327. Wien - New York: Springer. 45-94.
41. Nguyen, Q.S. and Petryk, H. (1991) A constitutive inequality for time-independent dissipative solids. *C. R. Acad. Sci. Paris, Serie II* **312**:7-12.

42. Nguyen, Q.S., and Radenkovic, D. (1976) Stability of equilibrium in elastic plastic solids. In *Lecture Notes in Mathematics, Vol.* 503. Berlin: Springer. 403-414.
43. Ogden, R.W. (1984a) On Eulerian and Lagrangean objectivity in continuum mechanics. *Arch. Mech.* **36**:207-218.
44. Ogden, R.W. (1984b) *Non-linear Elastic Deformations*. Chichester: Ellis Horwood.
45. Peirce, D. (1983) Shear band bifurcations in ductile single crystals. *J. Mech. Phys. Solids* **31**:133-153.
46. Petryk, H. (1982) A consistent energy approach to defining stability of plastic deformation processes. In Schroeder, F.H., ed., *Stability in the Mechanics of Continua, Proc. IUTAM Symp. Nümbrecht* 1981. Berlin - Heidelberg: Springer. 262-272.
47. Petryk, H. (1985a) On energy criteria of plastic instability. In *Plastic Instability, Proc. Considère Memorial.* Paris: Ecole Nat. Ponts Chauss. 215-226.
48. Petryk, H. (1985b) On stability and symmetry conditions in time-independent plasticity. *Arch. Mech.* **37**:503-520.
49. Petryk, H. (1989) On constitutive inequalities and bifurcation in elastic-plastic solids with a yield-surface vertex. *J. Mech. Phys. Solids* **37**:265-291.
50. Petryk, H. (1991a) On the second-order work in plasticity, *Arch. Mech.* **43**:377-397.
51. Petryk, H. (1991b) The energy criteria of instability in time-independent inelastic solids. *Arch. Mech.* **43**:519-545.
52. Petryk, H. (1992a) Material instability and strain-rate discontinuities in incrementally nonlinear continua. *J. Mech. Phys. Solids* **40**:1227-1250.
53. Petryk, H. (1992b) On stability of time-independent materials at finite strain. In Besdo, D. and Stein, E., eds., *Finite Inelastic Deformations - Theory and Applications*. Berlin - Heidelberg: Springer. 189-196.
54. Petryk, H. (1993a) Theory of bifurcation and instability in time-independent plasticity. In Nguyen, Q.S., ed., *CISM Lecture Notes No.* 327. Wien - New York: Springer. 95-152.
55. Petryk, H. (1993b) Stability and constitutive inequalities in plasticity. In Muschik, W., ed., *CISM Lecture Notes No.* 336. Wien - New York: Springer. 259-329.
56. Petryk, H. (1995) Thermodynamic stability of equilibrium in plasticity. *J. Non-Equil. Thermodyn.* **20**:132-149.
57. Petryk, H. (1997a) Instability of plastic deformation processes. In Tatsumi, T., et al., eds., *Theoretical and Applied Mechanics 1996, Proc. XIXth IUTAM Congress, Kyoto.* Amsterdam: Elsevier. 497-516.
58. Petryk, H. (1997b) Plastic instability: criteria and computational approaches. *Arch. Comp. Meth. Engng.* **4**:111-151.
59. Petryk, H. (1999a) On the micro-macro transition and hardening moduli in plasticity. In Bruhns, O.T. and Stein, E., eds., *Proc. IUTAM Symp. Micro- and Macrostructural Aspects of Thermoplasticity, Bochum 1997.* Dordrecht: Kluwer. 219-230.
60. Petryk, H. (1999b) (Unpublished).
61. Petryk, H. (2000) General conditions for uniqueness in materials with multiple mechanisms of inelastic deformation. *J. Mech. Phys. Solids* **48**:367-396.
62. Petryk, H., and Thermann, K. (1992) On discretized plasticity problems with bifurcations. *Int. J. Solids Structures* **29**:745-765.
63. Petryk, H., and Thermann, K. (1996) Post-critical plastic deformation of biaxailly stretched sheets. *Int. J. Solids Structures* **33**:689-705.
64. Raniecki, B. (1979) Uniqueness criteria in solids with non-associated plastic flow laws at finite deformations. *Bull. Acad. Polon. Sci., Sér. sci. techn.* **27**:391-399.
65. Raniecki, B. and Bruhns, O.T. (1981) Bounds to bifurcation stresses in solids with non-associated plastic flow law at finite strain. *J. Mech. Phys. Solids* **29**:153-172.
66. Rice, J.R. (1971) Inelastic constitutive relations for solids: an internal-variable theory and its application to metal plasticity. *J. Mech. Phys. Solids* **19**:433-455.
67. Rice, J.R. (1977) The localization of plastic deformation. In Koiter, W.T., ed., *Theoretical and Applied Mechanics*. Amsterdam: North- Holland. 207-220.
68. Rice, J.R. and Rudnicki, J.W. (1980) A note on some features of the theory of localization of deformation. *Int. J. Solids Structures* **16**:597-605.

69. Rizzi, E., Maier, G., and Willam, K. (1996) On failure indicators in multi-dissipative materials. *Int. J. Solids Structures* **33**:3187-3214.
70. Ryzhak, E.I. (1987) Necessity of Hadamard conditions for stability of elastic-plastic solids (in Russian) *Mekh. Tv. Tela No.* 4:101-104.
71. Sewell, M.J. (1972) A survey of plastic buckling. In Leipholz, H.H.E., ed., *Stability*. Waterloo, Ontario: Univ. of Waterloo. 85-197.
72. Steinmann, P. (1996) On localization analysis in multisurface hyperelasto-plasticity. *J. Mech. Phys. Solids* **44**:1691-1713.
73. Thermann, K. (2000) Post-critical plastic deformation pattern in incrementally nonlinear materials at finite strain. In Petryk H., ed., *CISM Lecture Notes No. 414*. Wien-New York: Springer-Verlag. 333-284.
74. Tomita, Y. (1994) Simulation of plastic instabilities in solid mechanics. *Appl. Mech. Rev.* **47**: no 6, part 1, 171-205.

Chapter 4
A Consistent Energy Approach to Defining Stability of Plastic Deformation Processes

Abstract A single energy-type definition of stability is used to study different aspects of plastic instability. The definition concerns stability of a quasi-static deformation process but applies also to an equilibrium state. The connections with the bifurcation theory as well as with energy criteria of stability of equilibrium are demonstrated. Hill's extremum principles are re-derived as necessary conditions for stability of a deformation process.

4.1 Introduction

Instability phenomena in plastic solids may manifest themselves in various ways. One of them is connected with loss of stability of an equilibrium configuration under constant loading. On the other hand, incipient buckling or necking in the plastic range in most cases occur under varying loads and without loss of stability of equilibrium. Such phenomena may be regarded as a symptom of instability of a deformation *process* [1]. Also the localization of plastic deformation into a thin band can be considered as an instability of plastic flow [2].

Several theoretical approaches to plastic instability problems exist in the literature. However, while for studying stability of equilibrium usually some version of the energy approach is used [3, 4, 5], the instability of a deformation process is examined in the other ways, e.g. by using Hill's bifurcation theory [3, 6], initial imperfection method or some intuitive approach (see [7, 8]). None of these approaches seems to be sufficiently universal to comprise different plastic instability aspects mentioned above.

In the previous paper [9] the author proposed a definition of stability which applies to quasi-static deformation processes as well as to an equilibrium state. The present paper is a continuation of that work. The definition may be regarded as extension

[0] Reprinted from: Stability in the Mechanics of Continua, Proc. IUTAM Symp., Nuembrecht 1981 (Ed. F.H.Schroeder), pp. 262-272, Springer-Verlag, Berlin Heidelberg 1982. H.Petryk, A consistent energy approach to defining stability of plastic deformation processes. Copyright © 1982.

H. Petryk, *Theory of Plastic Instability*,
https://doi.org/10.1007/978-3-031-98223-1_4

of the energy criterion of stability of equilibrium to processes of deformation under varying loading. The approach is based on the general theory of stability of motion. The motion is understood here as a quasi-static deformation process corresponding to some time-dependent boundary conditions. The term "stability" is used in the following sense: Let the theoretical deformation process be perturbed by an arbitrary disturbance from some class. The theoretical process is called stable in some time interval if the distance between this process and the perturbed one in this time interval is vanishingly small when the disturbance itself is vanishingly small. That definition has to be made precise by defining the class of disturbances considered and the measures of disturbance and of the distance between two processes.

The equations governing the behaviour of the body in a deformation process are never satisfied exactly in reality. This concerns not only the field equations but of course also the initial and boundary conditions as well as constitutive relations. Usually the stability with respect to disturbances of initial conditions only is considered. However, there is a priori no guarantee that a process stable for initial disturbances will also be stable against disturbances of another nature. In the present paper persistent disturbances of general character are taken into account. As the main attribute of all disturbances we assume the amount of energy supplied by a disturbance to the *system* consisting of both the deformed body and the loading device. That energy may be supplied at once or gradually during the process of deformation. Since energy may also be extracted from the system, the maximum value of the energy supplied by a disturbance is taken as a disturbance measure. The disturbances are thought to be represented by a set of additional body forces and surface tractions of arbitrary magnitude, in general varying in time and acting during an arbitrary period of time. By adding those perturbing forces, any kinematically admissible deformation process compatible with initial and boundary kinematical constraints may theoretically be realized. For stable unperturbed processes this always requires some additional energy to be supplied.

The whole approach is formally valid for any type of constitutive relations which do not contain a natural time. There is an open question, how wide is the class of solids or problems for which this approach leads to physically acceptable results. In this paper we are concerned with the conventional rigid–plastic or elastic-plastic solids obeying the normality flow rule. The loads acting on the body are assumed to be derivable from a potential, in general a time-dependent one. Under these assumptions the consequences of the stability definition are examined.

4.2 Definition of Stability

Let us consider a continuous plastic body subjected to an isothermal quasi-static proces of deformation in some time interval. We assume that the constitutive relations for the material do not contain a natural time. Denote by V and S the body volume and surface, respectively, in some reference configuration. Introduce Cartesian spatial coordinates (x_j) and the material coordinates (X_K), $j, K = 1, 2, 3$, which in the reference configuration coincide. The deformation process $\tilde{\chi}$ is defined by

$$x_j = \chi_j(X_K, t), \quad X_K \in V, \tag{4.1}$$

where t denotes a scalar parameter rather than a natural time, but for simplicity it is referred to as time.

The body surface is split onto two parts, S_{u} and S_{T}. Over S_{u} the displacements history is prescribed by the condition

$$x_j(X_K, t) = \bar{x}_j(X_K, t), \quad X_K \in S_{\mathrm{u}}, \tag{4.2}$$

where $\bar{x}_j$ are given functions. On S_{T} the nominal tractions T_j are given, which may depend not only on time but also on the place of a surface point in space. The tractions T_j are assumed to be derivable from a given potential ω,

$$T_j = -\frac{\partial \omega(X_K, x_i, t)}{\partial x_j}, \quad X_K \in S_{\mathrm{T}}. \tag{4.3}$$

In the particular case when T_j do not depend on the displacements, $\omega = -\bar{T}_j x_j + c$, where $\bar{T}_j$ and c are functions of X_K and t.

We will use the notion of the loading device, defined as a mechanical system by which the tractions (4.3) and the kinematical constraints (4.2) are applied to the body surface. That system may be elastic as well as "rigid" or "infinitely soft" in extreme cases. The loading device transmits energy from external sources to the deformed body, its current potential energy being

$$W^{\mathrm{e}} = \int_{S_{\mathrm{T}}} \omega \, \mathrm{d}S. \tag{4.4}$$

The initial configuration of the body at time $t = t_1$ is assumed to be known equilibrium configuration, compatible with the boundary conditions (4.2) and (4.3). In the following we will consider only such processes (4.1) which start from this initial configuration, satisfy at every moment the kinematical boundary conditions (4.2) and describe sufficiently regular deformations of the body. Such processes will be called kinematically admissible processes.

Let the functions $\bar{x}_j$ and ω be specified in some time interval $[t_1, t_2]$ (or $[t_1, \infty)$). We assume that there exists a kinematically admissible process χ^0 which corresponds to a theoretical quasi-static solution to the problem. That process, the stability of which is to be studied, is called fundamental process, and all corresponding functions or quantities are denoted by the superscript ^{0}. The stresses in the fundamental process, calculated from the constitutive relations, satisfy at every moment the equilibrium conditions (for simplicity we omit body forces)

$$
\begin{aligned}
s^0_{Kj,K} &= 0, && X_K \in V\,,\\
s^0_{Kj}\nu_K &= T^0_j, && X_K \in S\,,\\
T^0_j &= -\left.\frac{\partial\omega}{\partial x_j}\right|_{x_j=x^0_j}, && X_K \in S_{\mathrm{T}}\,,
\end{aligned}
\tag{4.5}
$$

where s_{Kj} is the (unsymmetric) nominal stress tensor, the comma denotes partial derivative with respect to the rectangular material coordinates X_K, ν_K is the unit normal to the surface S in the reference configuration, and the summation convention holds.

The fundamental process is thought to be perturbed by some set of additional body forces (per unit reference volume) $F^{\mathrm{d}}_j(X_K, t)$ and nominal surface tractions $T^{\mathrm{d}}_j(X_K, t)$. The perturbed forces F^{d}_j and T^{d}_j are of arbitrary magnitude and may act in the whole time interval $[t_1, t_2]$ (or $[t_1, \infty)$). The perturbed process $\tilde{\chi}^*$ is, by definition, any kinematically admissible process such that the stresses, related to the deformation history by the constitutive relations, satisfy at every moment the conditions (equations of perturbed motion)

$$
\begin{aligned}
S^*_{Kj,K} + F^{\mathrm{d}}_j &= 0, && X_K \in V\,,\\
S^*_{Kj,K}\nu_K &= T^*_j + T^{\mathrm{d}}_j, && X_K \in S\,,\\
T^*_j &= -\left.\frac{\partial\omega}{\partial x_j}\right|_{x_j=x^*_j}, && X_K \in S_{\mathrm{T}}\,,
\end{aligned}
\tag{4.6}
$$

where the superscript * always denotes quantities associated with the perturbed process $\tilde{\chi}^*$. The process $\tilde{\chi}^*$ is considered as a quasi-static one, possible inertial forces being included into perturbing forces. The perturbed process can also be considered as a dynamic motion, which will be discussed later.

The disturbance measure ρ is defined as

$$
\rho = \rho(\tilde{\chi}^*, \tilde{\chi}^0, t) = \sup_{\tau\in[t_1,t]} E^{\mathrm{d}}(\tilde{\chi}^*, \tilde{\chi}^0, \tau)\,,
\tag{4.7}
$$

where $E^{\mathrm{d}}(\tilde{\chi}^*, \tilde{\chi}^0, \tau)$ is the energy supplied by a disturbance during the perturbed process $\tilde{\chi}^*$ in the time interval $[t_1, \tau]$. This energy is assumed in the form

$$
E^{\mathrm{d}} = W^{\mathrm{d}} + (W^*_{\mathrm{ext}} - W^0_{\mathrm{ext}}) \equiv W^{\mathrm{d}} + \Delta W_{\mathrm{ext}}\,,
\tag{4.8}
$$

where

$$
W^{\mathrm{d}} = W^{\mathrm{d}}(\tilde{\chi}^*, \tau) = \int_{t_1}^{\tau}\left(\int_V F^{\mathrm{d}}_j v^*_j\,\mathrm{d}V + \int_S T^{\mathrm{d}}_j v^*_j\,\mathrm{d}S\right)\mathrm{d}t
\tag{4.9}
$$

is the work done directly by the perturbing forces in the interval $[t_1, \tau]$ ($v_j = \partial\chi_j/\partial t$ denotes velocity), and ΔW_{ext} is the supplement of the energy

$$
W_{\mathrm{ext}} = W_{\mathrm{ext}}(\tilde{\chi}, \tau) = \int_{t_1}^{\tau}\left(\int_S T_j v_j\,\mathrm{d}S + \frac{\mathrm{d}W^{\mathrm{e}}}{\mathrm{d}t}\right)\mathrm{d}t
\tag{4.10}
$$

taken in the interval $[t_1, \tau]$ from external sources by the *loading device* in the process $\tilde{\chi}^*$ with respect to that in the fundamental process $\tilde{\chi}^0$. Here and in the following the prefix Δ denotes the difference of any corresponding quantities in the perturbed and fundamental processes at the same moment of time. For given processes $\tilde{\chi}^*$ and $\tilde{\chi}^0$, E^{d} is a continuous function of time, so we may write $\rho(t) = \max_{\tau\in[t_1,t]} E^{\mathrm{d}}(\tau)$. $\rho(\mathrm{t})$ is a non-negative, continuous and non-decreasing function.

From the energy balance we have

$$E^{\mathrm{d}} = E^* - E^0 \equiv \Delta E, \tag{4.11}$$

where

$$E = E(\tilde{\chi}, \tau) = \int_{t_1}^{\tau} \int_V s_{Kj} v_{j,K}\, \mathrm{d}V\, \mathrm{d}t + [W^{\mathrm{e}}]_{t_1}^{\tau} \tag{4.12}$$

is the total energy which must be supplied to the system: {body + loading device} in the time interval $[t_1, \tau]$ in order to realize the process $\tilde{\chi}$ in a quasi-static manner. (4.11) presents an alternative definition of the energy supplied by a disturbance if dynamic effects are neglected. If the same perturbed process $\tilde{\chi}^*$ were considered as a dynamic motion, then on the right hand side of (4.11) the kinetic energy of the system at time τ would appear as an additional term. Since the kinetic energy is positive definite, the disturbance measure would be increased, the distance between the processes being unaffected. This means that including inertial forces into perturbing forces we did not overlook a possibility of dynamic instability.

The distance d between the perturbed and fundamental processes is assumed in the form

$$d = d(\tilde{\chi}^*, \tilde{\chi}^0, t) = \int_V (x_j^* - x_j^0)(x_j^* - x_j^0) dV. \tag{4.13}$$

Definition *The process $\tilde{\chi}^0$ is stable in the time interval $[t_1, t_2]$ if and only if for every number $\varepsilon > 0$ there is another number $\delta > 0$ such that for every perturbed process $\tilde{\chi}^*$ and all $t \in [t_1, t_2]$*

$$\rho(\tilde{\chi}^*, \tilde{\chi}^0, t) < \delta \quad \textit{implies} \quad d(\tilde{\chi}^*, \tilde{\chi}^0, t) < \varepsilon.$$

The process $\tilde{\chi}^0$ is called unstable if it is not stable.

This definition with the accompanying definitions of $\tilde{\chi}^*, \rho$ and d constitutes a stability postulate.

4.3 Necessary Conditions for Stability of a Deformation Process

In this section some implications of the stability postulate are examined. In the following we will consider the stability *in the small*, defined as the stability of the process $\tilde{\chi}^0$ in every sufficiently small time interval $[\tau, \tau + \delta\tau] \subset [t_1, t_2]$. Evidently, stability in the small is necessary for stability.

Our aim is to examine the particular case when the fundamental and perturbed processes coincide up to time τ and then diverge abruptly, in the sense that the velocity fields in these processes at time τ are different (as velocity at time τ we understand $\lim_{\delta t \to 0}(v_j|_{\tau+\delta t})$). We assume that processes are sufficiently regular in the interval $[\tau, \tau + \delta\tau]$; in particular, the velocities in both processes are continuous and continuously differentiable with respect to X_K.

The energy functional $\mathrm{E}(\tilde{\chi}, t)$ becomes, for a fixed process $\tilde{\chi}$, a real function of time. Expanding this function into Taylor series we obtain

$$E(\tau + \delta t) = E(\tau) + E_1 \delta t + E_2 (\delta t)^2 + o(\delta t)^2 ,$$

where

$$\begin{aligned} E_1 = \frac{\mathrm{d}E}{\mathrm{d}t} &= \int_V s_{ij} v_{j,i} \, \mathrm{d}V + \int_{S_\mathrm{T}} \left(\frac{\partial \omega}{\partial x_j} v_j + \frac{\partial \omega}{\partial t} \right) \mathrm{d}S \\ &= \int_V \sigma_{ij} \epsilon_{ij} \, \mathrm{d}V - \int_{S_\mathrm{T}} T_j v_j \, \mathrm{d}S + \int_{S_\mathrm{T}} \frac{\partial \omega}{\partial t} \, \mathrm{d}S , \end{aligned}$$

$$\begin{aligned} E_2 = \frac{1}{2} \frac{\mathrm{d}^2 E}{\mathrm{d}t^2} &= \frac{1}{2} \int_V \left(\dot{s}_{ij} v_{j,i} + s_{ij} \dot{v}_{j,i} \right) \mathrm{d}V \\ &+ \frac{1}{2} \int_{S_\mathrm{T}} \left(\frac{\partial \omega}{\partial x_j} \dot{v}_j + \frac{\partial^2 \omega}{\partial x_i \partial x_j} v_i v_j + 2 \frac{\partial^2 \omega}{\partial x_j \partial t} v_j + \frac{\partial^2 \omega}{\partial t^2} \right) \mathrm{d}S \\ &= \frac{1}{2} \int_V \dot{s}_{ij} v_{j,i} \, \mathrm{d}V - \int_{S_\mathrm{T}} \left(\frac{\partial T_j}{\partial t} + \frac{1}{2} \frac{\partial T_j}{\partial x_i} v_i \right) v_j \, \mathrm{d}S \\ &+ \frac{1}{2} \left[\int_V \sigma_{ij} \dot{v}_{j,i} \, \mathrm{d}V - \int_{S_\mathrm{T}} T_j \dot{v}_j \, \mathrm{d}S \right] + \frac{1}{2} \int_{S_\mathrm{T}} \frac{\partial^2 \omega}{\partial t^2} \, \mathrm{d}S . \end{aligned}$$

All quantities in the above expressions are calculated at time τ. The body configuration at time τ is taken as the reference one, the same indexes referring to both the space and material coordinates. σ_{ij} is the Cauchy stress tensor (equal to s_{ij} in the reference configuration), $\epsilon_{ij} = \frac{1}{2}(v_{i,j} + v_{j,i})$ is the strain rate tensor at time τ, and the dot denotes the material time derivative. It is convenient to introduce the velocity functionals

$$J_1(\tilde{v}) = \int_V \sigma_{ij} \epsilon_{ij} \, \mathrm{d}V - \int_{S_\mathrm{T}} T_j v_j \, \mathrm{d}S , \tag{4.14}$$

$$J_2(\tilde{v}) = \frac{1}{2} \int_V \dot{s}_{ij} v_{j,i} \, \mathrm{d}V - \int_{S_\mathrm{T}} \left(\frac{\partial T_j}{\partial t} + \frac{1}{2} \frac{\partial T_j}{\partial x_i} v_i \right) v_j \, \mathrm{d}S \tag{4.15}$$

appearing in the expressions for E_1 and E_2, respectively.

First, we consider the case of the rigid-plastic material, for which the stress σ_{ij} depends on the actual strain rate. Since $\partial\omega/\partial t$ at time τ is the same in both the fundamental and perturbed processes, we have

$$\Delta E|_{\tau+\delta t} = \Delta J_1|_{\tau}\, \delta t$$

to first order. If ΔJ_1 is negative at time τ for some velocity field $\tilde{v}^*$, then ΔE for the corresponding process $\tilde{\chi}^*$ is negative in a sufficiently small time interval $(\tau, \tau + \delta\tau)$. Thus the disturbance measure $\rho = 0$ at time $\tau + \delta\tau$, and the distance d at time $\tau + \delta\tau$ is a small but finite positive number since the body configuration at time $\tau + \delta\tau$ is different in both processes. This implies instability. Hence the necessary condition for stability of the process $\tilde{\chi}^0$ in the interval $[t_1, t_2]$ is

$$J_1(\tilde{v}^*) \geqslant J_1(\tilde{v}^0) = \int_{S_u} T_j^0 v_j \, \mathrm{d}S \tag{4.16}$$

for all $\tau \in [t_1, t_2]$ and for every admissible velocity field $\tilde{v}^*$ taking the values prescribed over S_u. But this is just the extremum principle due to Hill [10], satisfied automatically if the yield surface is convex and the normality flow rule is adopted. For velocity fields which are additionally compatible with the stress field σ_{ij}^0 we have $\Delta J_1 = 0$, and for *such* fields ΔE needs to be calculated to second order. We will not consider this problem here.

In the case of elastic-plastic solids we have $\sigma_{ij}^* = \sigma_{ij}^0$, and ΔJ_1 vanishes identically by Green's theorem and the condition $\Delta v_j = 0$ over S_u. Similarily, the term in square brackets in the expression for E_2 is the same for both processes, since $\Delta \dot{v}_j = 0$ over S_u. Thus

$$\Delta E|_{\tau+\delta t} = \Delta J_2|_{\tau}\, (\delta t)^2$$

to second order, since the remaining terms cancel. In a way analogical to presented above it is easy to show that $\Delta J_2 < 0$ implies instability. Hence the necessary condition for stability of the process $\tilde{\chi}^0$ in the interval $[t_1, t_2]$ is

$$J_2(\tilde{v}^*) \geqslant J_2(\tilde{v}^0) = \frac{1}{2}\left(\int_{S_u} \dot{T}_j^0 v_j \, \mathrm{d}S - \int_{S_T} \frac{\partial T_j}{\partial t} v_j^0 \, \mathrm{d}S \right) \tag{4.17}$$

for all $\tau \in [t_1, t_2)$ and for every admissible velocity field $\tilde{v}^*$ taking the values prescribed over S_u. The inequality (4.17) is not satisfied automatically, and presents a true stability criterion.

In the examined case of loading, $J_2(\tilde{v})$ is just the velocity functional considered by Hill [11, 6] in connection with the first-order rate problem. The condition (4.17) presents (without the equality sign) the extremum principle due to Hill [11] which holds if the sufficient condition for uniqueness is satisfied. Moreover, for many problems of interest it can be shown that the first moment τ^{cr} beyond which (4.17) ceases to hold coincides with the moment of primary bifurcation. The connections between the present approach and the bifurcation theory are apparent. However, the interpretation of the critical moment τ^{cr} is different. While bifurcation can take place not earlier than τ^{cr}, the process becomes unstable in the present case not later than τ^{cr}. Note that the functional $J_2(\tilde{v})$ plays here an analogical role as the potential energy functional in examining stability of elastic conservative systems ($J_2(\tilde{v})$ is stationary at $\tilde{v} = \tilde{v}^0$ [11]).

4.4 Stability of an Equilibrium Configuration

The definition of stable deformation process can be directly used to examine stability of a fixed[1] equilibrium configuration. For this purpose we consider the stability of a particular fundamental process in which the velocities vanish identically, the functions $\bar{x}_j$ and ω being independent of time. In that case we have $W^*_{\text{ext}} \equiv W^0_{\text{ext}} \equiv 0$ and $E^{\text{d}} \equiv W^{\text{d}} \equiv E^*$ at any moment. The disturbance measure reduces to the maximum value of the work W^{d} done by perturbing forces. The stability of equilibrium may be examined in infinite time interval $[t_1, \infty)$.

From (4.17) we obtain that the necessary condition for stability of a fixed equilibrium configuration of elastic-plastic body is

$$\int_V \dot{s}^*_{ij} v^*_{j,i} \, \mathrm{d}V - \int_{S_\mathrm{T}} \frac{\partial T_j}{\partial x_i} v^*_i v^*_j \, \mathrm{d}S \geqslant 0 \tag{4.18}$$

for every admissible velocity field $\tilde{v}^*$ vanishing on S_u. In the special case of dead loading, $\partial T_j / \partial x_i = 0$ and the second term in (4.18) vanishes. Then the condition (4.18) with strict inequality takes the form proposed by Hill [3] as a sufficient condition for stability. There is also correspondence with Drucker's postulate of stability in the small of a time-independent system [4].

The sufficient condition for stability of a fixed equilibrium configuration is

$$W^{\text{d}}(\tilde{\chi}^*, t) = E(\tilde{\chi}^*, t) \geqslant c_1 \, d(\tilde{\chi}^*, \tilde{\chi}^0, t) \quad \text{for } d \leqslant c_2 \tag{4.19}$$

for every perturbed process $\tilde{\chi}$ and all $t > t_1$, where c_1 and c_2 are fixed positive numbers. The proof follows directly from the stability definition. The condition (4.19) is also sufficient for dynamic stability against initial disturbances, in the sense of the definition used by Nguyen and Radenkovic [5].

Drucker [4] proposed a criterion of stability in the large of a time-independent system, which in the present notation becomes $W^{\text{d}}(\tilde{\chi}^*, t) > 0$ for every $\tilde{\chi}^*$ and all $t > t_1$. In general this is neither sufficient nor necessary for stability in the present sense.

We also mention that Krawietz [12] used the measure for initial disturbances which is equivalent to (4.7) for $W_{\text{ext}} \equiv 0$.

[1] The present approach has the advantage that the equilibrium configuration may also be considered in moving coordinate basis. Results are coordinate invariant.

References

1. Klushnikov, V.D.: Stability of elastic-plastic systems (in Russian). Nauka, Moscow 1980.
2. Rice, J.R.: The localization of plastic deformation. Proc. 14th IUTAM Congr. Theor. and Applied Mech. (ed. W.T.Koiter), North-Holland, Delft 1976, 207-220.
3. Hill, R.: A general theory of uniqueness and stability in elastic-plastic solids.J.Mech. Phys.Solids **6** (1958) 236-249.
4. Drucker, D.C.: On the postulate of stability of material in the mechanics of continua. J. Mec. **3** (1964) 235-249.
5. Nguyen, Q.S.; Radenkovic, D.: Stability of equilibrium in elastic plastic solids. Proc. IU-TAM/IMU Symp. Appl. Methods of Functional Analysis to Probl.Mech., Marseille 1975, 403-414.
6. Hill, R.: Aspects of invariance in solid mechanics. Advances in Appl. Mech. Vol. 18, Acad. Press, New York, 1978, 1-75.
7. Hutchinson, J.W.: Plastic buckling. Advances in Appl. Mech. Vol. 14, Acad. Press, New York 1974, 67-144.
8. Miles, J.P.: On necking phenomena and bifurcation solutions. Arch. Mech. **32** (1980) 909-931.
9. Petryk, H.: A stability postulate for quasi-static processes of plastic deformation. GAMM Tagung, Wurzburg 1981. (To be published in Arch. Mech.)
10. Hill, R.: On the state of stress in a plastic-rigid body at the yield point. Phil. Mag. **42** (1951) 868-875.
11. Hill, R.: Uniqueness criteria and extremum principles in self-adjoint problems of continuum mechanics. J. Mech. Phys. Solids **10** (1962) 185-194.
12. Krawietz, A.: Stabilitat des Gleichgewichts Mechanischer Systeme. ZAMM **61** (1981) H4, T46.

Chapter 5
On Energy Criteria of Plastic Instability

5.1 Introduction

An observation based on long-standing experience is that a plastic body deformed under conditions which could result in a quasi-static, macroscopically uniform deformation process, in certain circumstances begins to deform in a quite different way. The term instability is commonly used in such cases though more in an intuitive than in a specified sense. Typical examples of so-understood instability are the buckling, bulging and necking, phenomena as "geometric" instabilities, snap-through phenomena as "dynamic" instabilities, and shear band localization of deformation as a "local" instability. Similar phenomena may naturally appear in the course of non-uniform deformations.

The different forms of instability in plastic solids have been investigated in the literature by using various theoretical approaches. The basic tool in analysis of "geometric" instabilities is nowadays the bifurcation theory due to Hill [1]÷[4], supplemented by studies of post-bifurcation behaviour and imperfection sensitivity (cf. [5]÷[8]). Appearing of uncontrolled, dynamic deformations is usually attributed to an instability of equilibrium and investigated on the basis of an energy-type criterion of instability of equilibrium, as e.g. those in [1, 9, 10]. The localization of deformation, either in thin metal sheets or within a three-dimensional body, is usually examined by considering a possible bifurcation within a band [11, 12] or by studying the growth of initial imperfections as in Marciniak's approach [13] (cf. [14, 17]). In many cases, an intuitive instability criterion is formulated for the particular problem considered, as in [18, 19].

None of the approaches mentioned above allows to investigate all the "geometric", "dynamic" and "local" instabilities in a unified way. In this paper we explore a possibility that the observed distinct forms of plastic instability are merely different symptoms of *instability of the fundamental deformation process in the energy sense*.

[0] Reprinted from: Plastic Instability, Proc. Considère Memorial, pp. 215-226, Presses de l'ecole nationale des ponts et chaussées, Paris 1985. H.Petryk, On energy criteria of plastic instability. Copyright © 1985.

H. Petryk, *Theory of Plastic Instability*,
https://doi.org/10.1007/978-3-031-98223-1_5

Such approach, based on a single energy-type postulate of stability of a quasi-static deformation process, has been proposed recently by the author [20]÷[22]. In the assumed definition of stability the persistent disturbances are considered for which an energy measure is adopted. For our present purposes, we use a simplified and less restrictive version of the postulate, starting from a stability criterion which is a consequence of the basic stability definition and not from the definition itself. To make the paper self-contained, the stability criterion is introduced here as an independent hypothesis. *In the criterion it is assumed that a sufficiently small deformation increment in a stable fundamental process requires less energy to be supplied from external sources to the system consisting of the body and loading device than any other kinematically admissible deformation increment.* The deformations are caused by varying loading conditions; in particular, if the loading conditions do not vary in time then the criterion of stability of a deformation process reduces to the familiar concept of stability of equilibrium in the energy sense. The criterion is proposed here for loadings which are conservative in an overall sense, generalizing in this way the previous criteria [21, 22] formulated for the surface tractions which were conservative at each surface point separately.

In formulating the stability postulate, the mechanical properties of the material are assumed to be time-independent but are otherwise arbitrary. However, the postulate itself imposes certain symmetry restrictions on the general form of constitutive relations (which will be studied elsewhere). Implications of the assumed stability criterion will be examined below assuming a class of nonlinear constitutive rate equations which admit a potential provided they are expressed in terms of rates of work-conjugate stress and strain measures [23, 4]. This class of constitutive relations contains the rate equations for conventional elastic-plastic solids obeying the normality flow rule as a special case. Existence of potential assures that the incremental moduli have the needed principal symmetry property. The evolution equations for the moduli are left unspecified though their form is also expected to be restricted by the stability requirement.

It is not the aim of the present paper to analyze particular examples, rather, to provide a synthesis of the known results by specifying possibly general circumstances under which the proposed energy criterion yields the known criteria of plastic instability.

5.2 Definitions and Assumptions

5.2.1 Notation

We are concerned with isothermal, quasi-static deformations of a continuous body of time-independent material which in a fixed reference configuration occupied a space domain V bounded by a piecewise-regular surface S. $\mathrm{d}\xi$ and $\mathrm{d}a$ are infinitesimal elements of volume and of surface area, respectively, in the reference configuration in which space or surface integrations will be performed. The position vector of a material element in the reference or current configuration is denoted by $\boldsymbol{\xi}$ or $\mathbf{x}$, respectively. All vector or tensor components are for simplicity taken relative to a fixed rectangular basis and denoted by latin subscripts which range from 1 to 3. A natural time does not appear at all throughout the paper and its role plays a scalar parameter t which is called time for simplicity. A deformation process is described by the equation $\mathbf{x} = \mathbf{x}(\boldsymbol{\xi}, t)$. For any local quantity φ, the derivatives

$$\varphi_{,i} \equiv \frac{\partial \varphi(\boldsymbol{\xi}, t)}{\partial \xi_i}, \quad \dot{\varphi} \equiv \frac{\partial \varphi(\boldsymbol{\xi}, t)}{\partial t}$$

and likewise the higher-order ones, are assumed to be at least piecewise continuous functions of $(\boldsymbol{\xi}, t)$. A field defined on V will be distinguished from its value at $\boldsymbol{\xi}$ by a superimposed tilda, when needed.

$\mathbf{u} = \mathbf{x} - \boldsymbol{\xi}$, $\mathbf{v} = \dot{\mathbf{x}} = \dot{\mathbf{u}}$ and $\dot{\mathbf{F}} = \partial \mathbf{v}/\partial \boldsymbol{\xi}$ denote the displacement, velocity and vèlocity gradient in the reference configuration, respectively. The deformation and stress are measured by the deformation gradient tensor $\mathbf{F} = \partial \mathbf{x}/\partial \boldsymbol{\xi}$ and the nominal stress tensor $\mathbf{s}$, respectively, which form a work-conjugate pair in the sense that the deformation work in the body is equal to

$$W = \int_V \int s_{ij}\, \mathrm{d}F_{ji}\, \mathrm{d}\xi = \int_V \int s_{ij} v_{j,i}\, \mathrm{d}t\, \mathrm{d}\xi, \tag{5.1}$$

with the summation convention for repeated indices. $\mathbf{s}$ is related to the symmetric Cauchy stress $\boldsymbol{\sigma}$ by the formula $F_{ij} s_{jk} = \det(\mathbf{F}) \sigma_{ik}$. We assume that $\mathbf{u}$ is a continuous function of $(\boldsymbol{\xi}, t)$, $\mathbf{F}$ varies continuously in time while $\mathbf{v}$ is a continuous function of place.

5.2.2 Constitutive Relations

A general assumption is made that the mechanical properties of the material do not depend in any way on a natural time. Starting from the Sect. 5.3 we will assume that the constitutive relations for the material can be expressed in the general rate form proposed by Hill [1, 2]:

$$\dot{s}_{ij} = \frac{\partial U}{\partial (v_{j,i})}, \quad 2U = \dot{s}_{ij} v_{j,i}\,, \tag{5.2}$$

no matter what is their original (objective) form (for the formulae relating (5.2) to the constitutive rate equations written in terms of objective work-conjugate measures, see [4]). U is a continuous, continuously differentiable and piecewise-continuously twice differentiable function of $\dot{\mathbf{F}}$, dependent also on the deformation history. Since the material is time-independent, the potential U is homogeneous of degree two and can be thus written in the form $(5.2)_2$. A value of U will be denoted equivalently as $U(\dot{\mathbf{F}}) = U(v_{j,i}) = U[\mathbf{v}]$. The homogeneous relationship (5.2) between $\dot{\mathbf{s}}$ and $\dot{\mathbf{F}}$ can also be written, by the Euler theorem, as

$$\dot{s}_{ij} = C_{ijkl}(\dot{\mathbf{F}})\, v_{l,k}\,, \quad C_{ijkl} = C_{klij}\,, \tag{5.3}$$

provided the second derivatives of U:

$$\frac{\partial^2 U(\dot{\mathbf{F}})}{\partial\left(v_{l,k}\right)\partial(v_{j,i})} \equiv C_{ijk1}(\dot{\mathbf{F}}) \tag{5.4}$$

are continuous at $\dot{\mathbf{F}}$ so that the incremental moduli tensor $\mathbf{C}(\dot{\mathbf{F}})$ is well defined. The notation $\mathbf{C}(\dot{\mathbf{F}})$ indicates the dependence of the moduli on the actual rate of strain, present in any model of elastic-plastic response. The dependence is homogeneous of degree zero and in general only piecewise-continuous, for instance, piecewise constant when the relation (5.3) is piecewise-linear. We will denote $\mathbf{C}(\dot{\mathbf{F}})$ equivalently as $\mathbf{C}[\mathbf{v}]$. In general, the body may be inhomogeneous so that U and $\mathbf{C}$ are tacitly assumed below to depend also expticitly on $\boldsymbol{\xi}$ (in a piecewise smooth manner).

5.2.3 Loading Device

We consider the body placed in a loading device which applies body forces and surface tractions and constraints surface displacements in a manner dependent on a scalar loading parameter λ varying in time. The surface tractions are allowed to be configuration-dependent and may even be functionals (dependent on λ) of displacement field on the whole body surface. However, the surface loading is assumed to be *conservative* in the following overall sense (cf. [3, 24]): For each *fixed* value of λ, the total work done by the applied surface tractions vanishes for all closed virtual displacement paths which are compatible with geometric constraints and restore the material surface points to their starting positions. If we denote by $\mathbf{T}$ the nominal surface tractions (per unit reference area) then this assumption is written as

$$\int_S \oint T_j \, \mathrm{d}u_j \, \mathrm{d}a = 0, \quad \lambda = \text{ const}\,. \tag{5.5}$$

As examples of such loadings, we can mention (i) the tractions acting on the contact surface with a deformable hyperelastic continuum, (ii) the loading by uniform pressure on a surface part of fixed perimeter, (iii) the loading by smoothly distributed springs, or (iv) the dead loading. At a certain $\boldsymbol{\xi}$ and fixed λ, $\mathbf{T}$ is (i) a functional of the surface displacement field, (ii) a function of the displacement and its surface gradient at $\boldsymbol{\xi}$, (iii) a function of the displacement at $\boldsymbol{\xi}$, or (iv) independent of other factors, respectively. Surface displacement may be constrained by contact with a rigid tool, however, if slipping on the tool-material interface is allowed then it must occur without friction. The nominal body forces $\mathbf{b}$ (per unit reference volume) are taken for simplicity to depend only on $\boldsymbol{\xi}$ and λ, though conservative configuration-dependent body forces could be considered as well.

From the above assumptions it follows that the total work done by the body forces and surface tractions in a virtual motion compatible with geometric constraints and leading from $\tilde{\mathbf{u}}^1$ to $\tilde{\mathbf{u}}^2$ at *fixed* λ (the right hand side expression in (5.6)) is path-independent. (This work is path-dependent if λ is varying during the motion). Hence, for each λ we may define *the potential energy of the loading device*, Ω, as a functional defined on a class of admissible displacement fields $\tilde{\mathbf{u}}$ such that

$$\Omega[\tilde{\mathbf{u}}^1;\lambda]-\Omega[\tilde{\mathbf{u}}^2;\lambda] = \int_V \int_{\mathbf{u}^1}^{\mathbf{u}^2} b_j\, \mathrm{d}u_j\, \mathrm{d}\xi + \int_S \int_{\mathbf{u}^1}^{\mathbf{u}^2} T_j\, \mathrm{d}u_j\, \mathrm{d}a, \quad \lambda = \mathrm{const}\,. \tag{5.6}$$

Formally, Ω is defined in this way to within an additive function of λ which may be chosen arbitrarily with no influence on the stability criteria discussed below. The choice may be done once the loading device is specified in order to give the term "potential energy" a physical meaning. Note that if λ is varying along a deformation path then the increment of the value of Ω is still path-independent but is no longer equal to the work done by the body forces and surface tractions with negative sign.

5.2.4 General Stability Criterion

For a class of kinematically admissible deformation processes, introduce *the energy functional* E defined by

$$E = W + \Omega\,. \tag{5.7}$$

In general, E is a functional of the whole deformation history due to the path-dependence of the deformation work W. With an appropriate specification of Ω (cf. the remark following the formula (5.6)), the increment of the value of E in a quasi-static deformation process can be interpreted as the amount of energy which has to be supplied in that process from external sources to the *system* consisting of the body and the loading device.

We are concerned with stability of an idealized, isothermal, quasi–static deformation process, called the fundamental process, which is intended to describe sufficiently slow deformations of a real body subject to varying loading. The value of any quantity in the fundamental process will be distinguished by the superscript ^{0}. At every stage of the deformation along the fundamental path, we consider a class of kinematically admissible branching paths. Note that a branching path need not satisfy here the conditions of continuing equilibrium; when it does then the term "bifurcation" will be used. We will consider only such deformation paths along which the increment δE of the value of the energy functional (5.7) from that at the branching point can be developed into a Taylor series with respect to a small time increment δt at least to the second order,

$$\delta E = \dot{E}\delta t + \frac{1}{2}\ddot{E}(\delta t)^2 + \cdots\,, \tag{5.8}$$

where $\dot{E}$ and $\ddot{E}$ are the first and second time derivatives of E taken aloang a branching path at the instant of branching. We postulate now the following stability criterion (cf. [21, 22]).

Criterion 1 *The fundamental process is stable only if*

$$\delta E \geq \delta E^0 \tag{5.9}$$

for all kinematically admissible branching paths and for sufficiently small δt.

Note that this criterion gives only necessary condition for stability. If (5.9) does not hold for some branching mode then the fundamental process is unstable in the energy sense, and it is conjectured that this corresponds to some observable form of instability.

Suppose now that the stresses, surface tractions and body forces vary continuously in time in any deformation process. This assumption is generally not satisfied for rigid-plastic solids or for unilateral constraints, therefore both are excluded from now on. Denote by the prefix Δ the difference of any corresponding quantities in the branching and fundamental processes at the instant of branching. From (5.1), (5.6) and (5.7) we have

$$\dot{E} - \dot{E}^0 \equiv \Delta\dot{E} = \int_V (s^0_{ij}\Delta v_{j,i} - b_j \Delta v_j)\, \mathrm{d}\xi - \int_S T^0_j \Delta v_j \, \mathrm{d}a \tag{5.10}$$

for all $\Delta\tilde{\mathbf{v}}$ compatible with the kinematical constraints. Since we have excluded unílateral constraints, from (5.8) and the Criterion 1 it follows that the expression (5.10) vanishes in a stable fundamental process for all admissible $\Delta\mathbf{v}$. This is just the virtual work principle which is satisfied *a fortiori* if the fundamental process corresponds to a quasi-static solution to the problem. From (5.8) and the Criterion *l* we obtain the following

Criterion 2 *The fundamental process is stable only if*

$$\Delta\ddot{E} \geq 0 \tag{5.11}$$

for all kinematically admissible branching paths.

5.3 Minimum Principle for Velocities

To find an explicit form of (5.11), we specify now the surface data in an incremental, linearized form. The material body surface is split into two complementary parts, S_u and S_T. On the part S_u the displacement increments $\delta\mathbf{u}$ are given functions of λ. On the part S_T the first-order expression for the increment $\delta\mathbf{T}$ of the nominal traction from its starting value $\mathbf{T}^0$ is assumed to consist, as in [3, 4], of two parts

$$\delta \mathbf{T} = \frac{\partial \mathbf{T}}{\partial \lambda}\delta\lambda + \mathbf{f}[\delta \mathbf{u}] + \cdots . \tag{5.12}$$

The first term on the right hand side of (5.12) represents the part of the traction increment which is independent of the deformation increment, and the second term is the deformation-sensitive part. Both terms may depend on $\boldsymbol{\xi}$ and λ. $\mathbf{f}[\mathbf{w}]$ is assumed to be a linear homogeneous function of a space vector $\mathbf{w}$ and its surface gradient at the surface point considered ($\mathbf{f}[\mathbf{w}]$ could depend in the same way also on higher order surface gradients of $\mathbf{w}$ with no influence on the following considerations). The assumption (5.5) of conservative loading impliess Hill's "self-adjointness" condition [3, 4]

$$\int_{S_T} \left(f_j[\mathbf{v}^{(2)}]v_j^{(1)} - f_j[\mathbf{v}^{(1)}]v_j^{(2)} \right) \mathrm{d}a = 0\,, \tag{5.13}$$

where $\tilde{\mathbf{v}}^{(1)}$ and $\tilde{\mathbf{v}}^{(2)}$ are any pair of continuous and piecewise-continuously differentiable vector fields whose difference vanishes over S_u.

Identifying the displacement fields $\tilde{\mathbf{u}}^2$ and $\tilde{\mathbf{u}}^1$ in the expression (5.6) with those on the branching and fundamental paths and using (5.12) and (5.13), we obtain the second-order formula

$$-\hat{\Delta}\Omega = \int_V b_j \hat{\Delta} u_j \,\mathrm{d}\xi + \int_{S_T} \left(T_j^0 + \frac{1}{2} f_j[\hat{\Delta} u] \right) \hat{\Delta} u_j \,\mathrm{d}a + \cdots . \tag{5.14}$$

All quantities in (5.14), and likewise their differences denoted by $\hat{\Delta}$, are calculated at an instant following the instant of branching. Hence, at the instant of branching when $\Delta\tilde{\mathbf{u}} = \mathbf{0}$ we have

$$-\Delta\ddot{\Omega} = \int_V \left(2\dot{b}_j \Delta v_j + b_j \Delta \dot{v}_j \right) \mathrm{d}\xi + \int_{S_T} \left((2\dot{T}_j^0 + f_j[\Delta \mathbf{v}])\Delta v_j + T_j^0 \Delta \dot{v}_j \right) \mathrm{d}a\,. \tag{5.15}$$

In calculating the second time derivative of ΔW, we take into account the possibility of a velocity-gradient discontinuity across a moving surface whose image S_D in the reference configuration moves with a normal speed v_n. By using the transport theorem, we obtain

$$\ddot{W} = \int_V \left(\dot{s}_{ij} v_{j,i} + s_{ij} \dot{v}_{j,i} \right) \mathrm{d}\xi - \int_{S_D} s_{ij} [\![v_{j,i}]\!] v_n \,\mathrm{d}a, \tag{5.16}$$

where $[\![\cdot]\!]$ denotes a jump across S_D with the usual sign convention. Since $\tilde{\mathbf{v}}$ is and remains continuous, the standard compatibility conditions require that there is a jump in accelerations $\dot{\mathbf{v}}$ across S_D such that

$$[\![v_{j,i}]\!] v_n = -[\![\dot{v}_j]\!] n_i, \tag{5.17}$$

where $\mathbf{n}$ is the (appropriately directed) unit normal to S_D. By substituting (5.17) to (5.16) and using the equilibrium conditions for $\mathbf{s}^0$ and the divergence theorem, we obtain that at the instant of branching there is

$$\Delta \ddot{W} = \Delta \left(\int_V \dot{s}_{ij} v_{j,i} \, d\xi \right) + \int_V b_j \Delta \dot{v}_j \, d\xi + \int_{S_T} T_j^0 \Delta \dot{v}_j \, da \,. \tag{5.18}$$

Finally, by combining (5.15) and (5.18) and using (5.2), (5.12) and (5.13), we arrive at the result

$$\frac{1}{2} \Delta \ddot{E} = \Delta J[\tilde{\mathbf{v}}] \,, \tag{5.19}$$

where

$$J[\tilde{\mathbf{v}}] = \int_V \left(U[\mathbf{v}] - \dot{b}_j v_j \right) d\xi - \int_{S_T} \left(\frac{\partial T_j}{\partial \lambda} \dot{\lambda} + \frac{1}{2} f_j[\mathbf{v}] \right) v_j \, da \tag{5.20}$$

is a functional defined on the class of continuous and piecewise-continuously differentiable velocity fields $\tilde{\mathbf{v}}$ taking the values prescribed over S_u at the instant of branching. For convenience, from now on $\tilde{\mathbf{w}}$ will denote an arbitrary continuous and piecewise-continuously differentiable vector field on V which vanishes over S_u. Each velocity field admissible in (5.20) has now the form $\tilde{\mathbf{v}}^0 + \tilde{\mathbf{w}}$. By substituting (5.19) in the Criterion 2, we obtain the following minimum principle for velocities.

Criterion 3 *The fundamental process is stable only if*

$$J[\tilde{\mathbf{v}}] \geq J[\tilde{\mathbf{v}}^0] \quad \textit{for all} \quad \tilde{\mathbf{v}} = \tilde{\mathbf{v}}^0 + \tilde{\mathbf{w}} \,. \tag{5.21}$$

This is an extension of the criterion derived in [21] to a wider class of loadings and with velocity gradient discontinuities taken into account. The Criterion 3 reduces the problem of "detecting" instabilities to establishing the circumstances under which the fundamental velocity field $\tilde{\mathbf{v}}^0$ fails to minimize the value of the functional (5.20).

The functional (5.20) coincides with that considered by Hill [3, 4]. As shown by Hill, its first weak (Gateaux) variation vanishes at $\tilde{\mathbf{v}} = \tilde{\mathbf{v}}^0$ if and only if $\tilde{\mathbf{v}}^0$ is a solution to the actual first-order rate boundary value problem. This variational principle appears here as a necessary condition for (5.21) and thus for stability of the fundamental deformation process.

The minimum principle (5.21) (with strict inequality) was proved originally by Hill [3] under the additional assumption that a condition sufficient for uniqueness of the solution is satisfied. On the contrary, (5.21) has been derived here from the postulated stability Criterion 1 and holds for any solution $\tilde{\mathbf{v}}^0$, unique or not, which corresponds to a stable deformation process.

5.4 Discussion

Along a typical path of deformation, the condition (5.21) (or (5.11), equivalently) will be satisfied with strict inequality for all non-zero $\tilde{\mathbf{w}}$ up to a certain critical stage beyond which the functional (5.20) becomes indefinite. If instability did not take place earlier for another reason ((5.11) is only necessary for stability) then the critical stage marks the onset of instability of the fundamental deformation path. Suppose that this is the case. Now, it is essential in what way the functional (5.20) becomes indefinite and what properties it has *at* the critical stage since this strongly influences the post-critical behaviour of the body. Three typical cases are discussed below which are related to the distinct observable forms of plastic instability mentioned in Sect. 5.1.

5.4.1 "Geometric" Instabilities

Suppose that, at the critical stage, (5.21) still holds but with equality for some $\tilde{\mathbf{v}}^* \neq \tilde{\mathbf{v}}^0$. Then $\tilde{\mathbf{v}}^*$, likewise $\tilde{\mathbf{v}}^0$, renders the functional (5.20) a minimum and thus also a stationary value. From the variational principle mentioned above it follows that $\tilde{\mathbf{v}}^*$ is another solution to the first-order rate boundary value problem, that is, at the critical stage we have bifurcation in velocities. Hence, in this case the search for the onset of instability in the sense of the energy criterion is reduced to a search for the corresponding bifurcation point. If the secondary post-bifurcation path is stable then it may replace the fundamental path beyond the critical point, leading in that way to an observable "geometric" instability.

In general, a bifurcation point may precede the considered critical stage with no consequence for stability of the fundamental path. We specify below the circumstances under which such cases are excluded so that the primary bifurcation takes place exactly at the critical stage. Suppose first that along the fundamental deformation path the potential U has continuous second derivatives with respect to $\dot{\mathbf{F}}$ at $\dot{\mathbf{F}}^0$, except possibly on certain surfaces, e.g. on the elastic-plastic interface. In other words, let the incremental moduli tensor $\mathbf{C}^0$ relating $\mathbf{s}^0$ to $\dot{\mathbf{F}}^0$ (cf. (5.3) and (5.4)) be well defined almost everywhere in V. In that case the second weak variation of the functional (5.20) at $\tilde{\mathbf{v}} = \tilde{\mathbf{v}}^0$ exists and is equal to

$$\delta^2 J[\tilde{\mathbf{v}}^0; \tilde{\mathbf{w}}] \equiv \frac{1}{2}\frac{\mathrm{d}^2}{\mathrm{d}\gamma^2} J[\tilde{\mathbf{v}}^0 + \gamma\tilde{\mathbf{w}}]\Big|_{\gamma=0} = I^0[\tilde{\mathbf{w}}]\,, \tag{5.22}$$

where γ is a scalar and

$$I^0[\tilde{\mathbf{w}}] = \frac{1}{2}\int_V C^0_{ijkl} w_{j,i} w_{l,k}\,\mathrm{d}\xi - \frac{1}{2}\int_{S_T} f_j[\mathbf{w}] w_j\,\mathrm{d}a \tag{5.23}$$

is a quadratic functional defined on the class of $\tilde{\mathbf{w}}$ as above.

The proof follows at once from the standard theorem of the calculus of variations applied separately to each subdomain in V in which the needed regularity conditions are satisfied, provided that in each subdomain the moduli tensor $\mathbf{C}[\mathbf{w}^0 + \gamma\mathbf{w}]$ depends continuously on γ in an interval of γ containing 0 and independent of place. If this is not the case then we can separate a region $R(\hat{\gamma})$ in V such that $C[\mathbf{v}^0 + \gamma\mathbf{w}]$ is a continuous function of γ in $V \backslash R(\hat{\gamma})$ for $|\gamma| < \hat{\gamma}$ and apply the theorem in $V \backslash R(\hat{\gamma})$.

By the assumption introduced above and the assumed regularity of U, the volume of $R(\hat{\gamma})$ can be taken to tend to zero as $\hat{\gamma} \to 0$ and in the limit we obtain (5.22).

Non-negativeness of the second variation is necessary for a mimimum. Hence, when the moduli $\mathbf{C}^0$ are well defined then the Criterion 3 yields the

Criterion 4 *The fundamental process is stable only if*

$$I^0[\tilde{\mathbf{w}}] \geq 0 \quad \textit{for all } \tilde{\mathbf{w}}\,. \tag{5.24}$$

Suppose now that the moduli C^0 define the incrementally linear "comparison solid" [1, 2, 4] such that

$$\Delta\dot{s}_{ij}\Delta\dot{F}_{ji} \geq C^0_{ijkl}\Delta\dot{F}_{ji}\Delta\dot{F}_{lk} \quad \text{for all } \Delta\dot{\mathbf{F}}\,, \tag{5.25}$$

where $\Delta\dot{\mathbf{s}} = \dot{\mathbf{s}} - \dot{\mathbf{s}}^0$, $\Delta\dot{\mathbf{F}} = \dot{\mathbf{F}} - \dot{\mathbf{F}}^0$ and $\dot{\mathbf{s}}$ is related to $\dot{\mathbf{F}}$ by the constitutive equation (5.2). The condition (5.25) is a weakened form of the "relative convexity" property [2, 4] which requires (5.25) to hold also if $\dot{\mathbf{F}}^0$ is replaced by an arbitrary velocity gradient. For instance, (5.25) is satisfied for the elastic-plastic solids with piecewise-linear incremental response subject to the normality flow rule, either at a regular point or at a vertex on the yield surface, provided that the moduli $\mathbf{C}^0$ in the plastic zone correspond to the "fully" plastic branch [1, 5, 4].

From (5.25) by the same argument as in [4] we obtain that a bifurcation in velocities is excluded and (5.21) is satisfied as long as the functional I^0 is positive definite (i.e. $I^0[\tilde{\mathbf{w}}] > 0$ if $\tilde{\mathbf{w}} \neq \tilde{\mathbf{0}}$). In usual circumctances examined so far in many papers (cf. [5]÷[8]) this is so along the fundamental deformation path up to the stage when the primary bifurcation takes place. The corresponding eigenmode $\tilde{\mathbf{w}}^* = \tilde{\mathbf{v}}^* - \tilde{\mathbf{v}}^0$ renders the functional I^0 the zero stationary value. In general, beyond this stage the functional I^0 becomes indefinite; the Criterion 4 now shows that the instant of the primary bifurcation coincides with the onset of instability of the fundamental path in the energy sense. In those circumstances, the onset of buckling or necking found by using Hill's bifurcation theory results also from the Criteria 3 or 4 being now equivalent to each other. This is a generalization of the previous result [22] obtained for the conventional elastic–plastic solids with a smooth yield surface and for less general loading conditions. The Criterion 2 gives additionally an energy interpretation of the primary bifurcation as well as a reason for rejecting the fundamental post-bifurcation path as being unstable in the energy sense. The secondary post-bifurcation path *may* be stable; if there are more post-bifurcation paths then the stable path must minimize the increment δE of the energy functional (5.7), developed into a series up to the order needed.

5.4.2 "Dynamic" Instabilities

To show that the Criterion 2 (or 3) can also be used to find the onset of uncontrolled dynamic deformations, we derive from it the following

Criterion 5 *The fundamental process is stable only if*

$$I[\tilde{\mathbf{w}}] \equiv \int_V U[\mathbf{w}]\,\mathrm{d}\boldsymbol{\xi} - \frac{1}{2}\int_{S_T} f_j[\mathbf{w}]w_j\,\mathrm{d}a \geq 0 \quad \textit{for all } \tilde{\mathbf{w}}\,. \tag{5.26}$$

In proof, consider the quantity $\left(J[\tilde{\mathbf{v}}^0 + \gamma\tilde{\mathbf{w}}] - J[\tilde{\mathbf{v}}^0]\right)/\gamma^2$ and put $\gamma \to \infty$ while $\tilde{\mathbf{w}}$ is held fixed. By using homogeneity and continuity of U and $\mathbf{f}$, we obtain in the limit $I[\tilde{\mathbf{w}}]$. Thus, if $I[\tilde{\mathbf{w}}]$ is negative then $J[\tilde{\mathbf{v}}^0 + \gamma\tilde{\mathbf{w}}]$ does not satisfy (5.21) for γ large enough. Hence, $I[\mathbf{w}] \geq 0$ is necessary for stability, and the Criterion 5 has been proved.

Alternatively, the condition (5.26) can be obtained as a necessary condition for the *stability of equilibrium*, defined as the stability of the degenerate deformation process in which an equilibrium configuration of the body does not vary in time. The corresponding loading conditions cannot depend on time so that we must have $\dot{\lambda} = 0$ which implies $\dot{\mathbf{b}} = \mathbf{0}$ in V and $\dot{\mathbf{u}} = \mathbf{0}$ on S_u. Then (5.21) reduces simply to (5.26) (cf. [21, 22]). Under these conditions, $I[\delta\tilde{\mathbf{u}}]$ is the second-order expression for the amount of energy which has to be supplied to the system (body and loading device) along a direct virtual quasi-static deformation path leading from the equilibrium state to a neighbouring configuration defined by the displacement increment $\delta\tilde{\mathbf{u}}$ (cf. (5.8), (5.10), (5.19) and the remark following (5.7)). Relation to the familiar energy criteria of stability of equilibrium (cf. [1, 9]) is apparent, however, (5.26) is here necessary for stability rather than sufficient.

If (5.26) is not satisfied then there are certain modes of departure from the equilibrium state along which the energy is released from the system and may, at least in principle, be inverted into the kinetic energy since the modes require no quasi-statically applied increment of the loading parameter λ. Then, the system *may* start spontaneously to move dynamically; in that sense the Criterion 5 is a criterion for "dynamic" instabilities.

Consider now the stage on the deformation path at which there is an eigenmode $\tilde{\mathbf{w}}^* \neq \tilde{\mathbf{0}}$ which renders the functional I in (5.26) the stationary value necessarily equal to zero. Evidently, $\tilde{\mathbf{w}}^*$ is a solution to the first-order rate boundary value problem obtained for $\dot{\lambda} = 0$. If the functional I is positive definite before this stage then it becomes usually indefinite beyond this stage. Now, observe that that stage is necessarily reached *when the loading parameter λ attains its analytic extremum value versus a typical displacement on the fundamental deformation path*. For, $\dot{\lambda} = 0$ and $\tilde{\mathbf{v}}^0 \neq \tilde{\mathbf{0}}$ at the extremum point so that the fundamental mode $\tilde{\mathbf{v}}^0$ itself constitutes the eigenmode. This is a generalization of the classical maximum load criterion which is obtained here in the special case when the given loads vary monotonically with λ.

5.4.3 "Local" Instabilities

All the criteria discussed above involve an integral over the body volume. For such global criteria to be satisfied, a prerequisite is that certain local conditions hold. The local conditions discussed below involve the material mechanical properties at a single point only and are independent of the boundary conditions corresponding to the fundamental deformation process.

Our starting point is again the Criterion 3 as equivalent to the Criterion 2. Observe first that in a stable process the volume integral in (5.20) must be minimized by $\tilde{\mathbf{v}}^0$ within the class of all continuous and piece-wise-continuously differentiable velocity fields which coincide with $\tilde{\mathbf{v}}^0$ over the whole body surface S. Now, the theorem due to Graves [25] gives a local necessary condition for this (the condition (5.27) below) so that we obtain the following criterion for "local" instabilities.

Criterion 6 *The fundamental process is stable only if at every point in V*

$$U(v^0_{j,i} + g_j n_i) - U(v^0_{j,i}) - \dot{s}^0_{ij} g_j n_i \geq 0 \quad \textit{for all } \mathbf{g}, \mathbf{n}\,. \tag{5.27}$$

This criterion was derived in [22] (under less general assumptions) by considering a kinematically admissible mode of localization of deformation within a disc-like region whose thickness grew from zero at the initial instant of localization. It has been shown that if (5.27) does not hold then a localization mode of this type can be found which corresponds to the increment δE of the energy functional less that δE^0 by a third-order term.

Suppose now that (5.27) is satisfied in the body along the fundamental deformation path up to the critical stage and then ceases to hold, implying instability of the fundamental deformation process. Suppose also that (5.27) still holds at the critical stage, but with equality for some non–zero vectors $\mathbf{g}^*, \mathbf{n}^*$. Then the left hand side expression in (5.27) attains the minimum (zero) value for these $\mathbf{g}^*, \mathbf{n}^*$ and, consequently, its partial derivatives with respect to g_j vanish. This yields

$$(\dot{s}^*_{ij} - \dot{s}^0_{ij}) n^*_i = 0 \tag{5.28}$$

at the critical stage, where $\dot{s}^*_{ij}$ is related by the constitutive equation (5.2) to the velocity gradient $(v^0_{j,i} + g^*_j n^*_i)$. $\dot{\mathbf{s}}^*$ can be envisaged as the stress-rate within a vanishingly thin band of orientation $\mathbf{n}^*$ at the initial instant of localization since the velocity gradient within such bands is just of such a form. Now, (5.28) is the condition of continuing equilibrium across the band, that is, the condition for *bifurcation* within the band [12]. Note that the term "bifurcation" is in this context used in a local sense only and not necessarily in the sense of an exact bifurcation in a finite body. The condition (5.28), re-derived here from the general stability criterion, has been widely used in the literature (cf. [12], [14]÷[17]) as a bifurcation criterion for the onset of the shear band localization. It has been usually used for piecewise-linear constitutive rate equations which not necessarily admit a potential but under the assumption that the moduli inside and outside the band are the same.

The above considerations allow to interpret the Criterion 6 as the energy criterion for localization of deformation within narrow regions inside the body.

Consider a material point at which the moduli $\mathbf{C}^0$ are well defined. By substituting $\gamma\mathbf{g}$ in place of $\mathbf{g}$ in (5.27) and putting $\gamma \to 0$ or $\gamma \to \infty$, we obtain that the inequalities

$$C^0_{ijkl} g_j n_i g_l n_k \geq 0 \qquad \text{for all } \mathbf{g}, \mathbf{n}, \tag{5.29}$$

$$U(g_j n_i) \geq 0 \qquad \text{for all } \mathbf{g}, \mathbf{n}, \tag{5.30}$$

respectively, are necessary for the local stability. (5.29) is the known ellipticity condition, associated with the criterion $\det(C_{ijkl} n^*_i n^*_k) = 0$ which follows from (5.28) if the moduli within the band are equal to $\mathbf{C}^0$ [12].

The condition (5.30) can alternatively be obtained directly by applying Graves' theorem to (5.26) and is thus necessary also for stability of equilibrium. Hence, violation of (5.30) may lead to a *dynamic localization process* associated with an "internal snap-through" within the body. In this context it is perhaps worthwhile to mention that (5.30) is necessary and sufficient for the speed of all dynamic acceleration waves propagating into the material *being at rest* to be real. For, from the wave propagation condition (cf. [26]) and $(2)_2$ it follows that the expression in (5.30) with $|\mathbf{g}| = |\mathbf{n}| = 1$ is proportional to the square of wave speed.

Now, consider as a special case the conventional elastic-plastic solid for which the moduli $\mathbf{C}$ take constant values in each of two half-spaces obtained by dissection of the strain-rate space by the hyperplane tangent to the smooth yield surface in strain space. Then (5.29) is implied by (5.30) since each velocity gradient corresponds to the moduli $\mathbf{C}^0$ provided its sign is appropriately chosen. Consequently, in the conventional elastic-plastic solids the loss of ellipticity by the plastic moduli results in loss of the stability of equilibrium in the energy sense. In that case, a quasi-static study of post-bifurcation behaviour may be of no physical meaning.

Acknowledgement

A part of this work has been written when I stayed at the Department of Mechanical Engineering, Dortmund University, as a Fellow of the Alexander von Humboldt Foundation. The support of the AvH Foundation is gratefully acknowledged.

References

1. Hill, R., J. Mech. Phys. Solids, **6** (1958) 236.
2. Hill, R., J. Mech. Phys. Solids, **7** (1959) 209.
3. Hill, R., J. Mech. Phys. Solids, **10** (1962) 185.
4. Hill, R., Adv. Appl. Mech., Vol. 18 (1978) 1.

5. Sewell, M.J., Stability (edited by H. Leipholz), p.85. Univ. of Waterloo Press (1972).
6. Hutchinson, J.W., Adv. Appl. Mech., **14** (1974) 67.
7. Storåkers, B., Arch. Mech., **27** (1975) 821.
8. Miles, J.P., Arch. Mech., **32** (1980) 909.
9. Drucker, D.C., J. Méc., **3** (1964) 235.
10. Nguyen, Q.S. and Radenkovic, D., Lecture Notes in Mathematics, 503, p.403. Springer-Verlag (1976).
11. Stören, S. and Rice, J.R., J. Mech. Phys. Solids, **23** (1975) 421.
12. Rice, J.R., Proc. 14th Int. Cong. Theoretical and Applied Mechanics (edited by W.T. Koiter), p.207. North-Holland (1977).
13. Marciniak, Z. and Kuczynski, K., Int. J. Mech. Sci., **9** (1967) 609.
14. Needleman, A. and Rice, J.R., Mechanics of Sheet Metal Forming (edited by D.P. Koistinen and N.-M. Wang), p.237. Plenum Press (1978).
15. Hutchinson, J.W. and Tvergaard, V., Int. J. Solids Struct., **17** (1981) 451.
16. Asaro, R.J. and Rice, J.R., J. Mech. Phys. Solids, **25** (1977) 309.
17. Peirce, D., J. Mech. Phys. Solids, **31** (1983) 133.
18. Swift, H.W., J. Mech. Phys. Solids, **1** (1952) 1.
19. Hart, E.W., Acta Met., **15** (1967) 351.
20. Petryk, H., Arch. Mech., **35** (1983) 753.
21. Petryk, H., Stability in the Mechanics of Continua (edited by F.H. Schroeder), p.262. Springer-Verlag (1982).
22. Petryk, H., Plasticity Today: Modelling, Methods and Applications (edited by A. Sawczuk and L. Bianchi). Elsevier (1985).
23. Hill, R., J. Mech. Phys. Solids, **16** (1968) 229; 315.
24. Sewell, M.J., Arch. Rat. Mech. Anal., **23** (1967) 327.
25. Graves, L.M., Duke Math. J., **5** (1939) 656.
26. Hill, R., J. Mech. Phys. Solids, **10** (1962) 1.

Chapter 6
The Energy Criteria of Instability in Time-Independent Inelastic Solids

Abstract Conditions sufficient for instability of deformation are examined for a class of incrementally nonlinear, time independent inelastic solids. Instability of a quasi-static deformation process (path) at varying loading is distinguished from a narrower concept of an unstable equilibrium state. The energy criterion is extended to deformation paths under general assumptions which ensure that the incremental boundary value problem can be given a variational formulation. For a discretized problem, fulfillment of the instability condition along a path is shown to imply either instability of the traversed equilibrium states in a dynamic sense or persisting possibility of quasi-static bifurcation at varying loading. For a continuum, instability at varying loading is interpreted as sensitivity of the incremental deformation to arbitrarily small perturbing forces.

6.1 Introduction

The following question is investigated here: what processes of inelastic deformation cannot be practically realized in a physical system. A deformation process is thought to represent a theoretical quasi-static solution to an initial-boundary value problem formulated for an inelastic continuous body subject to *varying* loading. Real material properties and boundary conditions are assumed to be adequately modeled. The question posed above is related to the well known fact that there are simple and theoretically correct solutions, e.g. describing macroscopically uniform straining, which at sufficiently advanced deformations are no longer in accord with experimentally observed deformation modes. One of possible explanations is that this is a symptom of an instability of the solution beyond a certain critical stage.

Instability in inelastic solids at slowly varying loading may manifest itself in various ways, not necessarily involving observable dynamic effects. Familiar examples

[0] Reprinted from: Archives of Mechanics 43, 519-545, 1991. H.Petryk, The energy criteria of instability in time-independent inelastic solids. Copyright © 1991, with permission from IPPT PAN.

H. Petryk, *Theory of Plastic Instability*,
https://doi.org/10.1007/978-3-031-98223-1_6

are buckling or necking as "geometric" instabilities and strain localization within shear bands as a "material" instability. According to the theory of uniqueness and stability in time-independent inelastic solids developed by Hill [3, 4], preceded by Shanley's [23] pioneering observation of the essence of plastic column buckling, such phenomena need not be connected with loss of stability of *equilibrium* on the fundamental deformation path. Rather, they can be explained in many cases by appealing to the bifurcation theory or to sensitivity to initial imperfections (cf. e.g. [11, 16]).

Alternatively, the above mentioned phenomena can be consistently regarded as symptoms of instability provided that this term concerns not an equilibrium *state* (at constant loading) but rather a deformation *process* (or *path* – at varying loading). Traditionally, stability of a quasi-static deformation path is often identified with stability of equilibrium states traversed by the path. However, for path-dependent and incrementally nonlinear solids distinction must be made between these two kinds of instability. Stability of equilibrium means that the effects of disturbing influences become negligible when the strength of disturbance is vanishingly small *and* the value of the loading parameter is kept fixed. For path-dependent inelastic solids we generally cannot expect that after a transitory perturbed motion the initial equilibrium state will be restored exactly, even in an asymptotic sense. Now, if the final effect is very small but nonzero and if a quasi-static loading program is continued, there is no guarantee that the distance between the *subsequent* configurations along the perturbed and fundamental paths at the same loading levels will remain small; note that continuous dependence of this distance on initial data is necessarily broken e.g. at a bifurcation point. This observation alone indicates that stability of an equilibrium state and of a quasi-static process of inelastic deformation are essentially different concepts.[1]. The distinction is even more clear when the effect of application of small perturbing forces is studied under the assumption of a time-independent material. Depending on whether such forces are applied at constant loading or at varying loading, different branches of the incrementally nonlinear constitutive law can be activated. In the former case the straining "direction" is defined by the disturbance itself while in the latter, it is primarily determined by the fundamental velocity field; for elastic-plastic solids the "stiffiness" in the latter case can be much smaller (cf. the reduced and tangent modulus "stiffiness" of the Shanley column).

It can be concluded that the question of stability of a deformation process, and also its relation to bifurcation suggested by the above remarks, deserve more thorough studies. In particular, stability of post-bifurcation paths should be examined in order to establish which of them can have a practical meaning. In this paper, we will

[1] The scale of time and the meaning of a perturbed motion are also different: while in examining stability of equilibrium a natural time scale can be adopted and inertia effects should then be taken into account, for a quasi-static loading program the role of "time" is usually played by the loading parameter and it is then natural to neglect interia effects.

examine stability of an isothermal quasi-static process of rate-independent inelastic deformation under varying loading at presence of small disturbing influences.[2]

Although the idea itself is not new (it could be traced back to Considère) and became used in particular problems, a respective general theory is lacking. To the author's knowledge, general stability conditions for such processes were, in effect, only postulated (cf. e.g. [2, 13, 18, 19, 1]). On the contrary, in the present paper certain conditions for instability are not only postulated but also related to a defined kind of instability of motion for a broad class of problems.

The general assumptions accepted here are essentially the same as in the bifurcation theory developed by Hill [4, 5, 6, 8]. It is assumed that the constitutive rate equations for the material do not contain a natural time and admit a potential, and that the incremental loading is conservative (possibly in an overall sense). Incremental nonlinearity of the material is otherwise arbitrary: for instance, elastoplastic models obeying the normality flow rule, with or without the yield-surface vertex effect, as well as so-called softening materials are included. The above assumptions ensure that any solution to the first-order rate boundary value problem will be a critical point of a velocity functional [4, 6]. On the basis of an energy interpretation of the functional, the author proposed [18, 19] the energy criterion of instability of a quasi-static deformation process. The condition of instability of equilibrium (cf. [9]) was included as a special case. The previous work [19] established a definite connection between the onset of instability of the fundamental deformation path, found from the second-order energy criterion, and the primary bifurcation point. However, the relation to general concepts of instability of motion and the meaning of predicted instability remained to be clarified.

Disturbing influences are often thought as perturbations of initial conditions. In reality, any material body is subject to perpetual disturbing influences *during* a deformation process; in the general theory of stability of motion such influences are called *persistent disturbances* (cf. e.g. [15]). They can be represented, for instance, by small perturbing forces acting independently of the loading program. If application of arbitrarily small perturbing forces in any finite interval of a loading parameter can cause finite deviations from a theoretical, unperturbed deformation path, then such a path may be regarded as being unstable and, consequently, practically unrealizable in a physical system.

The above concept of instability is, of course, not the only possible one: there is a vast literature on various theories of stability of motion, in particular, on Lyapunov's theory where *initial* disturbances are considered. Moreover, the same solution can be stable or not depending on the chosen measures of the distance and disturbance: this is well known in the context of stability of equilibrium in elastic continua (cf. [14]). In absence of sufficiently strong reasons for assuming *a priori* a particular mathematical definition of stability or instability in inelastic continua, the following approach is adopted here: qualitative properties of certain post-critical solutions are first selected as *possible candidates* for conditions sufficient for instability, and then examined in

[2] Disturbing influences (called also disturbances) should be carefully distinguished from geometric or material imperfections: the former perturb the motion of a given material system while the latter change the system itself.

detail in order to learn whether (or to what extent) they do imply instability in a physically acceptable sense. More specifically, the purpose of the present paper is to establish a connection between the energy criterion of path instability and the instability for persistent disturbances mentioned above.

That approach is new and will be shown to lead to the conclusions which are in accord with but are not derivable from Hill's theory of bifurcation and stability. Roughly speaking (details are given in subsequent sections), it will be demonstrated that if the basic velocity functional [4, 6] is not minimized along a solution path by the actual velocity field, then that deformation process is unstable in the sense of sensitivity of the incremental deformation to arbitrarily small perturbing forces. This in turn will provide additional and essential information, not given by Hill's theory, which post-bifurcation paths can be stable and thus practically realizable. Usually, the initial imperfection approach is adopted to obtain such information, and in simplest cases the predictions of both approaches will be similar. However, the present approach gives definite advantages since no comparable general criterion for incrementally nonlinear solids has been established yet via the considerations of initial imperfections. Therefore, a numerical analysis of development of various initial inhomogeneities for more complex examples, being avoided here, is still required in the usual approach to eliminate the possibility that the conclusions are only valid for particular imperfections. Moreover, from a physical point of view it is not clear why instabilities which are observed e.g. at large plastic strains should always have a form decided by *initial* imperfections. Of course, there are other advantages of the initial imperfection method as e.g. assessment of the effect of geometrical inaccuracies on the buckling loads.

A finite solid body with specified boundary conditions is considered here; instability of uniform deformation at the level of a material element will be examined in a separate paper. Sect. 6.3 and 6.4 constitute the principal part of the present paper; for convenience of the reader, more important statements are distinguished in the text in italic and summarized in Sect. 6.6. The topic requires precise formulation of the assumptions and definitions; this is done in Sect. 6.2, along with certain extensions of Hill's results and their novel energy interpretation.

6.2 Preliminaries

6.2.1 Notation

Throughout the paper the standard symbolic notation is used. Vectors or tensors are denoted by boldface letters, and their Cartesian components are denoted by lower case Latin subscripts for which the summation convention is adopted. Contraction over two pairs of the subscripts is distinguished by a dot between two symbols, e.g. $A_{ij}B_{ji} = \mathbf{A} \cdot \mathbf{B}$, $C_{ijkl}B_{lk} = (\mathbf{C} \cdot \mathbf{B})_{ij}$, while $A_{ik}B_{kj} = (\mathbf{AB})_{ij}, p_i A_{ij} = (\mathbf{pA})_j$ or $p_i q_i = \mathbf{pq}$. The upper index T denotes a transpose, and $(A_{ij}A_{ij})^{1/2} = |\mathbf{A}|$.

In a reference configuration which is arbitrary but fixed, a continuous solid body, in general inhontogeneous, is assumed to occupy a spatial domain V bounded by a piecewise-regular surface S with a unit normal $\boldsymbol{\nu}$. $\boldsymbol{\xi}$ and $\mathbf{x}$ denote position vectors in the reference and current configuration, respectively, and $\mathbf{u} = \mathbf{x} - \boldsymbol{\xi}$ stands for displacement. A field defined on $\overline{V} = V \cup S$ is distinguished from its value at some $\boldsymbol{\xi}$ by a superimposed tilde. The symbol ∇ denotes the spatial gradient evaluated in the reference configuration. t stands for a scalar time-like parameter, called time for simplicity, identified with a natural time only when examining stability of an equilibrium state. A dot over a symbol denotes the rate of change with respect to t, understood as a material time derivative in the *right-hand* sense.

Finite strains and rotations are considered while their rates are assumed to be bounded. For the sake of conciseness, the formulae appearing in stability considerations are expressed in terms of rates of the deformation gradient $\mathbf{F} = \nabla\mathbf{x}$ and the (unsymmetric) nominal stress $\mathbf{N}$ (related to the symmetric Cauchy stress σ and Kirchhoff stress $\boldsymbol{\tau}$ by $\mathbf{FN} = \det(\mathbf{F})\sigma = \boldsymbol{\tau}$). The formulae can be transformed in the routine way to other stress and strain measures when needed.

The first weak variation (Gateaux differential) of a functional J at its argument value $\tilde{\mathbf{v}}$ in the direction $\tilde{\mathbf{w}}$ is denoted by

$$\delta J(\tilde{\mathbf{v}}, \tilde{\mathbf{w}}) \equiv \frac{\mathrm{d}}{\mathrm{d}\gamma} J(\tilde{\mathbf{v}} + \gamma\tilde{\mathbf{w}})\bigg|_{\gamma=0}, \tag{6.1}$$

where $\tilde{\mathbf{w}} = \delta\tilde{\mathbf{v}}$ stands for an admissible variation of $\tilde{\mathbf{v}}$ and γ is a real variable.

In the following, $\mathbf{v} = \dot{\mathbf{x}}$ stands for velocity.

6.2.2 Constitutive Rate Equations

A general assumption is made that the mechanical properties of the material in isothermal deformation do not depend in any way on a natural time. Under restriction to piecewise smooth deformation paths, it is assumed that the stresses vary in time in a continuous and piecewise smooth manner. At the current state of a material element, a (homogeneous) constitutive relationship between the (right-hand) rates of stress and strain, no matter in which variables it is originally formulated, can be expressed as

$$\dot{\mathbf{N}} = \dot{\mathbf{N}}(\dot{\mathbf{F}}) \tag{6.2}$$

The constitutive function $\dot{\mathbf{N}}(\cdot)$ itself is functionally dependent on the strain history prior to the current instant, however, this dependence (as well as a piecewise-smooth dependence on $\boldsymbol{\xi}$) will not be indicated explicitly for simplicity. Following Hill [4], we assume that Eq. (6.2) admits a continuously differentiable potential $U(\dot{\mathbf{F}})$ (necessarily homogeneous of degree two), viz.

$$\dot{\mathbf{N}} = \frac{\partial U}{\partial \dot{\mathbf{F}}^T} = \mathbf{C}(\dot{\mathbf{F}}) \cdot \dot{\mathbf{F}}, \quad U(\dot{\mathbf{F}}) = \frac{1}{2}\dot{\mathbf{N}}(\dot{\mathbf{F}}) \cdot \dot{\mathbf{F}}, \quad \mathbf{C} = \frac{\partial^2 U}{\partial \dot{\mathbf{F}}^T \partial \dot{\mathbf{F}}^T}. \tag{6.3}$$

The resulting dependence of the instantaneous moduli $C_{ijkl} = C_{klij}$ on the direction of strain rate is assumed to be piecewise smooth but may be arbitrarily nonlinear and discontinuous. Existence of the potential and symmetry of the moduli are preserved when Eq. (6.2) is expressed in terms of rates of any *work-conjugate* pair of stress and strain measures [8]. For elastic materials, this symmetry property is a consequence of existence of a strain energy potential. It is well known [3] that the form (6.3) is valid for conventional elastoplastic models (with a smooth yield surface and a linear relationship (6.2) for the loading branch) if and only if the normality flow rule holds in work-conjugate variables. Equation (6.3) is acceptable also at a yield surface vertex but then it is no longer a consequence of the normality flow rule alone. More generally, Eqs. (6.3) may be adopted as a constitutive restriction for solids other than elastic-plastic.

6.2.3 Boundary Value Problem and the Fundamental Deformation Process

We are concerned with isothermal deformations of the time-independent solid body subject to a quasi-static loading program. Let $\lambda = \lambda(t)$ denote a scalar loading parameter which varies continuously in "time" t (infinitely slowly with respect to a natural time); it is convenient to keep the parameters t and λ as being distinct since t is always monotonically increasing while $\lambda(t)$ need not be a monotonic function. The standard boundary value problem is considered below; an extension to a more general case of configuration-dependent conservative loading will be discussed later in Sect. 6.5. Assume that on a given nonzero part S_u of S the displacements $\mathbf{u} = \bar{\mathbf{u}}(\xi, \lambda)$ are controlled while on a complementary part S_T the nominal surface tractions $\mathbf{T} = \bar{\mathbf{T}}(\xi, \lambda)$ (per unit reference area) are controlled. Nominal body forces $\mathbf{b} = \bar{\mathbf{b}}(\xi, \lambda)$ (per unit reference volume) are prescribed in V; note that λ need not be a load multiplier. The loading functions are assumed to be at least piecewise smooth with respect to their arguments and to exhibit no strong discontinuities other than across material surfaces or lines. An initial equilibrium state of the body at a certain value of λ is assumed to be known.

Suppose that a theoretical quasi-static response of the body to the above loading program is known, at least in principle. The respective deformation process is described by $\mathbf{x} = \chi^0(\xi, t), \xi \in \bar{V}$, and called the fundamental process (path); the superscript ^{0} is used to distinguish the corresponding values of quantities of interest. We assume that the function χ^0 is continuous and at least piecewise continuously three times differentiable with respect to its arguments and, moreover, that the respective deformation gradient $\mathbf{F}^0$ varies continuously in time while the velocity $\mathbf{v}^0$ is a continuous function of place.

Let $\overline{\mathcal{W}}$ denote the linear space of admissible virtual velocity fields at any fixed λ (the fields from $\overline{\mathcal{W}}$ vanish on a given part S_u of the material surface S). Different regularity restrictions could be imposed at this moment; we assume that $\overline{\mathcal{W}}$ consists of all continuous and piecewise continuously twice differentiable vector fields on

$\overline{V}$ which vanish on S_u. At any λ, $\mathcal{V} = \mathcal{V}(\lambda)$ denotes the class of kinematically admissible velocity fields $\tilde{\mathbf{v}}$ of the form $\tilde{\mathbf{v}} = \tilde{\mathbf{v}}^0 + \tilde{\mathbf{w}}, \tilde{\mathbf{w}} \in \overline{\mathcal{W}}$ (so that $\mathbf{v} = \dot{\lambda}\frac{\partial \overline{\mathbf{u}}}{\partial \lambda}(\xi, \lambda)$ on S_u). Note that any functional defined on $\mathcal{V}$ can equivalently be regarded as a certain other functional defined on the *linear* space $\overline{\mathcal{W}}$. Since the zero element of $\overline{\mathcal{W}}$ will have to be excluded in a number of formulae, it will be convenient to consider also the class $\mathcal{W} = \{\tilde{\mathbf{w}} \in \overline{\mathcal{W}} : \tilde{\mathbf{w}} \neq \tilde{\mathbf{0}}\}$.

In absence of moving strong discontinuities in $\mathbf{N}$, $\mathbf{b}$ and $\mathbf{T}$, and when inertia forces are neglected, any solution $\tilde{\mathbf{v}} \in \mathcal{V}$ to the first-order rate boundary value problem satisfies the rate form of the virtual work principle, viz.

$$\int_V \dot{\mathbf{N}}(\nabla \mathbf{v}) \cdot \nabla \mathbf{w} \, \mathrm{d}V = \int_V \dot{\mathbf{b}}\mathbf{w} \, \mathrm{d}V + \int_{S_T} \dot{\mathbf{T}}\mathbf{w} \, \mathrm{d}S, \quad \text{for every } \tilde{\mathbf{w}} \in \overline{\mathcal{W}}. \tag{6.4}$$

The variational equality (6.4) can be taken as a definition of a first-order solution; under the regularity restrictions imposed above, Eq. (6.4) is equivalent, by the well known transformation, to the standard local conditions of continuing equilibrium in V and on S_T. In the fundamental process, Eq. (6.4) is satisfied by $\tilde{\mathbf{v}}^0(t)$ at every t; it is recalled that the rates are understood in the right-hand sense.

6.2.4 Basic Functionals

The following three functionals play an essential role in the analysis of uniqueness and stability (cf. [4]):

$$J(\tilde{\mathbf{v}}) \equiv \int_V (U(\nabla \mathbf{v}) - \dot{\overline{\mathbf{b}}}\mathbf{v}) \, \mathrm{d}V - \int_{S_T} \dot{\overline{\mathbf{T}}}\mathbf{v} \, \mathrm{d}S, \quad \tilde{\mathbf{v}} \in \mathcal{V}, \tag{6.5}$$

$$I^0(\tilde{\mathbf{w}}) \equiv \frac{1}{2}\int_V \nabla \mathbf{w} \cdot \mathbf{C}^0 \cdot \nabla \mathbf{w} \, \mathrm{d}V, \quad \tilde{\mathbf{w}} \in \overline{\mathcal{W}}, \tag{6.6}$$

$$I(\tilde{\mathbf{w}}) \equiv \int_V U(\nabla \mathbf{w}) \, \mathrm{d}V, \quad \tilde{\mathbf{w}} \in \overline{\mathcal{W}}. \tag{6.7}$$

The functionals are considered here at a certain stage of the fundamental deformation process. The quadratic functional I^0 is based on the so-called tangent moduli $\mathbf{C}^0 = \mathbf{C}(\nabla \mathbf{v}^0(\xi), \xi)$ corresponding to the fundamental continuation of deformation. The functional I^0 is well defined if the constitutive potential U is twice differentiable at $\nabla \mathbf{v}^0$ almost everywhere in V, i.e. except possibly in a region of zero volume only (on the elasticplastic interface, for instance). This is tacitly assumed below whenever the functional I^0 appears in considerations.

A velocity field $\tilde{\mathbf{v}} \in \mathcal{V}$ is a solution to the rate boundary value problem if and only if it assigns to the functional (6.5) a stationary value [4], that is

$$\delta J(\tilde{\mathbf{v}}, \tilde{\mathbf{w}}) = 0 \quad \text{for every } \tilde{\mathbf{w}} \in \overline{\mathcal{W}}. \tag{6.8}$$

Suppose now that a fundamental first-order solution $\tilde{\mathbf{v}}^0$ is known, which is common case in practice. Uniqueness of $\tilde{\mathbf{v}}^0$ is then ensured when [8]

$$\int_V (\dot{\mathbf{N}}(\nabla\mathbf{v}) - \dot{\mathbf{N}}^0) \cdot (\nabla\mathbf{v} - \nabla\mathbf{v}^0)\, \mathrm{d}V > 0 \quad \text{for every } \tilde{\mathbf{v}} \in \mathcal{V}, \quad \tilde{\mathbf{v}} \neq \tilde{\mathbf{v}}^0 . \tag{6.9}$$

Under the stronger condition that inequality (6.9) is valid if $\tilde{\mathbf{v}}^0$ is replaced by *any* admissible field from $\mathcal{V}$, it was proved in [4, 8] that a velocity solution $\tilde{\mathbf{v}}^0$ strictly minimizes the value of $J(\tilde{\mathbf{v}})$ in $\mathcal{V}$, viz.

$$J(\tilde{\mathbf{v}}) > J(\tilde{\mathbf{v}}^0) \quad \text{for every} \quad \tilde{\mathbf{v}} \in \mathcal{V}, \quad \tilde{\mathbf{v}} \neq \tilde{\mathbf{v}}^{\mathrm{v}} . \tag{6.10}$$

As an extension of this result, we shall prove the following proposition:

If $\tilde{\mathbf{v}}^0$ is a solution to Eq. (6.4) then (6.9) implies (6.10).

In *proof*, observe first that

$$\delta J(\tilde{\mathbf{v}}, \tilde{\mathbf{w}}) = \int_V (\dot{\mathbf{N}}(\nabla\mathbf{v}) - \dot{\mathbf{N}}^0) \cdot \nabla\mathbf{w}\, \mathrm{d}V, \tag{6.11}$$

on subtracting $\delta J(\tilde{\mathbf{v}}^0, \tilde{\mathbf{w}}) = 0$ from the expression for $\delta J(\tilde{\mathbf{v}}, \tilde{\mathbf{w}})$. It follows that (6.9) is equivalent to

$$\delta J(\tilde{\mathbf{v}}, \tilde{\mathbf{v}} - \tilde{\mathbf{v}}^0) > 0 \quad \text{for every } \tilde{\mathbf{v}} \in \mathcal{V},\ \tilde{\mathbf{v}} \neq \tilde{\mathbf{v}}^0 ; \tag{6.12}$$

in passing, we note that (6.12) excludes (6.8) except for $\tilde{\mathbf{v}} = \tilde{\mathbf{v}}^0$, implying uniqueness of the solution $\tilde{\mathbf{v}}^0$ on another route. From (6.73) or (6.74)[3] we obtain that if (6.12) holds then $J(\tilde{\mathbf{v}})$ is increasing along any straight "ray" in $\mathcal{V}$ with the origin at $\tilde{\mathbf{v}}^0$. This completes the proof of (6.10). It is worth mentioning that (6.12) does not imply convexity of the functional J.

Direct substitution shows that the uniqueness condition (6.9) holds if

$$I^0(\tilde{\mathbf{w}}) > 0 \quad \text{for every} \quad \tilde{\mathbf{w}} \in \mathcal{W}, \tag{6.13}$$

provided that I^0 is well defined and that the following constitutive inequality is satisfied for $\dot{\mathbf{F}}^0 \equiv \nabla\mathbf{v}^0$ [19]

$$(\dot{\mathbf{N}} - \dot{\mathbf{N}}^0) \cdot (\dot{\mathbf{F}} - \dot{\mathbf{F}}^0) \geq (\dot{\mathbf{F}} - \dot{\mathbf{F}}^0) \cdot \mathbf{C}(\dot{\mathbf{F}}^0) \cdot (\dot{\mathbf{F}} - \dot{\mathbf{F}}^0) \quad \text{for every } \dot{\mathbf{F}} . \tag{6.14}$$

This is an extension of the well known "comparison theorem" [4] based on the concept of relative convexity of the constitutive potential with respect to that for an incrementally linear comparison material. It is pointed out that (6.14) is generally a weaker restriction than the relative convexity property with respect to the "tangent" comparison material since $\dot{\mathbf{F}}^0$ in (6.14) is *fixed*. That straightforward mod-

[3] These formulae are given in Appendix in Sect. 6.7.

ification of Hill's theory brings a definite advantage since (6.14) can be derived from micromechanical considerations [20].

Comparison with the standard theorem of the calculus of variations suggests that

$$I^0(\tilde{\mathbf{w}}) = \delta^2 J(\tilde{\mathbf{v}}^0, \tilde{\mathbf{w}}) \tag{6.15}$$

whenever I^0 is well defined, where

$$\delta^2 J(\tilde{\mathbf{v}}^0, \tilde{\mathbf{w}}) \equiv \frac{1}{2} \lim_{\gamma \to 0} \left(\delta J(\tilde{\mathbf{v}}^0 + \gamma \tilde{\mathbf{w}}, \tilde{\mathbf{w}}) - \delta J(\tilde{\mathbf{v}}^0, \tilde{\mathbf{w}}) \right) / \gamma \tag{6.16}$$

is the second weak variation of J at $\tilde{\mathbf{v}}^0$ in the direction $\tilde{\mathbf{w}}$. However, the regularity assumptions of the theorem are not satisfied and the proof requires some modification. To prove (6.15) directly, substitute (6.11) into (6.16), which yields

$$\delta^2 J(\tilde{\mathbf{v}}^0, \tilde{\mathbf{w}}) = \frac{1}{2} \lim_{\gamma \to 0} \int_V \frac{1}{\gamma} \left(\dot{\mathbf{N}}(\nabla \mathbf{v}^0 + \gamma \nabla \mathbf{w}) - \dot{\mathbf{N}}(\nabla \mathbf{v}^0) \right) \cdot \nabla \mathbf{w} \, \mathrm{d}V \,. \tag{6.17}$$

If the tangent moduli $\mathbf{C}^0$ are well defined almost everywhere in V then the integrand in (6.17) tends, as $\gamma \to 0$, to the integrand in (6.6) almost everywhere in V. On account of the estimate (6.66) the integrand in (6.17) for fixed $\tilde{\mathbf{w}}$ is uniformly bounded so that we can use the known theorem on the limiting passage under the integral sign. This completes the proof of (6.15).

6.2.5 Energy Interpretation

The work of deformation in the body can be written down as

$$W = \int_V \int \mathbf{N} \cdot \, \mathrm{d}\mathbf{F} \, \mathrm{d}V, \tag{6.18}$$

where the stresses are determined pointwise by integration of the constitutive rate equations (6.2) along the deformation path. A potential energy of the loading device which applies nomial surface tractions $\overline{\mathbf{T}}$ and nomial body forces $\overline{\mathbf{b}}$ independently of the body configuration can be expressed as

$$\Omega = \Omega(\tilde{\mathbf{u}}, \lambda) = -\int_V \overline{\mathbf{b}}\mathbf{u} \mathrm{d}V - \int_{S_T} \overline{\mathbf{T}}\mathbf{u} \, \mathrm{d}S \,. \tag{6.19}$$

Introduce the *energy functional* [18, 19]

$$E = W + \Omega \tag{6.20}$$

defined for any kinematically admissible deformation process at varying or fixed λ. In general, E is a functional of the deformation history due to path-dependence of W. An increment of the value of E can be interpreted as the amount of energy which has to be supplied from external sources to the mechanical system consisting of the deformed body and the loading device in order to produce quasi-statistically a deformation increment, generally with the help of additional perturbing forces.[4] It is emphasized that an increment of the value of $(-\Omega)$ is generally *not* equal to the work done by the controlled loads unless $\overline{\mathbf{b}}$ and $\overline{\mathbf{T}}$ are constant in time.

By appealing to the virtual work principle it can be shown that the body is in equilibrium if and only if $\dot{E}(\tilde{\mathbf{v}})$, the first time derivative of the expression (6.20), has the same value for all kinematically admissible velocity fields from $\mathcal{V}$. The value of the second time derivative $\ddot{E}$ of E, when evaluated at an equilibrium state, does not depend on accelerations and differs from $2J(\tilde{\mathbf{v}})$ only by a state-dependent constant, that is [18, 19][5]

$$\frac{1}{2}\ddot{E}(\tilde{\mathbf{v}}_1) - \frac{1}{2}\ddot{E}(\tilde{\mathbf{v}}_2) = J(\tilde{\mathbf{v}}_1) - J(\tilde{\mathbf{v}}_2) \quad \text{for every} \quad \tilde{\mathbf{v}}_1, \tilde{\mathbf{v}}_2 \in \mathcal{V}. \tag{6.21}$$

The results quoted in the preceding subsection can now be given an energy interpretation, simply by replacing J by $\ddot{E}$ in the formulae (6.8), (6.10), (6.12), (6.15). Explicitly, the stationarity principle (6.8) valid for any solution $\tilde{\mathbf{v}}$ becomes

$$\delta\ddot{E}(\tilde{\mathbf{v}}, \tilde{\mathbf{w}}) = 0 \quad \text{for every} \quad \tilde{\mathbf{w}} \in \overline{\mathcal{W}}. \tag{6.22}$$

The functional I^0 which appears in the uniqueness condition (6.13) and in the path instability condition (6.30) below, has the following interpretation

$$I^0(\tilde{\mathbf{w}}) = \frac{1}{2}\delta^2\ddot{E}(\tilde{\mathbf{v}}^0, \tilde{\mathbf{w}}), \quad \tilde{\mathbf{w}} \in \mathcal{W}. \tag{6.23}$$

The minimum property (6.10), implied by (6.9), is equivalent to

$$\ddot{E}(\tilde{\mathbf{v}}) > \ddot{E}(\tilde{\mathbf{v}}^0) \quad \text{for every} \quad \tilde{\mathbf{v}} \in \mathcal{V}, \ \tilde{\mathbf{v}} \neq \tilde{\mathbf{v}}^0. \tag{6.24}$$

It is of interest to recall here the work interpretation of $J(\tilde{\mathbf{v}})$ which, unlike (6.21), is restricted to velocity *solutions*. If $\tilde{\mathbf{v}}$ satisfies (6.4) then we have the formula [5]

$$J(\tilde{\mathbf{v}}) = \frac{1}{2}\left(\int_{S_u} \dot{\mathbf{T}}\dot{\mathbf{u}}\, \mathrm{d}S - \int_V \dot{\overline{\mathbf{b}}}\mathbf{v}\, \mathrm{d}V - \int_{S_T} \dot{\overline{\mathbf{T}}}\mathbf{v}\, \mathrm{d}S\right), \tag{6.25}$$

where $\dot{\mathbf{T}} = \nu\dot{\mathbf{N}}(\nabla\mathbf{v})$ is the reaction-rate on S_u. When multiplied by the square of a small time increment, this is the second-order work done on controlled displacements minus the second-order work done by controlled loads.

If λ is fixed then an increment of E along any kinematically admissible path is equal to the work of deformation minus the work of the given loads and, for a deformation process starting from an equilibrium state with a velocity field $\tilde{\mathbf{w}}$, we

[4] It is possible to consider an enlarged system consisting of the mechanical system placed in an ideal heat reservoir. In a quasi-static deformation process, an increment of E can be identified with an increment of the total internal energy of the enlarged system.

[5] There is a misprint in the proof given in [19]: in Eq. (18) there should be $\Delta\ddot{W}$ instead of $\ddot{W}$.

have [3]

$$\frac{1}{2}\ddot{E}(\tilde{\mathbf{w}}) = I(\tilde{\mathbf{w}}) \quad \text{at} \quad \lambda = \text{const}, \quad \tilde{\mathbf{w}} \in \overline{\mathcal{W}}. \tag{6.26}$$

This is in agreement with (6.21) since at constant λ the functional J reduces to I. Contrary to that for the functionals J and I^0, the energy interpretation of I is well known.

6.2.6 Energy Criteria of Instability

As an object of the stability analysis we take a segment of the fundamental process χ^0 in some interval of t for a given loading program, generally at varying λ. In other words, we wish to investigate stability of a quasi-static motion of a *system* consisting of the deformed body and the loading device. Instability of equilibrium is regarded as a particular case of instability of the fundamental deformation process when λ is fixed. More precisely, at any deformation stage (at $t = t_1$, say), the loading program may be stopped so that λ is kept fixed while the parameter t is increasing further up to $t = t_2$, say; the loading program can then be continued with the shift in "time" t by $t_2 - t_1$. Under the assumption of quasi-static and rate-independent deformations, the fundamental deformation process is then described by $\mathbf{x} = \hat{\chi}^0(\boldsymbol{\xi}, t) = \chi^0(\boldsymbol{\xi}, \hat{t})$ where $\hat{t} = t$, $\hat{t} = t_1$ or $\hat{t} = t - (t_2 - t_1)$ for $t < t_1$, $t_1 \le t \le t_2$ or $t > t_2$, respectively. The distinction between the processes χ^0 and $\hat{\chi}^0$ is purely formal and they are thus identified with each other. Instability of the equilibrium state $\chi^0(\boldsymbol{\xi}, t_1)$ is now identified with instability of the process $\hat{\chi}^0$ on an interval $[t_1, t_2)$ which represents an infinite interval of a natural time. If we assume that a process χ^0 stable on some interval of t must also be stable on any subinterval, then instability of an equilibrium state (at $t = t_1$, say) implies instability (on any open interval containing t_1) of the process χ^0 going through this state.

Without assuming any particular definition of stability, we define now the concept of "instability in the energy sense". An equilibrium state is said to be unstable in the energy sense if

$$\ddot{E}(\tilde{\mathbf{w}}) < 0 \quad \text{for some} \quad \tilde{\mathbf{w}} \in \mathcal{W} \quad \text{at } \lambda = \text{const}. \tag{6.27}$$

This condition is equivalent to that given in [9] as a *part* of a set of sufficiency conditions for instability of equilibrium in a dynamic sense. Although (6.27) means that a spontaneous departure from the equilibrium state is energetically possible, no convincing general argument has been given yet, to the author's knowledge, that (6.27) *alone* is sufficient for instability of equilibrium in a dynamic sense.

The fundamental *process* χ^0 is said to be unstable in the energy sense at some stage of the deformation if the respective velocity solution $\tilde{\mathbf{v}}^0$ does not minimize the value of $\ddot{E}$ in $\mathcal{V}$, that is, if

$$\ddot{E}(\tilde{\mathbf{v}}) < \ddot{E}(\tilde{\mathbf{v}}^0) \quad \text{for some} \quad \tilde{\mathbf{v}} \in \mathcal{V}. \tag{6.28}$$

Alternatively, as a condition necessary for stability of the fundamental process in the energy sense, we take

$$\ddot{E}(\tilde{\mathbf{v}}) \geq \ddot{E}(\tilde{\mathbf{v}}^0) \qquad \text{for every} \quad \tilde{\mathbf{v}} \in \mathcal{V}\,. \tag{6.29}$$

This condition was proposed by the author [18, 19] as an intuitive criterion for predicting instability of a plastic deformation process.[6] In the special cases when $\lambda = \text{const}$ or when U is quadratic, (6.28) reduces to (6.27). Since at an equilibrium state the value of $\dot{E}(\tilde{\mathbf{v}})$ is independent of $\tilde{\mathbf{v}}$, the condition (6.28) (or (6.29)) can be written down in an equivalent incremental form as $\Delta E < \Delta E^0$ (or $\Delta E \geq \Delta E^0$)), where ΔE^0 and $\Delta E \equiv \dot{E}\Delta t + \frac{1}{2}\ddot{E}(\Delta t)^2$ are increments of E evaluated *to second order* along the fundamental path and along any kinematically admissible branching path, respectively.

On the basis of the identity (6.21) and with the help of the constitutive inequality (6.14), the onset of instability found from this energy criterion has been shown [19] to coincide with the primary bifurcation point for a class of plasticity problems. We recall briefly that (6.28) (but not (6.27)) is implied by

$$I^0(\tilde{\mathbf{w}}) < 0 \qquad \text{for some} \qquad \tilde{\mathbf{w}} \in \mathcal{W}\,, \tag{6.30}$$

on account of (6.23). In typical cases examined so far in the literature, (6.30) is met along the primary deformation path immediately beyond the instant of primary bifurcation. In turn, by comparison with the minimum property (6.24), the inequality (6.29) holds in the uniqueness range (6.9). If (6.14) is valid then (6.29) holds exactly as long as $I^0(\tilde{\mathbf{w}})$ is nonnegative on $\mathcal{W}$. This follows from the implication: (6.13) $\Rightarrow$ (6.9) $\Rightarrow$ (6.24) which remains true when ($>$) is replaced in each of the inequalities by ($\geq$).

By appealing to homogeneity and continuity of $U(\dot{\mathbf{F}})$ it can be shown that (6.27) implies (6.28) [19] but the converse need not be true. This is in accord with the relation between instability of equilibrium and instability of a deformation process, and allows to treat (6.28) as a single energy criterion applicable to a variety of problems of inelastic instability. Below we will examine the question whether fulfillment of the condition (6.28) (or of (6.27) in particular) is connected with an instability in a physically acceptable sense.

6.3 Discretized Problems

In this section, admissible variations of velocities (or of displacements) are restricted to the form

[6] The criterion was initially formulated [18] as a consequence of a more stringent postulate of stability of a deformation process at presence of persistent disturbances for which an energy measure was adopted. Later [19], the criterion in an incremental form was introduced as an independent hypothesis.

$$\mathbf{w}(\boldsymbol{\xi}) = \sum_{K=1}^{N} w_K \boldsymbol{\beta}_K(\boldsymbol{\xi}), \tag{6.31}$$

where $\boldsymbol{\beta}_K$ are given "shape functions" from the class $\mathcal{W}$ and w_K are arbitrary scalar coefficients. Their number N is finite so that the fields (6.31) form a finite-dimensional subspace $\overline{\mathcal{W}^N}$ of $\overline{\mathcal{W}}$; it will be convenient to define also a set $\mathcal{W}^N = \{\tilde{\mathbf{w}} \in \overline{\mathcal{W}^N} : \tilde{\mathbf{w}} \neq \tilde{\mathbf{0}}\}$. The definition of a solution is accordingly changed, following the standard discretization procedure.

6.3.1 Instability of Equilibrium

Throughout this subsection λ is kept constant, t is identified with a natural time, and inertia forces are taken into account. Let $\Delta\mathbf{N}(t) = \mathbf{N}(t) - \mathbf{N}^0$ denote a stress increment along a deformation path starting from an equilibrium state. By subtracting the virtual work expressions in the initial and current instants, the discretized equations of motion can be written in the form

$$\int_V (\Delta\mathbf{N} \cdot \nabla\mathbf{w} + \bar{\varrho}\mathbf{a}\mathbf{w})\,\mathrm{d}V = 0 \quad \text{for every} \quad \mathbf{w} \in \overline{\mathcal{W}^N}, \tag{6.32}$$

where $\bar{\varrho}$ is the material density in the reference configuration and $\mathbf{a} = \ddot{\mathbf{u}}$ denote accelerations.

We will examine the possibility of departure from an equilibrium configuration $\tilde{\mathbf{u}}^0$ along a *direct* path of deformation which, by definition, can be approximated by a straight path

$$\tilde{\mathbf{u}}(t) = \tilde{\mathbf{u}}^0 + A(t)\tilde{\mathbf{w}}, \quad A(0) = 0, \quad \dot{A} \geq 0, \tag{6.33}$$

for sufficiently small values of A and for some fixed $\tilde{\mathbf{w}} \in \mathcal{W}^N$. Along straight (direct) paths, the direction of velocity gradient remains (approximately) constant and the rates in (6.2) can be replaced by small increments of $\mathbf{N}$ and $\mathbf{F}$ with a negligible error. Hence, for a path (6.33) and small A we obtain

$$\Delta\mathbf{N}(t) = A(t)\dot{\mathbf{N}}_{\mathrm{eq}}(\nabla\mathbf{w}), \tag{6.34}$$

where $\dot{\mathbf{N}}_{\mathrm{eq}}(\cdot)$ denotes the constitutive function at the considered equilibrium state. The approximation involved in (6.34) is similar to that in the linear theory of elastic stability, however, the distinction is that in path-dependent inelastic solids (6.34) need not be a valid approximation for circuitous paths corresponding to variable $\tilde{\mathbf{w}}$.

Since $\mathcal{W}^N$ is finite-dimensional and U depends continuously on $\dot{\mathbf{F}}$, the functional (6.7) is continuous in $\mathcal{W}^N$ and, when constrained to a level set of the function

$$\Phi(\tilde{\mathbf{w}}) \equiv \frac{1}{2}\int_V \bar{\varrho}|\mathbf{w}|^2\,\mathrm{d}V, \tag{6.35}$$

attains a minimum at some $\mathbf{w}^* \in \mathcal{W}^N$. By the method of Lagrangian multipliers, there is a number μ such that

$$\delta I(\tilde{\mathbf{w}}^*, \tilde{\mathbf{w}}) = \mu\, \delta\Phi(\tilde{\mathbf{w}}^*, \tilde{\mathbf{w}}) \quad \text{for every} \quad \tilde{\mathbf{w}} \in \overline{\mathcal{W}^N}. \tag{6.36}$$

At the equilibrium state, this reads

$$\int_V \dot{\mathbf{N}}_{\mathrm{eq}}(\nabla\mathbf{w}^*) \cdot \nabla\mathbf{w}\, \mathrm{d}V = \mu \int_V \bar{\varrho}\mathbf{w}^*\mathbf{w}\, \mathrm{d}V \quad \text{for every} \quad \tilde{\mathbf{w}} \in \overline{\mathcal{W}^N}. \tag{6.37}$$

On multiplying both sides of (6.37) by A and using (6.34) we obtain that the stresses along the deformation path (6.33) satisfy (within the approximations involved in (6.34)) the discretized equations of motion (6.32) provided in (6.33) we take $\tilde{\mathbf{w}} = \tilde{\mathbf{w}}^*$ and $A(t)$ such that

$$\ddot{A} = -\mu A\,. \tag{6.38}$$

Clearly, the sign of μ decides whether there is a tendency to decrease or to increase the speed of departure from equilibrium.

By homogeneity of degree two of I and Φ, substitution of $\tilde{\mathbf{w}} = \tilde{\mathbf{w}}^*$ into (6.36) yields $I(\tilde{\mathbf{w}}^*) = \mu\Phi(\tilde{\mathbf{w}}^*)$ and

$$\mu = \min_{\tilde{\mathbf{w}} \in \mathcal{W}^N} \frac{I(\tilde{\mathbf{w}})}{\Phi(\tilde{\mathbf{w}})}\,. \tag{6.39}$$

Suppose that the equilibrium state is unstable in the energy sense (6.27), with $\mathcal{W}$ replaced by $\mathcal{W}^N$. From (6.26) we obtain that $I(\tilde{\mathbf{w}})$ takes negative values for some $\tilde{\mathbf{w}} \in \mathcal{W}^N$ so that $\mu < 0$. As an admissible function $A(t)$ in (6.33) satisfying (6.38) we can thus take

$$A(t) = (\varepsilon/\kappa)\,\mathrm{sh}(\kappa t), \quad \kappa = (-\mu)^{1/2}, \quad \varepsilon > 0\,, \tag{6.40}$$

which corresponds to a free inertial motion starting from the equilibrium state at $t = 0$ with initial velocities $\varepsilon\mathbf{w}^*$. Evidently, a given small finite distance from the initial equilibrium configuration is exceeded for t sufficiently large no matter how small ε is. We have thus proved the following theorem on instability in the first approximation:

If an equilibrium state of a discretized system is unstable in the energy sense (6.27) (with $\mathcal{W}$ replaced by $\mathcal{W}^N$) then it is also unstable in the dynamic sense for vanishingly small initial disturbances, under the approximation (6.34).[7]

An alternative to (6.40) is $A(t) = (\varepsilon/\kappa)(\cosh(\kappa t) - 1)$ which corresponds to another disturbance: initial velocities at $t = 0$ are not perturbed but rather a system

[7] It appears likely that under reasonable regularity restrictions on the variation of the constitutive response $\dot{\mathbf{N}}(\dot{\mathbf{F}})$ along a smooth deformation path, a dynamic path with initial velocities $\varepsilon\tilde{\mathbf{w}}^*$ at $\mu < 0$ will be sufficiently close to the straight path in order to demonstrate instability of a discretized system without any constitutive approximations.

of perturbing forces is applied. For that function $A(t)$ we have $\ddot{A} = -\mu A + \varepsilon\kappa$, and by an analogous argument as above we obtain that the perturbing forces corresponding to the path (6.33) with $\tilde{\mathbf{w}} = \tilde{\mathbf{w}}^*$ are constant in time and are vanishingly small when ε tends to zero. For $t \geq \tau$, say, we can take $A(t) = (\varepsilon/\kappa)\,(\cosh(\kappa t) - \cosh(\kappa(t-\tau)))$ which corresponds to a further motion free of perturbing forces since (6.38) is again satisfied. Hence, under the same approximations as above, instability of equilibrium in the energy sense (6.27) can be given the following interpretation: *if (6.27) holds then a finite distance from the equilibrium configuration can be reached in a dynamic motion caused by arbitrarily small perturbing forces.*

The method used here to demonstrate dynamic instability of equilibrium of an inelastic time-independent system is an extension of the well known "kinetic" method for linear elastic systems, where μ is defined as a square of the lowest natural frequency of vibrations. Of course, for inelastic solids μ defined by (6.39) has no longer this special interpretation. At the present generality, the above proof of instability seems to be new.

The limiting case when I is just positive semi-definite ($\mu = 0$) corresponds, under the approximation (6.34), also to dynamic instability of the considered equilibrium configuration which can be left with velocities $\varepsilon\tilde{\mathbf{w}}^*$ constant in time.[8]

Hence, under the approximation (6.34), a departure from equilibrium in a direction $\tilde{\mathbf{w}}^*$ is dynamically possible at vanishingly small disturbances unless

$$I(\tilde{\mathbf{w}}) > 0 \quad \text{for every} \quad \tilde{\mathbf{w}} \in \mathcal{W}^N . \tag{6.41}$$

On the other hand, (6.41) excludes such a departure in any direction from $\mathcal{W}^N$ by ensuring an insufficiency of energy [3, 4], as can be seen from the identity (6.26). The condition (6.41) is thus necessary and sufficient for a kind of stability which may be called *directional stability of equilibrium* (of a discretized system). It must be pointed out that due to path-dependence of the material, (6.41) does not guarantee stability for arbitrarily circuitous paths unless further assumptions are introduced (cf. [3, 4, 9, 17]).

6.3.2 Instability of a Deformation Process

Consider now the fundamental process χ^0 of quasi-static deformation at varying λ. Admissible velocity fields in the discretized problem have the form $\tilde{\mathbf{v}} = \tilde{\mathbf{v}}^0 + \tilde{\mathbf{w}}$ where $\tilde{\mathbf{w}} \in \mathcal{W}^N$; the class of such fields $\tilde{\mathbf{v}}$ is denoted by $\mathcal{V}^N$. A velocity field $\tilde{\mathbf{v}} \in \mathcal{V}^N$ is called a solution to the discretized (quasi-static) rate-problem if (6.4) holds for every $\tilde{\mathbf{w}} \in \mathcal{W}^N$, and not necessarily for every $\tilde{\mathbf{w}} \in \mathcal{W}$. In this subsection the fundamental velocity field $\tilde{\mathbf{v}}^0$ needs to be a solution in this weakened sense only. In analogy to (6.8), any solution $\tilde{\mathbf{v}}$ to the discretized rate-problem is characterized

[8] When higher-order terms are taken into account then, depending on the problem, we can expect either instability of equilibrium (at constant λ) or path bifurcation at varying λ (with $\lambda = 0$ initially), in analogy to the critical state of an elastic conservative system.

by the stationarity property

$$\delta J(\tilde{\mathbf{v}}, \tilde{\mathbf{w}}) = 0 \quad \text{for every} \quad \tilde{\mathbf{w}} \in \overline{\mathcal{W}^N}. \tag{6.42}$$

Since U is a continuously differentiable function of $\dot{\mathbf{F}}$, J is differentiable in $\mathcal{V}^N$ and it follows that a minimizer of J in $\mathcal{V}^N$ is automatically a solution. In general, J may be unbounded from below so that an absolute minimum need not be attained. However, this cannot happen if the equilibrium state at which the discretized rate-problem is formulated is directionally stable in the sense of (6.41). As shown in Subsect. A.3 of Appendix in Sect. 6.7, we have the following

Theorem 1 *If (6.41) holds then the discretized rate-problem has a solution which assigns to the functional J its absolute minimum value in $\mathcal{V}^N$.*

This existence theorem provides an interpretation for instability of the fundamental deformation process in the energy sense (6.28) when equilibrium states in the process are (directionally) stable. In the range of validity of (6.24), the solution guaranteed by the above theorem coincides with $\tilde{\mathbf{v}}^0$. However, if (6.28) holds true (with $\mathcal{V}$ replaced by $\mathcal{V}^N$), or equivalently, if

$$J(\tilde{\mathbf{v}}) < J(\tilde{\mathbf{v}}^0) \quad \text{for some} \quad \tilde{\mathbf{v}} \in \mathcal{V}^N, \tag{6.43}$$

then the solution guaranteed by the Theorem 1 must be distinct from the fundamental solution $\tilde{\mathbf{v}}^0$. Hence, we have the following:

Corollary *At any stage of deformation of a discretized system, if (6.41) holds and the fundamental process is unstable in the energy sense (6.28) (with $\mathcal{V}$ replaced by $\mathcal{V}^N$), then there is a velocity solution different from $\tilde{\mathbf{v}}^0$.*

Evidently, the solution in velocities is nonunique also when $\tilde{\mathbf{v}}^0$ does minimize J in $\mathcal{V}^N$ but the minimum is not strict.

If the fundamental moduli C^0 corresponding to the solution $\tilde{\mathbf{v}}^0$ are well defined almost everywhere in V then, by (6.15), a necessary condition for $\tilde{\mathbf{v}}^0$ to be a minimizer of J in $\mathcal{V}^N$ is that

$$I^0(\tilde{\mathbf{w}}) \geq 0 \quad \text{for every} \quad \tilde{\mathbf{w}} \in \mathcal{W}^N, \tag{6.44}$$

or equivalently, that the tangent stiffness matrix is positive definite or at least positive semi-definite (cf. [21]). Hence, we have proved the following theorem:

For a discretized system, there is a bifurcation in velocities at every point on the fundamental path along which (6.41) holds and (6.44) does not.

This generalizes the well known observation [23, 12, 10] that the incremental response of a straight plastic column under increasing compressive loading is nonunique between the tangent modulus load (above which (6.44) fails) and the reduced modulus load (at which (6.41) fails).

Contrary to the bifurcation at the typical critical instant at which I^0 is just positive semi-definite, in the range (6.43) the secondary solution $\tilde{\mathbf{v}}^*$ which minimizes $J(\tilde{\mathbf{v}})$ in $\mathcal{V}^N$ must correspond to a moduli field different from the fundamental moduli field $\widetilde{\mathbf{C}}^0$. For, if the moduli fields were the same, then straightforward transformations with the help of symmetry of the moduli would yield $J(\tilde{\mathbf{v}}^*) - J(\tilde{\mathbf{v}}^0) = I^0(\tilde{\mathbf{v}}^* - \tilde{\mathbf{v}}^0)$. This would contradict (6.43) since from (6.11) and (6.42) we would have $I^0(\tilde{\mathbf{v}}^* - \tilde{\mathbf{v}}^0) = \delta J(\tilde{\mathbf{v}}^*, \tilde{\mathbf{v}}^* - \tilde{\mathbf{v}}^0) = 0$.

We may assume that there exists a quasi-static deformation path initiated with the displacement rate field $\tilde{\mathbf{v}}^*$ (for at least one minimizer if the minimum is not strict). Since $\tilde{\mathbf{v}}^* \neq \tilde{\mathbf{v}}^0$, the bifurcating path deviates from the fundamental one on a finite distance after some time increment, without the need of any disturbances if we neglect inaccuracy of the solution due to discretization of the problem. If this were only possible at an isolated instant then this would not suffice to conclude about instability of the fundamental postbifurcation branch. However, as shown above, if (6.41) and (6.43) hold along a segment of the fundamental path then the bifurcation is possible at every point of the segment. This must be interpreted as an instability of the fundamental deformation process. For, if the velocity solutions at a bifurcation point have comparable chances to determine the actual continuation of the deformation then any segment of the considered fundamental path, no matter how short, has zero probability to be followed due to existence of infinitely many alternatives. Moreover, the secondary continuations of the deformation are energetically preferable, according to (6.21), the incremental form of (6.28) and the interpretation of an increment of E. From a physical point of view, this provides an additional argument for rejecting the fundamental solution.[9]

To summarize, if a discretized problem is considered then the energy inequality (6.28) with $\mathcal{V}$ replaced by $\mathcal{V}^N$, equivalent to (6.43), can be accepted as a sufficient condition for instability of the fundamental deformation process in the following sense. If (6.43) is satisfied when (6.41) fails (typically beyond a limit point) then at any fixed λ the current equilibrium state is unstable in the dynamic sense discussed. If (6.43) is satisfied in some interval of λ when (6.41) still holds (typically beyond a primary bifurcation point if (6.44) fails) then the fundamental deformation path can be left quasi-statically starting from any λ, even in absence of disturbances. This confirms the previous supposition that fulfillment of (6.29) is necessary for path stability.

It is worth pointing out that the instability condition (6.28) is applicable irrespectively of the character of the critical point itself (for instance, instability may be induced by a discontinuous drop of the incremental stiffness of the material, without singularity of the tangent stiffness matrix at the critical point). It can be applied also to a secondary post-bifurcation path since it is a matter of choice which path is regarded as fundamental. Using (6.29) as a necessary condition for path stability,

---

[9] Recently, Bazant [1] considered a possibility of deriving a criterion of path stability from thermodynamic considerations. Not going into a discussion on thermodynamic aspects of stability, we mention only that within the class of problems considered here the final criterion of choice of a stable path obtained by Bazant (Eq. (35) in [1]) represents, on account of (6.25) and (6.22), a particular case of the energy criterion (6.29) (cf. Eq. (17) in [18]).

Petryk and Thermann [21] have proposed a new computational method for crossing bifurcation points with automatic rejection of an unstable post-bifurcation branch.

6.4 Instability of Continuous Systems

The results obtained in the preceding section for a discretized problem cannot be automatically extended to infinite-dimensional continuous systems. An essential point in proving instability was that a minimum of the considered functionals was actually attained, which need not be the case when the functional space $\overline{\mathcal{W}}$ is infinite-dimensional.

Measures of the distance between two configurations of a continuous body are in general not equivalent to each other. Taking into account that rigid body displacements at fixed λ have been excluded, we will use the measure

$$\|\tilde{\mathbf{w}}\| \equiv \left(\int_V |\nabla \mathbf{w}|^2 \, dV\right)^{1/2} \tag{6.45}$$

which is a norm on the linear space $\overline{\mathcal{W}}$ (cf. also the estimates (6.70)).

6.4.1 Directional Stability of Equilibrium

If (6.27) holds and $\mathcal{W}^N$ approximates $\mathcal{W}$ sufficiently well then $\ddot{E}(\tilde{\mathbf{w}})$ takes negative values also in $\mathcal{W}^N$ so that μ defined by (6.39) is negative. Hence, instability of equilibrium in the dynamic sense discussed in the preceding section can be inferred from (6.27) for any sufficiently "fine" discretization of the problem. On this basis we may assume, as a hypothesis, that the condition (6.27) is sufficient for a dynamic instability of equilibrium also for a non-discretized continuum. This is confirmed when a minimum of $I(\tilde{\mathbf{w}})$ on a hypersurface $\Phi(\tilde{\mathbf{w}}) = \text{const}$ in $\mathcal{W}$ is actually attained so that the proof from Sect. 6.3, with $\mathcal{W}^N$ replaced by $\mathcal{W}$, remains valid. A general proof that (6.27) implies dynamic instability of a continuum is lacking at present; note that the functional $I(\tilde{\mathbf{w}})$ may be unbounded from below on a hypersurface $\Phi(\tilde{\mathbf{w}}) = \text{const}$ in an infinite-dimensional space.

For the discretized problem, fulfillment of the condition (6.41) ensured that the distance from equilibrium attainable on a straight (direct) path vanished with a vanishing input of the energy supplied by a disturbance. The same inequality in an infinite-dimensional space does not ensure this rigorously since $I(\tilde{\mathbf{w}})$ may approach zero in a non-trivial way without achieving it. A possible extension of (6.41) to the continuum problem is that $I(\mathbf{w})$ is *uniformly* positive, that is

$$I(\tilde{\mathbf{w}}) \geq a\|\tilde{\mathbf{w}}\|^2 \quad \text{for every} \quad \tilde{\mathbf{w}} \in \overline{\mathcal{W}}, \tag{6.46}$$

where a is a positive parameter, independent of place, which may be arbitrarily small. Equilibrium states of a continuum are said to be directionally stable if (6.46) holds; in a finite-dimensional space the conditions in (6.41) and (6.46) are equivalent to each other. As already remarked in Sect. 6.3, without further assumptions (6.46) cannot be regarded as being sufficient for dynamic stability of equilibrium for arbitrarily circuitous paths.

The parameter a can be defined as

$$a = \inf_{\tilde{\mathbf{w}} \in \mathcal{W}} \frac{I(\tilde{\mathbf{w}})}{||\tilde{\mathbf{w}}||^2} . \tag{6.47}$$

From (6.68) with $\tilde{\mathbf{v}} = \tilde{\mathbf{0}}$ it follows that a has always a finite value. We can expect that in typical circumstances a will vary along a deformation path and differ from zero with the exception of isolated points between the ranges where $a > 0$ and $a < 0$, that is, between the ranges of validity of (6.46) and (6.27). This means that in practice the condition (6.46) will frequently represent a negation of (6.27), possibly except at a critical point itself. Note that if the infimum in (6.47) is equal to zero and is actually reached in $\mathcal{W}$, then we have an *eigenstate* [9, 7].

6.4.2 Instability of a Deformation Process

Consider a certain stage of the fundamental process. We shall assume below that (6.46) holds for some constant $a > 0$. Then it can be proved (see Subsect. A.2 of Appendix in Sect. 6.7) that

$$J(\tilde{\mathbf{v}}) \to +\infty \quad \text{when} \quad ||\tilde{\mathbf{v}}|| \to \infty, \quad \tilde{\mathbf{v}} \in \mathcal{V} \tag{6.48}$$

and that $J(\tilde{\mathbf{v}})$ is bounded from below in $\mathcal{V}$.

Contrary to the finite-dimensional case, a minimum of J in $\mathcal{V}$ need not in general be attained, even if $\overline{\mathcal{W}}$ is enlarged to be a complete space (the Sobolev space, for instance). Therefore, we modify the argument from the preceding section and examine a minimizing sequence rather than a minimizer itself.

Let $(\tilde{\mathbf{v}}_n) \in \mathcal{V}, n = 1, 2, \ldots$, be a minimizing sequence, i.e. such that

$$J(\tilde{\mathbf{v}}_n) \to d = \inf_{\tilde{\mathbf{v}} \in \mathcal{V}} J(\tilde{\mathbf{v}}) \quad \text{as} \quad n \to \infty. \tag{6.49}$$

For instance, the minimizing sequence can be determined by the Ritz method and the velocity fields $\tilde{\mathbf{v}}_n$ can be the solutions to a sequence of discretized rate-problems. As proved in Subsect. A.4 of Appendix in Sect. 6.7, the sequence has the following property:

$$|\delta J(\tilde{\mathbf{v}}_n, \tilde{\mathbf{w}})| \leq \varepsilon_n ||\tilde{\mathbf{w}}|| \quad \text{for every } \tilde{\mathbf{w}} \in \overline{\mathcal{W}}, \quad \text{where } \varepsilon_n \to 0 \text{ as } n \to \infty . \tag{6.50}$$

Since the first variation of J vanishes at a solution field, its convergence to zero according to (6.50) means that inaccuracy of satisfying the equations of continuing equilibrium by the velocity fields $\tilde{\mathbf{v}}_n$ tends in a sense to zero. To clarify this, observe first that each velocity field $\tilde{\mathbf{v}}$ from $\mathcal{V}$ can be treated as a solution to the system of *perturbed* equations of continuing equilibrium, viz.

$$\begin{aligned} \operatorname{Div}\dot{\mathbf{N}}^T + \dot{\bar{\mathbf{b}}} + \dot{\mathbf{b}}^* = \mathbf{0} \quad \text{in } V\backslash S_D\,, \quad \boldsymbol{\nu}_D[\![\dot{\mathbf{N}}]\!] + \dot{\mathbf{T}}^* &= \mathbf{0} \quad \text{on } S_D\,, \\ \boldsymbol{\nu}\dot{\mathbf{N}} &= \dot{\bar{\mathbf{T}}} + \dot{\mathbf{T}}^* \quad \text{on } S_T\,, \end{aligned} \tag{6.51}$$

where the volume and surface densities $\dot{\mathbf{b}}^* = \dot{\mathbf{b}}^*(\tilde{\mathbf{v}}), \dot{\mathbf{T}}^* = \dot{\mathbf{T}}^*(\tilde{\mathbf{v}})$ of the rates of perturbing forces are uniquely defined (locally) for each $\tilde{\mathbf{v}}$ (for simplicity, their dependence on $\boldsymbol{\xi}$ is not indicated explicitly). S_D stands for a collection of the surfaces of discontinuity (with $\boldsymbol{\nu}_D$ being a unit normal in the reference configuration and $[\![\cdot]\!]$ denoting the jump) of the moduli (e.g. across an elastic/plastic interface) or of the velocity gradient. The first variation of J, after transtorming it by using the Green theorem and substituting (6.51), can be written down in the form

$$\delta J(\tilde{\mathbf{v}}, \tilde{\mathbf{w}}) = \int_V \dot{\mathbf{b}}^*(\tilde{\mathbf{v}})\mathbf{w}\,\mathrm{d}V + \int_{S_T \cup S_D} \dot{\mathbf{T}}^*(\tilde{\mathbf{v}})\mathbf{w}\,\mathrm{d}S\,. \tag{6.52}$$

Consider an arbitrary material subdomain $G \subseteq V$ with a piecewise regular boundary ∂G and denote by $\mathbf{P}^*(G)$ the total perturbing force acting on G, including the perturbing tractions acting over ∂G. Assuming that $S_D \cap \overline{G}$ does not change discontinuously in time and that $\mathbf{b}^*$ and $\mathbf{T}^*$ are initially zero, the rate of $\mathbf{P}^*(G)$ which results from $\dot{\mathbf{b}}^*(\tilde{\mathbf{v}})$ is expressed by

$$\dot{\mathbf{P}}^*(G, \tilde{\mathbf{v}}) = \int_G \dot{\mathbf{b}}^*(\tilde{\mathbf{v}})\,\mathrm{d}V + \int_{\overline{G}\cap(S_T \cup S_D)} \dot{\mathbf{T}}^*(\tilde{\mathbf{v}})\,\mathrm{d}S\,. \tag{6.53}$$

In Subsect. A.5 of Appendix in Sect. 6.7, it is proved that for a sequence $(\tilde{\mathbf{v}}_n)$ of velocity fields satisfying (6.50), the rates $\dot{\mathbf{P}}^*(G, \tilde{\mathbf{v}}_n)$ of perturbing forces acting on any G satisfy the inequality

$$\left|\dot{\mathbf{P}}^*(G, \tilde{\mathbf{v}}_n)\right| \leq \varepsilon_n \sqrt{3}\,|\partial G| \tag{6.54}$$

and tend thus to zero as $n \to \infty$ provided the domain G is fixed or varies with n such that its boundary measure $|\partial G|$ is bounded.

Perturbations of continuing equilibrium characterized by (6.50) may thus be regarded as vanishingly small when $n \to \infty$. In typical circumstances expected in the uniqueness range (6.9), such perturbations can cause only vanishingly small changes of the velocity field. To show this, suppose that

$$\delta J(\tilde{\mathbf{v}}, \tilde{\mathbf{v}} - \tilde{\mathbf{v}}^0) \geq b \left\|\tilde{\mathbf{v}} - \tilde{\mathbf{v}}^0\right\|^2 \quad \text{for every } \tilde{\mathbf{v}} \in \mathcal{V} \tag{6.55}$$

for some positive constant b. Strictly speaking, the condition (6.55) is stronger than (6.12), however, in usual circumstances the ranges of validity of (6.12) and (6.55) along a deformation path can be expected to coincide, with the exception of isolated instants (cf. the discussion of (6.46)). In particular, if the constitutive inequality (6.14) holds then (6.55) is ensured so long as $I^0(\tilde{\mathbf{w}}) \geq b\|\tilde{\mathbf{w}}\|^2$. On substituting $\tilde{\mathbf{v}}_n$ in place of $\tilde{\mathbf{v}}$ into (6.55) and using (6.50) we obtain that $\|\tilde{\mathbf{v}}_n - \tilde{\mathbf{v}}^0\| \leq \varepsilon_n/b$. It follows that *if (6.55) holds then any sequence* $(\tilde{\mathbf{v}}_n)$ *satisfying (6.50) tends to* $\tilde{\mathbf{v}}^0$ *in the sense that*

$$\|\tilde{\mathbf{v}}_n - \tilde{\mathbf{v}}^0\| \to 0 \quad \text{as } n \to \infty\,. \tag{6.56}$$

The situation is different when the process is unstable in the energy sense (6.28). As shown above, if (6.46) holds then (6.50) and thus also (6.54) are true for any sequence $(\tilde{\mathbf{v}}_n)$ minimizing $J(\tilde{\mathbf{v}})$ in $\mathcal{V}$. On account of (6.21), for a minimizing sequence $(\tilde{\mathbf{v}}_n)$ satisfying (6.49) we thus have $J(\tilde{\mathbf{v}}^0) - J(\tilde{\mathbf{v}}_n) > \varepsilon = \text{const} > 0$ for n sufficiently large. From (6.48) and (6.72) it follows that the distance (6.45) between the perturbed and fundamental velocity fields remains finite as the infimum of $J(\tilde{\mathbf{v}})$ is approached at $n \to \infty$, viz.

$$M_2 > \|\tilde{\mathbf{v}}_n - \tilde{\mathbf{v}}^0\| > M_1 \quad \text{for} \quad n > K \tag{6.57}$$

for some positive constants M_1, M_2 and K. Hence, we have proved the following instability theorem.

Theorem 2 *If (6.46) holds and the fundamental process is unstable in the energy sense (6.28) then finite deviations from the findamental velocity field, measured by (6.45), can be caused by vanishingly small perturbations of the equations of continuing equilibrium, in the sense of (6.50) and (6.54).*[10]

In short but less precisely (see below), we may say that if (6.28) holds simultaneously with (6.46) then the fundamental process is unstable in the sense of sensitivity of the incremental deformation to arbitrarily small perturbing forces.

If it happens that the infimum value in (6.49) is reached then the rates of perturbing forces can be exactly zero. In that case the path instability in the sense of Theorem 2 in some interval of λ is associated with persisting nonuniqueness in velocities, as in the finite-dimensional case considered in the preceding section.

Let (6.46) and (6.28) hold along the fundamental deformation path in a certain interval of λ. As shown above, at each equilibrium state of the body within that interval there exists a sequence of perturbed velocity fields $\tilde{\mathbf{v}}_n$ satisfying (6.49) and (6.57) while the rates of perturbing forces tend to zero in the sense of (6.54). Consider a perturbed deformation path initiated with a displacement rate field $\tilde{\mathbf{v}}_n$; we assume that if the *rates* of perturbing forces are kept fixed along the path then variations of the velocity field in a small interval of λ along any such path can be neglected. According to (6.57), the measure (6.45) of the distance between body configurations in the fundamental and perturbed processes grows then in time with a *finite* rate and achieves a given small but finite value after a small time increment. This distance

[10] In an equivalent formulation, if (6.46) and (6.28) hold simultaneously then there is $\varepsilon > 0$ such that for every $\gamma > 0$, however small, there is $\tilde{\mathbf{v}}_\gamma \in \mathcal{V}$ satistying $|\delta J(\tilde{\mathbf{v}}_\gamma, \tilde{\mathbf{w}})| < \gamma\|\tilde{w}\|$ for every $\tilde{\mathbf{w}} \in \mathcal{W}$ and $|\dot{\mathbf{P}}^*(G, \tilde{\mathbf{v}}_\gamma)| < \gamma|\partial G|$ for every G and such that $\|\tilde{\mathbf{v}}_\gamma - \tilde{\mathbf{v}}^0\| \geq \varepsilon$.

can be reached at perturbing forces as small as we please if n is taken sufficiently large, with the initial instant of their application being arbitrarily chosen from the considered interval of λ.

This is the final interpretation of the path instability implied by (6.28) when each equilibrium state separately is directionally stable. As remarked in the Introduction, that kind of instability corresponds to the instability of a process for so-called *persistent* disturbances in the absence of *initial* disturbances. However, it must be noted that the transition from the conclusion of Theorem 2 formulated in terms of the rates to the final statement for the *increments*, although it may seem intuitively obvious, is not mathematically rigorous in view of the additional assumption made. From a physical point of view, the conclusion about instability is strengthened by the fact that for n sufficiently large the perturbed continuations correspond to a smaller value of J and require thus incrementally less energy to be supplied to the system than the fundamental continuation. This follows from (6.21), (6.49), the incremental form of (6.28) and the interpretation of an increment of E.

6.5 Extension to Configuration-Dependent Conservative Loading

In the cases of an elastic support or fluid-pressure loading, the nominal surface tractions $\mathbf{T}$ on S_T are not only functions of $\boldsymbol{\xi}$ and λ but depend also on the actual displacements or their surface gradients. For one- or two-dimensional idealizations of the solid body, generalized body forces dependent in a similar way on the displacement field are induced by lateral surface tractions; other examples can also be given. Suppose thus (cf. [6, 8]) that the incremental loading consists not only of a controllable part (distinguished by a bar), as it was assumed above, but also of a deformation-sensitive part, that is

$$\dot{\mathbf{T}} = \dot{\bar{\mathbf{T}}} + \mathbf{f}(\tilde{\mathbf{v}}), \quad \dot{\mathbf{b}} = \dot{\bar{\mathbf{b}}} + \mathbf{g}(\tilde{\mathbf{v}}), \tag{6.58}$$

where $f(\tilde{\mathbf{v}})$ and $\mathbf{g}(\tilde{\mathbf{v}})$ are linear homogeneous expressions in the velocity $\mathbf{v}$ and its gradient $\nabla\mathbf{v}$ at the considered material point. Coefficients in the expressions may depend piecewise continuously on $\boldsymbol{\xi}$ and sufficiently smoothly on λ and $\tilde{\mathbf{u}}$. We show below in outline (cf. [19]) that the energy criteria of instability can be extended to the case of configuration-dependent loading corresponding to (6.58) provided the loading is conservative in an overall sense (cf. [6, 22]). Namely, it is assumed that when λ is *fixed* then the total work done by the body forces $\mathbf{b} = \mathbf{b}(\boldsymbol{\xi}, \lambda, \tilde{\mathbf{u}})$ and surface tractions $\mathbf{T} = \mathbf{T}(\boldsymbol{\xi}, \lambda, \tilde{\mathbf{u}})$ in any virtual motion compatible with the kinematic constraints and leading from a configuration $\tilde{\mathbf{u}}^0$ to any sufficiently close configuration $\tilde{\mathbf{u}}$ is *path-independent*, viz.

$$\int_V \int_{\mathbf{u}^0}^{\mathbf{u}} \mathbf{b} \mathrm{d}\mathbf{u}\, \mathrm{d}V + \int_{S_T} \int_{\mathbf{u}^0}^{\mathbf{u}} \mathbf{T} \mathrm{d}\mathbf{u}\, \mathrm{d}S = \Omega(\tilde{\mathbf{u}}^0, \lambda) - \Omega(\tilde{\mathbf{u}}, \lambda), \quad \lambda = \text{const.} \tag{6.59}$$

The functional $\Omega(\tilde{\mathbf{u}}, \lambda)$ is defined to within an additive function of λ which may be chosen arbitrarily. For a physically appropriate choice of that function, Ω can be identified with the potential energy of the loading device. When λ is *varying* then, in general, the left-hand expression in (6.59) is path-dependent and differs from the respective increment of $(-\Omega)$.

At a given body configuration $\tilde{\mathbf{u}}^0$ at certain λ, introduce the bilinear functional Q of velocities (or their variations) and the respective quadratic functional R, defined by

$$Q(\tilde{\mathbf{v}}_1, \tilde{\mathbf{v}}_2) = \int_V \mathbf{g}(\tilde{\mathbf{v}}_1)\mathbf{v}_2 \, \mathrm{d}V + \int_{S_T} \mathbf{f}(\tilde{\mathbf{v}}_1)\mathbf{v}_2 \, \mathrm{d}S \,, \quad R(\tilde{\mathbf{v}}) = -\frac{1}{2} Q(\tilde{\mathbf{v}}, \tilde{\mathbf{v}}) \,. \tag{6.60}$$

The first variation of $\mathbf{T}$ and $\mathbf{b}$ at fixed λ can be identified with $\mathbf{f}$ and $\mathbf{g}$, respectively. Path independence of the left-hand expression in (6.59) requires (cf. e.g. [24]) that $Q(\tilde{\mathbf{w}}_1, \tilde{\mathbf{w}}) = Q(\tilde{\mathbf{w}}, \tilde{\mathbf{w}}_1)$ for every $\tilde{\mathbf{w}}_1, \tilde{\mathbf{w}} \in \overline{\mathcal{W}}$. Substitution of the difference of two fields from $\mathcal{V}$ in place of $\tilde{\mathbf{w}}_1$ shows that the value of the expression: $Q(\tilde{\mathbf{v}}, \tilde{\mathbf{w}}) - Q(\tilde{\mathbf{w}}, \tilde{\mathbf{v}})$ is independent of $\tilde{\mathbf{v}} \in \mathcal{V}$. Since $\mathbf{f}$ and $\mathbf{g}$ depend only locally on $\tilde{\mathbf{v}}$ this expression vanishes if $\mathbf{v} = \mathbf{0}$ in V wherever $\mathbf{w} \neq \mathbf{0}$. Consequently, it must vanish, by a suitable choice of $\tilde{\mathbf{v}}$, if $\mathbf{w} = \mathbf{0}$ in some neighborhood of S_u, and hence for every $\tilde{\mathbf{w}} \in \overline{\mathcal{W}}$ by the limit transition. It follows that (6.59) implies

$$Q(\tilde{\mathbf{v}}_1, \tilde{\mathbf{v}}_2) = Q(\tilde{\mathbf{v}}_2, \tilde{\mathbf{v}}_1) \quad \text{for every } \tilde{\mathbf{v}}_1, \tilde{\mathbf{v}}_2 \in \mathcal{V} \cup \overline{\mathcal{W}} \,; \tag{6.61}$$

this is an extension of Hill's [6] "self-adjointness" condition for surface loading to the loading (6.58).

From (6.58), (6.60) and (6.61) we obtain the equality

$$\int_V \dot{\mathbf{b}}\mathbf{w} \, \mathrm{d}V + \int_{S_T} \dot{\mathbf{T}}\mathbf{w} \, \mathrm{d}S = \int_V \dot{\bar{\mathbf{b}}}\mathbf{w} \, \mathrm{d}V + \int_{S_T} \dot{\bar{\mathbf{T}}}\mathbf{w} \, \mathrm{d}S - \delta R(\tilde{\mathbf{v}}, \tilde{\mathbf{w}}) \tag{6.62}$$

valid for every $\tilde{\mathbf{v}} \subset \mathcal{V}$ and $\tilde{\mathbf{w}} \subset \overline{\mathcal{W}}$. By using this identity it can be shown (cf. [6, 19]) that all conclusions from the Sect. 6.2 remain valid when the basic functionals J, I^0, I are modified by adding to the right-hand expressions in (6.5) and (6.6), (6.7) the quadratic form $R(\tilde{\mathbf{v}})$ and $R(\tilde{\mathbf{w}})$, respectively. Accordingly, $R(\tilde{\mathbf{v}} - \tilde{\mathbf{v}}^0)$ and $R(\tilde{\mathbf{w}})$ have to be added to the integral expressions in (6.9) and (6.17), respectively, while $Q(\tilde{\mathbf{v}} - \tilde{\mathbf{v}}^0, \tilde{\mathbf{w}})$ has to be subtracted from the right-hand expression in (6.11). Instead of (6.19), by the trapezoid rule of quadrature we obtain the following second-order expression for the difference between the values of Ω in two body configurations close to each other:

$$\Omega(\tilde{\mathbf{u}}^0 + \tilde{\mathbf{w}}, \lambda) - \Omega(\tilde{\mathbf{u}}^0, \lambda) = -\int_V \bar{\mathbf{b}}\mathbf{w} \, \mathrm{d}V - \int_{S_T} \bar{\mathbf{T}}\mathbf{w} \, \mathrm{d}S + R(\tilde{\mathbf{w}}) + o(\|\tilde{\mathbf{w}}\|^2). \tag{6.63}$$

The other formulae in Sect. 6.2 remain unchanged, including the definition (6.20) of the energy functional and the basic identity (6.21) [19].

Similar modifications of the formulae in Sect. 6.3 and 6.4 are straightforward and hence need not be given in detail. By applying the Cauchy inequality and (6.70) to (6.60) it can be shown that the estimates in Appendix in Sect. 6.7 remain true (with modified constants). The theorems from Sect. 6.3 and 6.4 remain thus valid for the incremental loading (6.58) provided the loading is conservative in the sense of (6.59).

6.6 Conclusions

The principal results established above can be summarized as follows:

(i) If an equilibrium state of a discretized system is unstable in the energy sense (6.27) then it is also unstable in the dynamic sense, under the constitutive approximation (6.34).

(ii)If a process of quasi-static deformation of a discretized system is unstable in the energy sense (6.28) while the traversed equilibrium states are (directionally) stable then there is bifurcation in velocities at every such state, and the secondary solutions minimize the second-order increment of energy to be supplied to the system.

(iii)Under assumptions as in (ii) above but for a continuous system, the deformation process is unstable for persistent disturbances, in the sense of sensitivity of the incremental deformation to arbitrarily small perturbing forces.

More precise formulations have been given in preceding sections. The result (i) indicates that the familiar second-order energy inequality at constant loading (cf. (6.41) and (6.26)), usually interpreted as a sufficiency condition for stability of equilibrium, may be interpreted (with a sign $\geq$) as a condition necessary for stability in a dynamic sense. From (ii) we obtain that if the tangent stiffness matrix becomes indefinite (cf. (6.30)) along a deformation path without loss of stability of equilibrium, then a continuous nonuniqueness range is entered; moreover, the secondary solutions are energetically preferable. For continuous systems under analogous circumstances, the persisting possibility of path bifurcation is, according to (iii), replaced in general by the persisting possibility of switching to a secondary deformation path under action of arbitrarily small perturbing forces. Similarly to the case (ii), the secondary delormation modes are energetically preferable to the fundamental continuation.

These statements and conclusions at the present generality are believed to be new. The obtained results lend support to the hypothesis that, for a broad class of problems, the theoretical deformation processes characterized by (6.28) are unstable in a physically meaningful sense and hence cannot be practically realized in a physical system. Certainly, there are many aspects of instability in inelastic solids which still remain to be examined; as remarked in the Introduction, investigation in that direction has been rather limited.

Illustration

To illustrate the meaning of the instability condition (6.28) on a possibly simple and well-known example, consider the classical Shanley model of an elastic-plastic column with a central two-flange hinge [23]. The model has two degrees of freedom: the rotation angle θ of rigid arms and the relative vertical displacement u of the end points (Fig. 6.1a); $\tilde{\mathbf{v}}$ can thus be identified with $(\dot{\theta}, \dot{u})$. We will examine stability of the fundamental deformation process $\theta \equiv 0$ when the vertical compressive load $P \equiv \lambda(t)$ is a given increasing function of time, so that $\Omega = -Pu$. At a certain stage of the fundamental process, the rate of the force carried by an elastoplastic flange (K), $K = 1$ or 2, is given by $\dot{P}_K = L(\dot{e}_K)\dot{e}_K$ (no sum), where $\dot{e}_K$ is the shortening rate of the flange (taken positive for compression). $L(\dot{e})$ is equal to L_p (current tangent modulus) or to L_e (elastic modulus) for $\dot{e} > 0$ (loading) or $\dot{e} < 0$ (unloading), respectively, with $L_\mathrm{e} > L_\mathrm{p} > 0$. At an unbuckled configuration with the current dimensions as shown in Fig. 6.1a we have $\dot{e}_1 = \dot{u} + \dot{\theta}$, $\dot{e}_2 = \dot{u} - \dot{\theta}$ while $\ddot{u} = \frac{1}{2}(\ddot{e}_1 + \ddot{e}_2) + 2l\dot{\theta}^2$ (all variables are treated as nondimensional). Elementary calculations show that at an unbuckled equilibrium state we have

$$J(\dot{\theta}, \dot{u}) = \frac{1}{2}\ddot{E}(\dot{\theta}, \dot{u}) + \frac{1}{2}\ddot{P}u = \frac{1}{2}L(\dot{e}_1)\dot{e}_1^2 + \frac{1}{2}L(\dot{e}_2)\dot{e}_2^2 - \dot{P}\dot{u} - P\dot{\theta}^2 l. \tag{6.64}$$

The graph of $J(\dot{\theta}, \dot{u})$ consists of four quadrics joined smoothly along the lines $\dot{\theta} = \pm\dot{u}$. The uniqueness range (6.9) corresponds to $P < P_\mathrm{t}$, where $P_\mathrm{t} = L_\mathrm{p}/l$ is the tangent modulus load, and in this range the expression (6.64) is minimized by the fundamental continuation $\dot{\theta}^0 = 0, \dot{u}^0 = \dot{P}/2L_\mathrm{p}$. If L_p decreases continuously along the fundamental path then the uniqueness range terminates at $P = P_\mathrm{t}$; this is the primary bifurcation point at which an absolute minimum of $J(\dot{\theta}, \dot{u})$ is attained at any $\tilde{\mathbf{v}}$ from the "fan" of velocity solutions: $\dot{u} = \dot{u}^0, |\dot{\theta}| \leq \dot{u}^0$ (cf. Fig. 6.1a). However, if L_p and thus also P_t decrease discontinuously then the range $P > P_\mathrm{t}$ need not be preceded by $P = P_\mathrm{t}$. Within the range $P_\mathrm{t} < P < P_\mathrm{r}$, where $P_\mathrm{r} = 2L_\mathrm{e}L_\mathrm{p}/l(L_\mathrm{e} + L_\mathrm{p})$ is the reduced modulus load, the fundamental solution $(\dot{\theta}^0, \dot{u}^0)$ becomes a saddle point of J so that (6.28) is met while the condition (6.41) of stability of equilibrium is still satisfied. In accord with Theorem 1, along this segment of the fundamental path the absolute minimum value of J is attained at the secondary solution points (cf. Fig. 6.1b) which are easily found to be

$$\dot{\theta}^* = \pm\frac{\dot{P}}{2L_\mathrm{p}}z, \quad \dot{u}^* = \frac{\dot{P}}{2L_\mathrm{p}}(1 + (P/P_\mathrm{t} - 1)z), \quad \text{where } z \equiv \frac{1 - L_\mathrm{p}/L_\mathrm{e}}{2(1 - P/P_\mathrm{r})}. \tag{6.65}$$

As P is increasing, the column can start to buckle with these rates at *any* point along the segment, without any perturbing forces: this is the present interpretation of the path instability associated with (6.28). According to the energy interpretation of J, buckling initiated with $(\dot{u}^*, \dot{\theta}^*)$ requires incrementally less energy to be supplied to the *system* than the fundamental solution, i.e. $\Delta E^* < \Delta E^0$. If $P(t)$ is a gravitational force coming from a mass supplied with a prescribed rate then $(\Delta E^* - \Delta E^0)$ can

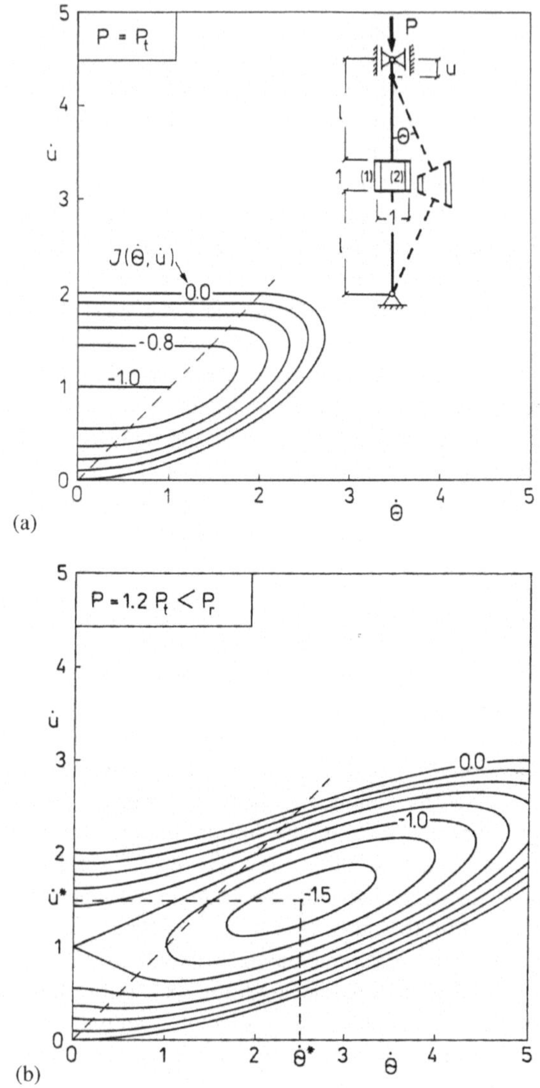

Fig. 6.1 Representation of the velocity functional (6.5) for the Shanley column (a) at the tangent modulus load, (b) between the tangent and reduced modulus loads. Numerical values correspond to $\dot{P}/2L_{\mathrm{p}} = 1$, $L_{\mathrm{p}}/L_{\mathrm{e}} = 0.5$, $L_{\mathrm{p}} = 1$. Contours of $J(\dot{\theta}, \dot{u}) > 0$ are not shown.

be identified with the decrease of the work needed to transport the mass; note that $\dot{u}^* > \dot{u}^0$.

For $P > P_{\mathrm{r}}$ the condition (6.28) is satisfied simultaneously with (6.27), and J is unbounded from below. In this range, instability of the fundamental process is interpreted as instability of the traversed equilibrium states in a dynamic sense.

As indicated in Sect. 6.3, the qualitative picture presented above can be extended to much more complicated finite-dimensional systems. This is illustrated elsewhere [21] on the example of necking under plane strain tension, calculated by using the finite element method.

The picture may change if a continuous system is considered: from one side, the absolute minimum of the functional J in the range where (6.28) and (6.46) hold need not be attained among ordinary functions, and from another side, the number of solution paths emanating from a bifurcation point may be even infinite. Both cases can be illustrated e.g. on the example of necking in a biaxially stretched, infinitely thin sheet of finite in-plane dimensions, by assuming specific boundary conditions. In the second case, the stability condition (6.29) can be applied to *secondary* post-bifurcation branches to select the incipient width of a necking band which is undetermined by the bifurcation theory alone; details will be given elsewhere.

6.7 Appendix

A.1. Certain estimates and identities

The function $\dot{\mathbf{N}}(\cdot)$ has been assumed to be positively homogeneous of degree one, continuous and piecewise smooth (outside $\dot{\mathbf{F}} = \mathbf{0}$). It follows that for any pair $\dot{\mathbf{F}}_1, \dot{\mathbf{F}}_2$ of velocity gradients we have the estimate

$$\left|\dot{\mathbf{N}}(\dot{\mathbf{F}}_1) - \dot{\mathbf{N}}(\dot{\mathbf{F}}_2)\right| \leq C \left|\dot{\mathbf{F}}_1 - \dot{\mathbf{F}}_2\right|, \tag{6.66}$$

where C is a positive constant, independent of place if it is taken as a supremum over V of the local constant. From (6.66) and the identity

$$U(\dot{\mathbf{F}}_1) - U(\dot{\mathbf{F}}_2) - \frac{1}{2}\dot{\mathbf{N}}(\dot{\mathbf{F}}_2) \cdot (\dot{\mathbf{F}}_1 - \dot{\mathbf{F}}_2) + \frac{1}{2}(\dot{\mathbf{N}}(\dot{\mathbf{F}}_1) - \dot{\mathbf{N}}(\dot{\mathbf{F}}_2)) \cdot \dot{\mathbf{F}}_1,$$

we obtain

$$\left|U(\dot{\mathbf{F}}_1) - U(\dot{\mathbf{F}}_2)\right| \leq \frac{1}{2} C \left|\dot{\mathbf{F}}_1 - \dot{\mathbf{F}}_2\right| \left(\left|\dot{\mathbf{F}}_1\right| + \left|\dot{\mathbf{F}}_2\right|\right). \tag{6.67}$$

Let the definition (6.7) of the functional I be formally extended to the class $\mathcal{V}$. For any pair $\tilde{\mathbf{v}}, \tilde{\mathbf{w}}$ of vector fields from $\mathcal{V} \cup \overline{\mathcal{W}}$, from (6.67), (6.45) and the Cauchy inequality for integrals we obtain the estimate

$$|I(\tilde{\mathbf{v}}) - I(\tilde{\mathbf{w}})| \leq \frac{1}{2} C \|\tilde{\mathbf{v}} - \tilde{\mathbf{w}}\| (\|\tilde{\mathbf{v}}\| + \|\tilde{\mathbf{w}}\|). \tag{6.68}$$

Similarly, for any fields $\tilde{\mathbf{v}}_1, \tilde{\mathbf{v}}_2$ and $\tilde{\mathbf{w}}$, from (6.1), (6.5) and (6.66) it follows that

$$|\delta J(\tilde{\mathbf{v}}_1, \tilde{\mathbf{w}}) - \delta J(\tilde{\mathbf{v}}_2, \tilde{\mathbf{w}})| \leq C \, \|\tilde{\mathbf{v}}_1 - \tilde{\mathbf{v}}_2\| \, \|\tilde{\mathbf{w}}\|. \tag{6.69}$$

Since $\tilde{\mathbf{w}} \in \overline{\mathcal{W}}$ means that $\mathbf{w} = \mathbf{0}$ on a finite part S_u of the boundary of V, it is well known that

$$\int_{S_T} \mathbf{w}\mathbf{w}\, \mathrm{d}S \leq k_1 \|\tilde{\mathbf{w}}\|^2, \quad \int_V \mathbf{w}\mathbf{w}\, \mathrm{d}V \leq k_2 \|\tilde{\mathbf{w}}\|^2 \quad \text{for every } \tilde{\mathbf{w}} \in \overline{\mathcal{W}}, \tag{6.70}$$

where k_1 and k_2 are positive constants, dependent on V. Since $\dot{\bar{\mathbf{b}}}$ and $\dot{\bar{\mathbf{T}}}$ in (6.5) are bounded, after standard transformations with the use of the Cauchy inequality and (6.70), we obtain

$$\left| \int_V \dot{\bar{\mathbf{b}}}\mathbf{w}\, \mathrm{d}V + \int_{S_T} \dot{\bar{\mathbf{T}}}\mathbf{w}\, \mathrm{d}S \right| \leq k_3 \|\tilde{\mathbf{w}}\|, \quad \tilde{\mathbf{w}} \in \overline{\mathcal{W}}, \tag{6.71}$$

where k_3 is another positive constant.

Let $\tilde{\mathbf{v}}_1, \tilde{\mathbf{v}}_2$ be any pair of velocity fields from $\mathcal{V}$ so that $\tilde{\mathbf{v}}_1 - \tilde{\mathbf{v}}_2 \in \mathcal{W}$. From (6.5), (6.71) and (6.68) we arrive at the estimate

$$|J(\tilde{\mathbf{v}}_1) - J(\tilde{\mathbf{v}}_2)| \leq \|\tilde{\mathbf{v}}_1 - \tilde{\mathbf{v}}_2\| \left(k_3 + \frac{1}{2} C(\|\tilde{\mathbf{v}}_1\| + \|\tilde{\mathbf{v}}_2\|) \right). \tag{6.72}$$

From the definition (6.1) it follows that at $\gamma > 0$ we have the identity

$$\frac{\mathrm{d}}{\mathrm{d}\gamma} J(\tilde{\mathbf{v}} + \gamma\tilde{\mathbf{w}}) = \delta J(\tilde{\mathbf{v}} + \gamma\tilde{\mathbf{w}}, \tilde{\mathbf{w}}) = \frac{1}{\gamma} \delta J(\tilde{\mathbf{v}} + \gamma\tilde{\mathbf{w}}, \gamma\tilde{\mathbf{w}}), \tag{6.73}$$

which yields

$$\begin{aligned} J(\tilde{\mathbf{v}} + \tilde{\mathbf{w}}) - J(\tilde{\mathbf{v}}) &= \int_0^1 \delta J(\tilde{\mathbf{v}} + \gamma\tilde{\mathbf{w}}, \tilde{\mathbf{w}})\, \mathrm{d}\gamma \\ &= \delta J(\tilde{\mathbf{v}}, \tilde{\mathbf{w}}) + \int_0^1 (\delta J(\tilde{\mathbf{v}} + \gamma\tilde{\mathbf{w}}, \tilde{\mathbf{w}}) - \delta J(\tilde{\mathbf{v}}, \tilde{\mathbf{w}}))\, \mathrm{d}\gamma. \end{aligned} \tag{6.74}$$

From (6.69) we obtain the estimate

$$\int_0^1 |(\delta J(\tilde{\mathbf{v}} + \gamma\tilde{\mathbf{w}}, \tilde{\mathbf{w}}) - \delta J(\tilde{\mathbf{v}}, \tilde{\mathbf{w}}))|\, \mathrm{d}\gamma \leq \int_0^1 C\|\gamma\tilde{\mathbf{w}}\|\|\tilde{\mathbf{w}}\|\, \mathrm{d}\gamma = \frac{1}{2} C\|\tilde{\mathbf{w}}\|^2.$$

On substituting it into (6.74), we arrive at:

$$|J(\tilde{\mathbf{v}} + \tilde{\mathbf{w}}) - J(\tilde{\mathbf{v}}) - \delta J(\tilde{\mathbf{v}}, \tilde{\mathbf{w}})| \leq \frac{1}{2} C\|\tilde{\mathbf{w}}\|^2. \tag{6.75}$$

A.2. Proof of property (6.48)

Let $\tilde{\mathbf{v}}'$ be a fixed field from $\mathcal{V}$, $\tilde{\mathbf{w}}$ be any field from $\mathcal{W}$ and $\tilde{\mathbf{v}} = \tilde{\mathbf{v}}' + \tilde{\mathbf{w}}$. From (6.68) and the triangle inequality it follows that

$$I(\tilde{\mathbf{v}}) \geq I(\tilde{\mathbf{w}}) - \frac{1}{2}C\,\|\tilde{\mathbf{v}}'\|\,(\|\tilde{\mathbf{v}}'\| + 2\|\tilde{\mathbf{w}}\|) \tag{6.76}$$

On substituting (6.76) into (6.5), rearranging and using (6.71), we obtain

$$J(\tilde{\mathbf{v}}) \geq I(\tilde{\mathbf{w}}) - k'\|\tilde{\mathbf{w}}\| - c', \tag{6.77}$$

where the numbers k' and c' depend on $\tilde{\mathbf{v}}'$ but are independent of $\tilde{\mathbf{w}}$. If (6.46) holds then (6.77) implies that

$$J(\tilde{\mathbf{v}}) \longrightarrow +\infty \quad \text{as} \quad \|\tilde{\mathbf{w}}\| \longrightarrow \infty\,. \tag{6.78}$$

$\tilde{\mathbf{v}}'$ is fixed so that $\|\tilde{\mathbf{w}}\| \longrightarrow \infty$ is equivalent to $\|\tilde{\mathbf{v}}\| \longrightarrow \infty$. Since every $\tilde{\mathbf{v}} \in \mathcal{V}$ can be expressed as $\tilde{\mathbf{v}}' + \tilde{\mathbf{w}}$, the property (6.48) has been proved.

From (6.78) it follows that for $\|\tilde{\mathbf{w}}\| \geq M$ with M sufficiently large we have $J(\tilde{\mathbf{v}}) > J(\tilde{\mathbf{v}}')$. For $\|\tilde{\mathbf{w}}\| \leq M, J(\tilde{\mathbf{v}})$ is bounded on account of (6.72). Hence, $J(\tilde{\mathbf{v}})$ is bounded from below in $\mathcal{V}$.

A.3. Proof of Theorem 1

Let a be defined by (6.47) with $\mathcal{W}$ replaced by $\mathcal{W}^N$. Since $\mathcal{W}^N$ is finite-dimensional and I is homogeneous of degree two and continuous in $\mathcal{W}^N$, the infimum value is reached at some nonzero $\tilde{\mathbf{w}}^*$. Fron (6.41) it follows that $I(\tilde{\mathbf{w}}^*) > 0$ so that $a > 0$. By the argument given in Sect. A.2 (with $\mathcal{V}$ and $\mathcal{W}$ replaced by $\mathcal{V}^N$ and $\mathcal{W}^N$, respectively), we obtain that $J(\tilde{\mathbf{v}}) > J(\tilde{\mathbf{v}}')$ for $\|\tilde{\mathbf{v}} - \tilde{\mathbf{v}}'\| \geq M$, where $\tilde{\mathbf{v}}, \tilde{\mathbf{v}}' \in \mathcal{V}^N$, $\tilde{\mathbf{v}}'$ is fixed and M is a sufficiently large number. Since $\mathcal{V}^N$ is finite dimensional and J is continuous in $\mathcal{V}^N$, the absolute minimum of J in the ball $\|\tilde{\mathbf{v}} - \tilde{\mathbf{v}}'\| < M$, and thus in $\mathcal{V}^N$, is attained at some $\tilde{\mathbf{v}} = \tilde{\mathbf{v}}^*$. Since J is Gateaux differentiable, (6.42) must be satisfied at the minimum point so that $\tilde{\mathbf{v}}^*$ is a solution to the discretized rate-problem. This completes the proof of Theorem 1.

A.4. Proof of property (6.50)

Consider a sequence $(\tilde{\mathbf{v}}_n)$ in $\mathcal{V}$ such that (6.49) holds, and a sequence of numbers γ_n. On substituting $\tilde{\mathbf{v}}_n$ and $\gamma_n\tilde{\mathbf{w}}$ into (6.75) in place of $\tilde{\mathbf{v}}$ and $\tilde{\mathbf{w}}$, respectively, and rearranging, we obtain

$$-\gamma_n\,\delta J(\tilde{\mathbf{v}}_n, \tilde{\mathbf{w}}) - \frac{1}{2}C\gamma_n^2\|\tilde{\mathbf{w}}\|^2 \leq J(\tilde{\mathbf{v}}_n) - J(\tilde{\mathbf{v}}_n + \gamma_n\tilde{\mathbf{w}})\,. \tag{6.79}$$

Let $\|\tilde{\mathbf{w}}\| = 1$ and $\gamma_n = -\delta J(\tilde{\mathbf{v}}_n, \tilde{\mathbf{w}})/C$. On taking into account that $J(\tilde{\mathbf{v}}_n + \gamma_n\tilde{\mathbf{w}}) \geq d$, we arrive at the result

$$(\delta J(\tilde{\mathbf{v}}^n, \tilde{\mathbf{w}}))^2 \leq 2C(J(\tilde{\mathbf{v}}_n) - d) \quad \text{for every } \tilde{\mathbf{w}} \in \overline{\mathcal{W}}, \ \|\tilde{\mathbf{w}}\| = 1 .$$

This can be rewritten as

$$|\delta J(\tilde{\mathbf{v}}^n, \tilde{\mathbf{w}})| \leq \varepsilon_n \|\tilde{\mathbf{w}}\| \quad \text{for every } \tilde{\mathbf{w}} \in \overline{\mathcal{W}},$$

where

$$\varepsilon_n = (2C)^{1/2}(J(\tilde{\mathbf{v}}_n) - d)^{1/2} .$$

Since $J(\tilde{\mathbf{v}}_n)$ tends to d, $\varepsilon_n \to 0$ as $n \to \infty$. The property (6.50) has been proved.

A.5. Proof of property (6.54)

Consider a domain $G \subset V$ with a regular boundary ∂G. Let $h(\boldsymbol{\xi})$ denote the distance from a point $\boldsymbol{\xi}$ to the set G, i.e. $h(\boldsymbol{\xi}) = |\boldsymbol{\xi} - \boldsymbol{\xi}'|$, where $\boldsymbol{\xi}'$ is a point from $\overline{G}$ closest to $\boldsymbol{\xi}$. Denote by G_H a spatial set surrounding G and consisting of $\boldsymbol{\xi}$ such that $0 < h(\boldsymbol{\xi}) < H$.

We will consider the following trial scalar fields $\widetilde{w}$ defined on $\overline{V}$:

$$w(\boldsymbol{\xi}) = \begin{cases} H & \text{if } \boldsymbol{\xi} \in \overline{G}, \\ H - h(\boldsymbol{\xi}) & \text{if } \boldsymbol{\xi} \in G_H, \\ 0 & \text{if } h(\boldsymbol{\xi}) \geq H. \end{cases}$$

Let $w_1 = w$ and $w_2 \equiv w_3 \equiv 0$. ∂G may contain a part of S_T but we assume that $\partial G \cap \overline{S_u} = \emptyset$. ∇h at typical $\boldsymbol{\xi} \in G_H$ exists and is equal to a unit vector (normal to ∂G at $\boldsymbol{\xi}'$). For H sufficiently small, S_u does not intersect G_H and $\tilde{\mathbf{w}}$ just defined belongs to $\overline{\mathcal{W}}$. We have then $w_{1,i}w_{1,i} = 1$ in G_H and $\|\tilde{\mathbf{w}}\| = |G_H|$, where $|G_H|$ denotes the volume of G_H. On substituting (6.52) with that field $\tilde{\mathbf{w}}$ into (6.50) and rearranging, we obtain

$$|\dot{P}_1^*(G, \tilde{\mathbf{v}}_n)| \equiv \left| \int_G \dot{b}_1^*(\tilde{\mathbf{v}}_n) \,\mathrm{d}V + \int_{(S_T \cup S_D) \cap \overline{G}} \dot{T}_1^*(\tilde{\mathbf{v}}_n) \,\mathrm{d}S \right|$$
$$\leq \varepsilon_n |G_H| / H + \left| \int_{G_H} \dot{b}_1^*(\tilde{\mathbf{v}}_n)(1 - h/H) \,\mathrm{d}V + \int_{(S_T \cup S_D) \cap G_H} \dot{T}_1^*(\tilde{\mathbf{v}}_n)(1 - h/H) \,\mathrm{d}S \right| .$$

In the limit as $H \to 0$, the set G_H becomes vanishingly narrow and the integrals on the right-hand side vanish while $|G_H| / H \to |\partial G|$, the surface measure of the boundary ∂G. By repeating the above procedure for the remaining coordinates, we arrive at (6.54) for any regular domain $G \subset V$ such that $\partial G \cap \overline{S_u} = \emptyset$.

A domain $G \subset V$ with a piecewise regular boundary disjoint with $\overline{S_u}$ can be approximated by a sequence (G_k) of regular domains considered above. If this is done, as is always possible, in such a manner that $|\partial G \backslash \partial G_k| \to 0$ then in the limit of approximation we obtain (6.54). The same argument shows that (6.54) remains valid when $\partial G \cap \overline{S_u}$ is not empty but is of zero surface measure. If the common part of S_u

and ∂G is of finite area then in the limit expression the rate of perturbing tractions acting on this common surface part cannot be taken into account. However, $\dot{\mathbf{T}}^* = \mathbf{0}$ on S_u (any reaction on S_u is statically admissible), so that (6.54) is finally valid for any domain $G \subseteq V$ with a piecewise regular boundary.

References

1. Z. P. Bazant, Stable states and stable paths of propagation of damage zones and interactive fractures, in: Cracking and Damage, J. Mazars and Z. P. Bazant [Eds.], pp. 183-206, Elsevier, London and New York 1989.
2. D. C. Drucker, A definition of stable inelastic material, J. Appl. Mech., **26**, 101-106, 1959.
3. R. Hill, A general theory of uniqueness and stability in elastic-plastic solids, J. Mech. Phys. Solids, **6**, 236-249, 1958.
4. R. Hill, Some basic principles in the mechanics of solids without a natural time, J. Mech. Phys. Solids, **7**, 209-225, 1959.
5. R. Hill, Bifurcation and uniqueness in non-linear mechanics of continua, in: Problems of Continuum Mechanics, N.I. Muskhelishwili Anniversary Volume, pp. 155-164, S.I.A.M., Philadelphia 1961.
6. R. Hill, Uniqueness criteria and extremum principles in self-adjoint problems of continuum mechanics, J. Mech. Phys. Solids, **10**, 185-194, 1962.
7. R. Hill, Eigenmodal defomations in elastic-plastic continua, J. Mech. Phys. Solids, **15**, 371-386, 1967.
8. R. Hill, Aspects of invariance in solid mechanics, Adv. in Appl. Mech., Vol. 18, 1-75, Acad. Press, New York 1978.
9. R. Hill and M. J. Sewell., A general theory of inelastic column failure-II, J. Mech. Phys. Solids, **8**, 112-118, 1960.
10. R. Hill and M. J. Sewell, A general theory of inelastic column failure-III, J. Mech. Phys. Solids, **10**, 285-300, 1962.
11. J. W. Hutchinson, Plastic buckling, Adv. in Appl. Mech., Vol. 14, pp. 67-144, Academic Press, New York 1974.
12. T. von Kármán, Discussion on Shanley's paper: Inelastic column theory, J. Aero. Sci., **14**, 267-268, 1947.
13. V. D. Klushnikov, Stability of elastic-plastic systems [in Russian], Nauka, Moscow 1980.
14. R. J. Knops and E. W. Wilkes, Theory of elastic stability, Handbuch der Physik VI a/3, Springer, Berlin 1973.
15. N. N. Krasovskii, Stability of motion, Stanford Univ. Press, California 1963.
16. A. Needleman and V. Tvergaard, Aspects of plastic postbuckling behavior, in: Mechanics of Solids, The Rodney Hill 60th Anniversary Volume, H. G. Hopkins and M. J. Sewell [Eds.], pp. 453-497, Pergamon Press, Oxford 1982.
17. Q. S. Nguyen and D. Radenkovic, Stability of equilibrium in elastic plastic solids, Springer Lecture Notes in Mathematics, vol. 503, pp. 403-414, 1975.
18. H. Petryk, A consistent energy approach to defining stability of plastic deformation processes, in: Stability in the Mechanics of Continua, F. H. Schroeder [Ed.], Proc. IUTAM Symp., Nümbrecht 1981; pp. 262-272, Springer, Berlin, Heidelberg 1982.
19. H. Petryk, On energy criteria of plastic instability, in: Plastic Instability, Proc. Considère Memorial, pp. 215-226, Ecole Nat. Ponts Chauss. Press, Paris 1985.
20. H. Petryk, On constitutive inequalities and bifurcation in elastic-plastic solids with a yield-surface vertex, J. Mech. Phys. Solids, **37**, 265-291, 1989.
21. H. Petryk and K. Thermann, On discretized plasticity problems with bifurcations, Int. J. Solids Structures, **29** (1992) 745-765.
22. M.J. Sewell, On configuration-dependent loading, Arch. Rat. Mech. Anal., **23**, 327-351, 1967.

23. F.R. Shanley, Inelastic column theory, J. Aero. Sci., **14**, 261-267, 1947.
24. M. M. Vainberg, Variational methods for the study of nonlinear operators [in Russian], G.I.T.T.L., Moscow 1956 (English transl.: Holden-Day, San Francisco 1964).

Chapter 7
Material Instability and Strain-Rate Discontinuities in Incrementally Nonlinear Continua

Abstract A systematic analysis of theoretical aspects of material instability is presented for a general class of timeindependent, incrementally nonlinear solids which admit a velocity-gradient potential. The second-order energy criteria of instability of uniform straining and of equilibrium are derived for a material element embedded in a quasi-statically deforming continuum. Several theorems concerning propagation of acceleration waves in an incrementally nonlinear material are proved, with particular reference to stationary discontinuities. Initiation of shear band formation either by a dynamic growth of an initial disturbance or by a quasi-static bifurcation is studied as a symptom of material instability. The derived condition for the instability of uniform quasi-static straining does not coincide with that for instability of equilibrium and both differ in general from the familiar condition of strong ellipticity loss formulated for the tangent moduli. This is illustrated on an example of a non-proportional deformation path of material obeying the J_2 corner theory of plasticity.

7.1 Introduction

In classical elastic-plastic solids, the incremental constitutive relationship at a given state is assumed to be linear in each of the two (loading and unloading) branches of the material response. It seems nowadays to be generally agreed that such "bi-linear"[1] models overestimate the material stiffness against an abrupt change of the direction of straining; this becomes an essential shortcoming in bifurcation calculations. A micromechanical analysis of a time-independent macroscopic response of a structural polycrystalline metal leads to the conclusion [15, 21] that the actual incremental constitutive law for such a material involves a nonlinear range of partial unloading, related to the formation of a pointed yield-surface vertex at the current loading point.

[0] Reprinted from: Journal of the Mechanics and Physics of Solids 40, 1227-1250, 1992. H.Petryk, Material instability and strain-rate discontinuities in incrementally nonlinear continua. Copyright © 1992, with permission from Elsevier.

[1] The term "bi-linear" is used here in the sense different from that usually employed in mathematics.

H. Petryk, *Theory of Plastic Instability*,
https://doi.org/10.1007/978-3-031-98223-1_7

Specific models of that type have been developed (e.g. [5]) and effectively used in finite-element computations. Inadequacy of the classical model is evident for metal crystals deforming by multislip [26, 14, 19]. Also for soils the need to introduce incremental nonlinearity of non-classical type into the constitutive relations has been indicated (cf. e.g. [24, 6] and the references quoted therein).

Incrementally nonlinear models of a time-independent material which are not simply "bi-linear" are thus of more than academic interest. Nevertheless, essential aspects of instability of material type in such solids have not been sufficiently explored in the literature. For instance, although the phenomenon of localization of deformation into shear bands (a possible symptom of material instability) has been studied in many papers during the last 20 years, "truly nonlinear" bifurcation studies where the constitutive branch activated inside the band at the bifurcation point was not prescribed in advance (cf. [35, 24, 4, 25]) were infrequent and limited either to general remarks or to a particular material law. It has not been clarified what is the correct generalization of the condition of (strong) ellipticity loss when the constitutive rate equations are arbitrarily nonlinear. Hill's general theory of bifurcation and instability [12, 17] concerns a finite solid body, and its relation to material instability problems is not evident. It may be recalled that the status of Drucker's postulate ([7, 8] as a condition of material stability is not free of ambiguities when changes in geometry are taken into account [16]. In a geometrically exact formulation, there remains the problem of proper definition of the loading environment of a material element, cf. [17]. It may also be mentioned that the energy criterion for local instabilities formulated in [30] was not given a convincing justification. This brief recapitulation, albeit incomplete, does indicate that a theory of material instability in incrementally nonlinear continua has not reached a satisfactory stage of development.

In the present paper theoretical aspects of material instability are examined afresh for a wide class of solids whose constitutive rate equations are arbitrarily nonlinear but do not contain a natural time and admit a velocity-gradient potential [12]. Thermal effects are disregarded, while changes in geometry are fully taken into account. The analysis deals with three topics closely related to each other: (a) the second-order energy criteria of instability of uniform straining and of equilibrium in a material element embedded in a quasi-statically deforming continuum; (b) dynamic propagation of acceleration waves into material which is either at rest or deforming, with particular reference to stationary discontinuities; (c) initiation of shear band formation either by a quasi-static bifurcation or by a dynamic localization process. The theorems, their corollaries and interpretations given below provide an extension of the theory developed for incrementally linear or bi-linear solids by Hadamard [10], Thomas [37], Hill [13] and Mandel [27]. The present results for an incrementally nonlinear material turn out to be qualitatively different in several aspects. In particular, the derived condition for the instability of uniform quasi-static straining does not coincide with that for instability of equilibrium and both differ in general from the familiar condition of strong ellipticity loss formulated for the tangent moduli. This will be illustrated in Sect. 7.6 on an example of non-proportional deformation path of the material obeying the so-called J_2 corner theory of plasticity developed by Christoffersen and Hutchinson [5].

The present analysis of instabilities in a material element provides a counterpart to a recent study by the author [33] concerned with a finite solid body with specified boundary conditions. In these papers, a unified approach to various bifurcation and instability problems is developed which is based on the concept of instability of a deformation process and on the relevant energy criterion [30].

Throughout the paper the standard symbolic notation is employed where vectors or tensors are denoted by boldface letters. Lower-case Latin subscripts denote vector or tensor components on a fixed rectangular basis, with the summation convention adopted for repeated indices. A dot between two symbols indicates the full contraction, e.g. $A_{ij}B_{ji} = \mathbf{A} \cdot \mathbf{B}$, $(a_i A_{ij}) = \mathbf{a} \cdot \mathbf{A}$, while $(a_i b_j) = \mathbf{a} \otimes \mathbf{b}$. The transpose and norm of a second-order tensor $\mathbf{A}$ are denoted by $\mathbf{A}^T$ and $|\mathbf{A}| = (A_{ij}A_{ij})^{1/2}$, respectively. A dot over a symbol denotes the forward rate understood as a material time derivative in the right-hand sense.

7.2 Constitutive Rate Equations

It will be convenient to use the deformation gradient $\mathbf{F}$ and the unsymmetric nominal stress $\mathbf{N}$ (related to the symmetric Cauchy stress $\boldsymbol{\sigma}$ by $F_{ik}N_{kj} = \det(\mathbf{F})\sigma_{ij}$) as a specific pair of measures of the deformation and stress. We shall assume that constitutive rate equations $\dot{\mathbf{N}} = \dot{\mathbf{N}}(\dot{\mathbf{F}})$ at a given state of a material element do not depend in any way on a natural time and admit a velocity-gradient potential introduced by Hill [12]:

$$\dot{\mathbf{N}} = \frac{\partial U}{\partial \dot{\mathbf{F}}^T} = \mathbf{C}(\dot{\mathbf{F}}) \cdot \dot{\mathbf{F}}, \quad U = \frac{1}{2}\dot{\mathbf{N}}(\dot{\mathbf{F}}) \cdot \dot{\mathbf{F}}, \quad \mathbf{C} = \frac{\partial^2 U}{\partial \dot{\mathbf{F}}^T \partial \dot{\mathbf{F}}^T}. \tag{7.1}$$

In the indical notation, $\dot{N}_{ij} = \partial U / \partial \dot{F}_{ji} = C_{ijkl}(\dot{\mathbf{F}})\dot{F}_{lk}$; of course, the rate $\dot{\mathbf{F}}$ of deformation gradient coincides with the velocity gradient $\nabla\mathbf{v}$ evaluated in the *reference* configuration. It is nowadays a standard task to determine the actual form of the potential U and moduli C_{ijkl}, if the former exists and the latter are well defined, from the constitutive law expressed in other constitutive variables; the needed formulae can be found, e.g. in [17]. The potential U is a continuously differentiable and homogenous function of degree two of $\dot{\mathbf{F}}$ (with $\mathbf{C}(\dot{\mathbf{F}}/|\dot{\mathbf{F}}|)$ piecewise continuous for $\dot{\mathbf{F}} \neq \mathbf{0}$) which is itself dependent on the deformation history (and explicitly dependent on the current stress). Nonlinearity of $U(\dot{\mathbf{F}})$ is otherwise arbitrary (e.g. U may be piecewise quadratic with $\dot{\mathbf{N}}(\dot{\mathbf{F}})$ piecewise linear or it may describe a thoroughly nonlinear material response at a pointed yield-surface vertex), and any path-dependent hardening or softening characteristics are allowed; on the other hand, bi-linear models with the nonassociative flow rule of plasticity are excluded since (7.1) implies $C_{ijkl} = C_{klij}$. At a given state of an inhomogeneous or nonuniformly stressed body we have $U = U(\boldsymbol{\xi}, \dot{\mathbf{F}})$, where $\boldsymbol{\xi}$ is a position vector in the reference configuration; the dependence of U on $\boldsymbol{\xi}$ is assumed to be at least piecewise differentiable.

The above assumptions are adopted throughout this paper. On the contrary, the constitutive inequality discussed below shall not be used to obtain the results but will be useful in their interpretation. It has recently been shown [31] that an incrementally nonlinear, macroscopic constitutive law for an elastic-plastic heterogeneous material which at a micro-level obeys the normality, piecewise linearity and symmetry postulates satisfies the following constitutive inequality:

$$\int \dot{\mathbf{F}}^* \cdot \mathrm{d}\mathbf{N} \geqslant \int \dot{\mathbf{N}}^* \cdot \mathrm{d}\mathbf{F}\,; \tag{7.2}$$

F and **N** can be replaced by any other pair of work-conjugate measures of strain and stress. The inequality (7.2) holds for all segments of every piecewise-smooth deformation path while the starred rates are virtual and may vary arbitrarily along the path. It is understood that $\dot{\mathbf{N}}^*$ is related to $\dot{\mathbf{F}}^*$, similarly as $\mathrm{d}\mathbf{N}$ is to $\mathrm{d}\mathbf{F}$, by the currently valid incremental constitutive law which varies along the path but is unaffected in any way by variations of the starred rates (so that the starred (virtual) and unstarred (actual) rates in (7.2) cannot be interchanged). The validity of (7.2) has been examined [29] for a class of materials obeying the maximum dissipation principle and extended, in particular, to an elastic material with a number of propagating planar Griffith micro-cracks. An interpretation of (7.2) in terms of the second-order work of deformation is given in [32].

We recall briefly the consequences of (7.2) which are of importance for the study of bifurcation and material stability aspects:

– (7.2) implies the existence of the constitutive potential (7.1);

– along a regular segment of a deformation path with the actual rates $\dot{\mathbf{F}}^0$ and $\dot{\mathbf{N}}^0$, (7.2) reduces to

$$\dot{\mathbf{N}}^0 \cdot \dot{\mathbf{F}}^* - \dot{\mathbf{N}}^* \cdot \dot{\mathbf{F}}^0 \geqslant 0 \quad \text{for every } \dot{\mathbf{F}}^*, \quad \dot{\mathbf{N}}^* = \partial U / \partial \dot{\mathbf{F}}^T (\dot{\mathbf{F}}^*). \tag{7.3}$$

These consequences of (7.2) are obtained under certain regularity assumptions [31] which are omitted here for brevity. (7.3) has the interpretation that the actual rates do not correspond to abrupt unloading at the micro-level. If the tangent moduli $C_{ijkl}(\dot{\mathbf{F}}^0) = \partial^2 U / \partial \dot{F}_{ji} \partial \dot{F}_{lk} (\dot{\mathbf{F}}^0)$ (cf. (7.1)) are well defined then (7.3) is equivalent to [31]

$$(\dot{\mathbf{N}}(\dot{\mathbf{F}}) - \dot{\mathbf{N}}^0) \cdot (\dot{\mathbf{F}} - \dot{\mathbf{F}}^0) \geqslant (\dot{\mathbf{F}} - \dot{\mathbf{F}}^0) \cdot \mathbf{C}(\dot{\mathbf{F}}^0) \cdot (\dot{\mathbf{F}} - \dot{\mathbf{F}}^0) \quad \text{for every } \dot{\mathbf{F}}\,; \tag{7.4}$$

this may be regarded as a weakened form (since one member of the pair $\dot{\mathbf{F}}, \dot{\mathbf{F}}^0$ in (7.4) is fixed) of the relative convexity property introduced by Hill [12]. From (7.1) it is evident that (7.4) implies

$$U(\dot{\mathbf{F}}) \geqslant \frac{1}{2} \dot{\mathbf{F}} \cdot \mathbf{C}(\dot{\mathbf{F}}^0) \cdot \dot{\mathbf{F}} \quad \text{for every } \dot{\mathbf{F}} \tag{7.5}$$

as a limit case when $|\dot{\mathbf{F}}^0| / |\dot{\mathbf{F}}| \to 0$.

7.3 Energy Criteria of Material Instability

7.3.1 Integral Conditions of Material Instability

Consider a material element embedded in a quasi-statically deforming continuous body which in general is inhomogeneous and of finite dimensions. Body forces are omitted for simplicity,[2] and the external surface loading is assumed at the moment to be conservative (kinematical constrains are allowed) to make contact with the global energy criteria of instability; that assumption is otherwise not necessary here and will soon be dropped. In order to confine attention to instabilities of material type only, we shall consider a particular class of perturbed deformation modes which coincide with the fundamental mode except within a small *interior*[3] subdomain G of the reference body domain V (all fields are considered in the reference configuration). This means that the motion of (i.e. velocity and acceleration on) the boundary ∂G of G is not perturbed while the deformation within G is perturbed *non-uniformly*; that approach allows the formulation of conditions for material instability independently of any interaction rule between the configuration of the material element and its loading environment (cf. the discussion by Hill [17]). This clearly distinguishes the kind of material instability investigated here from that considered e.g. by Drucker [7, 8].

Denote by $\tilde{\mathbf{v}}$ a continuous and piecewise smooth velocity field such that $\mathbf{v} = \mathbf{v}^0$ outside a subdomain G; the superscript ^{0} is used to distinguish quantities evaluated in the fundamental (unperturbed) deformation process and the tilde over a symbol indicates a spatial field. Under the assumption that $\dot{\mathbf{v}} = \dot{\mathbf{v}}^0$ outside G, the condition of instability of the fundamental deformation *process*, in the energy sense of Petryk [30, 33], is reduced for such trial fields to

$$\ddot{W}(\tilde{\mathbf{v}}) < \ddot{W}(\tilde{\mathbf{v}}^0) \text{ for some } \tilde{\mathbf{v}} \text{ with } \mathbf{v} = \mathbf{v}^0 \text{ outside some } G, \tag{7.6}$$

where $\ddot{W}(\tilde{\mathbf{v}})$ stands for the second-order rate of the deformation work W, evaluated for the velocity field $\tilde{\mathbf{v}}$ at a given equilibrium state of the body. At an equilibrium state and for $\mathbf{v} = \mathbf{v}^0$ and $\dot{\mathbf{v}} = \dot{\mathbf{v}}^0$ over ∂V, the first-order rate $\dot{W}$ of work is independent of $\tilde{\mathbf{v}}$, while $\ddot{W}$ is independent of $\tilde{\dot{\mathbf{v}}}$ and can be expressed as [30][4]

$$\ddot{W}(\tilde{\mathbf{v}}) = \int_V 2U(\boldsymbol{\xi}, \nabla\mathbf{v}(\boldsymbol{\xi}))\,\mathrm{d}\boldsymbol{\xi} + \int_{\partial V} \mathbf{T}^0 \cdot \dot{\mathbf{v}}^0\,\mathrm{d}S\,,$$

where $\mathbf{T}^0$ are nominal surface tractions, regarded as given. As a special case of (7.6) when $\mathbf{v}^0$ and $\dot{\mathbf{v}}^0$ vanish identically, we obtain

[2] It can be shown that the conditions of material instability formulated below do not change when the first-order rate of nominal body forces is not zero but is prescribed or dependent locally on the velocity.

[3] Local instabilities at the body surface (cf. e.g. [18, 22, 2]) are not investigated in this paper.

[4] There is a misprint in equation (18) in this reference: $\ddot{W}$ should be replaced there by $\Delta\ddot{W}$.

$$\ddot{W}(\tilde{\mathbf{v}}) < 0 \quad \text{for some } \tilde{\mathbf{v}} \text{ with } \mathbf{v} = \mathbf{0} \text{ outside some } G, \tag{7.7}$$

which is a reduced form of the condition of instability of *equilibrium* in the energy sense. Fulfilment of (7.6) or (7.7) implies global instability of the fundamental deformation process or of equilibrium, respectively, in the sense discussed in [33] provided the external surface loading is conservative. The latter restriction is not needed in a discussion of instabilities at the level of a material element and can from now on be dropped.

Since $\tilde{\mathbf{v}}^0$ satisfies the conditions of continuing equilibrium by assumption, $\ddot{W}(\tilde{\mathbf{v}}^0)$ evaluated within G is equal to the second-order rate of work of the tractions acting on ∂G from the side of the surrounding material. Hence, fulfilment of the conditions (7.7) or (7.6) implies that mechanical work can be extracted from the material occupying the subdomain G (by imposing suitable perturbing body forces within G) while the material outside G remains at rest or is deforming as in the fundamental deformation mode, respectively.

On substituting the expression for $\ddot{W}$, the condition (7.6) takes the form

$$\int_G U(\boldsymbol{\xi}, \nabla\mathbf{v}(\boldsymbol{\xi}))\mathrm{d}\boldsymbol{\xi} < \int_G U(\boldsymbol{\xi}, \nabla\mathbf{v}^0(\boldsymbol{\xi}))\mathrm{d}\boldsymbol{\xi} \quad \text{for some } \tilde{\mathbf{v}} \text{ with } \mathbf{v} = \mathbf{v}^0 \text{ outside some } G. \tag{7.8}$$

Note that (7.8) is excluded if the global uniqueness condition holds since then $\tilde{\mathbf{v}}^0$ is characterized by the known minimum property [12] stronger than a negation of (7.8).

Let G be a neighborhood of a regular interior point $\bar{\boldsymbol{\xi}}$ in V at which the dependence of U and $\dot{\mathbf{F}}^0$ on $\boldsymbol{\xi}$ is continuous. It is then intuitively clear that within sufficiently small G the variations of $\dot{\mathbf{F}}^0$ and the *explicit* dependence of U on $\boldsymbol{\xi}$ become negligible. As a matter of fact, it can be rigorously proved (cf. Morrey [28], Ball [1]) that (7.8) (or (7.6), equivalently) is implied by

$$\int_{\hat{G}} U(\bar{\boldsymbol{\xi}}, \dot{\mathbf{F}}^0(\bar{\boldsymbol{\xi}}) + \nabla\mathbf{w}(\mathbf{X}))\, \mathrm{d}\mathbf{X} < |\hat{G}|\, U(\bar{\boldsymbol{\xi}}, \dot{\mathbf{F}}^0(\bar{\boldsymbol{\xi}})) \quad \text{for some } \tilde{\mathbf{w}} \text{ with } \mathbf{w} = 0 \text{ on } \partial\hat{G}\,. \tag{7.9}$$

The integral in (7.9) is taken now over a bounded open subset $\hat{G}$ of the Euclidean space $\mathbb{R}^3$ (and no longer over a subset of V), $\mathbf{X}$ is a variable position vector within $\hat{G}$ while $\bar{\boldsymbol{\xi}}$ is an arbitrary regular point in V which is regarded as fixed, $\tilde{\mathbf{w}}$ is a continuous and piecewise smooth field on $\hat{G} \cup \partial\hat{G}$, $\nabla\mathbf{w} = \partial\mathbf{w}/\partial\mathbf{X}$, and $|\hat{G}|$ denotes the volume of $\hat{G}$. A finite domain $\hat{G}$ in $\mathbb{R}^3$ may be identified with a vanishingly small region G in V but considered in a different scale of observation; this is illustrated schematically in Fig. 7.1. It has been shown [28, 1] that if (7.9) is satisfied for one $\hat{G}$ then it can be satisfied for any other $\hat{G}$ by suitable scaling of the field $\tilde{\mathbf{w}}$; this means that the size and shape of $\hat{G}$ in (7.9) are inessential. In the terminology introduced by Morrey [28], fulfilment of (7.9) means that U as a function of $(\boldsymbol{\xi}, \dot{\mathbf{F}})$ is *not quasi-convex at* $(\bar{\boldsymbol{\xi}}, \dot{\mathbf{F}}^0(\bar{\boldsymbol{\xi}}))$; the condition of quasi-convexity appears to be not considered previously in the published articles concerning plasticity. For a homogeneous and uniformly deforming material, the conditions (7.8) and (7.9) are equivalent to each other.

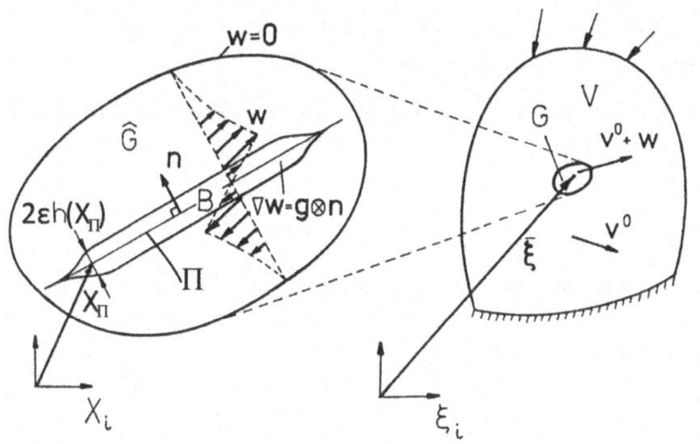

Fig. 7.1 Local peturbations of the fundamental velocity field.

In the particular case when $\dot{\mathbf{F}}^0 \equiv \mathbf{0}$, as a condition sufficient for (7.7) we analogously obtain

$$\int_{\hat{G}} U(\bar{\boldsymbol{\xi}}, \nabla\mathbf{w}(\mathbf{X}))\,\mathrm{d}\mathbf{X} < 0 \quad \text{for some } \tilde{\mathbf{w}} \text{ with } \mathbf{w} = \mathbf{0} \text{ on } \partial\hat{G}. \tag{7.10}$$

Since U is a continuous and homogeneous function of degree two of $\dot{\mathbf{F}}$, we have the implications (7.7) ⇒ (7.6) and (7.10) ⇒ (7.9) [30].

The condition (7.9) has the following interpretation: for a finite, homogeneous material element occupying a reference domain $\hat{G}$ in which the actual deformation, stress and mechanical properties are uniform and the same as at a point $\bar{\boldsymbol{\xi}}$ of V, there is a kinematically admissible deformation mode which satisfies the kinematical boundary conditions over $\partial\hat{G}$ compatible with the uniform incremental solution $\dot{\mathbf{F}} \equiv \dot{\mathbf{F}}^0$ in $\hat{G}$, and corresponds to a smaller value of the second-order work increment than the uniform mode. Moreover, if $\hat{G}$ is identified with an *interior* subdomain of the homogeneous element then the boundary conditions compatible with uniform straining (at zero body forces) can be arbitrary and not just purely kinematical. This means (cf. the above discussion of (7.6)) that if (7.9) holds then mechanical work can be extracted from the *deforming* material element *embedded in a continuum*, by perturbations vanishing outside that element. Hence, (7.9) may be regarded as a condition sufficient for instability, at least in the energy sense, of the uniform straining with the velocity gradient $\dot{\mathbf{F}}^0$ of the material element at $\bar{\boldsymbol{\xi}}$.

Analogously, if (7.10) holds then energy can be extracted from the homogeneous material element with a rigidly constrained boundary. Hence, (7.10) may be regarded as a condition sufficient for instability (in the energy sense) of equilibrium of the material element at $\bar{\boldsymbol{\xi}}$ *embedded in a non-deforming continuum* (under the current stress).

It is recalled that since the trial perturbation mode $\tilde{\mathbf{w}}$ vanishes over the element boundary, (7.9) and (7.10) interpreted as conditions *sufficient* for a material instability are derived independently of the actual interaction rule between the configuration of the material element and its loading environment.

The argument $\bar{\xi}$ is henceforth omitted to simplify the notation; it will be understood that the material element under consideration is arbitrary but fixed.

7.3.2 Pointwise Conditions of Instability

Although (7.9) is expressed in terms of material properties at one point only, it is not an explicit pointwise condition on the function $U(\dot{\mathbf{F}})$ or its derivatives. It can be reduced to such a form if the class of considered instability modes $\tilde{\mathbf{w}}$ is further restricted. We shall discuss the possibility that the instability mode $\nabla\mathbf{w}$ interpreted as a perturbation of the deformation rate in $\hat{G}$ is effectively concentrated in a disk-like volume B of a (nonuniform) thickness $2\varepsilon h$ which is small compared with the dimensions of $\hat{G}$ (Fig. 7.1); ε denotes here a positive small parameter. A similar deformation mode was used to draw the conclusion [10, 13] that fulfilment of the Legendre-Hadamard (or semi-strong ellipticity) condition is necessary for stability of equilibrium in an *incrementally linear* solid ($\mathbf{C}$ independent of $\dot{\mathbf{F}}$); for incrementally nonlinear materials the result will not be the same.

The shape function $h = h(\mathbf{X}_\Pi), \mathbf{X}_\Pi \in \Pi$, defined on the central plane Π of the disk B is assumed to be smooth and to vanish in some neighbourhood of the boundary $\partial\hat{G}$ of $\hat{G}$; h may be constant over the major part of the disk B. Denote by $\mathbf{n}$ the unit normal to Π, and let $\mathbf{g}$ be an arbitrary nonzero vector. Let $\tilde{\mathbf{w}}$ of the properties assumed in (7.9) be specified as a vector field dependent on ε such that $|\nabla\mathbf{w}|$ is of order ε outside $\overline{B} = B \cup \partial B$ while $\nabla\mathbf{w}$ inside B is independent of ε and equal to $\mathbf{g}\otimes\mathbf{n}$ (that is, $\partial w_j/\partial X_i = g_j n_i$ in B). Such a field can easily be constructed; an explicit example is given in the Subsect. Al of Appendix in Sect. 7.8.

Introduce the Weierstrass "excess" function $\mathscr{E}$ associated with U, defined by

$$\mathscr{E}(\dot{\mathbf{F}}^0, \dot{\mathbf{F}}) = U(\dot{\mathbf{F}}) - U(\dot{\mathbf{F}}^0) - \dot{\mathbf{N}}^0 \cdot (\dot{\mathbf{F}} - \dot{\mathbf{F}}^0), \quad \dot{\mathbf{N}}^0 = \frac{\partial U}{\partial \dot{\mathbf{F}}^T}(\dot{\mathbf{F}}^0)\,. \tag{7.11}$$

From the Green theorem and the boundary condition $\mathbf{w} = \mathbf{0}$ on $\partial\hat{G}$ it follows that[5]

$$\int_{\hat{G}} (U(\dot{\mathbf{F}}^0 + \nabla\mathbf{w}(\mathbf{X})) - U(\dot{\mathbf{F}}^0))\,\mathrm{d}\mathbf{X} = \int_{\hat{G}} \mathscr{E}(\dot{\mathbf{F}}^0, \dot{\mathbf{F}}^0 + \nabla\mathbf{w}(\mathbf{X}))\,\mathrm{d}\mathbf{X}\,. \tag{7.12}$$

Note that the latter integrand is of the order of ε^2 if $|\nabla\mathbf{w}|$ is of the order of ε, as can be seen from (7.11) or from (7.19) below. For a field $\tilde{\mathbf{w}}$ of the properties specified above, in particular for the field $\tilde{\mathbf{w}}$ defined in Subsect. Al of Appendix in Sect. 7.8 with $\nabla\mathbf{w}$ given by (7.35), we thus obtain

---

[5] An analogous identity is obviously valid for an inhomogeneous region where U and $\dot{\mathbf{F}}^0$ are dependent on $\mathbf{X}$ provided the associated stress-rate $\dot{\mathbf{N}}^0$ satisfies the conditions of continuing equilibrium.

$$\int_{\hat{G}} U(\dot{\mathbf{F}}^0 + \nabla \mathbf{w}(\mathbf{X}))\, \mathrm{d}\mathbf{X} - |\hat{G}|\, U(\dot{\mathbf{F}}^0) = |B|\, \mathscr{E}(\dot{\mathbf{F}}^0, \dot{\mathbf{F}}^0 + \mathbf{g} \otimes \mathbf{n}) + O(\varepsilon^2), \quad (7.13)$$

where the volume $|B|$ of the disk is proportional to ε. Hence, (7.9) is satisfied for ε sufficiently small by the field $\tilde{\mathbf{w}}$ if at a regular point $\bar{\xi}$ at the current state we have

$$\mathscr{E}(\dot{\mathbf{F}}^0, \dot{\mathbf{F}}^0 + \mathbf{g} \otimes \mathbf{n}) < 0 \quad \text{for some vectors } \mathbf{g}, \mathbf{n}\,. \quad (7.14)$$

The statement just proved is a corollary of the theorem due to Graves [9] who has shown that, in the present context, a necessary condition for the left-hand integral in (7.8) to be minimized by $\tilde{\mathbf{v}}^0$ is that

$$\mathscr{E}(\dot{\mathbf{F}}^0, \dot{\mathbf{F}}^0 + \mathbf{g} \otimes \mathbf{n}) \geqslant 0 \quad \text{for all } \mathbf{g}, \mathbf{n} \quad (7.15)$$

at every regular point in G. The above construction of the instability mode provides a mechanical interpretation for (7.14) and will be useful later in the discussion of shear band initiation. In the terminology of Ball [1], if (7.15) holds then U is said to be *rank* 1 *convex at* $\dot{\mathbf{F}}^0$.

In particular, the condition (7.10) (and thus also (7.7)) is satisfied by the field $\tilde{\mathbf{w}}$ just discussed for ε sufficiently small if

$$U(\mathbf{g} \otimes \mathbf{n}) < 0 \quad \text{for some } \mathbf{g}, \mathbf{n}\,. \quad (7.16)$$

In passing, it may be remarked that on substituting $(7.1)_2$ in the condition (7.16) and transforming with the help of standard formulae one easily shows that (7.16) is equivalent to

$$\dot{\sigma}_{ij} D_{ij} < 0 \quad \text{for some } \dot{\mathbf{F}} = \mathbf{g} \otimes \mathbf{n}\,, \quad (7.17)$$

where $\mathbf{D}$ is the Eulerian strain rate and σ is the constitutively related material time derivative of the Cauchy stress, both evaluated for a given $\dot{\mathbf{F}}$ of the form indicated. In spite of some analogy (with reversed inequality) to Drucker's condition of material stability [7, 8], the present constraint on the velocity gradient makes the conditions clearly distinct and removes ambiguities raised when the material spin is unspecified.

The inequalities (7.14) and (7.16) represent the final energy conditions for instability of uniform straining and of equilibrium, respectively, of a homogeneous material element in an arbitrarily "stiff" loading environment. Equivalent criteria have been formulated in [30] but not discussed in detail. Justification for (7.14) and (7.16) as conditions sufficient for a material instability (without referring to the energy criteria of global instability) and the nature of predicted instability are studied in the subsequent sections. It is an open problem in which cases the conditions (7.14) and (7.16) are different from (7.9) and (7.10), respectively.

The above inequalities, as well as all related inequalities discussed below, are invariant under transformation to another reference configuration. This follows from the invariance in value of the expression $\dot{\mathbf{N}} \cdot \dot{\mathbf{F}}/\rho$, where ρ is the reference density and $\dot{\mathbf{N}}$ and $\dot{\mathbf{F}}$ need not be constitutively related to each other.

7.3.3 Mutal Relations Between the Instability Conditions

On using (7.1) the left-hand expression in (7.16) can be transformed to

$$U(\mathbf{g}\otimes\mathbf{n}) = \frac{1}{2}(\mathbf{g}\otimes\mathbf{n})\cdot\mathbf{C}(\mathbf{g}\otimes\mathbf{n})\cdot(\mathbf{g}\otimes\mathbf{n}) \equiv \frac{1}{2}C_{ijkl}(\mathbf{g}\otimes\mathbf{n})n_i n_k g_j g_l \qquad (7.18)$$

provided the moduli tensor $\mathbf{C} = \partial^2 U/\partial\dot{\mathbf{F}}^T\partial\dot{\mathbf{F}}^{\mathrm{T}}$ is well defined at $\dot{\mathbf{F}} = \mathbf{g}\otimes\mathbf{n}$. For comparison, the left-hand expression in (7.14) can be equivalently expressed as (the elementary proof is given for completeness in Subsect. A.2 of Appendix in Sect. 7.8)

$$\mathscr{E}(\dot{\mathbf{F}}^0, \dot{\mathbf{F}}^0+\mathbf{g}\otimes\mathbf{n}) = \frac{1}{2}(\mathbf{g}\otimes\mathbf{n})\cdot\overline{\mathbf{C}}\cdot(\mathbf{g}\otimes\mathbf{n}), \quad \overline{\mathbf{C}} = 2\int_0^1 \mathbf{C}(\dot{\mathbf{F}}^0+s\mathbf{g}\otimes\mathbf{n})(1-s)\mathrm{d}s \qquad (7.19)$$

while negation of the familiar Legendre-Hadamard condition at $\dot{\mathbf{F}}^0$ reads

$$(\mathbf{g}\otimes\mathbf{n})\cdot\mathbf{C}(\dot{\mathbf{F}}^0)\cdot(\mathbf{g}\otimes\mathbf{n}) < 0 \quad \text{for some } \mathbf{g}, \mathbf{n}\,, \qquad (7.20)$$

where $\mathbf{C}(\dot{\mathbf{F}}^0)$ is assumed to be well defined. By substituting $\gamma\mathbf{g}$ in place of $\mathbf{g}$ in (7.14) and using homogeneity and continuity of $U(\dot{\mathbf{F}})$ it becomes evident that each of (7.16) or (7.20) implies (7.14) when γ tends to infinity or to zero, respectively. The converse implications need not be true so that in general the three criteria are distinct (cf. the example given in Sect. 7.6).

For convenience, the obtained instability conditions are ordered in the following diagram

$$\begin{array}{ccc|ccc|} \cline{4-6} \text{instability of deformation process} & \Leftarrow (7.6) \Leftarrow & & (7.9) & \Leftarrow (7.14) & \Leftarrow (7.20) \\ & \Uparrow & & \Uparrow & \Uparrow & \\ \text{instability of equilibrium} & \Leftarrow (7.7) \Leftarrow & & (7.10) & \Leftarrow (7.16) & \\ \cline{4-6} \end{array} \qquad (7.21)$$

where the conditions in the box refer to "intrinsic" instabilities in a homogeneous and uniformly deformed material element.

For the classical (bi-linear) elastic-plastic response with the normality flow rule (in work-conjugate variables) relative to a smooth yield surface, the incremental constitutive relationship is of the form (7.1) with U composed of two different quadratic forms corresponding to loading and unloading. For such solids all the conditions in the box in (7.21) are equivalent to each other provided $\dot{\mathbf{F}}^0$ corresponds to the loading regime. For, the inequality (7.4) is then satisfied as a consequence of Hill's relative convexity property [11, 17], and fulfilment of (7.5) implies equivalence of (7.9), (7.14) and (7.20) (see below). Moreover, (7.20) implies then (7.16) since for any velocity gradient of the form $\mathbf{g}\otimes\mathbf{n}$ we can have $\mathbf{C}(\mathbf{g}\otimes\mathbf{n}) = \mathbf{C}(\dot{\mathbf{F}}^0)$ provided the sense of a vector $\mathbf{g}$, immaterial in (7.20), is appropriately selected. This completes the proof of equivalence of the conditions in the box. Evidently, this equivalence takes place also in the special case when U is *quadratic*, i.e. if the material is incrementally linear, and then $\mathscr{E}(\dot{\mathbf{F}}^0, \dot{\mathbf{F}}) = U(\dot{\mathbf{F}} - \dot{\mathbf{F}}^0)$.

On the contrary, instability of the process of uniform straining of an incrementally nonlinear material, predicted by the energy criterion (7.14), can appear even if the actual tangent moduli $\mathbf{C}(\dot{\mathbf{F}}^0)$ are strongly elliptic (i.e. the left-hand expression in (7.20) is positive for all nonzero $\mathbf{g}, \mathbf{n}$). However, that possibility is excluded if the constitutive inequality (7.5) is satisfied, e.g. as a consequence of (7.2). For, if (7.5) holds and $\mathbf{C}(\dot{\mathbf{F}}^0)$ is well defined then we have the equivalence (7.9) $\Leftrightarrow$ (7.14) $\Leftrightarrow$ (7.20). To prove this, we observe that if (7.9) holds for the actual potential U and (7.5) is valid, then the inequality in (7.9) will be satisfied for the same $\tilde{\mathbf{w}}$ also if U is replaced by the *quadratic* potential generated by the fixed moduli $\mathbf{C}(\dot{\mathbf{F}}^0)$. The needed implication (7.9) $\Rightarrow$ (7.20) follows then from the theorem of van Hove [20].

In turn, (7.20) implies (7.16) if some velocity gradient $\mathbf{g} \otimes \mathbf{n}$ satisfying (7.20) happens to lie in the constitutive domain of the moduli $\mathbf{C}(\dot{\mathbf{F}}^0)$. This need not be so in general, contrary to the case of classical bi-linear response. The criteria (7.14) and (7.16) of instability of uniform straining and of equilibrium, respectively, can thus be distinct even if (7.4) holds.

7.4 Instability and Acceleration Waves

In a deforming material continuum, consider a geometric surface Σ across which the velocity gradient $\nabla\mathbf{v} = \dot{\mathbf{F}}$ and nominal stress-rate $\dot{\mathbf{N}}$ are discontinuous while $\mathbf{v}, \mathbf{F}$ and $\mathbf{N}$ are continuous. For convenience but with no loss of generality, the reference configuration is taken to coincide momentarily with the current configuration. Inertia effects are taken into account. Following Hadamard [10] and Hill [13], if Σ propagates relative to the material and carries thus a jump in acceleration then it is called an acceleration wave, and if the speed of propagation is zero (with no acceleration jump) then it is called a stationary discontinuity. The well-known compatibility and wave propagation conditions read [13]

$$[\![\dot{\mathbf{F}}]\!] = \mathbf{g} \otimes \mathbf{n}, \quad \mathbf{n} \cdot [\![\dot{\mathbf{N}}]\!] = \rho c^2 \mathbf{g}, \tag{7.22}$$

where $[\![\cdot]\!]$ denotes a jump across Σ, $\mathbf{n}$ is the unit normal to Σ, $\mathbf{g}$ is some nonzero vector, $\rho > 0$ is the material density, c is the normal speed of propagation, and $\dot{\mathbf{N}}$ is constitutively related to $\dot{\mathbf{F}}$ on each side of Σ. The other compatibility conditions need not be considered here. It will be convenient to take into account also solutions to (7.22) with $c^2 < 0$; they will be called waves of imaginary speed.

A qualitative analysis of solutions to (7.22) is presented below separately for the two cases when the velocity gradient at one side of Σ vanishes or not.[6]

[6] The author was unable to find in the literature any comprehensive qualitative analysis of solutions to (7.22) for an incrementally nonlinear constitutive law of the general form (7.1).

7.4.1 Wave Propagation and Instability of Equilibrium

Throughout this subsection it is assumed that the velocity gradient at one side of Σ vanishes (possibly momentarily), so that the jumps in (7.22) can be replaced by the values at the other side of Σ. From (7.1) it follows that the left-hand side of $(7.22)_2$ can be expressed as $\partial U(\mathbf{g}\otimes\mathbf{n})/\partial\mathbf{g}$. For any fixed $\mathbf{n}$, (7.22) is thus satisfied for $\mathbf{g} = \mathbf{g}^*$ if and only if

$$\left.\frac{\partial U(\mathbf{g}\otimes\mathbf{n})}{\partial\mathbf{g}}\right|_{\mathbf{g}=\mathbf{g}^*} = \mu\mathbf{g}^*, \quad \mu = \rho c^2, \tag{7.23}$$

where μ may be of either sign or vanish for a stationary discontinuity. By the method of Lagrangian multipliers, (7.23) is satisfied for some μ if $\mathbf{g}^*$ corresponds to a relative extremum point for $U(\mathbf{g}\otimes\mathbf{n})$ constrained to a sphere $|\mathbf{g}|$ = constant. Hence, *for any* $\mathbf{n}$ *(7.22) have at least two distinct solutions*[7] $(\mathbf{g}_1, c_1)$, $(\mathbf{g}_2, c_2)$ (with $\mathbf{g}_1/|\mathbf{g}_1| \neq \mathbf{g}_2/|\mathbf{g}_2|$), associated with the minimum and maximum points, respectively, of $U(\mathbf{g}\otimes\mathbf{n})$ on a sphere $|\mathbf{g}| = 1$. On account of the homogeneity of U, the value of μ is determined by

$$2U(\mathbf{g}^*\otimes\mathbf{n}) = \mu|\mathbf{g}^*|^2 \tag{7.24}$$

and has thus the sign of $U(\mathbf{g}^*\otimes\mathbf{n})$. It follows that a solution to (7.22) describes a real wave, a wave of imaginary speed or a stationary discontinuity according to whether it assigns to $U(\mathbf{g}\otimes\mathbf{n})$ a positive, negative or zero value, respectively. In particular, for the solution $(\mathbf{g}_1, c_1)$ we have $c_1^2 < 0$ if $U(\mathbf{g}\otimes\mathbf{n})$ takes negative values for some $\mathbf{g}$. If $U(\mathbf{g}\otimes\mathbf{n})$ is never negative then $c^2 \geqslant 0$ for *any* solution to (7.23).

We have thus proved the following statement which provides an interpretation of the instability related to (7.16):[8] *The condition (7.16) is necessary and sufficient for the existence of waves of imaginary speed of an orientation* $\mathbf{n}$ *provided the velocity gradient vanishes at one side of the "wave".*

In analogy to the linear case [27], we will show in Sect. 7.5 that the existence of such non-real waves is associated with a specific material instability related to the possibility of spontaneous *dynamic localization of deformation.*

Turning to stationary discontinuities in a velocity gradient (we recall that the velocity gradient at one side of Σ is assumed to vanish), for their admissibility it is necessary that

$$U(\mathbf{g}\otimes\mathbf{n}) = 0 \quad \text{for some } \mathbf{n}, \mathbf{g} \neq \mathbf{0} \tag{7.25}$$

as can be seen, e.g. from (7.24) with $\mu = 0$. This result can also be deduced directly from Hill's [13] analysis and the identity (7.18). Contrary to the case of incrementally linear solids, singularity of the matrix $(C_{ijkl}(\mathbf{g}\otimes\mathbf{n})n_i n_k)$ does *not* imply in general that a stationary discontinuity is admissible for that orientation $\mathbf{n}$.

We shall show that (7.25) is sufficient for the existence of an admissible stationary discontinuity of *some* orientation $\mathbf{n}^*$ provided that there exists a nonzero vector $\mathbf{n}_s$

[7] For comparison, if U is *quadratic* then the existence of *three* distinct solutions with the "characteristic segments" $\mathbf{g}_i$ mutually orthogonal is ensured [13].

[8] An equivalent statement has been mentioned in [30] without providing a detailed proof.

such that

$$U(\mathbf{g} \otimes \mathbf{n}_s) \geqslant 0 \quad \text{for every } \mathbf{g}\,. \tag{7.26}$$

Observe first from (7.23) that a solution to (7.22) for $c = 0$ is associated with $\partial U(\mathbf{g} \otimes \mathbf{n})/\partial \mathbf{g} = \mathbf{0}$. It suffices thus to prove that for some $\mathbf{n} = \mathbf{n}^*$ the minimum value of $U(\mathbf{g} \otimes \mathbf{n})$ on a sphere $|\mathbf{g}| = 1$ is equal to zero. If (7.25) and (7.26) hold then for some $\mathbf{n}$ the minimum value is nonpositive and for some (other) $\mathbf{n} = \mathbf{n}_s$ is non-negative, respectively. On account of continuity of U the minimum value must vanish for certain $\mathbf{n}^*$, which completes the proof. It is evident that the conclusion would be valid if the inequality sign in (7.26) were reversed.

The above proof remains valid without any changes when the condition (7.25) is replaced by (7.16) (of course, for $\mathbf{n} \neq \mathbf{n}_s$). We note also that (7.26) means that (7.16) does not hold for some $\mathbf{n} = \mathbf{n}_s$. Hence, existence of admissible stationary discontinuities is proved if (7.16) holds but not for all $\mathbf{n}$.

For convenience, the obtained results are summarized in the following statement.

Theorem 1 *Let the velocity gradient vanish at one side of a discontinuity surface* Σ. *Then:*
(i) for any orientation $\mathbf{n}$ *of* Σ, *(7.22) have at least two distinct solutions (which may be associated with imaginary speed of wave propagation),*
(ii) the condition (7.16) is necessary and sufficient for the existence of waves of imaginary speed of an orientation $\mathbf{n}$,
(iii) the condition (7.16) is sufficient for the existence of admissible stationary discontinuities provided it is satisfied for not all orientations $\mathbf{n}$,
(iv) the condition (7.25) is necessary and sufficient for the existence of admissible stationary discontinuities provided $U(\mathbf{g} \otimes \mathbf{n}_s)$ *does not change sign for some fixed* $\mathbf{n}_s \neq \mathbf{0}$,
(v) for any given orientation $\mathbf{n}$ *of* Σ, *all solutions to (7.22) describe acceleration waves propagating with nonzero real speed (i.e. stationary discontinuities and waves of imaginary speed are simultaneously excluded) if and only if*

$$U(\mathbf{g} \otimes \mathbf{n}) > 0 \quad \textit{for every} \;\; \mathbf{g} \neq \mathbf{0}. \tag{7.27}$$

7.4.2 Wave Propagation and Instability of Uniform Straining

Throughout this subsection we shall assume that at one side of Σ the velocity gradient has a given nonzero value $\dot{\mathbf{F}}^0$ associated with the nominal stress-rate $\dot{\mathbf{N}}^0$. In (7.22) we thus substitute $[\![\dot{\mathbf{F}}]\!] = \dot{\mathbf{F}} - \dot{\mathbf{F}}^0$ and $[\![\dot{\mathbf{N}}]\!] = \dot{\mathbf{N}} - \dot{\mathbf{N}}^0$. The analysis runs on somewhat similar lines as in the preceding subsection, but the results turn out to be essentially different in certain aspects.

From (7.11) and (7.1) it follows that the left-hand side of $(7.22)_2$ can be expressed as $\partial \mathscr{E}(\dot{\mathbf{F}}^0, \dot{\mathbf{F}}^0 + \mathbf{g} \otimes \mathbf{n})/\partial \mathbf{g}$. For any fixed $\mathbf{n}$, (7.22) is thus satisfied for $\mathbf{g} = \mathbf{g}^*$ if and only if

$$\frac{\partial}{\partial \mathbf{g}} \mathscr{E}(\dot{\mathbf{F}}^0, \dot{\mathbf{F}}^0 + \mathbf{g} \otimes \mathbf{n})\bigg|_{\mathbf{g}=\mathbf{g}^*} = \mu \mathbf{g}^*, \quad \mu = \rho c^2, \tag{7.28}$$

where μ may be of either sign or vanish for a stationary discontinuity. Unlike in the special cases when $\dot{\mathbf{F}}^0 = \mathbf{0}$ or when U is quadratic, the direction of $\mathbf{g}^*$ satisfying (7.28) varies in general with $|\mathbf{g}^*|$. By the method of Lagrangian multipliers, (7.28) holds if $\mathbf{g}^*$ corresponds to a relative extremum point of $\mathscr{E}(\dot{\mathbf{F}}, \dot{\mathbf{F}}^0 + \mathbf{g} \otimes \mathbf{n})$ constrained to a sphere $|\mathbf{g}| =$ constant. Hence, *for any* $\mathbf{n}$ *(7.22) have at least two infinite families of solutions*, associated with the minimum and maximum values of $\mathscr{E}(\dot{\mathbf{F}}^0, \dot{\mathbf{F}}^0 + \mathbf{g} \otimes \mathbf{n})$ on each sphere $|\mathbf{g}| = R =$ constant; the former value is denoted by $\mathscr{E}_{\min}(R)$. From $(7.22)_2$ and (7.28) we find that a solution $(\mathbf{g}^*, c)$ to (7.22) satisfies the equality

$$\mu |\mathbf{g}^*|^2 = [\![\dot{\mathbf{N}}]\!] \cdot [\![\dot{\mathbf{F}}]\!] = \mathbf{g}^* \cdot \frac{\partial}{\partial \mathbf{g}} \mathscr{E}(\dot{\mathbf{F}}^0, \dot{\mathbf{F}}^0 + \mathbf{g} \otimes \mathbf{n})\bigg|_{\mathbf{g}=\mathbf{g}^*}. \tag{7.29}$$

Hence, a solution to (7.22) describes a real wave, a wave of imaginary speed or a stationary discontinuity according to whether it assigns to the quantity (7.29) a positive, negative or zero value, respectively. Only the first possibility remains if U is *strictly rank* 1 *convex*, that is, if for every $\dot{\mathbf{F}}$:

$$\mathscr{E}(\dot{\mathbf{F}}, \dot{\mathbf{F}} + \mathbf{g} \otimes \mathbf{n}) > 0 \quad \text{for all nonzero } \mathbf{g}, \mathbf{n}\,. \tag{7.30}$$

By the well known properties of convex functions, fulfilment of (7.30) for every $\dot{\mathbf{F}}$ is equivalent to $[\![\dot{\mathbf{N}}]\!] \cdot [\![\dot{\mathbf{F}}]\!] > 0$ for all pairs of $\dot{\mathbf{F}}$ whose nonzero difference $[\![\dot{\mathbf{F}}]\!]$ is of the form $\mathbf{g} \otimes \mathbf{n}$.

Observe that the curve $\mathscr{E}_{\min}(R)$ represents the lower envelope of all (smooth) curves $\mathscr{E}(|\mathbf{g}|)$, each drawn for a fixed direction and sense of $\mathbf{g}$ ($\dot{\mathbf{F}}^0$ and $\mathbf{n}$ being prescribed in advance). This means that the (right- or left-hand) derivative $\mathrm{d}\mathscr{E}_{\min}(R)/\mathrm{d}R$ multiplied by R is equal to the last expression in (7.29) evaluated at (one of the set of) $\mathbf{g}^*$ associated with $\mathscr{E}_{\min}(R)$. It follows that the existence of waves of imaginary speed for a given $\mathbf{n}$ is implied by the existence of an interval of R, where $\mathscr{E}_{\min}(R)$ is decreasing. This is the case if $\mathscr{E}(\dot{\mathbf{F}}^0, \dot{\mathbf{F}}^0 + \mathbf{g} \otimes \mathbf{n})$ takes negative values for this $\mathbf{n}$ and for some $\mathbf{g}$. $\mathscr{E}_{\min}(R)$ is then negative for some R and since it vanishes with vanishing R and is continuously dependent on R, it must somewhere decrease with increasing R.

We have thus proved that *fulfilment of the condition (7.14) is sufficient for existence of waves of imaginary speed of the orientation* $\mathbf{n}$ (provided the velocity gradient at one side of the "wave" is equal to $\dot{\mathbf{F}}^0$).

Direct connection of such non-real waves with a material instability is less clear than in the case $\dot{\mathbf{F}} = \mathbf{0}$. If equilibrium is stable then instability associated with (7.14) can rather be interpreted in relation to the possibility of stationary discontinuities (and shear band bifurcations, discussed in Sect. 7.5). Suppose that (7.27) holds for some fixed $\mathbf{n}$. This implies (cf. an analogous property (7.25) in [33]) that $\mathscr{E}(\dot{\mathbf{F}}^0, \dot{\mathbf{F}}^0 + \mathbf{g} \otimes \mathbf{n})$ increases unboundedly as $|\mathbf{g}| \to \infty$. As a continuously differentiable function of $\mathbf{g}$, $\mathscr{E}(\dot{\mathbf{F}}^0, \dot{\mathbf{F}}^0 + \mathbf{g} \otimes \mathbf{n})$ attains then at some $\mathbf{g} = \mathbf{g}^*$ an absolute minimum and its gradient vanishes at the minimum point. If (7.14) holds for that $\mathbf{n}$ then the minimum value is negative and $\mathbf{g}^*$ must thus be different from $\mathbf{0}$. By comparison with (7.28) we obtain that the minimum point corresponds to an admissible stationary discontinuity.

If (7.15) holds with the equality sign for some nonzero $\mathbf{g}^*, \mathbf{n}^*$ then the pair $(\mathbf{g}^*, \mathbf{n}^*)$ corresponds to a minimum point, and hence to a stationary point of $\mathscr{E}(\dot{\mathbf{F}}^0, \dot{\mathbf{F}}^0 + \mathbf{g} \otimes \mathbf{n})$ at fixed $\dot{\mathbf{F}}^0$. It follows again that the pair $(\mathbf{g}^*, \mathbf{n}^*)$ defines an admissible stationary discontinuity.

For convenience, the results obtained in this subsection are summarized in the following statement.

Theorem 2 *Let the velocity gradient at one side of a discontinuity surface* Σ *be equal to* $\dot{\mathbf{F}}^0$.
Then:
(i) for any orientation $\mathbf{n}$ *of* Σ*, (7.22) have at least two infinite families of solutions (possibly associated with imaginary speed of wave propagation),*
(ii) a stationary discontinuity of an orientation $\mathbf{n}$ *is admissible if any only if the Weierstrass function* $\mathscr{E}(\dot{\mathbf{F}}^0, \dot{\mathbf{F}}^0 + \mathbf{g} \otimes \mathbf{n})$ *is stationary with respect to* $\mathbf{g}$ *at* $\mathbf{g} \otimes \mathbf{n} = [\![\dot{\mathbf{F}}]\!]$,
(iii) the condition (7.14) is sufficient for the existence of waves of imaginary speed of the orientation $\mathbf{n}$,
(iv) the condition (7.14) is sufficient for the existence of admissible stationary discontinuities of the orientation $\mathbf{n}$ *provided (7.27) is satisfied,*
(v) in order that all solutions to (7.22) describe acceleration waves propagating with nonzero real speed (i.e. that stationary discontinuities and waves of imaginary speed are simultaneously excluded),
– it is necessary that (7.30) holds for $\dot{\mathbf{F}} = \dot{\mathbf{F}}^0$,
– it is sufficient that (7.30) holds for all $\dot{\mathbf{F}}$.

It is interesting to observe that the condition (7.20) or its negation did not appear explicitly in the foregoing analysis based on a general nonlinear constitutive law of the form (7.1). Of course, by using the mutual connections between (7.20) and (7.14), (7.16) discussed in Sect. 7.3, a number of corollaries of Theorems 1 and 2 can be formulated with the use of that condition. For instance, as a corollary from Theorem 2(iii) and (iv) we find that (7.20) is sufficient for the existence of waves of imaginary speed of the orientation $\mathbf{n}$, and also of admissible stationary discontinuities of that orientation provided (7.27) is satisfied.

7.5 Material Instability and Shear Band Formation

7.5.1 Dynamic Localization of Deformation Into a Planar Band

In Sect. 7.4 it has been shown (Theorem 1(ii)) that fulfilment of the energy condition (7.16) for instability of equilibrium of a material element with a clamped boundary is equivalent to the existence of waves of imaginary speed if the material on one side of the "wave" has been assumed to be non-deforming. In analogy to the linear case considered by Mandel [27], we show now that existence of such nonreal waves is associated with the possibility of *dynamic localization of deformation* into a planar band, under arbitrarily small disturbances. In a homogeneous, uniformly stressed infinite continuum in $\mathbb{R}^3$ being initially at rest, consider a band of an orientation $\mathbf{n}$ limited by the material planes $X = -h$ and $X = +h$, where $X = \mathbf{X} \cdot \mathbf{n}$. Introduce the approximation that (7.1) remains valid when the rates of $\mathbf{F}$ and $\mathbf{N}$ are replaced by respective small increments reached along a *straight* deformation path (a similar assumption would be unacceptable for arbitrarily circuitous paths). Assuming that the band boundaries move as rigid planes and the tractions along them remain constant in time, a motion induced by an initial disturbance can be taken in the form [27]

$$\begin{aligned} \mathbf{u}(\mathbf{X}, t) &= A(t)\,\mathbf{g}^* \sin(mX), \quad A(t) = \varepsilon \sinh(\kappa t), \\ m &= \frac{\pi}{2h}, \quad \kappa = m(-\mu/\rho)^{1/2}, \quad |X| \leqslant h, \end{aligned} \tag{7.31}$$

where $\mathbf{g}^*$ and $\mu < 0$ are fixed and satisfy (7.23), $\mathbf{u}$ is a displacement vector from the initial equilibrium configuration, t denotes time and ε is a small positive parameter. It is not difficult to check that under the constitutive approximation introduced above, the equations of motion are satisfied within the band. A finite deformation increment is produced at t sufficiently large no matter how small ε is, and an instability of equilibrium in the band can be inferred. It is recalled that the existence of a solution to (7.23) with $\mu < 0$ is a consequence of (7.16).

The function $A(t)$ in (7.31) corresponds to an initial perturbation of the velocity field at $t = 0$. Alternatively, we can consider the effect of imposing a system of small perturbing forces in an arbitrarily short initial time interval (cf. [33]) by assuming that $A(t) = \varepsilon(\cosh(\kappa t) - 1)$ for $0 \leqslant t \leqslant \tau$. For $t \geqslant \tau$ we can then take $A(t) = \varepsilon(\cosh(\kappa t) - \cosh(\kappa(t - \tau)))$ which corresponds to further motion free of perturbing forces. It can easily be seen that if $\mu < 0$ then finite deviations from equilibrium are caused by perturbing forces as small as we please if ε is chosen to be sufficiently small, implying instability of equilibrium in the band also for that kind of disturbance.

It must be noted that this is not a rigorous proof of instability in the continuum since the rigid body motion outside the band (which satisfies all compatibility conditions on the band boundaries) is not a free inertial motion; also the constitutive law has been taken in an approximate form. Nevertheless, the above argument does indicate a possible symptom of the material instability associated with fulfilment of (7.16). One may expect that the dynamic localization of deformation approximately described by (7.31) within a band can be associated with slight elastic unloading outside the band and that the latter is negligible if the band is sufficiently narrow.

If the incremental constitutive law varies smoothly along a deformation path then the instability range associated with fulfilment of (7.16) is entered through the critical state at which (7.25) is met with $U(\mathbf{g} \otimes \mathbf{n})$ still non-negative for all $\mathbf{g}, \mathbf{n}$. As shown in Sect. 7.4 (Theorem 1(iv)), at that instant a stationary discontinuity in the velocity gradient becomes first possible under the assumption that the material at one side of the discontinuity surface is not deforming; this is the critical stage discussed by Kolymbas [24]. The pair $(\mathbf{g}, \mathbf{n})$ satisfying (7.25) at the critical instant defines the incipient mode of localization of deformation into a band of orientation $\mathbf{n}$; depending on higher-order terms, this may be expected to initiate either an uncontrolled dynamic localization as discussed above, or a quasi-static localization at varying external boundary conditions as in the case of the bifurcations examined below.

7.5.2 Shear Band Bifurcation

Contrary to the above analysis, throughout this subsection the inertia effects are disregarded. There is a well-known connection between admissibility of stationary discontinuities in a dynamic problem and nonuniqueness of a quasi-static solution in velocities within a planar band in an infinite, homogeneous, uniformly stressed continuum: the first-order compatibility conditions are formally the same in both cases, with the distinction that the right-hand term in $(7.22)_2$ is in a quasi-static analysis disregarded at the outset (on taking $\rho = 0$, say) and c need *not* vanish, cf. [13, 18, 36, 35]. The existence of a nontrivial solution to these equations is for simplicity referred to as a shear band bifurcation, although the terminology concerns, strickly speaking, the particular case when $\mathbf{g}$ is orthogonal to $\mathbf{n}$ (for simplicity, the current configuration is again taken as reference).

From Theorem 2(iv) we obtain the following corollary:

Corollary 1 *Fulfilment of the condition (7.14) is sufficient for bifurcation in a shear band of the orientation* $\mathbf{n}$ *in the infinite continuum deforming with the velocity gradient* $\dot{\mathbf{F}}^0$ *provided (7.27) holds for that* $\mathbf{n}$.

Since $U(\dot{\mathbf{F}})$ and $\mathscr{E}(\dot{\mathbf{F}}^0, \dot{\mathbf{F}})$ are continuously dependent on $\dot{\mathbf{F}}$, fulfilment of (7.27) and (7.14) for one $\mathbf{n}$ implies that these conditions are satisfied for each $\mathbf{n}$ from some continuous range. This means that when the conditions in Corollary 1 are satisfied then there is *a continuous range of the band orientations* for which a shear band bifurcation can occur. In turn, since for a fixed $\mathbf{n}$ the conditions in Corollary 1 can be met at every instant along a segment of the deformation path, in general there is *a continuous range of the bifurcation points* for a *given* reference orientation of the band. An example is given in the next section.

From Theorem 2(ii) (or from (7.28)) it follows that for the bifurcation in velocity gradient in a shear band of orientation $\mathbf{n}$ it is necessary and sufficient that $\mathscr{E}(\dot{\mathbf{F}}^0, \dot{\mathbf{F}}^0 + \mathbf{g} \otimes \mathbf{n})$ is stationary with respect to $\mathbf{g}$ at $\mathbf{g} = \mathbf{g}^*$, where $\dot{\mathbf{F}}^0$ and $\dot{\mathbf{F}}^0 + \mathbf{g}^* \otimes \mathbf{n}$ denote the velocity gradient outside and inside the band. The conditions in Corollary 1 ensure the existence of such a stationary point as a minimum point, but neither (7.14) nor (7.27) is necessary for a shear band bifurcation. However, if (7.14) does not hold then formation of shear bands is not energetically preferable to further uniform straining in the sense discussed below. The case when the condition (7.27) fails for some $\mathbf{n}$ has been discussed in the preceding subsection. It seems that in the associated range of unstable equilibrium the quasi-static bifurcations at varying external boundary conditions are of less practical significance.

The shear band bifurcation mode appropriate for an infinite continuum need not be compatible with the boundary conditions at the intersection of a band with the body surface. If it is not compatible and the exact solution for a finite body is considered then the band at the moment of bifurcation must be vanishingly narrow, cf. [18]. Alternatively, the shear band bifurcation mode can always be modified as discussed in Sect. 7.3 in order to satisfy exactly any boundary conditions corresponding to uniform straining but at the cost of approximate fulfilment of the conditions of continuing equilibrium. As shown in the Subsect. A.1 of Appendix in Sect. 7.8, the pointwise error (identified with the volume or surface density of the rate of perturbing forces) can be made as small as we please for ε sufficiently small, i.e. if the band thickness becomes sufficiently small when compared with the dimensions of $\hat{G}$. As shown in Sect. 7.3, if the band is sufficiently narrow then superposition of such a modified shear band mode of bifurcation associated with the pair $(\mathbf{g}, \mathbf{n})$ satisfying (7.14) leads to a smaller increment of the deformation work than the uniform straining with $\dot{\mathbf{F}} \equiv \dot{\mathbf{F}}^0$. This provides the following interpretation (cf. Theorem 2 in [33]) for the instability associated with (7.14) at a stable equilibrium state of a *finite* homogeneous material body:

If (7.14) is satisfied then the incremental deformation corresponding to formation of shear bands of the orientation $\mathbf{n}$ *is energetically preferable to further uniform straining with the velocity gradient* $\dot{\mathbf{F}}^0$, *and is possible at arbitrarily small perturbations of incremental equilibrium provided (7.27) also holds.*

Let us discuss now the *linearized* condition for shear band bifurcation, formulated in terms of the tangent moduli $C_{ijkl}(\dot{\mathbf{F}}^0)$ which henceforth are assumed to be well defined. Since (7.14) is implied by (7.20) which in turn is implied by (7.32) below, another corollary of Theorem 2 (iv) can be formulated.[9]

Corollary 2 *Fulfilment of the condition*

$$\det \mathbf{A}^0(\mathbf{n}) < 0, \quad A^0_{jl}(\mathbf{n}) = C_{ijkl}(\dot{\mathbf{F}}^0) n_i n_k\,, \tag{7.32}$$

is sufficient for bifurcation in a shear band of the orientation $\mathbf{n}$ *in the infinite continuum deforming with the velocity gradient* $\dot{\mathbf{F}}^0$ *provided (7.27) holds for that* $\mathbf{n}$.

If $\mathbf{C}(\dot{\mathbf{F}}) = \mathbf{C}(\dot{\mathbf{F}}^0)$ for directions of $\dot{\mathbf{F}}$ sufficiently close to that of $\dot{\mathbf{F}}^0$ then the well-known condition (cf. [13, 35])

$$\det \mathbf{A}^0(\mathbf{n}) = 0 \tag{7.33}$$

is sufficient for bifurcation in a shear band of the orientation $\mathbf{n}$. If the incremental constitutive law is not piecewise linear then (7.33) is in general not sufficient for a

[9] It can be remarked that in spite of apparent resemblance of the bifurcation condition, the present statement is not related to that formulated recently by Borre and Maier [3]: the latter concerns only incrementally bi-linear elastic-plastic materials while the former is just empty for such materials since then (7.32) and (7.27) contradict each other.

shear band bifurcation *in velocities* but does imply a "smooth" bifurcation in quasi-static *accelerations* [31]. Hence, if the term "shear band bifurcation" is extended to such bifurcations then the Corollary 2 remains valid if " $<$ " is replaced by " $\leq$ ".

From Corollary 2 it becomes evident that the bifurcation associated with (7.33), usually considered in the literature, is merely a special case of shear band bifurcation possible in an incrementally nonlinear continuum. This conclusion may be of particular interest when bifurcation is induced by a discontinuous drop of the tangent stiffness of the material along a deformation path. Moreover, shear band bifurcations may precede fulfilment of (7.33) since the conditions in Corollary 1 do not exclude that $\mathbf{A}^0(\mathbf{n})$ is positive definite for all nonzero $\mathbf{n}$.

If the constitutive inequality (7.2) is satisfied and the fundamental deformation path is regular so that (7.3) holds at each instant then the latter remark does not apply. From (7.4) it follows that positive definiteness of $\mathbf{A}^0(\mathbf{n})$ implies that $(\dot{\mathbf{N}}-\dot{\mathbf{N}}^0)\cdot(\dot{\mathbf{F}}-\dot{\mathbf{F}}^0) > 0$ for $\dot{\mathbf{F}}-\dot{\mathbf{F}}^0 = \mathbf{g}\otimes\mathbf{n} \neq \mathbf{0}$ and hence excludes a bifurcation into a shear band of the orientation $\mathbf{n}$ [31]. In those circumstances, the condition (7.33) is a correct condition for determining the *critical* point at which the *continuous* range of shear band bifurcations along the fundamental path is entered (provided the conditions in Corollary 2 are satisfied beyond that point). According to the preceding discussion, further uniform straining beyond that point is connected with fulfilment of (7.14) and may be regarded as unstable. Discussion of nonuniform post-bifurcation deformations is beyond the scope of the present paper.

7.6 Critical Points on a Uniform Deformation Path – an Example

In an illustrative example, we will consider an elastic-plastic model of a polycrystalline metal, taking into account the effect of formation of a vertex on the yield surface at the current loading point, cf. [15]. The incremental response of the material at a yield-surface vertex is specified by using a finite-strain version of the socalled J_2 corner theory developed by Christoffersen and Hutchinson [5]. This material model has been used in a number of numerical studies (e.g. [5, 23, 38, 34]), and the reader is referred to these papers for a detailed description of the constitutive equations and their basic features. Under plane strain conditions, consider two paths, I and II, of uniform deformation of a material element. The path I corresponds to monotonic compression ($\dot{F}^0_{11} < 0$) from an unstressed state. The path II coincides with I up to the engineering strain $\Delta F^0_{11} = F^0_{11} - 1 = -0.25$, beyond which the material is subject to further compression simultaneously with superimposed simple shearing ($\dot{F}^0_{21} > 0, \dot{F}^0_{12} = 0$) such that $\dot{F}^0_{21}/\dot{F}^0_{11} = \text{constant} = -0.5$. In Fig. 7.2 the deformation paths are drawn, and several critical points are indicated which have been calculated numerically[10] in a manner discussed below. To simplify the calculations, slight compressibility of the material has been neglected ($F^0_{11}F^0_{22} = 1$). The

[10] For this purpose, the subroutines worked out by Petryk and Thermann [34] have been employed. In particular, the stress variations along the path II have been calculated incrementally with accuracy to second-order terms in each increment.

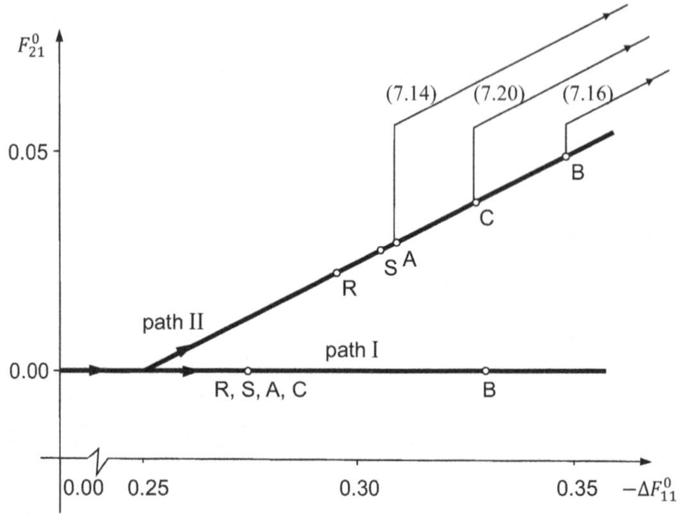

Fig. 7.2 Critical points on proportional and non-proportional deformation paths for a material element obeying the J_2 corner theory of plasticity.

critical points in Fig. 7.2 are determined for the power strain-hardening law with an exponent 0.1 and for a relatively blunt conical corner formed on the yield surface, characterized by an external cone angle $\beta_c = 110°$. Other material parameters and the transition function which defines the type of nonlinearity of the constitutive potential U have been taken to be the same as in calculations by Tvergaard et al. [38] and by Petryk and Thermann [34], except that the formula (7.15) in [5] has been adopted without modification.

Under the assumptions of incompressibility and of plane strain, for a given nonzero $\mathbf{n}$ the direction (but not the sense) of $\mathbf{g}$ in any expression discussed above is uniquely defined by the requirements $\mathbf{g} \cdot \mathbf{n} = 0$ and $g_3 = 0$. On a circle $|\mathbf{n}| = 1$, minimum values of the left-hand expressions in the conditions (7.14), (7.16) and (7.20) have been calculated for $|\mathbf{g}| = 1$ in the second and third cases and for variable $|\mathbf{g}|$ in the first case. This has been done at successive deformation stages in order to find when these minimum values become negative. In that way the critical points A, B and C on the deformation path have been determined beyond which the conditions (7.14), (7.16) and (7.20), respectively, start to hold. On the path II the points A, B and C are clearly separate which illustrates the fact that the respective conditions are distinct. It has been found that in other circumstances the point C can even lie beyond B; on the other hand, from the diagram (7.21) it can be seen that neither B nor C can precede A. For comparison, two other critical points are also indicated. The point R marks the onset of ellipticity loss for the moduli from the so-called total loading cone (which are defined by the formulae of the deformation theory of plasticity): beyond the point R the condition (7.30) necessarily fails for $\dot{\mathbf{F}}$

corresponding to that cone. Beyond the point S a local minimum of $\mathscr{E}(\dot{\mathbf{F}}^0, \dot{\mathbf{F}}^0+\mathbf{g}\otimes\mathbf{n})$ as a function of $\mathbf{g}$ can be found at $|\mathbf{g}| \neq 0$ for some range of $\mathbf{n}$; this means that the stationary discontinuities with $\dot{\mathbf{F}}^0$ as a velocity gradient at its one side (as well as waves of imaginary speed) become admissible. Up to the point A this local minimum value is positive, which shows that fulfilment of (7.15) (with strict inequality for $\mathbf{n}, \mathbf{g} \neq \mathbf{0}$) does not exclude such discontinuties.

The above results can be re-interpreted in terms of shear band bifurcations. Starting from the point R on the path II, a shear band bifurcation *could* take place if the direction of the velocity gradient $\dot{\mathbf{F}}^0$ outside the band were abruptly changed to fall into the total loading cone. For the actual direction of $\dot{\mathbf{F}}^0$ tangent to the path II, shear band bifurcations are possible beyond the point S, although they become energetically preferable to further uniform straining (in the sense discussed in Sect. 7.3 and 7.5) only beyond the point A. The critical point C found from the linearized bifurcation condition (7.32) or (7.33) lies inside the actual range of shear band bifurcations. As the point B is approached, the maximal magnitude of an admissible jump in velocity gradient tends to infinity; beyond the point B shear bands can be formed due to a dynamic growth of arbitrarily small disturbances, as discussed in Subsection 7.5.1.

On the path I the points R, S, A, C coincide. This follows from the fact that for proportional loading the tangent moduli are the same as the "total loading" moduli and the inequality (7.3) is satisfied implying (7.4) and (7.5): cf. [31] and the discussion at the end of Sect. 7.3. Consequently, the onset of shear band bifurcations on that path is correctly predicted by the linearized criterion (7.33) or (7.32).

The general conclusion from Sect. 7.5 concerning the existence of a continuous range of bifurcation points and shear band angles is illustrated in Fig. 7.3 for the path I of plane strain compression. *Every point in the shaded domain corresponds to bifurcation* in a band whose orientation (in Fig. 7.3 in the currrent configuration) satisfies the sufficiency conditions in Corollary 2 (and thus also in Corollary 1). From the condition (7.33) only the limiting solid line can be found as a representation of posssible band bifurcations in the *linearized* solid.

A point on the broken line inside the shaded area corresponds to an absolute minimum of $\mathscr{E}(\dot{\mathbf{F}}^0, \dot{\mathbf{F}} + \mathbf{g} \otimes \mathbf{n})$ at a fixed state. That particular bifurcation band may be regarded in a sense as energetically "most preferable" since, according to (7.13) and Fig. 7.1, a negative value of $\mathscr{E}(\dot{\mathbf{F}}^0, \dot{\mathbf{F}}^0 + \mathbf{g} \otimes \mathbf{n})$ is, in the limit, proportional to the amount of work extracted from the deforming element due to superposition of the shear rate $\mathbf{g} \otimes \mathbf{n}$ in a vanishingly thin band. Fig. 7.4 shows how the bifurcation amplitude $g = |\mathbf{g}||\mathbf{n}|/|\dot{F}^0_{11}|$ for that band increases and tends to infinity when the compressive strain increases to the critical value (point B in Fig. 7.2) which marks the onset of instability of equilibrium in the uniform plane strain compression under all-round kinematical boundary conditions.

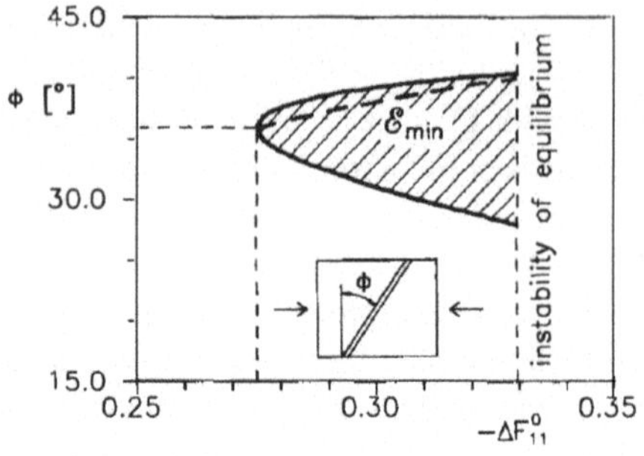

Fig. 7.3 Continuous range of bifurcation points and shear band angles in plane strain compression of the incrementally nonlinear material.

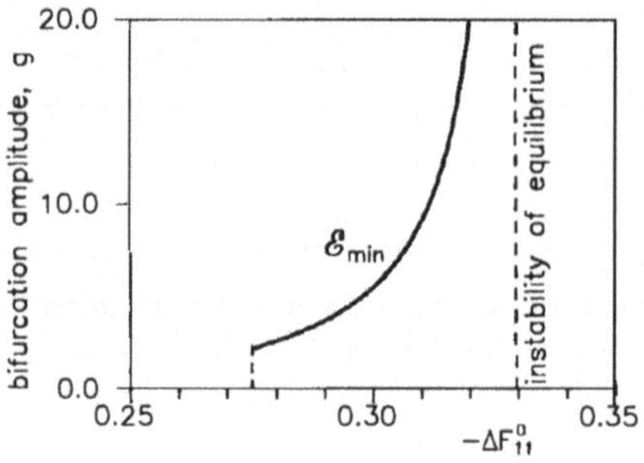

Fig. 7.4 Magnitude of superimposed shear rate in the bifurcation band corresponding to an absolute minimum of $\mathscr{E}(\dot{\mathbf{F}}^0, \dot{\mathbf{F}}^0 + \mathbf{g} \otimes \mathbf{n})$.

7.7 Concluding Remarks

A systematic analysis of theoretical aspects of material instability has been presented for a general class of time-independent, incrementally nonlinear solids which admit a velocity-gradient potential. The condition (7.14) is a central point of this analysis. It has been derived in Sect. 7.3 from energy considerations as a sufficiency condition for instability of uniform straining with the velocity gradient $\dot{\mathbf{F}}^0$. If $\dot{\mathbf{F}}^0$ vanishes then (7.14) reduces to the sufficiency condition (7.16) for instability of equilibrium. Sect. 7.4 has shown how the instability predicted by (7.14) can be interpreted in terms of vanishing or imaginary speed of propagation of acceleration waves; in Sect. 7.5 this has been re-interpreted in relation to the possibility of shear band formation. The instability has a different character according to whether the equilibrium is stable or unstable in the restricted sense of (7.27) or (7.16), respectively. In the former case the shear band formation can be initiated as a quasi-static deformation branching (possibly at the presence of arbitrarily small perturbing forces if a finite body is considered) at varying external boundary conditions. In the latter case a spontaneous dynamic localization of deformation with a band at fixed external loading can rather be expected.

In comparison with the related earlier work, the obtained results provide extensions in several directions. To avoid repetitions of what has already been stated in the Introduction and subsequent sections, we restrict ourselves here to a few remarks. In Sect. 7.3 the concept of instability of a deformation process (path) in the energy sense [30, 33] has been specified for a material element embedded in a deforming continuum. The analysis of acceleration waves presented in Sect. 7.4 may be treated as an independent section. Nothing significant can be found there for incrementally linear or bi-linear solids (cf. [13]), but in the case of essentially nonlinear constitutive rate equations the method used to attack the problem and the obtained results appear to be new. Analogous remarks apply to the analysis of shear band initiation given in Sect. 7.5. The obtained sufficiency conditions for bifurcation in a shear band complement the well-known condition (7.33) (cf. [13, 36, 35]) and confirm its validity as a *critical* condition for shear band formation in incrementally nonlinear solids *provided* that the constitutive inequality (7.2) is valid and the deformation path is regular. If the inequality (7.4) is violated then shear band bifurcations may occur in the strongly elliptic range of the tangent moduli associated with $\dot{\mathbf{F}}^0$, as illustrated on the example given in Sect. 7.6.

7.8 Appendix

A.1. Localized deformation mode in a finite domain $\hat{G}$

Let an auxiliary scaler function $\beta = \beta(\mathbf{X})$ be defined on $\mathbb{R}^3$ as a continuous function satisfying $\beta = 0$ ouside $\hat{G}$ and $\beta = 1$ in B for all sufficiently small ε (cf. Fig. 7.1), and being smooth outside $\overline{B}$. Let $\tilde{\mathbf{w}}$ of the properties required in (7.9) be specified by

$$\mathbf{w}(\mathbf{X}) = \mathbf{g}\beta(\mathbf{X})X_B, \quad X_B = \begin{cases} X & \text{if } \mathbf{X} \in B\,, \\ \text{sh}(\mathbf{X}_\Pi)\,\text{sign}(X) & \text{if } \mathbf{X} \notin B\,, \end{cases} \tag{7.34}$$

where $\mathbf{g}$ is a fixed nonzero vector, $\mathbf{X}_\Pi$ is the orthogonal projection of $\mathbf{X}$ on Π and $X \equiv (\mathbf{X} - \mathbf{X}_\Pi) \cdot \mathbf{n}$ is the distance from Π with (+) or (−) sign. Accordingly,

$$\nabla\mathbf{w}(\mathbf{X}) = \begin{cases} \mathbf{g} \otimes \mathbf{n} & \text{if } |X| < \varepsilon h(\mathbf{X}_\Pi), \\ \varepsilon\,\text{sign}(X)(h(\mathbf{X}_\Pi)\mathbf{g} \otimes \nabla\beta(\mathbf{X}) + \beta(\mathbf{X})\mathbf{g} \otimes \nabla_\Pi h(\mathbf{X}_\Pi)) & \text{if } |X| > \varepsilon h(\mathbf{X}_\Pi). \end{cases} \tag{7.35}$$

It follows that $\nabla\mathbf{w}$ has the properties used in the derivation of (7.13): within the disk B, $\nabla\mathbf{w} = \mathbf{g} \otimes \mathbf{n}$ is independent of ε, and outside $\overline{B}$, $|\nabla\mathbf{w}|$ tends uniformly to zero as $\varepsilon \to 0$.

Let the field $\tilde{\mathbf{w}}$ be a perturbation of the fundamental velocity field of a constant gradient $\dot{\mathbf{F}}^0$. Let $\beta(\mathbf{X})$ outside $\overline{B}$ and $h(\mathbf{X}_\Pi)$ on Π have bounded second gradients.

From (7.35) and the continuity and piecewise differentiability of the constitutive relationship $\dot{\mathbf{N}}(\dot{\mathbf{F}})$ it follows that the perturbed velocity field satisfies the conditions of continuing equilibrium with a pointwise error of order ε (i.e. $\dot{N}_{ij,i} = O(\varepsilon)$ and $\upsilon_i[\![\dot{N}_{ij}]\!] = O(\varepsilon)$, where $[\![\cdot]\!]$ denotes a jump across a surface of the reference unit normal $\boldsymbol{\nu}$) everywhere except possibly across the boundary ∂B of B. Suppose now that $\mathbf{g} \otimes \mathbf{n}$ constitutes an admissible mode of bifurcation (when added to $\dot{\mathbf{F}}^0$) within a planar band of the reference normal $\mathbf{n}$ in an infinite continuum deforming with the velocity gradient $\dot{\mathbf{F}}^0$. Then, from the condition of continuing equilibrium on the band boundary and from (7.35) it follows that $\upsilon_i[\![\dot{N}_{ij}]\!] = O(\varepsilon)$ also on ∂B, where υ_i approximates $\pm n_i$ with an error of the order of ε. We thus conclude that if $\mathbf{g} \otimes \mathbf{n}$ defines the exact bifurcation mode for an infinite continuum then the field $\tilde{\mathbf{w}}$ defined by (7.34) is an approximate bifurcation mode for a finite domain $\hat{G}$, with the pointwise error in fulfilment of the incremental equilibrium condition being of the order of ε everywhere.

A.2. Proof of (7.19)

For arbitrary but fixed $\dot{\mathbf{F}}^0$, $\mathbf{g}$ and $\mathbf{n}$, consider an auxiliary function of a real variable γ, defined by

$$f(\gamma) = \mathscr{E}(\dot{\mathbf{F}}^0, \dot{\mathbf{F}}^0 + \gamma \mathbf{g} \otimes \mathbf{n}) .$$

The second derivative of f reads

$$f''(\gamma) = (\mathbf{g} \otimes \mathbf{n}) \cdot \mathbf{C}(\dot{\mathbf{F}}^0 + \gamma \mathbf{g} \otimes \mathbf{n}) \cdot (\mathbf{g} \otimes \mathbf{n}) .$$

Since f is a continuously differentiable function of γ satisfying $f(0) = 0$ and $f'(0) = 0$, we have the elementary formula

$$f(\gamma) = \int_0^\gamma f''(s)(\gamma - s)\mathrm{d}s .$$

On substituting the expression for f'' and taking $\gamma = 1$, we obtain (7.19).

Acknowledgement

This work was partially supported by the Committee for Research (KBN), Poland under the project No. 310829101.

References

1. Ball, J. M. (1977) Arch. Rat. Mech. Anal. **63**, 337.

2. Benellal, A., Billardon, R. and Geymonat, G. (1990) C. R. Acad. Sci. Paris **310**, S. II, 679.
3. Borre, G. and Maier, G. (1989) Meccanica **24**, 36.
4. Chambon, R. and Desrues, J. (1985) Plastic Instability, Proc. Considère Memorial, p. 101. Ecole Nat. Ponts Chauss., Paris.
5. Christofferson J. and Hutchinson, J. W. (1979) J. Mech. Phys. Solids **27**, 465.
6. Darve, F., Flavigne, E. and Rojas, E. (1986) Comp. Geotechnics **2**, 43.
7. Drucker, D. C. (1950) Q. Appl. Mech. **7**, 411.
8. Drucker, D. C. (1951) Proc. 1st U.S. National Congress of Applied Mechanics, p. 487. A.S.M.E.
9. Graves, L. M. (1939) Duke Math. J. **5**, 656.
10. Hadamard, J. (1903) Lesons sur la Propagation des Ondes et les Equations de l'Hydrodynamique, Hermann, Paris.
11. Hill, R. (1958) J. Mech. Phys. Solids **6**, 236.
12. Hill, R. (1959) J. Mech. Phys. Solids **7**, 209.
13. Hill, R. (1962) J. Mech. Phys. Solids **10**, 1.
14. Hill, R. (1966) J. Mech. Phys. Solids **14**, 95.
15. Hill, R. (1967) J. Mech. Phys. Solids **15**, 79.
16. Hill, R. (1968) J. Mech. Phys. Solids **16**, 229, 315.
17. Hill, R. (1978) Advances in Applied Mechanics, Vol. 18, p. 1. Academic Press, New York.
18. Hill, R. and Hutchinson, J. W (1975) J. Mech. Phys. Solids **23**, 239.
19. Hill, R. and Rice, J. R. (1972) J. Mech. Phys. Solids **20**, 401.
20. van Hove, L. (1947) Proc. Kon. Ned. Akad. Wet. **50**, 18.
21. Hutchinson, J. W. (1970) Proc. Roy. Soc. Lond. A **319**, 247.
22. Hutchinson, J. W. and Tvergaard, V. (1980) Int. J. Mech. Sci. **22**, 339.
23. Hutchinson, J. W. and Tvergaard, V. (1981) Int. J. Solids Struct. **17**, 451.
24. Kolymbas, D. (1981) Ing.-Archiv **50**, 131.
25. Loret, B. (1987) J. Méc. théor. appl. **6**, 423.
26. Mandel, J. (1965) Int. J. Solids Struct. **1**, 273.
27. Mandel, J. (1966) In Rheology and Soil Mechanics (edited by J. Kravtchenko), p. 58. Springer, Berlin.
28. Morrey, C. B. (1952) Pacific J. Math. **2**, 25.
29. Nguyen, Q. S. and Petryk, H. (1991) C. R. Acad. Sci. Paris **312**, S. II, 7.
30. Petryk, H. (1985) In Plastic Instability, Proc. Considère Memorial, p. 215. Ecole Nat. Ponts Chauss., Paris.
31. Petryk, H. (1989) J. Mech. Phys. Solids **37**, 265.
32. Petryk, H. (1991a) Arch. Mech. **43**, 377.
33. Petryk, H. (1991b) Arch. Mech. **43**, 519.
34. Petryk, H. and Thermann, K. (1992) Int. J. Solids Struct. **29**, 745.
35. Rice, J. R. (1977) In Theoretical and Applied Mechanics (edited by W. T. Koiter), p. 207. North-Holland, Amsterdam.
36. Rudnicki, J. W. and Rice, J. R. (1975) J. Mech. Phys. Solids **23**, 371.
37. Thomas, T. Y. (1961) Plastic Flow and Fracture in Solids. Academic Press, New York.
38. Tvergard, V., Needleman, A. and Lo, K.K. (1981) J. Mech. Phys. Solids **29**, 115.

Chapter 8
On Stability of Time-Independent Materials at Finite Strain

Abstract Stability of plastic deformation at the level of a homogeneous material element imbedded in a deforming continuum is discussed. Distinction is made between instability of equilibrium and instability of uniform straining of the element. Under the general assumption that constitutive rate equations admit a potential, pointwise energy criteria of instability are formulated and related to propagation of acceleration waves and to incipient localization of deformation. For incrementally nonlinear materials they are shown to differ in general from the familiar condition of ellipticity loss. Existence of a continuous range of shear band bifurcations is demonstrated. In a numerical example, a finite strain version of the Christoffersen-Hutchinson model of material behaviour at a yield-surface vertex is used.

8.1 Introduction

We are concerned here with stability of quasi-static, isothermal deformation of an incrementally nonlinear continuum, at the level of a material element and at unrestricted strains. The restriction is made that the constitutive law does not contain a natural time and, in the rate form, admits a velocity-gradient potential. No spatial discontinuities in displacements or velocities are considered: fracture, slip discontinuities or cavitation are thus excluded from consideration. A certain class of macroscopic instabilities, primarily in metals, is intended to be so modeled. Physical aspects of plastic instabilities, especially at the level of dislocation motion, are not discussed here.

The concept of an "intrinsic" stability or instability of the material in the plastic range was propounded in the mechanics of continua in various contexts, but no general agreement about its precise meaning has been worked out yet. Definitions of Drucker's type [1] implicitly involve an arbitrary choice of the measure of strain (or

[0] Reprinted from: Finite Inelastic Deformations - Theory and Applications, Proc. IUTAM Symp. Hannover 1991 (Ed. D.Besdo and E.Stein), pp. 189-196, Springer-Verlag, Berlin Heidelberg 1992. H.Petryk, On stability of time-independent materials at finite strain. Copyright © 1992.

H. Petryk, *Theory of Plastic Instability*,
https://doi.org/10.1007/978-3-031-98223-1_8

stress-rate) when geometric changes are not disregarded [2]. In turn, the measure-invariant requirement based on comparison of the internal and external work for a virtual *uniform* deformation increment involves an arbitrary specification of the interaction rule between the configuration of a material element and the tractions on the element boundary applied by its environment [3]. An analogous difficulty is met if material instability is understood in the sense of "intrinsic eigenstates" associated with constitutive branching [4].

In another approach, used recently in many investigations of shear band formation, instability of the material is understood in the sense of possible bifurcation in a planar band in an infinite homogeneous continuum [5]. Admissibility of quasi-static bifurcation in a band is strictly related to vanishing speed of dynamic propagation of acceleration waves [6] in the *deforming* material. If the constitutive domain of the currently active "tangent" moduli in strain-rate space coincides with the half-space (classical elastoplasticity) then the onset of instability in the above sense can be determined from the familiar condition of (strong) ellipticity loss, as in elastic or incrementally linear solids [7][6]. Moreover, uniform post-critical deformations exhibit then still another kind of instability, namely, an intrinsic instability of equilibrium in the energy sense (*op.cit.*) and also in a dynamic sense [8], related to imaginary speed of wave propagation with respect to the *non-deforming* material.

For an incrementally nonlinear response, predicted e.g. by micromechanical studies [9][10] and regarded as more realistic, the situation is more complex since the critical points associated with shear band bifurcation or loss of stability of equilibrium or ellipticity loss for tangent moduli need not coincide with each other in general. The question thus arises, not fully explored yet, what is the correct extension of the condition of ellipticity loss to an incrementally nonlinear case. That question is discussed here by starting from the concept of local instability of a quasi-static deformation process in the energy sense [11]. Further results presented below are extracted from the recent work [12] but illustrated on another numerical example.

Following Hill [13], constitutive rate equations will be expressed in terms of the (forward) rates of the deformation gradient $\mathbf{F}$ and the nominal stress $\mathbf{N}$ as

$$\dot{N}_{ij} = \frac{\partial U}{\partial \dot{F}_{ji}} = C_{ijkl}(\dot{\mathbf{F}}, \mathcal{H})\dot{F}_{lk}, \qquad U = U(\dot{\mathbf{F}}, \mathcal{H}) = \frac{1}{2}\dot{N}_{ij}(\dot{\mathbf{F}}, \mathcal{H})\dot{F}_{ji}, \tag{8.1}$$

where the subscripts (obeying the summation convention) denote tensor components on a fixed rectangular basis, a dot over a symbol denotes right-hand differentiation by a time-like parameter t,

$$C_{ijkl} = \frac{\partial^2 U}{\partial \dot{F}_{ji} \partial \dot{F}_{lk}} = C_{klij} \tag{8.2}$$

are the instantaneous "stiffness" moduli dependent (piecewise-continuously) on the direction of the strain-rate, and $\mathcal{H}$ indicates the influence of the deformation history prior to the current instant; the symbolic parameter $\mathcal{H}$ will be omitted below for simplicity. U is a positively homogeneous of degree two and continuously differentiable function of $\dot{\mathbf{F}}$, being arbitrary otherwise. Constitutive laws for hyperelastic solids or for classical elastoplastic solids obeying the normality flow rule are included in (8.1) as particular cases.

8.2 Pointwise Conditions Sufficient for Instability in the Energy Sense

Consider a material element which constitutes a small part of a quasi-statically deforming continuum. We are concerned with stability of a fundamental deformation process $\mathbf{F}^0(t)$ in the element; the process is regarded as known and spatially homogeneous in the limit as the element size tends to zero. Perturbations of the deformation increment are considered which preserve continuity of the material and are *non-uniform inside and vanish outside the element.* For stability of the fundamental process in the energy sense it is required [11] that among all virtual deformations of that type the work of deformation in the element, evaluated to *second* order, is minimized by the uniform mode. It can be shown by appealing to Graves' [14] theorem that the above minimum property is necessarily violated if

$$\mathcal{E}(\dot{\mathbf{F}}^0, \dot{\mathbf{F}}) < 0 \qquad \text{for some } \dot{F}_{ji} = \dot{F}^0_{ji} + g_j n_i, \tag{8.3}$$

where $\mathcal{E}$ is the Weierstrass function associated with the constitutive potential $U(\dot{\mathbf{F}})$:

$$\mathcal{E}(\dot{\mathbf{F}}^0, \dot{\mathbf{F}}) = U(\dot{\mathbf{F}}) - U(\dot{\mathbf{F}}^0) - \dot{N}^0_{ij}(\dot{F}_{ji} - \dot{F}^0_{ji}), \qquad \dot{N}^0_{ij} = \frac{\partial U}{\partial \dot{F}_{ji}}(\dot{\mathbf{F}}^0). \tag{8.4}$$

A direct derivation which provides a mechanical interpretation to (8.3) as the condition that incipient localization of deformation in a band becomes energetically preferable to uniform deformation is given in [12]. Other interpretations of (8.3) as a condition *sufficient for instability of uniform straining with the rate* $\dot{\mathbf{F}}^0$ will be discussed in the next section.

A *possible* critical stage associated with (8.3) is that $\mathcal{E}(\dot{\mathbf{F}}^0, \dot{\mathbf{F}})$ vanishes non-trivially for some $\dot{F}_{ji} = \dot{F}^0_{ji} + g_j n_i$ while (8.3) is still not satisfied; obviously, this corresponds to $\partial\mathcal{E}/\partial g_j = 0$ at such $(\dot{\mathbf{F}}^0, \dot{\mathbf{F}})$. At that instant we thus have [11]

$$(\dot{N}_{ij}(\dot{\mathbf{F}}) - \dot{N}^0_{ij})n_i = 0 \qquad \text{for some } \dot{F}_{ji} = \dot{F}^0_{ji} + g_j n_i \tag{8.5}$$

which are just the statical and kinematical first-order compatibility conditions on a stationary discontinuity (of normal n_i) in a dynamic problem [6] or on a boundary of a bifurcation band in a quasi-static problem [5].

We shall assume below that the tangent moduli $C^0_{ijkl} = C_{ijkl}(\dot{\mathbf{F}}^0)$ are well defined whenever they appear in the considerations. If the material is incrementally linear ($\mathbf{C}$ independent of $\dot{\mathbf{F}}$) then (8.3) is equivalent to

$$(C^0_{ijkl} n_i n_k) g_j g_l < 0 \qquad \text{for some } n_i, g_j \tag{8.6}$$

and (8.5) reduces to

$$\det(C^0_{ijkl} n_i n_k) = 0. \tag{8.7}$$

For the classical elastic-plastic solids obeying normality, (8.3) is still equivalent to (8.6) provided the moduli C^0_{ijkl} correspond to the loading branch. However, if the material is incrementally nonlinear then in general (8.3) may hold even if C^0_{ijkl} satisfy the strong ellipticity condition, i.e. if (8.6) and (8.7) are both excluded. In turn, (8.6) implies (8.3) as can be seen e.g. from the Taylor series expansion of $\mathcal{E}(\dot{\mathbf{F}}^0, \dot{\mathbf{F}})$ at $\dot{\mathbf{F}} = \dot{\mathbf{F}}^0$. The condition (8.6) may be regarded as a specification of (8.3) for the *linearized* constitutive law.

If the work argument leading to (8.3) is applied to a material element imbedded in a *non-deforming* continuum ($\dot{\mathbf{F}}^0 = \mathbf{0}$) then it leads to the following condition *sufficient for instability of equilibrium*:

$$U(\dot{\mathbf{F}}) < 0 \qquad \text{for some } \dot{F}_{ji} = g_j n_i. \tag{8.8}$$

The instability is meant here in the following energy sense: (8.8) implies that energy can be extracted from the material element without perturbing the equilibrium *state* of the surrounding continuum. Interpretation of a typical critical stage associated with (8.8) is analogous to (8.5) with the restriction that now the material at one side of the discontinuity surface (or outside the band) is *non-deforming*. It can be seen [11] that for the classical elastoplastic solids with normality (and obviously also for incrementally linear materials), the condition (8.8) is equivalent to (8.3) and also to (8.6) for C^0_{ijkl} taken from the loading branch, but in a general nonlinear case we only have the implication (8.8) $\Longrightarrow$ (8.3).

8.3 Interpretations of Instability

To understand better what kind of material instability is predicted by the above energy criteria, their connection with propagation of acceleration waves and formation of shear bands is now briefly discussed. A more detailed study is presented in [12] where the proofs of the statements quoted below are given.

For the speed of dynamic propagation of all acceleration waves with respect to the *non-deforming* material to be real it is necessary and sufficient that (8.8) is not satisfied. The speed is always real and nonzero if and only if

$$U(\dot{\mathbf{F}}) > 0 \qquad \text{for every nonzero } \dot{F}_{ji} = g_j n_i. \tag{8.9}$$

The inequality (8.8) may also be interpreted as the condition sufficient for initiation of *dynamic localization of deformation* in a planar band, at a fixed value of the external loading parameter. Namely, if (8.8) holds then, in the first approximation, a dynamic growth of an arbitrarily small disturbance within a band can take place under constant loading along the (non-deforming) band boundaries.

Non-fulfillment of (8.3) is necessary (but not sufficient) for the speed of dynamic propagation of all acceleration waves into the material *deforming* with the rate $\dot{\mathbf{F}}^0$ to be real. A general connection between (8.3) and instability in a dynamic sense is less clear than in the case when (8.8) holds. However, if (8.9) is satisfied then uniform straining with $\dot{\mathbf{F}}^0 \neq \mathbf{0}$ satisfying (8.3) should be regarded as unstable in the *quasi-static* sense. For, it can be proved, also if n_i is fixed in advance, that validity of (8.5), and hence existence of admissible *stationary discontinuities*, is implied by (8.3) so long as (8.9) holds. This means that quasi-static *bifurcation* in a band can take place (in the conventional formulation of the problem for an infinite continuum) at *every* point on a deformation path $\mathbf{F}^0(t)$ along which (8.3) is satisfied simultaneously with (8.9). In turn, at a given instant from that range the bifurcation can take place for a continuous "fan" of band orientations; an illustration is given below.

It can be seen that in a general case of an incrementally nonlinear material, loss of ellipticity for the *tangent* moduli, in the form (8.6) or (8.7), is not necessary for localization of deformation to be initiated, neither in dynamic nor quasi-static way. Validity of the criterion of ellipticity loss as a *critical* condition for shear band formation can be, however, re-established for material models which satisfy a constitutive inequality recently derived from micromechanical considerations [15]. The inequality is satisfied automatically for classical elastic-plastic solids with normality, and also for the material model examined below *provided* the tangent moduli coincide with "total loading" moduli.

8.4 Example

The general results are now illustrated on an incrementally nonlinear model of time-independent material behaviour at a yield-surface vertex, of the form proposed in [16] for polycrystalline metals as the so-called J_2 corner theory of plasticity. More essential material parameters adopted in the calculations are: a power-law hardening exponent $n = 0.1$, an external angle of the (rather sharp) yield-surface corner $\beta_c = 135^o$, an angle of total loading cone $\theta_0 = 0$ (thoroughly nonlinear plastic response). The material is assumed to be practically incompressible, with the Poisson ratio for elastic strains taken as 0.499. The transition function is described by eq. (2.46) in [16] with $m = 3$, the Zaremba-Jaumann flux of Kirchhoff stress is substituted for a stress-rate, and the deformation theory moduli in the hyperelastic version are used as total loading moduli.

A particular family of non-proportional fundamental deformation paths $\mathbf{F}^0(t)$ is considered (Fig. 8.1). Along each path, plane-strain compression up to a fixed engineering strain $F^0_{11} - 1 = -0.25$ is followed by further compression accompanied by additional shear F^0_{21} which increases with a constant ratio $\dot{F}^0_{21}/|\dot{F}^0_{11}|$ from an interval $[0, 0.5]$. In a manner described in more detail in [12], for several such paths the range has been calculated where the inequalities (8.3), (8.6) or (8.8) are satisfied. To complete the picture, two further curves have also been determined. It is pointed out that due to strong path- and parameter-sensitivity, the results shown in

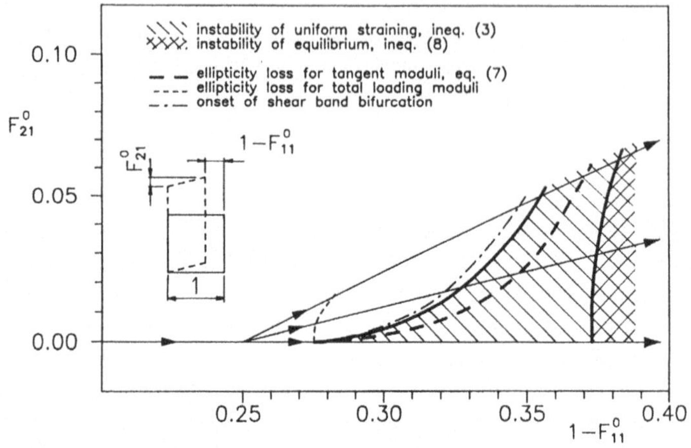

Fig. 8.1 Instability ranges on non-proportional deformation paths for an incrementally nonlinear material obeying the J_2 corner theory of plasticity

Fig. 8.1 are only valid for the specified family of deformation paths for the assumed model; it is not discussed here whether the quantitative predictions are realistic or not. For any path with $\dot{F}^0_{21} \neq 0$, the order of events is the following. First, the limit of ellipticity is reached for the total loading moduli; this corresponds to the onset of some "latent" instability which *would* manifest itself *if* the direction of the fundamental deformation rate $\dot{\mathbf{F}}^0$ were abruptly changed to satisfy (8.6). For $\dot{\mathbf{F}}^0$ as prescribed above, the critical condition (8.7) is met at a later stage which, however, is preceded by two other critical points. The second critical point on the path is found when $\mathcal{E}(\dot{\mathbf{F}}^0, \dot{\mathbf{F}})$ with $\dot{F}_{ji} = \dot{F}^0_{ji} + g_j n_i$ has for the first time a nontrivial stationarity point (in fact, a local minimum point was searched numerically), which corresponds to (8.5) and hence to the onset of shear band bifurcation. However, such bifurcations are not energetically preferable (in the sense discussed above) to the fundamental continuation of deformation unless the third critical point is crossed and the range of validity of (8.3) is entered. Finally, the range of unstable equilibrium is reached and uncontrolled dynamic localization of deformation in a shear band becomes possible. It can be seen that fulfillment of the condition (8.7) does not correspond to the onset of instability in any sense discussed here. However, one can still relate (8.7) to a critical deformation stage when a *finite* growth of an *infinitesimal* initial imperfection becomes possible without bifurcation (cf. [5][17]).

In the limiting case $\dot{F}^0_{21} = 0$ (plane-strain compression), all the critical points coincide except the one related to loss of stability of equilibrium. In that particular case, to be examined from now on, the actual tangent moduli coincide with the total loading moduli, and (8.3) reduces to (8.6). The general result that (8.3) and (8.9) satisfied simultaneously along a path imply existence of a *continuous* range of shear band bifurcations is visualized in Fig. 8.2, where a domain of admissible band angles (in the current configuration) is indicated. The solid line represents band normals n_i

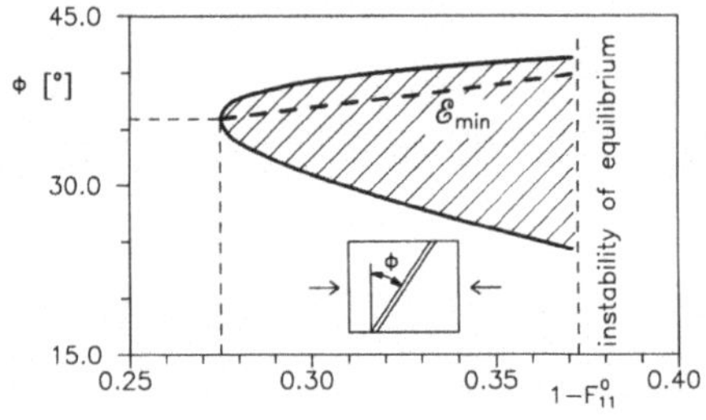

Fig. 8.2 Continuous range of shear band bifurcations in plane strain compression

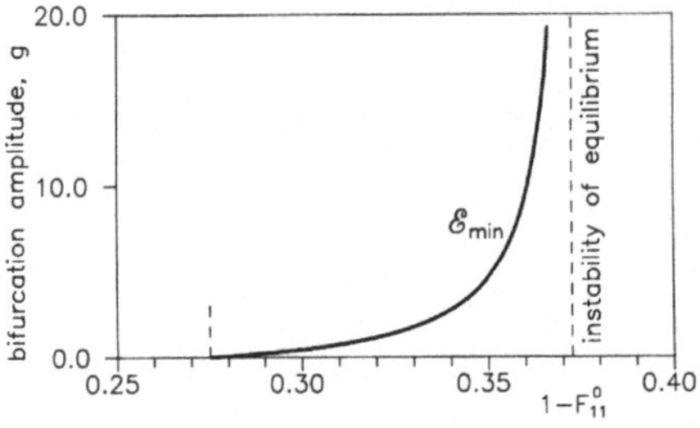

Fig. 8.3 Magnitude of superimposed shear rate in the energetically most preferable bifurcation band

satisfying (8.7), while *any* band orientation from the shaded area, at the state attained at a certain compressive strain, satisfies the conditions (8.6), (8.3) *and* the bifurcation condition (8.5). A point on the broken line inside this area corresponds to an absolute minimum of $\mathcal{E}(\dot{\mathbf{F}}^0, \dot{\mathbf{F}})$ among all $\dot{F}_{ji} = \dot{F}^0_{ji} + g_j n_i$ at a fixed state. That bifurcation band may be regarded as energetically most preferable since $\mathcal{E}(\dot{\mathbf{F}}^0, \dot{\mathbf{F}})$ is proportional to the second-order work difference due to superposition of the bifurcation mode in a vanishingly thin band [12]. The bifurcation amplitude $g = (g_j g_j n_i n_i)^{1/2}/|\dot{F}^0_{11}|$ for that band grows from 0 at the instant of ellipticity loss (we recall that the plastic response is here thoroughly nonlinear) and tends to infinity when the critical strain corresponding to transition from the range (8.9) to (8.8) is approached (Fig. 8.3). Shear band bifurcations are possible also beyond that instant but appear to be not of much interest in view of the loss of stability of equilibrium.

Acknowledgement

This work was supported by the Committee for Research (KBN), Poland under the project No. 3 1082 91 01.

References

1. Drucker, D.C.: On the postulate of stability of material in the mechanics of continua. *J. Méc.* **3** (1964) 235-249.
2. Hill, R.: On constitutive inequalities for simple materials. *J. Mech. Phys. Solids* **16** (1968) 229-242, 315-322.
3. Hill, R.: Aspects of invariance in solids mechanics. *Advances in Applied Mechanics, Vol.* **18**, pp.1-75, Acad. Press, New York 1978.
4. Hill, R.: On intrinsic eigenstates in plasticity with generalized variables. *Math. Proc. Camb. Phil. Soc.* **93** (1983) 177-189.
5. Rice, J.R.: The localization of plastic deformation. *Theoretical and Applied Mechanics* (W.T. Koiter, ed.), pp.207-220, North- Holland, Amsterdam 1977.
6. Hill, R.: Acceleration waves in solids. *J. Mech. Phys. Solids* **10** (1962) 1-16.
7. Hadamard, J.: *Leçons sur la Propagation des Ondes et les Equations de l'Hydro-dynamique,* Hermann, Paris 1903.
8. Mandel, J.: Conditions de stabilité et postulat de Drucker. *Rheology and Soil Mechanics* (J. Kravtchenko, ed.), pp.58-67, Springer, Berlin 1966.
9. Hill, R.: The essential structure of constitutive laws for metal composites and polycrystals. *J. Mech. Phys. Solids* **15** (1967) 79-95.
10. Hutchinson, J.W.: Elastic-plastic behaviour of polycrystalline metals and composites. *Proc. Roy. Soc. Lond.* **A 319** (1970) 247-272.
11. Petryk, H.: On energy criteria of plastic instability. *Plastic Instability, Proc. Considère Memorial,* pp.215-226. Ecole Nat. Ponts Chauss., Paris 1985.
12. Petryk, H.: Material instability and strain-rate discontinuities in incrementally nonlinear continua. *J. Mech. Phys. Solids* **40** (1992) 1227-1250.
13. Hill, R.: Some basic principles in the mechanics of solids without a natural time. *J. Mech. Phys. Solids* **7** (1959) 209-225.
14. Graves, L.M.: The Weierstrass condition for multiple integral variation problems. *Duke Math. J.* **5** (1939) 656-660.
15. Petryk, H.: On constitutive inequalities and bifurcation in elastic-plastic solids with a yield-surface vertex. *J. Mech. Phys. Solids* **37** (1989) 265-291.
16. Christoffersen, J.; Hutchinson, W: A class of phenomenological corner theories of plasticity. *J. Mech. Phys. Solids* **27** (1979) 465-487.
17. Hutchinson, J.W.; Tvergaard, V.: Shear band formation in plane strain. *Int. J. Solids Structures* **17** (1981) 451-470.

Chapter 9
Plastic Instability: Criteria and Computational Approaches

Abstract General criteria of instability in time-independent elastic-plastic solids and the related computational approaches are reviewed. The distinction between instability of equilibrium and instability of a deformation process is discussed with reference to instabilities of dynamic, geometric or material type. Comparison is made between the bifurcation, energy and initial imperfection approaches. The effect of incremental nonlinearity of the constitutive law, associated with formation of a yield-surface vertex, on instability predictions is examined. A survey of the methods of post-critical analysis is presented.

9.1 Introduction

The meaning of the term 'plastic instability' does not appear to be commonly agreed. It was earlier used in connection with buckling or necking phenomena in the plastic range, predominantly under uniaxial stress, and also in connection with a peak on a load - displacement curve. Nowadays, it is frequently employed when discussing an abrupt change of smoothly distributed deformation into a localized mode, in the form of shear bands in solid bodies or of localized necking in thin sheets. It is well known that in plastic solids the concept of instability of *equilibrium* is too narrow to account for such a variety of instability phenomena. In the present article, plastic instability is understood in a wider sense, concerning a quasi-static deformation *process* (path), including the above mentioned phenomena as particular symptoms of instability.

Reference is made to a theoretical deformation process, called the fundamental process (or the fundamental path), which represents a *possible* response of a plastic body to a given program of slowly varying loading. A typical example is the process of macroscopically uniform deformation under appropriate boundary data, while a single equilibrium state may be regarded as a degenerate process under constant

[0] Reprinted from: Archives of Computational Methods in Engineering 4, 111-151, 1997. H.Petryk. Plastic instability: Criteria and computational approaches. Copyright ©1997 by CIMNE, Barcelona (Spain), with permission from Springer.

H. Petryk, *Theory of Plastic Instability*,
https://doi.org/10.1007/978-3-031-98223-1_9

loading. It is assumed that the boundary conditions and mechanical properties of the material are modeled with the accuracy sufficient for actual engineering purposes. Nevertheless, in the course of deformation in a physical system there are always some disturbing influences which are not taken into account in the theoretical description. [1] If the behaviour of the deformed body tends, with decreasing strength of such disturbing influences, to the fundamental process then no reference to instability is to be made. If, on the other hand, the perturbed deformation process deviates from the fundamental process on a finite distance, no matter how accurately the theoretical conditions are satisfied, then the fundamental process is treated as unstable and, consequently, as practically unrealizable in a physical system. It can be seen that instability is meant here more in a physical sense than in numerical one, [2] and concerns a whole deformation path, or at least its segment, rather than a single equilibrium state. For computational purposes, an instability *criterion* is needed which can be postulated or derived from a definition of stability; in particular, the bifurcation criterion or the energy criterion can be used. An alternative approach is to examine the effect of small initial imperfections. The distinction is that in the stability analysis a perturbed motion of a given material system is investigated while the initial imperfections change the system itself.

'Plastic' instability refers naturally to such cases of instability where plasticity plays a crucial role. Plasticity is meant below in a narrower sense, excluding visco-plasticity and thermo-plasticity, so that viscoplastic and thermoplastic instabilities (e.g. adiabatic shear bands) are not discussed. On the other hand, a general constitutive framework of incrementally nonlinear plasticity is used to include the important effect of formation of a yield-surface vertex.

This article contains a review of computational approaches to plastic instability phenomena, which are discussed in connection with the underlying instability criteria. As a rule, approaches will be discussed that have a sound theoretical foundation, are sufficiently general to be applicable to a wide class of problems, and have been effectively used in calculations. We will omit intuitive stability criteria formulated for particular problems, which may be useful for engineering purposes but have not yet been extended to provide a rational basis for a computational approach of a more general scope. The attention will be focused on issues belonging to mechanics; methods developed to overcome difficulties of a strictly numerical origin are outside the scope of this article.

Plastic instability manifests itself usually beyond a certain critical stage of the deformation, and then some more or less abrupt deviation from the fundamental process is observed. According to the character of such deviation, it is convenient to distinguish three kinds of plastic instability: of dynamic, geometric or material

[1] Specification of the class of disturbances and of the measure of their strength is required if a precise definition of stability is to be formulated; for the purposes of this survey, the concept of stability of a quasi-static path is described only in outline.

[2] A physical instability is likely to cause numerical troubles, but this need not always be the case. Numerical instabilities as such are not discussed here.

type. In the case of dynamic instabilities [3], we loose control over the process which can no longer be quasi-static in reality. Geometric instabilities manifest themselves by changes in shape of the deformed plastic body: buckling and necking are the best known examples. Material instabilities influence the deformation pattern inside the body, e.g. through the formation of shear bands, and are not removable by increasing the rigidity of boundary constraints. Of course, mixed kinds of instability are also possible: geometric or material instabilities may be accompanied by dynamic effects. The classification is only auxiliary and not absolute: for instance, localized necking in thin sheets may equally well be regarded as a symptom of the geometric or material instability according to whether the sheet is treated as a three- or two-dimensional body, respectively.

The paper is organized as follows. First, the incremental constitutive framework for time-independent simple materials is given, and a general first-order rate boundary value problem is reduced to a system of nonlinear algebraic equations. The criteria for dynamic, geometric and material instabilities and the respective computational approaches are subsequently reviewed in the next three sections. Methods of the analysis of a post-critical behaviour are discussed later, first in the absence of local instabilities and next in the case of strain localization into narrow bands as a consequence of material instability. This article may be regarded as complementary to the recent surveys [176, 181, 252] which addressed similar topics but were organized in a different manner.

9.2 Preliminaries

Constitutive Framework

Unless stated otherwise, we shall assume that mechanical properties of the material in isothermal deformation do not depend in any way on a natural time (*a time-independent material*) and depend locally on the deformation-gradient history (*a simple material*). It is assumed that the stresses vary along a path in a continuous and piecewise smooth manner, so that for rigid-plastic solids only a restricted class of deformation paths is considered. The basic constitutive equation takes the form of a *continuous and homogeneous of degree one* relationship between the forward (right-hand) rates of stress and strain. No matter in which variables it is originally formulated to ensure objectivity of the material description, the constitutive rate-equation can be expressed in the form

$$\dot{\mathbf{S}} = \dot{\mathbf{S}}(\dot{\mathbf{F}}, \mathcal{H}) = \mathbf{C}(\dot{\mathbf{F}}, \mathcal{H})\cdot\dot{\mathbf{F}}, \qquad \mathbf{C} \equiv \frac{\partial \dot{\mathbf{S}}}{\partial \dot{\mathbf{F}}}, \tag{9.1}$$

[3] The meaning of this term adopted here should not be confused with instability under dynamic loads; in the present article, only quasi-statically applied loadings are considered.

where $\mathbf{S}$ is the first (non-symmetric) Piola-Kirchhoff stress (i.e. the transpose of the nominal stress), related to the symmetric Cauchy stress $\boldsymbol{\sigma}$ and Kirchhoff stress $\boldsymbol{\tau}$ by $\mathbf{S}\mathbf{F}^T = \det(\mathbf{F})\,\boldsymbol{\sigma} = \boldsymbol{\tau}$, $\mathbf{F}$ is the deformation gradient, and the superscript 'T' denotes a transpose. A dot over a symbol denotes the *right-hand* material time derivative, [4] so that $\dot{\mathbf{F}} = \nabla\mathbf{v} = \partial\mathbf{v}/\partial\boldsymbol{\xi}$ is the velocity gradient evaluated in the reference configuration where a position vector is denoted by $\boldsymbol{\xi}$. A dot between two tensor symbols indicates full contraction, in the sense that $(\mathbf{C}\cdot\dot{\mathbf{F}})_{ij} = C_{ijkl}\dot{F}_{kl}$ and $\dot{\mathbf{S}}\cdot\dot{\mathbf{F}} = \dot{S}_{ij}\dot{F}_{ij}$; throughout this article, the summation convention is applied to repeated Latin indices which denote vector or tensor components on a fixed orthonormal basis $\{\mathbf{a}_1, \mathbf{a}_2, \mathbf{a}_3\}$ in $\mathcal{R}^3$. The symbol $\mathcal{H}$ (omitted later for simplicity) indicates a set of parameters by means of which the influence of the deformation-gradient history from a given initial state up to the current configuration (including the current stress and deformation) is taken into account. Arbitrary 'hardening' or 'softening' characteristics are allowed, noting that these concepts are influenced by arbitrary choice of the stress-rate in the current configuration [91], or alternatively, of the conjugate pair of strain and stress measures [94, 96]. The moduli tensor $\mathbf{C}$ may depend on the *direction* of the strain rate in a discontinuous or nonlinear manner.

Transformation from or to constitutive variables other than $\dot{\mathbf{S}}$ and $\dot{\mathbf{F}}$ is nowadays a standard task. For instance, let the constitutive rate equations in the current state (i.e. at given $\mathcal{H}$) be expressed as

$$\overset{\nabla}{\boldsymbol{\tau}} = \mathbf{L}(\mathbf{D}) \cdot \mathbf{D}, \qquad \mathbf{L} \equiv \frac{\partial \overset{\nabla}{\boldsymbol{\tau}}}{\partial \mathbf{D}}, \tag{9.2}$$

where $\overset{\nabla}{\boldsymbol{\tau}}$ is the corotational (Zaremba-Jaumann) flux of Kirchhoff stress and $\mathbf{D}$ is the Eulerian strain-rate. Then, we have (cf. [96])

$$\det(\mathbf{F}^{-1})F_{ip}F_{kq}C_{jplq} = L_{ijkl} - \frac{1}{2}(\sigma_{jk}\delta_{il} - \sigma_{ik}\delta_{jl} + \sigma_{il}\delta_{jk} + \sigma_{jl}\delta_{ik}), \tag{9.3}$$

where δ_{ij} is the Kronecker symbol. The additive term on the right-hand side, which represents the incremental effect of changes in geometry and is linear in Cauchy stress, in plasticity can be of the same order of magnitude as L_{ijkl}. It follows that in studies of plastic instability and bifurcation (where the use of the variables in (9.1) is particularly convenient [86]), this term should not be disregarded in advance, irrespectively of whether the strains are assumed to be 'small' or not [85].

It will be later of fundamental importance whether or not the relationship (9.1) admits a potential [86], $U(\dot{\mathbf{F}}, \mathcal{H})$ say, such that

$$\dot{\mathbf{S}} = \frac{\partial U}{\partial \dot{\mathbf{F}}}, \qquad \mathbf{C} = \frac{\partial^2 U}{\partial \dot{\mathbf{F}} \partial \dot{\mathbf{F}}}, \qquad U = \frac{1}{2}\dot{\mathbf{S}}\cdot\dot{\mathbf{F}}. \tag{9.4}$$

This is equivalent to imposing on (9.1) the symmetry restriction

$$C_{ijkl} = C_{klij} \qquad \Longleftrightarrow \qquad L_{ijkl} = L_{klij}\,. \tag{9.5}$$

---

[4] In a quasi-static process, the role of a natural time is played by a time-like parameter.

The constitutive relationship of classical elastoplasticity, generalized to include incremental changes in geometry [85, 91], represents a special case of (9.2) with $\mathbf{L}(\mathbf{D})$ constant on each side of a hyperplane in the strain-rate space, viz.

$$\mathbf{L} = \begin{cases} \mathbf{L}^p \text{ if } \boldsymbol{\lambda} \cdot \mathbf{D} > 0 & \text{(loading)} \\ \mathbf{L}^e \text{ if } \boldsymbol{\lambda} \cdot \mathbf{D} < 0 & \text{(unloading)}, \end{cases} \tag{9.6}$$

with the continuity restriction for $\overset{\nabla}{\boldsymbol{\tau}}(\mathbf{D})$:

$$\mathbf{L}^p = \mathbf{L}^e - \frac{1}{g}\boldsymbol{\lambda}^p \otimes \boldsymbol{\lambda}, \tag{9.7}$$

where $\mathbf{L}^e$ stands for the elastic moduli tensor, obeying the usual symmetries as in the small strain theory, $\boldsymbol{\lambda}$ is the outward normal to a smooth yield surface in *strain* space (based on the current configuration as reference), $\boldsymbol{\lambda}^p$ defines the direction of the plastic part of stress-rate, $\otimes$ denotes the tensor product, and g is a positive scalar parameter. The normality flow rule is obtained when

$$\boldsymbol{\lambda}^p = \boldsymbol{\lambda} \tag{9.8}$$

which implies the symmetry property (9.5). If $\mathbf{L}^e$ is invertible and $h = g - \boldsymbol{\lambda} \cdot (\mathbf{L}^e)^{-1} \cdot \boldsymbol{\lambda}^p$ (the hardening modulus for the pair $(\overset{\nabla}{\boldsymbol{\tau}}, \mathbf{D})$) is positive then the familiar inverted expression for $\mathbf{D}$ as a function of $\overset{\nabla}{\boldsymbol{\tau}}$ can be obtained. Of course, (9.6) imposes no restriction on the geometry of a smooth yield surface; its best known form is defined by the Huber-Mises yield condition. The respective version of J_2 flow theory of metal plasticity obeying normality is a widely used specification of (9.6) with (9.7) and (9.8). The normality condition (9.8) is usually not assumed for geomaterials, and for metals it is not used if the effect of void nucleation [79] or non-Schmid effects on the crystal level [11] is incorporated.

It is nowadays well recognized that the classical elastoplastic models obeying (9.6) with (9.7) are likely to overestimate the material stiffness against an abrupt change in the direction of straining. While this may remain unnoticed during simulation of smooth deformation paths, it becomes an essential shortcoming in calculations of bifurcation or instability points. The classical assumptions that the yield surface is smooth and that the direction of the plastic strain-rate is independent of the stress-rate are not supported by micromechanical considerations, cf. [17, 229, 92, 108, 225, 201]. At a micro-level, many internal mechanisms of plastic deformation can be distinguished that obey different loading-unloading criteria in terms of the macroscopic stress. If all rheological effects are disregarded then this leads to the formation of a pointed *vertex* on the macroscopic yield surface at the current loading point at which the individual yield surfaces for active mechanisms during plastic flow intersect. [5] The moduli tensor $\mathbf{C}$ in (9.1) depends then on the current direction

[5] It may be noted that the theoretical macroscopic yield surface corresponds to a vanishingly small plastic strain increment, while only finite increments can be measured experimentally; this can explain why experimental yield surfaces frequently exhibit rounded-off corners and not sharp vertices [140].

of strain-rate in a more complex fashion than in the classical elastoplasticity. At the vertex, three constitutive domains can generally be distinguished [92], in the strain-rate or stress-rate space: the cone of fully active loading, the cone of total elastic unloading, and the intermediate range of partial unloading which may be piecewise-linear or thoroughly nonlinear; a cone may degenerate to a single ray. The resulting incrementally nonlinear law can be defined indirectly by using a certain number of plastic deformation mechanisms, each obeying the classical law, and specifying the manner in which they interact; cf. [126, 147, 90, 103, 235, 96] for a general discussion, and e.g. [195, 10] for more specific models of metal crystals. For metal polycrystals, phenomenological relationships, e.g. of type (9.2) or its converse, with the nonlinearity defined directly seem to be more convenient; such models were discussed already long ago, e.g. in [45, 86, 123, 158]. In numerical calculations of plastic instability problems, the J_2 corner theory of plasticity proposed in [53] has been frequently used; cf. also [75, 255]. [6] Incrementally nonlinear models were also introduced to describe the behaviour of geomaterials, e.g. [60, 48, 59, 128, 47, 129]. A common feature of all such models is that the direction of the plastic part of strain-rate (or of stress-rate) is no longer fixed in the current state as in the classical plasticity but becomes dependent on the loading direction. This is connected with a significant decrease of the material stiffness (given by elastic moduli in the classical description) in the directions orthogonal to the fundamental mode of plastic flow. As discussed later in more detail, realistic estimates of the onset of plastic instability are then available without the need of appealing to initial imperfections or to physical softening; the incremental changes in geometry may give a softening effect sufficient to overcome small intrinsic hardening of the material.

Boundary-Value Problem for Velocities

The basic boundary value problem examined here has *the forward rates* of displacements and stresses as the principal unknowns. [7] In the assumed absence of *moving* strong discontinuities in stresses and in distributed loads, and neglecting inertia effects, we may start from the equations of continuing equilibrium expressed in the rate form of the virtual power principle, viz.

$$\int_V \dot{\mathbf{S}} \cdot \nabla \mathbf{w} \, dV = \int_V \dot{\mathbf{b}} \mathbf{w} \, dV + \int_S \dot{\mathbf{T}} \mathbf{w} \, dS \qquad \forall \, \tilde{\mathbf{w}} \in \mathcal{W}. \tag{9.9}$$

Here, V and S denote the body volume and surface, respectively, in an arbitrary but fixed reference configuration. $\dot{\mathbf{b}}$ and $\dot{\mathbf{T}}$ are forward rates of *nominal* body forces and *nominal* surface tractions per unit volume and area, respectively, in the *reference*

[6] It should be added, however, that a phenomenological corner model whose behaviour would be consistent with micromechanics along *arbitrary* paths has not been proposed yet, to the writer's knowledge.

[7] For time integration purposes, the basic incremental problem is frequently formulated and solved in terms of small but finite increments of displacements rather than in terms of velocities. The influence of this change on bifurcation and instability conditions is not discussed in this article.

configuration. A tilde over a symbol is used to distinguish a spatial field over V from its value at a material point, and the symbol $\forall$ means 'for every'. The current state of the deformed material body, represented by the fields of stresses, displacements and material parameters, is assumed to be known.

$\mathcal{W}$ stands for a linear space [8] of continuous and finite *variations* of a velocity field which are *defined* as kinematically admissible. Let $\tilde{\boldsymbol{\phi}}_\alpha$, $\alpha \in \mathcal{A}$, be linearly independent, continuous and piecewise sufficiently smooth vector functions which form a basis in the linear space $\mathcal{W}$, so that

$$\tilde{\mathbf{w}} = w_\alpha \tilde{\boldsymbol{\phi}}_\alpha \,, \qquad \tilde{\mathbf{w}} \in \mathcal{W}\,, \quad \alpha \in \mathcal{A}\,, \quad w_\alpha \in \mathcal{R}\,, \tag{9.10}$$

where $\mathcal{R}$ is the set of real numbers; the summation over $\mathcal{A}$ for repeated Greek subscripts is henceforth assumed. The set $\mathcal{A}$, of natural numbers, say, is taken to be infinite or finite, for the non-discretized continuum problem or for a spatially discretized problem, respectively. We shall focus attention mainly on a finite dimensional problem, but allowing the set $\mathcal{A}$ to be infinite we can consider, at least formally, also the non-discretized continuum problem within a unified framework. The fundamental questions concerning error estimates or the convergence of discretized solutions to an exact solution are beyond the scope of this article. On the other hand, it is not excluded that a solution to the discretized problem is simultaneously an exact one.

It is assumed that any member of the class $\mathcal{V}$ of kinematically admissible velocity fields $\tilde{\mathbf{v}}$ has the following representation

$$\tilde{\mathbf{v}} = \tilde{\mathbf{v}}^k + v_\alpha \tilde{\boldsymbol{\phi}}_\alpha \,, \qquad \tilde{\mathbf{v}} \in \mathcal{V}\,, \quad \alpha \in \mathcal{A}\,, \quad v_\alpha \in \mathcal{R}\,. \tag{9.11}$$

Here, $\tilde{\mathbf{v}}^k$ is a given (continuous and at least piecewise smooth) velocity field which is compatible with the kinematical constraints, for instance, takes prescribed values on the part S_u of S where $\mathbf{w}$ vanishes. If the prescribed velocities are nonzero then $\tilde{\mathbf{v}}^k$ is not contained in $\mathcal{W}$, and $\mathcal{V}$ is not a linear space. In a displacement-based finite element formulation, v_α correspond to a *difference* between the velocity components at nodal points for $\tilde{\mathbf{v}}$ and $\tilde{\mathbf{v}}^k$, but other discretization schemes are allowed as well.

On substituting the constitutive rate equations (9.1) into (9.9) and following the standard discretization procedure in Lagrangian description, a field $\tilde{\mathbf{v}} \in \mathcal{V}$, defined by a vector $[v]$ of components v_α , is called a solution in velocities if and only if it satisfies the system of nonlinear algebraic equations:

$$\dot{Q}_\alpha(\tilde{\mathbf{v}}) = \dot{P}_\alpha, \qquad \alpha \in \mathcal{A}, \quad \tilde{\mathbf{v}} \in \mathcal{V}\,, \tag{9.12}$$

where

$$\dot{Q}_\alpha(\tilde{\mathbf{v}}) = \int_V \dot{\mathbf{S}}(\nabla\mathbf{v})\cdot\nabla\boldsymbol{\phi}_\alpha \, dV \tag{9.13}$$

and

$$\dot{P}_\alpha = \int_V \dot{\mathbf{b}}\boldsymbol{\phi}_\alpha \, dV + \int_S \dot{\mathbf{T}}\boldsymbol{\phi}_\alpha \, dS \tag{9.14}$$

[8] Unilateral constraints are excluded from consideration.

may be interpreted as components of the rates of internal and external forces, respectively. The system of equations (9.12) pertains to both the non-discretized and discretized problems; the difference is in the number (infinite or finite) of the equations. The solution to (9.12) along a fundamental path will be denoted by $\widetilde{\mathbf{v}}^0$, and the respective quantities will also be distinguished by a superscript '0'.

The tangent stiffness matrix $[\bar{K}]$ for the deformed body, with $\dot{\mathbf{b}}$ and $\dot{\mathbf{T}}$ treated as prescribed, is defined by

$$\bar{K}_{\alpha\beta}(\widetilde{\mathbf{v}}) \equiv \frac{\partial \dot{Q}_\alpha}{\partial v_\beta}(\widetilde{\mathbf{v}}) = \int_V \nabla\boldsymbol{\phi}_\alpha \cdot \mathbf{C}(\nabla\mathbf{v}) \cdot \nabla\boldsymbol{\phi}_\beta \, dV, \qquad \alpha, \beta \in \mathcal{A}. \tag{9.15}$$

The dependence of $[\dot{Q}]$ and $[\bar{K}]$ on the velocity field $\widetilde{\mathbf{v}}$ is homogeneous of degree one and zero, respectively. However, it should be noted that, in general, $\dot{Q}_\alpha(\widetilde{\mathbf{v}}) \neq \bar{K}_{\alpha\beta}(\widetilde{\mathbf{v}})v_\beta$ unless $\widetilde{\mathbf{v}}^k = \widetilde{0}$. If the instantaneous moduli exhibit the symmetry property (9.5), i.e. if (9.4) holds, then $[\bar{K}]$ is symmetric.

If the loading is configuration-dependent (cf. [89, 234, 246]) then $\dot{P}_\alpha$ consists not only of the prescribed part, $\dot{\bar{P}}_\alpha$ say, but also of a deformation-sensitive part:

$$\dot{P}_\alpha = \dot{P}_\alpha(\widetilde{\mathbf{v}}) = \dot{\bar{P}}_\alpha - k_{\alpha\beta}v_\beta, \qquad \alpha, \beta \in \mathcal{A}. \tag{9.16}$$

For instance, elastic support or fluid-pressure loading can lead to expressions of type (9.16). We shall assume that, in a given state, the matrix $[k]$ is independent of $\widetilde{\mathbf{v}}$. The vector $[\bar{P}]$ may be regarded as a given function of a loading parameter, Λ say, which is directly controlled and varies in time in a quasi-static manner.

The global tangent stiffness matrix for the mechanical *system* {body + loading device}, which appears in the analysis of bifurcation and instability, reads

$$[K] = [\bar{K}] + [k]. \tag{9.17}$$

Consequently, the stiffness of the loading device can have a significant influence on the occurrence of instabilities of dynamic or geometric type.

Hill's [86, 89] general condition sufficient (but not necessary) for uniqueness of a solution in velocities in time-independent plasticity becomes, in the present notation,

$$\Delta\dot{Q}_\alpha \Delta v_\alpha + k_{\alpha\beta}\Delta v_\alpha \Delta v_\beta > 0 \qquad \forall\, \Delta\widetilde{\mathbf{v}}, \tag{9.18}$$

where the prefix Δ denotes the difference of the quantities corresponding to any pair of two distinct velocity fields from $\mathcal{V}$. The uniqueness condition (9.18) can be weakened by taking the fundamental solution $\widetilde{\mathbf{v}}^0$, if it is known, as one member of all the pairs [96, 201].

If the configuration-dependent incremental loading is *conservative* in an overall sense [89, 234, 202] then $[k]$ is symmetric. If (9.4) holds then the symmetry of $[K]$ is implied, and then more can be said about the solutions to (9.12). Every solution $\widetilde{\mathbf{v}}$ to (9.12), unique or not, is characterized by Hill's stationarity principle [86, 89], viz.

$$\frac{\partial J(\widetilde{\mathbf{v}})}{\partial v_\alpha} = 0, \qquad \alpha \in \mathcal{A}\,, \tag{9.19}$$

where

$$J(\widetilde{\mathbf{v}}) = \int_V U(\nabla \mathbf{v})\, dV + \frac{1}{2} k_{\alpha\beta} v_\alpha v_\beta - \dot{\bar{P}}_\alpha v_\alpha\,. \tag{9.20}$$

The existence of a solution to (9.19) for a *discretized* problem is ensured if the condition (9.21) below holds since a solution can then be found as an absolute minimizer of J in $\mathcal{V}$ [202, 209].

9.3 Onset of Dynamic Instability

We begin with the conditions for 'dynamic' instabilities associated with the loss of stability of *equilibrium*, although often this is not the first type of plastic instability encountered along a deformation path. Dynamic instability means here that a finite distance from an equilibrium state can be reached in a dynamic motion following an arbitrarily small disturbance, at a *fixed* value of the controlled loading parameter Λ. This represents a particular type of instability of the fundamental deformation path: it can be left in the time scale in which the variations of the loading parameter are disregarded.

The corresponding critical point (and the instability mode) might in principle be determined by performing a direct analysis of a perturbed dynamic motion, following an *arbitrary* disturbance, in vicinity of an equilibrium state of a loaded elastic-plastic body. On account of the path-dependence and incremental non-linearity of the governing equations in time-independent elastoplasticity, such a procedure can hardly be effective in general, although it can sometimes be applied to simple systems, e.g. [125]. Dynamic analyses bases on an incrementally *linear* model can be adequate if the instability mode happens to activate a single constitutive branch, as e.g. the shear-band mode [148].

The common approach to determining the onset of dynamic instabilities is to use the energy criterion of stability of equilibrium, formulated in the literature with a various degree of exactness. The criterion is obtained by comparison of the deformation work done along non-equilibrium paths starting from the equilibrium state with the respective work done by external loads. Following Hill [85, 89] and Drucker [69], we obtain the following condition

$$I(\widetilde{\mathbf{w}}) > 0 \qquad \forall\, \widetilde{\mathbf{w}} \in \mathcal{W}, \quad \widetilde{\mathbf{w}} \neq \widetilde{\mathbf{0}}, \tag{9.21}$$

where

$$2I(\widetilde{\mathbf{w}}) \equiv \dot{Q}_\alpha(\widetilde{\mathbf{w}}) w_\alpha + k_{\alpha\beta} w_\alpha w_\beta = K_{\alpha\beta}(\widetilde{\mathbf{w}}) w_\alpha w_\beta\,, \qquad \widetilde{\mathbf{w}} \in \mathcal{W}, \tag{9.22}$$

as a condition sufficient for 'directional' stability of equilibrium, in the following sense: If (9.21) holds then a departure from equilibrium on a straight (direct) path in a dynamic motion induced by a vanishingly small disturbance (so-called divergence instability) is excluded by imposing an energy barrier in all directions.

If plasticity is described with the help of internal state variables then the left-hand expression can be split into two parts, one dependent solely on a reversible response of the system and the second expressed in terms of a dissipation function [180]; cf. also [72, 179, 206, 207].

If $[K]$ is finite-dimensional and symmetric and if

$$I(\widetilde{\mathbf{w}}) < 0 \qquad \text{for some } \widetilde{\mathbf{w}} \in \mathcal{W} \tag{9.23}$$

then the existence of a straight dynamic path of departure from the equilibrium state induced by a vanishingly small disturbance can be demonstrated, in the first approximation [202].

To ensure stability of equilibrium for arbitrary circuitous paths, (9.21) must be accompanied by further assumptions (cf. [85, 86, 105, 182, 180, 206]). In particular, in the classical plasticity without normality where $[K]$ is not symmetric, the possibility of cyclic or 'flutter' instability of equilibrium with a sequence of loading-unloading loops should be taken into account, cf. [290, 199, 230]. This problem requires further study.

The critical state beyond which equilibrium becomes unstable can be searched as a primary *eigenstate* [93, 96] in which, by definition, the equation

$$K_{\alpha\beta}(\widetilde{\mathbf{w}})w_\beta = 0, \qquad \widetilde{\mathbf{w}} \in \mathcal{W}, \tag{9.24}$$

has a non-zero solution (eigenmode). (9.24) corresponds to the *homogeneous* rate boundary- value problem, obtained form (9.12) when $\widetilde{\mathbf{v}}^k$ and the *given* part $[\dot{P}]$ of the loading rate vanish, or simply when $\dot{\Lambda} = 0$. Of course, the solution can be freely scaled down by an arbitrary non-negative factor. The range of (9.21) for symmetric $[K]$ terminates in a natural way at an eigenstate, while for non-symmetric $[K]$, (9.21) is associated with another critical condition

$$K^{\mathrm{s}}_{\alpha\beta}(\widetilde{\mathbf{w}})w_\beta = 0, \qquad \widetilde{\mathbf{w}} \in \mathcal{W}, \qquad 2K^{\mathrm{s}}_{\alpha\beta} = K_{\alpha\beta} + K_{\beta\alpha}\,. \tag{9.25}$$

While (9.25) can be met earlier than (9.24), its mechanical interpretation is less clear.

On account of the dependence of the matrix $[K]$ on the unknown eigenmode, the problem (9.24) is nonlinear. This corresponds to determining the 'reduced modulus load' for buckling instability of a compressed elastic-plastic column (cf. the survey in [235]). In general, the analysis of eigenstates in elastic-plastic solids cannot be reduced to a linear problem and can thus be connected with difficulties. However, if eigenstates in homogeneous solids are examined under the boundary conditions that allow uniform eigenmodes then remarkable conclusions can be drawn from a pure analysis, e.g. [93, 155, 96, 219, 98, 99]. It can be added that under predominantly tensile loading, it is possible to eliminate the material spin from the stability condition [93].

Fortunately, eigenstates of a particular but practically most important type are found in a natural way on a general (uniform or non-uniform) deformation path at a point where the *controlled* loading parameter Λ (which may be, in particular, an external pressure or a displacement of the loading device) attains its *analytic* extremum value versus a typical displacement within the deformed body. Then $\dot{\Lambda} = 0$ so that $\widetilde{\mathbf{v}}^k$ and $[\dot{P}]$ vanish, and the fundamental solution to (9.12), when normalized to be bounded, represents itself an eigenmode, viz.

$$K_{\alpha\beta}(\widetilde{\mathbf{v}}^0)v^0_\beta = 0, \qquad \dot{\Lambda} = 0. \tag{9.26}$$

This is nothing else as a straightforward generalization of the concept of a limit point, most familiar when a controlled load attains a maximum. From (9.16) it can be seen that (9.26) is associated with

$$\dot{P}^0_\alpha = -k_{\alpha\beta}v^0_\beta \qquad \text{at } \dot{\Lambda} = 0\,, \tag{9.27}$$

so that the placement of the generalized limit point on the fundamental path can be strongly influenced by the stiffness of the loading device. In numerical step-by-step analysis, the incremental solution blows up as the critical state (9.26) is approached, unless the step size in Λ is decreasing to zero sufficiently fast. To calculate the eigenstate with sufficient accuracy, special numerical procedures can be used; cf. e.g. [285] and the references quoted therein. Calculations of post-critical deformation will be discussed later.

It must be noted that, in general, fulfillment of (9.26) may be preceded by an earlier eigenstate. Dynamic instability can also be induced by a discontinuous drop in the material stiffness (e.g. when the yield point for a previously inactive mechanism of plastic deformation is reached), the range of instability of equilibrium being not entered at an eigenstate. In particular, this can happen when the drop in the tangent stiffness leads abruptly to violation of the energy condition (9.21) by the fundamental mode $\widetilde{\mathbf{v}}^0$ substituted in place of $\widetilde{\mathbf{w}}$, which is possible if the prescribed boundary velocities vanish. This corresponds to the familiar case of a sharp peak on the load-displacement curve when the load is controlled.

9.4 Onset of Geometric Instability

9.4.1 Theoretical Foundations

Before the traversed equilibrium states become unstable, the fundamental deformation path of a plastic body may be left in a manner which is fundamentally distinct from that considered in the preceding section. Namely, the deviation from the fundamental path, as the path itself, may be *quasi-static* in character and may require simultaneous variation of the loading parameter Λ, that is, of prescribed external loads or displacements. If a finite distance from the fundamental path can be reached in that way along a perturbed path corresponding to vanishingly small (and physically admissible) disturbing influences then the fundamental path may be regarded as unstable. In the cases where this is associated with changes in the external shape of the deformed body or of its part (buckling, necking, bulging, wrinkling, etc.), reference to 'geometric' instability is made.

Directly related to path instability in the above sense is the possibility of *bifurcation* of a deformation path, in the absence of any disturbances, if the incremental response of the loaded body becomes nonuniquely defined. In the first instance, the bifurcation problem is reduced to that in velocities, i.e. to the nonuniqueness of a solution to (9.12). Due to the incremental nonlinearity of an elastic-plastic constitutive law, the constitutive branches activated by an eigenmode in (9.24) at $\dot{\Lambda} = 0$ and by a bifurcation mode at $\dot{\Lambda} \neq 0$ are different in general. The material moduli activated at bifurcation may be 'less stiff', so that bifurcation may take place before instability of equilibrium. This concept was introduced by Shanley [237, 238] in the context of buckling of a compressed elastic-plastic column, and was extended by Hill [85] to a general case.

Starting from the primary bifurcation point, the deforming plastic body *may* follow a secondary solution branch rather than the fundamental path. Strictly speaking, this means only a possibility, but on the basis of the accumulated knowledge about the behaviour of imperfect systems, the fundamental post-bifurcation branch is usually assumed to have no physical meaning. The circumstances in which this conclusion can be drawn from the energy criterion of *path* instability are specified below. The approaches to path stability in [68, 125, 18, 20] lead also to a similar conclusion.

In this section we discuss the methods used to determine the instant and the mode of primary bifurcation. The theory is much better developed under the assumptions which lead to a symmetric tangent stiffness matrix; this case is examined first.

Symmetric Tangent Stiffness Matrix

We shall assume now the symmetry of the moduli, cf. (9.4) and (9.5), and of the matrix $[k]$. A general theory of bifurcation in elastic-plastic solids under such restrictions was developed by Hill [85, 86, 87, 89]; for a more recent account and extensions, cf. [96, 205]. For the purposes of computation of the primary bifurcation point, one element of this theory is particularly useful, that is, the concept of the *incrementally linear comparison solid* whose moduli in (9.1), $\mathbf{C}^L$ say, do not depend on the current strain-rate.

For the classical elastic-plastic solids with a smooth yield surface and the normality flow rule, obeying (9.6) with (9.7) and (9.8), the comparison solid is defined by replacing the actual incremental stiffness moduli $\mathbf{L}$ in (9.2) by $\mathbf{L}^p$, independently of the actual strain-rate direction. In the elastic zone no modification is needed, of course. Then, for any pair $(\dot{\mathbf{F}}, \dot{\mathbf{F}}^*)$ and for the constitutively related pair $(\dot{\mathbf{S}}, \dot{\mathbf{S}}^*)$ determined from (9.1) through the equations of the classical elastoplasticity, we have [85]

$$(\dot{\mathbf{N}} - \dot{\mathbf{F}}^*) \cdot (\dot{\mathbf{F}} - \dot{\mathbf{F}}^*) \geq (\dot{\mathbf{F}} - \dot{\mathbf{F}}^*) \cdot \mathbf{C}^L \cdot (\dot{\mathbf{F}} - \dot{\mathbf{F}}^*) \qquad \forall\, \dot{\mathbf{F}}, \dot{\mathbf{F}}^* . \tag{9.28}$$

(9.28) has the interpretation that the constitutive potential U in (9.4) is 'more convex' than the quadratic potential U^L for the comparison solid, in the sense that $(U - U^L)$ is a convex function of $\dot{\mathbf{F}}$ [86, 87, 96].

On substituting (9.28) into (9.18), we obtain the condition which excludes bifurcation in velocities, in the form

$$K^L_{\alpha\beta} w_\alpha w_\beta > 0 \qquad \forall\, [w] \neq [0]\,, \tag{9.29}$$

where K^L is obtained from (9.17) on substituting $\mathbf{C} = \mathbf{C}^L$ into (9.15). It can be seen that (9.29) with (9.28) imply, under the symmetry restrictions (9.5) and $[k] = [k]^T$, that the functional J in the stationarity principle (9.19) is strictly convex in $\mathcal{V}$, which ensures uniqueness of the solution (cf. [82]). Validity of (9.29) as an exclusion condition for bifurcation in velocities has been recently extended to include non-local effects by using gradient-dependent plasticity with isotropic hardening [161, 29].

Along a given fundamental deformation path, the respective velocity solution $\widetilde{\mathbf{v}}^0$ is known. Suppose that $\widetilde{\mathbf{v}}^0$ corresponds to loading within the region being stressed to the yield point (the plastic zone), except possibly on certain interfaces which makes no contribution to volume integration. This means that unloading along the fundamental path may take place gradually, through propagation of elastic-plastic interfaces, but not abruptly. At such regular instants, the comparison moduli $\mathbf{C}^L$ can be defined, almost everywhere within the body, as the instantaneous tangent stiffness moduli $\mathbf{C}^0 \equiv \mathbf{C}(\nabla \mathbf{v}^0)$. Then the uniqueness condition (9.29) for velocities can be expressed through the tangent stiffness matrix for the fundamental mode:

$$2I^0(\widetilde{\mathbf{w}}) \equiv K^0_{\alpha\beta} w_\alpha w_\beta > 0 \qquad \forall\, [w] \neq [0]\,, \qquad [K^0] \equiv [K(\widetilde{\mathbf{v}}^0)]\,. \tag{9.30}$$

Henceforth, we shall assume that the matrix $[K^0]$ is well defined whenever it appears in the considerations.

It has been shown that (9.30) or (9.29) excludes also higher-order bifurcations of a regular type, in a linearized [125, 259] and nonlinear [208, 52] formulation. A similar result was obtained in [183] for the constitutive description of perfect plasticity extended to include internal variables.

We have considered so far the classical model of elastoplastic response. For a class of piecewise-linear constitutive rate equations at a pyramidal vertex on the yield surface, obeying the normality and symmetry postulates, the relative convexity property (9.28) still holds if the comparison moduli $\mathbf{C}^L$ are defined as the 'total loading' moduli, i.e. those from the current range of fully active loading [235]. This refers, in particular, to a class of models of elastic-plastic crystals deformed by multislip, cf. [147, 90, 103]. Similarly as above, at a regular point on the fundamental deformation path, the uniqueness condition reduces to (9.30).

This conclusion has been extended in [201] to a general (unspecified) incrementally nonlinear response of a polycrystal at a yield-surface vertex under the normality and symmetry assumptions adopted at the level of a single crystal (micro-level). The proof was based on the constitutive inequality

$$\dot{\mathbf{S}}^0 \cdot \dot{\mathbf{F}} - \dot{\mathbf{S}} \cdot \dot{\mathbf{F}}^0 \geq 0 \qquad \forall\, \dot{\mathbf{F}}, \tag{9.31}$$

which expresses the lack of abrupt unloading at the micro-level at a *regular* point along the fundamental path. Here, $\dot{\mathbf{S}}^0$ and $\dot{\mathbf{F}}^0$ are the fundamental stress and deformation rates, and $\dot{\mathbf{S}}$ is constitutively related to $\dot{\mathbf{F}}$ through (9.4). (9.31) was transformed to a form similar to but less restrictive than (9.28) with $\mathbf{C}^L = \mathbf{C}^0$, and then uniqueness was concluded from (9.30) without the need for the convexity of the functional (9.20). As a corollary, it follows that (9.30) and (9.31) imply (9.21) [201, 205].

The constitutive inequality (9.31) was derived from micromechanical considerations and, similarly to (9.28) with $\mathbf{C}^L = \mathbf{C}^0$, may not be satisfied by an incrementally nonlinear, phenomenological model of plasticity. In such cases, the use of (9.30) as the exclusion condition for bifurcation in velocities may be incorrect.

With the reservations indicated above, (9.30) may serve as a basis for determining the range where the uniqueness of an incremental solution is at least expected (if not ensured rigorously). Under certain additional assumptions, also a kind of path stability can be demonstrated in this range [206]; cf. also [125, 177, 239].

The next question is to what extent (9.30) is also *necessary* for incremental uniqueness and for path stability. For a *discretized* problem it has been proved [202, 209] that there is a bifurcation in velocities at *every* point on the segment of a fundamental path along which

$$K^0_{\alpha\beta} w_\alpha w_\beta < 0 \qquad \text{for some } [w] \tag{9.32}$$

holds true simultaneously with the condition (9.21) of stability of equilibrium. Contrary to incrementally linear (e.g. elastic) solids, the spectrum of bifurcation points, defined by the *inequalities* (9.32) and (9.21), is *continuous* along the path, without the need of singularity of $[K^0]$. For the simplest systems as the Shanley column [238] this result has been known for many years; the proof given in [202] (and in another version in [209]) is valid for any discretized problem under the assumptions leading to a symmetric tangent stiffness matrix $[K]$.

The secondary solutions in the continuous range of bifurcations were shown to assign to the functional J defined by (9.20) a minimum value in $\mathcal{V}$. The fundamental solution $\widetilde{\mathbf{v}}^0$ cannot be a minimizer there since the second variation of J at $\widetilde{\mathbf{v}}^0$ is (at most) indefinite on account of (9.32). On the other hand, in the range of validity of the uniqueness condition (9.18), the velocity solution $\widetilde{\mathbf{v}}^0$ always minimizes J in $\mathcal{V}$ [86, 96], also when (9.18) is weakened by taking $\widetilde{\mathbf{v}}^0$ as one member of all the pairs [202]. This makes contact with the *energy criterion* of plastic instability, *extended to deformation processes*, formulated as follows [198, 200]. For any kinematically admissible deformation path, define the energy functional as

$$E = W + \Omega, \tag{9.33}$$

where W is the work of deformation in a deformation process up to the current instant, and Ω is the potential energy of the loading device, which can be defined at least to second-order terms on account of the assumed symmetry of $[k]$. E can be interpreted as the internal energy of a thermally insulated, compound thermodynamic system that consists of the deformed body placed in a loading device and surrounded by a heat reservoir. The first-order rate of E is independent on $\widetilde{\mathbf{v}} \in \mathcal{V}$ on account of (9.9). The instability criterion states that the following inequality in the second-order rate of E:

$$\ddot{E}(\widetilde{\mathbf{v}}) < \ddot{E}(\widetilde{\mathbf{v}}^0) \qquad \text{for some } \widetilde{\mathbf{v}} \in \mathcal{V} \tag{9.34}$$

implies instability of the fundamental process associated with $\widetilde{\mathbf{v}}^0$. The energy criterion is related to the bifurcation criterion through the identity

$$\frac{1}{2}\ddot{E}(\widetilde{\mathbf{v}}) - \frac{1}{2}\ddot{E}(\widetilde{\mathbf{v}}^0) = J(\widetilde{\mathbf{v}}) - J(\widetilde{\mathbf{v}}^0) \qquad \forall\, \widetilde{\mathbf{v}} \in \mathcal{V}\,, \tag{9.35}$$

with the help of which (9.34) has been shown to be implied by (9.32) [200]. By (9.35) and the existence theorem for a discretized solution in velocities [202], at *every* instant along the fundamental post-critical path where (9.34) holds simultaneously with (9.21), there is a secondary incremental solution which is energetically preferable to the fundamental mode. This is the interpretation of the path instability predicted by the energy criterion (9.34) for a discretized problem; in the non-discretized continuum, the infimum of J need not be attained among ordinary functions, but the conclusion about path instability can still be drawn [202]. If (9.21) fails then dynamic instability of equilibrium is expected instead; cf. the preceding section.

To summarize, the problem of determining the onset of bifurcations and of path instability of 'geometric' type can be reduced, if the constitutive inequality (9.31) holds or at least in the first instance, to finding the earliest state beyond which (9.32) is met on a fundamental deformation path.

A typical (not always present) critical state that separates the ranges of (9.30) and (9.32) for a symmetric and continuously varying $[K^0]$ arises when the linear eigenvalue problem for $[K^0]$:

$$K^0_{\alpha\beta} w_\beta - \lambda w_\alpha = 0 \tag{9.36}$$

admits a nontrivial solution $\widetilde{\mathbf{w}}^*$ associated with a zero eigenvalue $\lambda^* = 0$, while no eigenvalue is negative. In a discretized problem, in this critical state we obviously have

$$\det[K^0] = 0\,. \tag{9.37}$$

The eigenmode can be found from the variational condition [86] (cf. (9.30))

$$\delta I^0 = 0 \qquad \text{at } \widetilde{\mathbf{w}} = \widetilde{\mathbf{w}}^* \tag{9.38}$$

implying $I^0(\widetilde{\mathbf{w}}^*) = 0$, where δ stands for the first weak variation in $\mathcal{W}$. $\widetilde{\mathbf{w}}^*$ represents a solution to the *linearized homogeneous* boundary value problem, obtained from the actual problem by fixing the stiffness moduli ($\mathbf{C} \equiv \mathbf{C}^0$) and by setting $\widetilde{\mathbf{v}}^k$ and $[\dot{P}]$ to zero. An arbitrary multiple of $\widetilde{\mathbf{w}}^*$ can be added to $\widetilde{\mathbf{v}}^0$ to generate another solution to the inhomogeneous but linearized problem (i.e. for $\dot{\Lambda} \neq 0$ but still with $\mathbf{C} \equiv \mathbf{C}^0$). The question arises whether this corresponds to a genuine bifurcation in the actual, incrementally nonlinear solid. The following three kinds of the critical state can be distinguished:

(B1) The condition (9.21) of directional stability of equilibrium still holds, and there exists a constant c *independent of place* such that at each point of the body at which the moduli $\mathbf{C}^0$ associated with $\tilde{\mathbf{v}}^0$ are well defined, we have

$$\mathbf{C}(\dot{\mathbf{F}}) = \mathbf{C}^0 \qquad \text{if } |\dot{\mathbf{F}} - \dot{\mathbf{F}}^0| < c\,. \tag{9.39}$$

Then the velocity field $\tilde{\mathbf{v}}^* = \tilde{\mathbf{v}}^0 + \gamma\tilde{\mathbf{w}}^*$ also corresponds to the moduli field $\mathbf{C}^0$ provided $|\gamma|$ is sufficiently small. It follows (cf. [87, 104]) that $\tilde{\mathbf{v}}^*(\gamma)$ represents an infinite family of velocity solutions to the actual first-order rate boundary value problem. This does not imply that infinitely many post-bifurcation branches are initiated in that way at the single equilibrium state under consideration; we return to this question when discussing post-critical deformations.

(B2) The condition (9.21) still holds but no constant c satisfying (9.39) can be found. This is a rule if the incremental constitutive law is thoroughly nonlinear but is possible also in the classical elastic-plastic solids, cf. [208]. In general, bifurcation in velocities is not possible at the critical instant, but a higher-order bifurcation can be inferred. For instance, the eigenmode $\tilde{\mathbf{w}}^*$ with an arbitrary multiplier may be added to the fundamental solution in quasi-static *accelerations* to generate another solution to the (linear) second-order rate boundary value problem [208]. The actual order of path bifurcation follows from the analysis of a post-bifurcation expansion and may be different from two [259, 178, 183].

(B3) The condition (9.24) for an eigenstate is met so that (9.21) ceases to hold simultaneously with (9.30). This takes place if the 'direction' of the eigenmode $\tilde{\mathbf{w}}^*$ corresponds to the tangent moduli field, in the sense that $\mathbf{C}(\nabla\mathbf{w}^*) = \mathbf{C}^0$ almost everywhere in the body. If $\dot{\Lambda} \neq 0$ then we have bifurcation in velocities. If $\dot{\Lambda} = 0$ then a generalized limit point has been reached (cf. (9.26)), so that the (normalized) fundamental solution $\tilde{\mathbf{v}}^0$ represents itself an eigenmode satisfying both (9.24) and (9.36) at $\lambda = 0$. Other eigenmodes may simultaneously exist which corresponds to bifurcation.

In the non-discretized continuum problem there is a possibility that the critical state is not itself an eigenstate but represents a limit of an infinite spectrum of eigenstates, e.g. [35, 101]. This corresponds to surface of material instabilities, to be considered later.

The discussion of the critical state which corresponds to the onset of geometric instabilities was limited so far to a smooth fundamental path along which the tangent stiffness matrix $[K^0]$ varied continuously. In general, $[K^0]$ can become abruptly indefinite, being not singular in the critical state. This can be caused by a discontinuous drop in the tangent stiffness associated with a non-smooth stress-strain curve. If this does not lead to the violation of (9.21) (cf. the preceding section) then the critical state is a bifurcation point of the type met within the continuous range discussed above. According to the energy criterion (9.34) and identity (9.35), the primary bifurcation mode which initiates a possibly stable post-bifurcation branch can then be determined as the minimizer of $\ddot{E}$ and J in $\mathcal{V}$ [200, 202, 209].

Non-Symmetric Tangent Stiffness Matrix

If the tangent stiffness matrix is not symmetric, e.g. in the classical plasticity without normality, then the first critical state for the functional in (9.30) cannot be identified with the primary bifurcation point [96]. The search for the genuine primary bifurcation state may be difficult and can be replaced by a search for upper and lower bounds to this state [219].

A natural approach is to determine the primary bifurcation point for the incrementally linear solid defined by the actual tangent moduli $\mathbf{C}_0$ along the fundamental path, disregarding in this way possible earlier bifurcations where instantaneous moduli from other constitutive branches could be activated. The critical state and the respective nonzero eigenmode are determined from the condition that

$$K^0_{\alpha\beta} w_\beta = 0 \qquad \text{or} \qquad \det[K^0] = 0\,. \tag{9.40}$$

If (9.39) is satisfied then the bifurcation mode in velocities for the actual elastic-plastic solid can be constructed as a sum of the fundamental solution and a multiple of the eigenmode, analogously to the symmetric case. Since earlier bifurcations have not been excluded, this leads to an upper bound to the primary bifurcation state along the fundamental path.

A lower bound to the onset of bifurcation along the fundamental path with non-symmetric $[K]$ can be found by determining the range in which the general uniqueness condition (9.18) holds. Most work in this direction concerned the classical non-associated plasticity, frequently neglecting the incremental effect of changes in geometry, e.g. [157, 158, 146]. In the geometrically exact formulation, the approach based on the inequality (9.28) has been found effective. It has been proved [218, 219] that (9.28) is satisfied in the classical elastoplasticity without normality if the comparison moduli $\mathbf{C}^L$ in the plastic zone are defined through (9.3) with $\mathbf{L} = \mathbf{L}^L$ of the form

$$\mathbf{L}^L = \mathbf{L}^e - \frac{1}{4gr}(\boldsymbol{\lambda}^p + r\boldsymbol{\lambda}) \otimes (\boldsymbol{\lambda}^p + r\boldsymbol{\lambda})\,, \qquad r > 0. \tag{9.41}$$

The lower bound can then be determined as the critical point at which (9.29) fails (*op. cit.*), in a manner analogous to that discussed the preceding section. However, contrary to the classical plasticity with normality, $\mathbf{C}^L \neq \mathbf{C}^0$ so that there is now a gap between the lower and upper bounds.

There is the possibility that the eigenvalue problem (9.36) for a non-symmetric matrix $[K^0]$ admits complex eigenvalues. If the moduli $\mathbf{C}^0$ were fixed then this could be associated with exponential growth of oscillatory perturbations (flutter), as remarked by Rice [221] in the context of material instability. However, the actual constitutive cones can change during such oscillations, and the physical meaning of 'flutter' instability in plasticity constitutes at present an open problem.

The above discussion was based on the constitutive description in the form (9.1) or (9.2) where stress-rates in a given state are expressed as functions of strain-rates. An alternative theory of uniqueness and stability has been developed by Q.S. Nguyen for materials obeying the principle of maximum dissipation power, and involves the rates of internal variables as unknown quantities. The theory concerned earlier a restricted class of materials with a fixed convex elastic domain in the space of thermodynamic forces [177, 216, 178, 183]. More recently, the theory was extended to a more general case of the elastic domain varying with the inelastic deformation [179, 180], which leads in general to a non-symmetric tangent stiffness matrix. In that approach, plasticity represents one of possible sources of dissipation; friction and fracture may be alternative sources, cf. the survey article [181].

9.4.2 Computational Approaches

If the fundamental path describes uniform deformation of a homogeneous material body of a simple shape then the linear eigenvalue problem (9.36) can in many cases be solved by using analytical tools. Similar situation can be met in problems with radial nonuniformity (tubes, shells).

The separation-of-variables technique, supported by a series expansion of unknown functions, has been used to determine the critical state associated with (9.38) or (9.40) for various materials and boundary data, e.g. [104, 233, 7, 70, 51, 155, 236, 44, 113, 101, 156, 245, 168, 287, 173, 246, 266, 267, 169, 14, 258, 49, 273, 13]. Analytic solutions, if available, are of fundamental importance for understanding the complex phenomena related to plastic instability. In particular, they illustrate the effect of material parameters and of specimen geometry on the onset and form of geometric instability.

From the above list of references it can be seen that one of the thoroughly investigated examples is the linearized bifurcation problem under plane strain. The general methods to solve this problem analytically, with the classification of possible regimes, were developed in [35, 101, 97]. Typical primary eigenstates for the tangent comparison solid on the fundamental quasi-static path lie inside the elliptic regime (changed to hyperbolic in a dynamic problem). The respective eigenmodes are of a sinusoidal form in certain directions, with the wavelength proportional to an overall dimension of a specimen. This corresponds to global geometric instabilities such as buckling or diffuse necking, whose appearance is dependent of the stress state and stiffness moduli and also on geometric factors such as the length-to-thickness ratio. The analysis has been recently extended to include non-local effects by using gradient-dependent plasticity with isotropic hardening [29].

The critical eigenmodes can be of a similar type as above but of a vanishingly short wavelength; this corresponds to a surface (or interface) instability being local in character, i.e. independent on the specimen dimensions. The critical point for the actual finite body represents then not an eigenstate but a point of accumulation of eigenstates [35, 101, 287, 169]. The critical point is recovered as an eigenstate when the surface instabilities are searched for a homogeneous and uniformly stressed infinite half-space. The eigenmode is of the form of a sinusoidal stationary wave which decays exponentially with the increasing distance from the free surface (or interface). The surface or interfacial instabilities were examined for various incrementally linear constitutive relationships, cf. [35, 7, 70, 101, 287, 173, 169, 14, 116, 106, 271, 242, 119, 27, 12, 64, 65, 170, 249, 144]. In the recent paper [144] the conditions for surface instability of flutter type have also been examined. It can be noted that the amplitude of short-wavelength eigenmodes may not decay with depth, e.g. [169]; this is related to the material instability examined in the next section.

In more complex problems, one must be satisfied with numerical solutions. In a problem discretized in space and time, the critical stage (9.40) is to be found numerically with some approximation, by detecting the change in sign of the lowest eigenvalue, determinant or pivots of the (factorized) tangent stiffness matrix. The accuracy can be increased by using special numerical techniques, cf. e.g. [243, 285] and the references therein. Then, the respective (right-hand) eigenvector corresponding to the (approximately) zero eigenvalue of $[K^0]$ can be determined, of course, with accuracy to an arbitrary multiplier. That approach was used e.g. in [164, 165, 166, 46, 54, 256, 120, 257, 253, 121, 37, 243, 12, 273]. The bifurcation mode in the actual elastic-plastic solid is defined, following Hill's theory, as a sum of the fundamental mode and of the eigenvector whose magnitude must be such that the resulting strain-rate lies everywhere within the constitutive cone where $\mathbf{C}^0$ applies, or on the cone boundary. A lower bound to the primary bifurcation state in a classical plastic solid without normality was determined in [43, 271, 120, 121] by using Raniecki's comparison solid (9.41).

Another and frequently used approach to determining approximately the onset of instability of the fundamental path is to perform a numerical step-by-step analysis of the behaviour of the system with a small initial imperfection in the body shape, material properties or loading conditions. If an initially small imperfection of a magnitude unavoidable in practice starts to grow considerably during the deformation process then this may be regarded as the onset of practical instability of the fundamental path. A number of examples of geometric instability at finite strain have been calculated in that way, e.g. [50, 153, 168, 264, 173, 184, 267, 120, 257, 254] and others; further references will be given later when discussing material instability and post-bifurcation behaviour, and earlier work on buckling instability is reviewed in [235, 111, 265]. The critical point is frequently defined as an instant when unloading begins. By using this approach, not only the onset of instability but also the post-critical behaviour (to be discussed below in a separate section), can be simulated by using the same numerical tools as in the absence of any instabilities. The results may depend on the form and magnitude of the imperfection which in many cases have to be assumed arbitrarily.

An alternative way to determine the onset of geometric instability and its sensitivity to imperfections is to study the growth of a small imperfection by using the regular perturbation method applied below the critical load, e.g. [110, 112, 168, 172, 173, 266, 1, 174]. In general, path-dependence of elastic-plastic materials is reflected by the presence of certain integrals in the equation for the displacement perturbation growth. In numerical calculations, the sensitivity can be summed-up incrementally along the fundamental path, cf. e.g. [263, 122, 284]. If the equation for the perturbation growth becomes singular at some level of the loading then the fundamental path beyond this point may be regarded as unstable. Of course, imperfection sensitivity in plasticity is an important topic in itself, also in the absence of any instabilities.

A different approach to determining the onset and mode of plastic instability in discretized problems with symmetric $[K]$ follows from the energy criterion (9.34). A solution to the *nonlinear* system of equations (9.12) for velocities is searched by *minimizing* the value of $\ddot{E}$, and thus of the functional J defined by the expression (9.20), within the set $\mathcal{V}$ of kinematically admissible velocity fields. [9] If (9.23) is met, indicating instability of equilibrium, then J decreases unboundedly [200], and if (9.21) holds then a minimizer can be found, as already discussed. The instant when the minimizer ceases to coincide with the fundamental solution marks the onset of instability of the fundamental path, and the minimizer defines the instability mode, being the bifurcation mode just *beyond* the primary bifurcation point; cf. [209, 210, 211, 212]. This is the crucial step in the computational procedure for crossing bifurcation points with automatic selection of the post-bifurcation branch, to be discussed later.

The next issue to be discussed is the significance of the choice of the *incremental* constitutive law. Let us begin with the classical *rigid-plastic* solids obeying a smooth plastic potential identified with a yield function. The direction of a non-zero strain-rate is then prescribed, which imposes severe restrictions on the admissible velocity fields. Under the conditions leading to spatial or radial uniformity of this direction, analytic expressions for a difference of two admissible velocity fields are available [217, 83, 154]. Such fields can be substituted into the rigid-plastic version of Hill's uniqueness condition [83] [10] to find the critical value of the hardening modulus above which bifurcation in velocities is excluded. [11] Analytic velocity functions can also be used to construct a secondary solution, e.g. [56, 248, 101, 62]. In certain problems, the critical hardening state obtained from Hill's sufficiency condition coincides with the actual primary bifurcation point, cf. e.g. [248, 101], as it does for the Shanley column [238] and other structures where any allowed strain-rate has *a priori* the admissible direction. In other problems, however, bifurcation solutions were not determined for the classical rigid-plastic solid since the severely restricted velocity functions were found to be incompatible with the kinematic boundary conditions, except for the trivial fundamental mode. This reveals an essential shortcoming of using classical plasticity models in bifurcation calculations, at least when elastic compliancies are disregarded.

One might expect at a first glance that the limitation disappears in the classical *elasto*-plasticity since then the direction of the *total* strain-rate is no longer constrained. However, the deviation of the strain-rate direction from that for an admissible rigid-plastic mode must be accommodated by an elastic strain-rate. The

---

[9] This can be contrasted with the usual procedure which reduces to the search of a *stationary* point of J in $\mathcal{V}$, not required to be a minimum point.

[10] The rigid-plastic version can be obtained from the uniqueness condition for the classical elastic-plastic solids in the limit when the elastic moduli are indefinitely increased [85].

[11] Similar procedure can be applied to determine the range of stability of equilibrium by using *the same* condition but with the narrower class of admissible functions that make both the direction and sense of the strain-rate compatible with the flow rule [84, 105].

latter activates the elastic stiffness moduli which for practical problems are much higher than the plastic hardening modulus. If the needed 'elastic' correction is small in comparison with the plastic part of the bifurcation mode, as in the case of uniaxial tension of a cylinder, then acceptable results are obtained [113]. In other cases, the elastic contribution can be substantial, and the calculated critical stresses (e.g. for J_2 flow theory for metals) can be of order of the elastic shear modulus, cf. [44, 156, 173, 116], which is unrealistic. A well known example is the torsional buckling of a cruciform column under axial compression, cf. [111] and the references quoted therein. The respective eigenmode involves only elastic (shear) component tangent to a smooth yield surface, and the critical bifurcation load is precisely the same as for a purely elastic column, contrary to experimental observations.

The above indicated limitations of using classical plasticity models in plastic instability calculations have been known for many years. It is also well known that the limit loads or strains can often be significantly reduced by the presence of *finite* initial imperfections. In the example of cruciform column, the magnitude of the imperfections needed to obtain realistic predictions of the buckling load was found to be extremely small [186, 112]. However, if a new problem is to be solved where the expected instability mode induces a considerable change of the straining direction with respect to the fundamental mode then the required imperfection magnitude is unknown in advance. This is not advantageous from the point of view of a predictive power of the computational approach.

The disagreement between experiments and the bifurcation predictions based on the classical plasticity models is not surprising when confronted with the 'essential structure' of constitutive laws for metal polycrystals, expected on micromechanical grounds [92]. It was indicated above that micromechanical time-independent elastic-plastic models of a polycrystal invariably predict the formation of a vertex on the theoretical yield surface at the current loading point during plastic flow. The tangent moduli tensor $\mathbf{C}^0$ used to construct $[K^0]$ is then substantially different from its classical counterpart, although the load-displacement curves along the fundamental path can be the same. As concluded with reference to the constitutive inequality (9.31), under the normality and symmetry assumptions at the micro-level and along a regular fundamental path, the knowledge of $\mathbf{C}^0$ is sufficient to determine the primary bifurcation point and the onset of path instability, without the need of full specification of the incrementally nonlinear constitutive law [201]. The problem remains how to define the tensor $\mathbf{C}^0$ in a satisfactory manner.

Along a (nearly) proportional loading path of an initially isotropic metal polycrystal, the tangent moduli $\mathbf{C}^0$ are frequently defined as the moduli of J_2 deformation theory of plasticity, extended to finite strain [95, 247, 114] when needed. Arguments have been given that this choice is admissible and may be treated as a reasonable first approximation [16, 17, 229, 45, 124, 108, 53]. Earlier work showed that the buckling-load predictions for plates based on the deformation theory moduli are in better agreement with experiment than those based on J_2 flow theory; cf. [111] and the references therein. Also in other cases, the difference in bifurcation predictions was found to be to the advantage of the deformation theory moduli, e.g. [173, 116, 269, 273], which should be interpreted as the correction due to the vertex effect. To simulate instability phenomena within the classical plasticity framework, sufficiently strong initial imperfections, material softening or deviation from normality were frequently introduced, especially to study the strain localization phenomenon, to be discussed below.

9.5 Onset of Material Instability

The instabilities investigated in the preceding sections concerned a finite solid body and were influenced by the actual boundary conditions. In this section we will focus attention on determining the onset of 'material' instability, understood as 'intrinsic' instability of the material at the level of an infinitesimal element of a continuum. This concept is not so clear as that of a global bifurcation or eigenstate at specified boundary conditions, and was given various interpretations in the literature.

The postulate of material stability 'in the small' in the sense of Drucker [66, 67, 69] is

$$\mathrm{d}\boldsymbol{\sigma} \cdot \mathrm{d}\boldsymbol{\varepsilon} > 0 \qquad \text{or} \qquad \mathrm{d}\boldsymbol{\sigma} \cdot \mathrm{d}\boldsymbol{\varepsilon}^{p} > 0\,, \tag{9.42}$$

where $\mathrm{d}\boldsymbol{\sigma}$ and $\mathrm{d}\boldsymbol{\varepsilon}$ are infinitesimal increments of stress and strain, respectively, in the small strain framework. The meaning of the postulate is not free of ambiguities when changes in geometry are not disregarded [94]. In turn, it is well known that the *incremental* effect of geometry changes is essential in investigations of bifurcations and instability [85], no matter whether the strains are 'small' or not. Nevertheless, (9.42) can still be met in the literature as a basis for material stability studies.

In a geometrically exact formulation of the inequalities analogous to (9.42) but expressed in terms of the rates of work-conjugate variables, their meaning is dependent on an arbitrary choice of the measure of strain (or of stress-rate, in the current state) [94]. This might be not clear at a first glance since such inequalities can be derived from a measure-invariant postulate where the internal work for a virtual increment of uniform deformation is compared with the respective external work. However, calculation of the latter (to second order terms) involves an arbitrary specification of the interaction rule between the configuration of a material element and the tractions on the element boundary applied by its environment [96]. An analogous problem occurs when defining 'intrinsic eigenstates' [98, 99]. Such inequalities and critical states fall in the categories examined in the preceding sections if six dimensionless geometric variables (strain components in particular) that specify the current shape of an embedded material cell during homogeneous straining are taken as generalized coordinates q_α [96]. Variations of $\dot{q}_\alpha$ can then be identified with our w_α, the matrix of generalized moduli can be identified with $[\bar{K}]$, while the matrix $[k]$ characterizes, as before, the stiffness of a loading device. *Both* the matrices change under the transition to another definition of q_α, so that if $[k] = [0]$ in one set of coordinates then $[k]$ is nonzero in another set for *the same* loading device [96]. It follows that the critical states of the material depend either on the chosen set of coordinates if

$[k] = [0]$ is assumed, or alternatively, on the loading conditions used to define $[k]$ in an invariant formulation. [12]

To eliminate the influence of boundary conditions, a material element may be treated, in another scale but under the current uniform stress, as an infinite homogeneous continuum. The loading conditions at infinity are imagined to be such that uniform and constitutively compatible quasi-static rate fields $\dot{\mathbf{F}}^0, \dot{\mathbf{S}}^0$ represent the fundamental solution. Then, non-trivial solutions that correspond to a material instability in the form of bifurcation within a planar band can be investigated [101, 225, 221]. The gradient $\nabla\mathbf{w}$ of admissible variations in velocities is assumed to vanish outside the band whose unit normal in the reference configuration is denoted by $\mathbf{n}$. Within the band, $\mathbf{w}$ is assumed constant on planes $\boldsymbol{\xi} \cdot \mathbf{n} = \text{const}$ parallel to the band boundaries. The traction rates applied along the band boundaries result from the uniform fundamental solution $(\dot{\mathbf{F}}^0, \dot{\mathbf{S}}^0)$. From the spatial continuity of $\widetilde{\mathbf{w}}$ and from the conditions of continuing equilibrium, we obtain the kinematic and static conditions to be satisfied by a pair $(\dot{\mathbf{F}}, \dot{\mathbf{S}})$ within the bifurcation band, e.g. [221], viz.

$$\dot{\mathbf{F}} - \dot{\mathbf{F}}^0 = \mathbf{g} \otimes \mathbf{n}, \qquad (\dot{\mathbf{S}} - \dot{\mathbf{S}}^0)\mathbf{n} = \mathbf{0}\,, \tag{9.43}$$

respectively, where $\mathbf{g} \in \mathcal{R}^3$ is some function of $(\boldsymbol{\xi} \cdot \mathbf{n})$. The existence of a nontrivial solution to (9.43) is referred to as the shear band bifurcation, although, strictly speaking, the terminology concerns the special case when $\mathbf{g} \otimes \mathbf{n}$ represents simple shearing. Without loss of generality (for simple materials), (9.43) can be reformulated at a fixed value of $\boldsymbol{\xi} \cdot \mathbf{n}$ as a particular case of the discretized problem (9.12), viz.

$$\dot{Q}_i^{(\mathbf{n})}(\dot{\mathbf{F}}) = \dot{P}_i^{(\mathbf{n})}\,, \qquad \dot{\mathbf{F}} = \dot{\mathbf{F}}^0 + \mathbf{g} \otimes \mathbf{n}\,, \qquad i = 1, 2, 3 \tag{9.44}$$

with $\mathbf{g}$ as the unknown vector in place of $[v]$, and

$$\dot{Q}_i^{(\mathbf{n})}(\dot{\mathbf{F}}) = \dot{S}_{ij}n_j, \qquad \dot{P}_i^{(\mathbf{n})} = \dot{S}^0_{ij}n_j\,. \tag{9.45}$$

The role of the tangent stiffness matrix $[K]$ plays now the acoustic tensor $\mathbf{K}^{(\mathbf{n})}$ associated with $\mathbf{C}$, with the components

$$K_{ik}^{(\mathbf{n})}(\dot{\mathbf{F}}) = C_{ijkl}(\dot{\mathbf{F}})n_j n_l\,. \tag{9.46}$$

If $\mathbf{C}$ is diagonally symmetric then $\mathbf{K}^{(\mathbf{n})}$ is symmetric, and then (9.44) admits a variational formulation (cf. (9.19) and [204])

$$\frac{\partial J^{(\mathbf{n})}}{\partial g_i} = 0, \qquad J^{(\mathbf{n})}(\mathbf{g}) = U(\dot{\mathbf{F}}^0 + \mathbf{g} \otimes \mathbf{n}) - \dot{P}_i^{(\mathbf{n})} g_i\,. \tag{9.47}$$

It is clear that the theoretical results from the preceding two sections apply also to the present problem. A number of corollaries can be formulated; to avoid repetitions, only some of them are stated explicitly below. In particular, from (9.40) we recover

[12] The special case $[k] = [0]$ is implicitly assumed in (9.42), but for unspecified measures of strain and stress.

the *linearized* bifurcation criterion [101, 225, 221]

$$K_{ij}^{(\mathbf{n})0} g_j = 0 \qquad \text{or} \qquad \det(\mathbf{K}^{(\mathbf{n})0}) = 0\,, \qquad \mathbf{K}^{(\mathbf{n})0} \equiv \mathbf{K}^{(\mathbf{n})}(\dot{\mathbf{F}}^0) \tag{9.48}$$

from which the *onset* of the localization of deformation into a band (or bands) of orientation $\mathbf{n}$ can be calculated, with analogous limitations as previously. If $\mathbf{C}^0$ is varying continuously along the fundamental path then the critical state can be determined by selecting such $\mathbf{n}$ for which (9.48) is met first. This instant corresponds to the loss of ellipticity of the rate equations of continuing equilibrium, linearized at $\dot{\mathbf{F}}^0$ [88].

The conditions (9.43) pertain also to the quantities on both sides of a stationary discontinuity, i.e. of an acceleration wave of vanishing speed of propagation when inertia effects are taken into account [80, 251, 88]. (9.48) represents thus a criterion for the existence of stationary discontinuities in the incrementally linear solid with the moduli $\mathbf{C}^0$ [88]. By Hill's argument from the general bifurcation theory, (9.48) is also a condition for bifurcation within a band, and for a stationary discontinuity, in an incrementally piecewise-linear solid *deforming* with the rate $\dot{\mathbf{F}}^0$. For, $\mathbf{g}$ may be scaled down such that $\dot{\mathbf{F}}$ in (9.43) falls within the constitutive cone where the currently active tangent moduli $\mathbf{C}^0$ apply. [13]

Consider now the case when this constitutive cone coincides with the loading half-space, i.e. the classical elastoplasticity. Circumstances in which (9.48) is met in the classical elastoplasticity were examined, following [101, 225, 221], in a number of papers, e.g. [11, 171, 286, 71, 31, 189, 33, 197, 244], although frequently with neglecting the incremental effect of changes in geometry. Without this effect and for the normality flow rule and $\mathbf{L}^e$ positive definite, (9.48) cannot be satisfied in the range of positive hardening modulus [225, 221], while in a geometrically exact formulation the shear band bifurcation may take place at a small but positive hardening rate [101, 225]. If normality does not apply then strain localization into a shear band may start at a positive hardening modulus, also if the effect of geometry changes is disregarded, cf. [148, 225] and others. Nevertheless, if the plastic part of the fundamental deformation rate does not contain a non-deforming plane then, as explained by Rice [221], the criterion (9.48) for localization is controlled to a certain extent by elastic moduli, which results in a strong softening rate necessary for the fulfillment of (9.48) [225].

If the yield-surface vertex effect is taken into account then the material stiffness moduli involved in (9.48) need no longer be of the order of elastic moduli, and more realistic bifurcation predictions can be expected (cf. the discussion in the preceding section). The shear band bifurcation analyses in [101, 225, 221, 171, 9, 169, 5, 278, 117, 6, 118, 195, 194, 282], among others, give a deeper insight into the role played by various constitutive parameters that influence the moduli $\mathbf{C}^0$ used in the analysis of shear band bifurcation. Frequently, the deformation theory moduli were

[13] In the case of thoroughly nonlinear constitutive rate equations, (9.48) can be interpreted as the condition for a stationary discontinuity in the *second* gradient of velocity, and for the second-order bifurcation in the band in the quasi-static formulation [88, 201].

substituted in (9.46) to determine the onset of material instability on a proportional straining path, e.g. [225, 221, 171, 5, 117, 282].

If $\mathbf{K}^{(\mathbf{n})0}$ is not symmetric then two eigenvalues of $\mathbf{K}^{(\mathbf{n})0}$ may be complex conjugates. For a *linear* constitutive law this leads to 'flutter' instability [221], whose physical significance in elastic-plastic solids, with due account for the possibility of unloading, is yet to be clarified. The circumstances in which flutter instability can or cannot occur have been examined in [145, 143, 190, 4, 142, 34, 30, 144].

The condition (9.48) has been derived under the assumption that the moduli within the bifurcation band coincide with the fundamental moduli $\mathbf{C}^0$. Since the elastoplastic constitutive rate equations are nonlinear (at least, piecewise-linear with two branches), there is in general the possibility that moduli from another constitutive cone are activated at the primary bifurcation point, which may thus precede that found from (9.48). Arguments have been given that this cannot happen in the classical plasticity, without assuming normality [222, 36, 189, 32, 33]. With normality, the conclusion is well known to follow from the inequality (9.28). Of course, these remarks concern the case $\mathbf{L}^0 = \mathbf{L}^p$.

Consider now the effect of vertex formation on the yield surface, as predicted by micromechanical models. If the constitutive inequality (9.31) holds then (9.48) is valid as a critical condition for the material instability associated with the shear band bifurcation, without any assumption about the constitutive branch activated within the band [201, 204]. $\mathbf{C}^0$ exhibits then the properties of the total loading moduli tensor in a micromechanical model, which need not be the case for a phenomenological model. For instance, in the J_2 corner theory of plasticity [53] the current moduli differ from the total loading moduli if the loading path deviates sufficiently from proportional loading, and shear band bifurcations preceding the fulfillment of (9.48) do exist [204, 203]. In this case and in general, determining the onset of the shear band bifurcation in an incrementally nonlinear solid requires a nonlinear analysis with the use of (9.43). Such analysis has been also performed for a class of so-called hypoplastic models of soils [127, 48, 141, 61, 130].

A special but important critical state can be determined from the condition for bifurcation within a band under the restriction that the remaining part of the continuum undergoes only rigid translation [127]. That is, a non-trivial solution to (9.43) is searched after substituting $\dot{\mathbf{F}}^0 = \mathbf{0}$, which leads to the condition (cf. (9.24))

$$K_{ij}^{(\mathbf{n})}(\dot{\mathbf{F}})g_j = 0 \qquad \text{for some } \dot{\mathbf{F}} = \mathbf{g} \otimes \mathbf{n} \neq \mathbf{0}\,. \tag{9.49}$$

On account of the homogeneity of the constitutive relationship, this is equivalent to finding a solution to (9.43) with $|\dot{\mathbf{F}}| \to \infty$ at finite $\dot{\mathbf{F}}^0$. For symmetric $\mathbf{K}^{(\mathbf{n})}$, this critical point may be interpreted as the onset of intrinsic instability of *equilibrium*, related to imaginary speed of wave propagation with respect to the *non-deforming* material, cf. [200, 204]. The critical states determined from (9.48) and (9.49) differ in general since, contrary to the case of classical plasticity, the eigenmode ($\mathbf{g}\otimes\mathbf{n}$) found from (9.48) need not lie, with either sign, in the constitutive domain of applicability of $\mathbf{K}^{(\mathbf{n})0}$. In the classical plasticity, with or without normality, (9.48) implies (9.49), and if the matrix $\mathbf{K}^{(\mathbf{n})0}$ has a negative real eigenvalue then a dynamic instability of a 'divergence' type can be demonstrated [148].

If the constitutive rate equations admit a potential (9.4) then the onset of material instability can be determined from the local form of the energy criterion (9.34) [200, 204], viz.

$$\mathcal{E}(\dot{\mathbf{F}}^0, \dot{\mathbf{F}}) < 0 \qquad \text{for some } \dot{\mathbf{F}} = \dot{\mathbf{F}}^0 + \mathbf{g} \otimes \mathbf{n}\,, \tag{9.50}$$

where $\mathcal{E}$ is the Weierstrass function associated with the constitutive potential $U(\dot{\mathbf{F}})$:

$$\mathcal{E}(\dot{\mathbf{F}}^0, \dot{\mathbf{F}}) = U(\dot{\mathbf{F}}) - U(\dot{\mathbf{F}}^0) - \dot{\mathbf{S}}^0 \cdot (\dot{\mathbf{F}} - \dot{\mathbf{F}}^0), \qquad \dot{\mathbf{S}}^0 = \frac{\partial U}{\partial \dot{\mathbf{F}}}(\dot{\mathbf{F}}^0). \tag{9.51}$$

(9.50) has the interpretation that the incipient localization of deformation within a band of orientation $\mathbf{n}$ is energetically preferable to uniform straining with the velocity gradient $\dot{\mathbf{F}}^0$. Relation of (9.50) to other approaches to material instability has been investigated in [204]. In particular, (9.48) and (9.49) are the critical conditions associated with the *linearized* local condition necessary for path instability:

$$C^0_{ijkl} n_j n_l g_i g_k \geq 0 \qquad \forall\, \mathbf{g}, \mathbf{n} \tag{9.52}$$

and with the local condition necessary for stability of equilibrium:

$$U(\mathbf{g} \otimes \mathbf{n}) \geq 0 \qquad \forall\, \mathbf{g}, \mathbf{n}\,, \tag{9.53}$$

respectively. (9.52) and (9.53) must hold if (9.50) is to be avoided when the modulus $|\mathbf{g}\otimes\mathbf{n}|$ tends to zero or to infinity, respectively [200]. In general, the onset of instability in the energy sense (9.50) along a deformation path can precede the critical states corresponding to (9.52) and (9.53), and can be preceded by shear band bifurcations that are not energetically preferable to uniform straining [204, 203].

If the symmetry condition (9.5) does not hold then the function $U(\dot{\mathbf{F}})$ can still be defined by $(9.4)_3$, although it is no longer a potential in the sense of $(9.4)_1$ and $(9.4)_2$. Then, by appealing to the Graves theorem as in [200], we obtain that (9.53) is necessary for the fulfillment of the energy condition (9.21) for stability of equilibrium of a non-discretized continuum, under any boundary conditions. [14] Similarly, by the classical Legendre-Hadamard condition, (9.52) is necessary for the fulfillment of the linearized uniqueness condition (9.30) for a non-discretized continuum with any boundary data. In turn, from the van Hove theorem [107] it follows that (9.52) is *sufficient* for the positive semi-definiteness of $[\bar{K}]$ for a homogeneous body with the kinematic constraint $\mathbf{w} \equiv \mathbf{0}$ over the whole boundary. The conclusion can be extended to the case when only the normal component of $\mathbf{w}$ vanishes over the surface of a rectangular homogeneous orthotropic specimen [227]. The above results indicate how the conditions for material instability are related to the instability conditions for a finite body.

---

[14] In the classical elastoplasticity, the necessity condition reduces to the semi-strong ellipticity of the moduli $\mathbf{C}$ from both branches; cf. [200, 226].

The conditions for material instability in three dimensions, the condition (9.48) in particular, can be applied to the problem of localized necking in a thin metal sheet treated as a two-dimensional continuum subject to in-plane straining. The mathematical framework remains the same, with the difference that the dimensionality of all vectors and matrices is reduced by one. Certain aspects of 'sheet material' instability become more transparent than in 3D-case. The consequences of using the classical plasticity model are best seen in the case of biaxial stretching of a rigid-plastic perfect sheet. In the absence of directions of zero rate of extension in the sheet plane, the bifurcation criterion gives no prediction for the localized necking instability, contrary to experimental evidence and in spite of the geometric softening due to a decreasing sheet thickness. On the other hand, the instability can take place if such directions exist [81]. If the vertex effect is taken into account then realistic predictions for the necking band bifurcation in a perfect sheet at any stretching ratio become available [247]. The condition (9.48) with the plane stress moduli substituted in place of **C** can be used for this purpose, with the reservations discussed above. Calculations of the respective critical strains have been done, e.g. in [247, 114, 76, 77]. The effect of a finite sheet thickness on localized necking was also investigated, e.g. in [173, 115, 260, 274, 212].

So far we have discussed material instability by assuming a perfectly homogeneous continuum. The analysis of a continuum with initial imperfections represents a widely used alternative. A simple and effective approach (known as MK-approach) to localized necking is to introduce an initial imperfection in the form of a weakened (e.g. thinner) band within which the strain-rate successively increases as the fundamental deformation path is followed outside the band [149, 150]. The critical strains at the forming limit are determined from the condition that the rate of deformation within the band becomes infinite relative to the fundamental strain rate. If the imperfection (in geometry or in material properties) is not too small then realistic critical strains can be predicted (for a power-hardening law) by the classical theory of plasticity also in the case of biaxial stretching. The process of strain concentration can be triggered by an increased curvature of the yield surface in the region traversed by the loading point, cf. [268, 269, 220, 272], and by anisotropic hardening [100, 102]. The results are sensitive to the imperfection magnitude and orientation, and are influenced by other factors; cf. [150, 241, 151, 168, 114, 171, 192, 268, 15, 269, 220, 272, 139] and others.

The analog of MK-approach has been also applied to 3D continua, with the imperfection concerning various material properties [286, 117, 228, 152, 224, 272, 275]. These studies illustrate well the sensitivity of the critical strains reached outside the band to the assumed initial orientation of a weakened band, imperfection magnitude and details of the constitutive description. If the MK-approach is applied to the material obeying the corner theory of plasticity then the limit strains in the sense of MK-approach in general do not tend with vanishing imperfection to those determined from (9.48) for a perfect continuum, cf. [53, 117]. Rather, by analogy to the limit point approach, MK-limit strains correspond to the onset of instability of *equilibrium* within the localization band; cf. [212].

The type of material instability discussed above where velocities and strains are continuous in space and time is, of course, not the only possible one. There are other types of material instability like propagation of Lüders bands or phase transformation phenomena where the assumptions of strain continuity in time and of spatial continuity in velocities are too strong. If slip discontinuities or cavitation instabilities are investigated then also displacements are not continuous and infinite strain jumps are allowed. Such instabilities could not be analyzed within the incremental constitutive framework adopted in this article, and are not discussed here.

9.6 Post-Critical Deformations

The discussion in the preceding sections was limited to the problem of determining the critical state associated with the onset of plastic instability of various types. The related topic, to be discussed from now on, is the post-critical behaviour of the plastic body. The post-critical analysis is connected with additional difficulties when local (material or surface) instabilities, which can occur in an arbitrarily small element of the continuum, come into play. Therefore, in the first subsection we will consider the cases where local instabilities are absent, while the approaches to post-critical strain localization related to material instability are reviewed separately in the next subsection.

9.6.1 Post-Critical Deformations Without Local Instability

The fundamental quasi-static deformation path in the range of unstable equilibrium, typically beyond a (generalized) limit point, may be regarded as unrealizable in a physical system and therefore as being of no practical interest. If no other possibility to continue quasi-static deformation exists (in the absence of bifurcation) then, to simulate the real post-critical behaviour, inertia effects (and perhaps the rate sensitivity and thermal effects) should be incorporated in the analysis; such analysis is not discussed here. If the post-critical fundamental path is still of interest (for instance, it could be realized after a change of the control parameter) then the mixed finite element - Rayleigh Ritz method [264] can be used, or the numerical techniques developed for crossing limit points in elastic solids (cf. e.g. [223, 57, 289] and the survey paper [58]) can be adopted.

In elastic-plastic solids just beyond the primary bifurcation point, stability of equilibrium is frequently preserved along any deformation path, so that quasi-static analysis of post-critical deformations is fully applicable. As already discussed, the fundamental post-critical path may again be regarded as unrealizable in practice, and a secondary path is to be followed. The approach due to Hutchinson [109, 111], applied and developed in other papers, e.g. [167, 276, 172, 266, 174, 259, 178, 183, 136], is to construct a singular asymptotic expansion along a secondary quasi-static post-bifurcation path, starting from the primary bifurcation point. From the infinite family of *velocity* solutions at the primary bifurcation point found according to Hill's theory, higher-order conditions of continuing equilibrium usually eliminate all except three: the fundamental mode and the two associated with the limits of the interval of the eigenmode multiplier such that the corresponding strain-rates lie within the constitutive cone of the tangent moduli $\mathbf{C}^0$. This means that the leading (linear) term in the asymptotic expansion is determined by the requirement that loading (or fully active loading in vertex models) occurs everywhere in the current plastic zone, except at one point or along one curve where neutral loading occurs. A characteristic feature of the post-bifurcation behaviour is the growth of a zone of unloading, either total (classical elastoplasticity) or partial (vertex models), starting from that single point or line. This strongly nonlinear effect leads to fractional power terms in the typical asymptotic expression for the post-critical load [109, 111]. If the admissible interval of the eigenmode multiplier is reduced to a single value (e.g. for a thoroughly nonlinear solid, but possibly also in the classical elastoplasticity) then the primary bifurcation takes place not in velocities but in higher-order rates of displacements, cf. [125, 174, 259, 208]. The leading term in the asymptotic expansion defines the actual order of the bifurcation, cf. [259, 183].

Another possibility to study the post-bifurcation behaviour is, of course, the numerical analysis performed e.g. by using the finite element method. If the primary bifurcation point and eigenmode have been determined then the secondary incremental solution can be calculated by using the above-mentioned neutral loading condition to find the eigenmode multiplier. Once the post-bifurcation branch has been so initiated then it can be followed further, e.g. [164, 165, 46, 256, 121, 37, 243], in many cases without numerical difficulties which could be attributed to plastic instability. This is possible due to the increase in the overall stiffness caused by the above-mentioned growth of the (total or partial) unloading zone, so that the secondary post-bifurcation solution up to a next critical point can exhibit analogous qualitative properties as the fundamental solution before the primary bifurcation.

The above numerical approach, resulting in a natural way from Hill's bifurcation theory, requires implementation of numerical techniques other than those used along a regular path without bifurcations. Moreover, a difficulty is met if the secondary branch emanates at the primary bifurcation point tangentially to the fundamental path (higher-order bifurcation), so that only the trivial (zero) value of the eigenmode multiplier is admissible. The approach fails also when the bifurcation is induced by a discontinuous drop in the incremental stiffness of the material, the tangent stiffness matrix being not singular at any instant.

To overcome those difficulties, a different method for crossing bifurcation points in numerical step-by-step analysis was proposed in [209] for the problems with a symmetric tangent stiffness matrix $[K]$. A solution in velocities to the actual, *nonlinear* rate-problem is determined as a *minimizer* of the functional J defined by (9.20), and not only as a stationarity point of J according to (9.19). This makes no difference in the range where uniqueness is ensured but the distinction becomes essential just after the primary bifurcation point. As soon as the tangent stiffness matrix $[K^0]$ ceases to be positive definite and becomes indefinite, the fundamental solution path is left, in accord with the energy criterion (9.34) of path instability, since $\widetilde{\mathbf{v}}^0$ can then no longer assign to J a minimum value. The secondary solution in velocities is determined if a minimizer of J in $\mathcal{V}$ is found, whose existence is ensured provided the condition (9.21) of directional stability of equilibrium still holds. A secondary solution path is initiated in this way *automatically*, irrespectively of the type of primary bifurcation and without changing the numerical technique used along the whole path. The use of this approach for determining the onset of diffuse necking was illustrated in [209, 212]. The approach is well suited for crossing multiple bifurcation points, cf. [211].

The approach which is commonly in use is to analyze numerically a system with a small initial imperfection such that the secondary post-bifurcation path is approximated by a perturbed path which does not exhibit bifurcations. Geometric inaccuracies, material inhomogeneities or additional perturbing forces can be used for that purpose. It was demonstrated on examples (cf. [110, 111, 167, 276, 266, 269, 174, 257, 243]) that with a decreasing amplitude of the initial imperfection, the post-critical path tends towards the secondary post-bifurcation path emanating at the primary bifurcation point for a body without imperfections. However, no general and rigorous proof of such a convergence has been published yet, to the writer's knowledge. An unconditional convergence is not expected for such incrementally nonlinear constitutive laws where a primary bifurcation may occur in the range of a positive definite tangent stiffness matrix corresponding to the fundamental velocity field. The initial imperfection approach has also been used without intention to follow as closely as possible the post-bifurcation path for a perfect system. As already discussed, when the classical elastoplastic flow rule is used then unrealistically high values of critical strains or loads at the primary bifurcation point can be predicted. In such cases, the imperfections can be introduced to compensate the high stiffness of the material model against an abrupt change of the straining direction.

9.6.2 Plastic Strain Localization Due to Material Instability

The localization of plastic deformation has been intensively investigated during the last two decades. It is not the purpose here to discuss the localization phenomenon as such. We will review only those of various aspects of this important phenomenon which are most closely related to the material instability in time-independent solids. It is recalled that localization phenomena where dynamic loading, viscous or thermal effects are essential, as e.g. the formation of adiabatic shear bands, are beyond the scope of this article. Situations where the concentration of plastic strain is due to particular boundary conditions or to strong initial nonuniformities, and not to the material properties themselves, do not also fall in the category examined here. Our attention will be focused on the localization of plastic deformation into narrow bands as a symptom or consequence of the loss of intrinsic stability of the material, in the sense discussed in the preceding section. However, we will consider now a finite material body with specified boundary conditions that vary quasi-statically in time.

The shear band bifurcation mode appropriate for an infinite continuum need not be compatible with the boundary conditions at the places where the band intersects the body surface. If it is not compatible and the exact solution for the non-discretized continuum body is searched then the band at the moment of bifurcation must be vanishingly narrow [101]. Alternatively, the shear band bifurcation mode can be modified to fit the boundary conditions exactly but at the cost of an arbitrarily small error in the fulfillment of the interior conditions of continuing equilibrium [204]. In this context, it may be remarked that the instability of a deformation process need not be identified with the occurrence of exact bifurcations. In numerical computations, the compatibility conditions at the boundary, cf. e.g. [28, 170], may lead to the kinks in the shear band near the body surface, e.g. [138]. A band may also be reflected from a boundary or symmetry plane, e.g. [271]. If the material stiffness decreases gradually during numerical calculations then the emergence of shear bands is frequently preceded by a diffuse instability and a surface instability, e.g. [277, 131, 262, 271, 255, 209]. The resulting non-uniformities, similarly as initial imperfections, can grow in certain directions, leading to the formation of shear bands at places of more intense straining. In a discretized problem, the strain localization into shear bands can thus start without bifurcation. The *onset* of localization and the band orientation can be predicted by the material stability analysis based on the condition (9.48). This information can help to select the finite element mesh (consisting e.g. of 'crossed-triangles') such that it can accommodate the shear band formation, cf. [277, 131, 195, 271, 175]. Appropriate remeshing [134, 188, 193, 196], the use of enhanced elements [187, 24, 73, 163] or spectral overlay methods [23, 26] can increase accuracy of the calculations in the presence of narrow regions of high strain. However, there is the fundamental difficulty: when the material model does not imply an internal length scale then the *final* width of the localization zone can be arbitrarily small in the continuum description, and is set by the mesh size in finite element calculations, cf. [277, 195, 270, 175, 37]. To eliminate the mesh-size dependence, a localization band of a width that is regarded as a material property may be embedded in the finite elements [213]. Recently, another approach is being developed where finite elements are enriched to capture strong discontinuities in displacements [133, 240, 8, 132, 185].

If the classical elastoplastic models (usually with strain softening) are used to simulate the localization process then (9.48) and (9.49) are met simultaneously, i.e. the loss of ellipticity of the fundamental moduli $\mathbf{C}^0$ can be identified with the onset of local instability of *equilibrium* along the fundamental path. A quasi-static analysis beyond that stage is connected with entering the non-elliptic regime and with considerable numerical difficulties. The situation changes if the localization is calculated by using a corner theory of plasticity. As shown recently in [212, 211] on the example of biaxial stretching of a sheet material of finite in-plane dimensions, the post-critical deformation path can exist which does not penetrate the inside of the non-elliptic range. Rather, on account of the vertex stiffening associated with the change of the straining direction, the path can return to the elliptic range or move along its boundary. Such paths have been calculated by rejecting other deformation modes with the help of the energy criterion (9.34) or (9.50). The respective localization zone has turned out to have a well-defined *finite incipient width* [212], although it may consist of many (or even infinitely many) bands of indeterminate individual widths [211]. The total width was shown to tend during further deformation theoretically to zero as the limit of stability of equilibrium was approached. The fundamental difficulty connected with the mesh-dependence of the *final* width of a localization zone persists, but may perhaps be viewed from another perspective. This issue requires further study.

As stated in the introductory sections, this survey is in principle restricted to time-independent *simple* materials. In the last decade, regularization methods that incorporate the effect of a material length scale have been developed for a wide spectrum of material behaviour, including elasticity, plasticity or damage in geomaterials or metals. The modified constitutive laws allow strain localization to develop within a band of a finite width, and without loss of ellipticity of the governing equations. One way of regularization is to abandon the assumption of a simple material by using a *non-local* constitutive model. This can be done by using a volume integral representation of material properties, e.g. [19, 214, 21, 22, 215, 135, 283], or by introducing higher-order gradients terms in the constitutive description, e.g. [2, 55, 232, 261, 3, 288, 25, 231, 161, 280, 39, 41, 281, 137, 29]. In applications, the non-local formulation has been usually restricted to a part of the constitutive description. Another way to introduce a characteristic length at the material level is to use a *polar* (or Cosserat) *continuum*, e.g. [159, 162, 279, 38, 42, 78, 63, 191, 40, 74, 250, 196]. A more detailed discussion of these approaches, as well as e.g. of viscous regularization in dynamic problems, is beyond the scope of this article.

9.7 Concluding Remarks

Plastic instability is a topic which has been vital through decades and is currently being examined in many directions. In this survey article, attention has been focused on general criteria of instability in time-independent elastic-plastic solids, and on computational approaches based on such criteria. To place various existing methods of attacking diverse problems of plastic instability within a possibly unified framework, the concept of stability of a quasi-static deformation process (path) has been found useful. A notable feature of instability in time-independent plasticity is that vanishingly small perturbations of the conditions in which a fundamental process takes place can cause a finite deviation from it in a *finite* interval of the loading parameter. This must be distinguished from dynamic instability of equilibrium where the deviation takes place at a fixed value of the loading parameter and requires, in the limit, an *infinite* interval of a natural time. We have seen that different approaches to path instability correspond to different understanding of 'vanishingly small perturbations'. In the bifurcation approach no perturbations are assumed, and the deviation is possible due to the nonuniqueness of the incremental solution. In the energy approach, the additional amount of energy supplied by a disturbing agency to the system is taken to define the disturbance strength, under the assumptions leading to a symmetric tangent stiffness matrix. In the initial imperfection approach, material properties or shape of the deformed body are perturbed, and the use of finite perturbations is analogous to examining so-called practical stability. The conditions have been discussed under which the critical points determined by different approaches coincide and when they are distinct.

The calculated onset of plastic instability can be strongly influenced by details of the constitutive description, also by those which are irrelevant along the fundamental path. In particular, the distinction between classical and corner theories of plasticity is fundamental; the non-normality effect within the classical plasticity framework can also be essential. This has been illustrated in the reviewed literature on numerous examples of geometric or material instability. For simulation of post-critical behaviour in the presence of material instabilities, the use of material models that introduce an internal length scale is of primary importance, although the effect of incremental nonlinearity also deserves further study. Micromechanical analysis of plastic deformation can be helpful in determining the structure of constitutive relations.

Finally, it is recalled that although a wide class of plastic instability phenomena can be investigated by using the approaches reviewed here, the constitutive assumptions (excluding viscoplastic and thermoplastic instabilities) and continuity restrictions (excluding phase transformation, slip discontinuity and cavitation instabilities) result in a limited scope of this survey.

References

1. R. Abeyratne and N. Triantafyllidis, The emergence of shear bands in plane strain, *Int. J. Solids Structures* 17, 1113-1134, 1981.
2. E.C. Aifantis, On the microstructural origin of certain inelastic models. *ASME J. Engng. Mat. Techn.* 106, 326-330, 1984.
3. E.C. Aifantis, The physics of plastic deformation. *Int. J. Plasticity* 3, 211-247, 1987.
4. L. An and D. Schaeffer, The flutter instability in granular flow. *J. Mech. Phys. Solids* 40, 683-698, 1992.
5. L. Anand and W.A. Spitzig, Initiation of localized shear bands in plane strain. *J. Mech. Phys. Solids* 28, 113-128, 1980.
6. L. Anand and W.A. Spitzig, Shear band orientations in plane strain. *Acta metall.* 30, 553-561, 1982.
7. S.T. Ariaratnam and R.N. Dubey, Some cases of bifurcation in elastic-plastic solids in plane strain. *Q. appl. Math.* 3, 349-358, 1969.
8. F. Armero, K. Garikipati, Recent advances in the analysis and numerical simulation of strain localization in inelastic solids. In: *Computational Plasticity: Fundamentals and Applications* (D.R.J. Owen and E. Oñate, Eds.) Vol. I, pp. 547-573, Pineridge Press, Swansea 1995.
9. R.J. Asaro, Geometrical effects in the inhomogeneous deformation of ductile single crystals. *Acta metall.* 27, 445-453, 1979.
10. R.J. Asaro, Micromechanics of crystals and polycrystals. *Adv. Appl. Mech.* 23, pp. 1-115, Acad. Press, New York 1983.
11. R.J. Asaro and J.R. Rice, Strain localization in ductile single crystals. *J. Mech. Phys. Solids* 25, 309-338, 1977.
12. J.P. Bardet, Finite element analysis of surface instability in hypoelastic solids. *Comp. Meth. Appl. Mech. Engng.* 78, 273-296, 1990.

13. J.P. Bardet, Analytical solutions for plane strain bifurcation of compressible solids, *ASME J. Appl. Mech.* 58, 651-657, 1991.
14. J.L. Bassani, D. Durban and J.W. Hutchinson, Bifurcations at a spherical hole in an infinite elastoplastic medium. *Math. Proc. Camb. Phil. Soc.* 87, 339-356, 1980.
15. J.L. Bassani, J.W. Hutchinson, K.W. Neale, On the prediction of necking in anisotropic sheets. In: *Metal Forming Plasticity* (H. Lippmann, Ed.), pp. 1-13, Springer-Verlag, Berlin Heidelberg 1979.
16. S.B. Batdorf, Theories of plastic buckling. J. Aeronaut. Sci. 16, 405-408, 1949.
17. S.B. Batdorf and B. Budiansky, A mathematical theory of plasticity based on the concept of slip. *NACA TN* No. 1871, 1949.
18. Z.P. Bažant, Stable states and paths of structures with plasticity or damage. *ASCE J. Engng. Mech.* 114, 2013-2034, 1988.
19. Z.P. Bažant, T. Belytschko and T.P. Chang, Continuum theory for strain softening. *J. Eng. Mech. ASCE* 110, 1666-1692, 1984.
20. Z.P. Bažant and L. Cedolin, *Stability of Structures: Elastic, Inelastic, Fracture and Damage Theories*, Oxford Univ. Press, New York 1991.
21. Z.P. Bažant and F.-B. Lin, Non-local yield limit degradation. *Int. J. Num. Meth. Engng.* 26, 1805-1823, 1988.
22. Z.P. Bažant and G. Pijaudier-Cabot, Nonlocal continuum damage, localization instability and convergence. *ASME J. Appl. Mech.* 55, 287-293, 1988.
23. T. Belytschko, J. Fish and A. Bayliss, The spectral overlay on finite elements for problems with high gradients. *Comp. Meth. Appl. Mech. Engng.* 81, 71-89, 1990.
24. T. Belytschko, J. Fish and B. Engelmann, A finite element with embedded localization zones. *Comp. Meth. Appl. Mech. Engng.* 70, 59-89, 1988.
25. T. Belytschko, D. Lasry, A study of localization limitiers for strain softening in statics and dynamics, *Comput. Struct.* 33, 707-715, 1989.
26. T. Belytschko and Y.Y. Lu, A curvilinear spectral overlay method for high gradient problems. *Comp. Meth. Appl. Mech. Engng.* 95, 383-396, 1992.
27. A. Benallal, R. Billardon and G. Geymonat, Conditions de bifurcation à l'intérieur et aux frontières pour une classe de matériaux non-standards. *C. R. Acad. Sci. Paris, Série II, t.* 308, 893-898, 1989.
28. A. Benallal, R. Billardon and G. Geymonat, Phénomènes de localisation à la frontière d'un solide. *C. R. Acad. Sci. Paris, Série II, t.* 310, 679-684, 1990.
29. A. Benallal and V. Tvergaard, Nonlocal continuum effects on bifurcation in the plane strain tension-compression test. *J. Mech. Phys. Solids* 43, 741-770, 1995.
30. D. Bigoni, On flutter instability in elastoplastic constitutive models. *Int. J. Solids Structures* 32, 3167-3189, 1995.
31. D. Bigoni and T. Hueckel, Uniqueness and localization - Part I and II. *Int. J. Solids Structures* 28, 197-213, 215-224, 1991.
32. D. Bigoni and D. Zaccaria, Loss of strong ellipticity in non-associative elastoplasticity. *J. Mech. Phys. Solids* 40 1313-1331, 1992.
33. D. Bigoni and D. Zaccaria, Strong ellipticity of comparison solids in elastoplasticity with volumetric non-associativity. *Int. J. Solids Structures* 29, 2123-2136, 1992.
34. D. Bigoni and D.Zaccaria, On the eigenvalues of the acoustic tensor in elastoplasticity. *Eur. J. Mech. A//Solids* 13, 621-638, 1994.
35. M.A. Biot, *Mechanics of Incremental Deformations*. John Wiley & Sons, New York, 1965.
36. G. Borre and G. Maier, On linear versus nonlinear flow rules in strain localization analysis, *Meccanica* 24, 36-41, 1989.
37. R. de Borst, Bifurcations in finite element models with a non-associated flow law. *Int. J. Num. Anal. Methods Geomech.* 12, 99-116, 1988.
38. R. de Borst, Simulation of strain localization: a reappraisal of Cosserat continuum. *Eng. Comput.* 8, 317-332, 1991.
39. R. de Borst, Gradient-dependent plasticity: formulation and algorithmic aspects. *Int. J. Num. Meth. Engng.* 35, 521-539, 1992.

40. R. de Borst, A generalization of J_2-flow theory for polar continua. *Comp. Meth. Appl. Mech. Engng.* 103, 347-362, 1993.
41. R. de Borst and H.-B. Mühlhaus, Gradient-dependent plasticity: formulation and algorithmic aspects. *Int. J. Num. Meth. Engng.* 35, 521-539, 1992.
42. R. de Borst and L.J. Sluys, Localisation in a Cosserat continuum under static and dynamic loading conditions. *Comp. Meth. Appl. Mech. Engng.* 90, 805-827, 1991.
43. O.T. Bruhns, Bounds to the critical stresses in bifurcation of cylindrical specimens in the case on non-associated flow laws. *Stability in the Mechanics of Continua* (F.H. Schroeder, Ed.), pp. 46-56, Springer, Berlin 1982.
44. O. Bruhns and K. Thermann, Stabilitätsprobleme elastoplastischer Kontinua. ZAMM 54, T73-T75, 1974.
45. B. Budiansky, A reassessment of deformation theories of plasticity. *ASME J. Appl. Mech.* 26, 259-264, 1959.
46. M.A. Burke and W.D. Nix, A numerical study of necking in the plane tension test. *Int. J. Solids Structures* 15, 379-393, 1979.
47. R. Chambon, Une classe de lois de comportement incrémantalement nonlinéaires pou les sols non visqueux - résolution de quelques problèmes de cohérence. *C. R. Acad. Sci. Paris, Série II, t.* 308, 1571-1576, 1989.
48. R. Chambon and J. Desrues, Bifurcation par localization et non linéarité incrémentale: un example heuristique d'analyse complète. *Plastic Instability, Proc. Considère Memorial*, pp. 101-113, Ecole Nat. Ponts Chauss. Press, Paris 1985.
49. K.-T. Chau and J.W. Rudnicki, Bifurcations of compressible pressure-sensitive materials in plane strain tension and compression. *J. Mech. Phys. Solids* 38, 875-898, 1990.
50. W.H. Chen, Necking of a bar, *Int. J. Solids Structures* 7, 685-717, 1971.
51. S.Y. Cheng, S.T. Ariaratnam and R.N. Dubey, Axisymmetric bifurcation in an elastic-plastic cylinder under axial load and lateral hydrostatic pressure. *Quart. Appl. Math.* 29, 41-51, 1971.
52. Y.S. Cheng and W.D. Lu, Uniqueness and bifurcation in elastic-plastic solids. *Int. J. Solids Structures* 30, 3073-3084, 1993.
53. J. Christoffersen and J.W. Hutchinson, A class of phenomenological corner theories of plasticity. *J. Mech. Phys. Solids* 27, 465-487, 1979.
54. C.-C. Chu, Bifurcation of elastic-plastic circular cylindrical shells under internal pressure. *ASME J. Appl. Mech.* 46, 889-894, 1979.
55. B.D. Coleman and M.L. Hodgon, On shear bands in ductile materials. *Arch. Rat. Mech. Anal.* 90, 219-247, 1985.
56. G.R. Cowper, E.T. Onat, The initiation of necking and buckling in plane plastic flow. *Proc. 4th U.S. Nat. Congr. Appl. Mech.*, 2, pp. 1023-1029, ASME, New York 1962.
57. M.A. Crisfield, A fast incremental//iterative solution procedure that handles 'snap-through', *Computers & Structures* 13, 55-62, 1981.
58. M.A. Crisfield and J. Shi, A review of solution procedures and path-following techniques in relation to the non-linear finite element analysis of structures. In: *Nonlinear Computational Mechanics* (P. Wriggers and W. Wagner, Eds.), pp. 47-68, Springer, Berlin 1991.
59. F. Darve, E. Flavigne and E. Rojas, A class of incrementally non-linear constitutive relations and applications to clays. *Comp. Geotechnics* 2, 43-66, 1986.
60. F. Darve, S. Labanieh, Incremental constitutive law for sands and clays: simulations of monotonic and cyclic tests. *Int. J. Numer. Anal. Methods Geomech.* 6, 243-275, 1982.
61. J. Desrues and R. Chambon, Shear band analysis for granular materials: the question of incremental nonlinearity. *Ing.-Archiv* 59, 187-196, 1989.
62. L. Dietrich, Analysis of necking in cylindrical bar of hardening materials. *Arch. Mech.* 32, 389-402, 1980.
63. A. Dietsche, P. Steinmann and K. Willam, Micropolar elastoplasticity and its role in localization analysis. *Int. J. Plasticity* 8, 361-384, 1992.
64. M. A. Dowaikh and R.W. Ogden, On surface waves and deformations in a pre-stressed incompressible elastic solid. *IMA J. Appl. Math.* 44, 261-284, 1990.
65. M. A. Dowaikh and R.W. Ogden, Interfacial waves and deformations in pre-stressed elastic media. *Proc. R. Soc. London* A 433, 313-328, 1991.

66. D.C. Drucker, Some implications of work hardening and ideal plasticity. *Q. Appl. Math.* 7, 411-418, 1950.
67. D.C. Drucker, A more fundamental approach to plastic stress-strain relations. *Proc.* 1-*st U.S. Nat. Congr. Appl. Mech.*, pp. 487-491, ASME 1951.
68. D.C. Drucker, A definition of stable inelastic material. *J. Appl. Mech.* 26, 101-106, 1959.
69. D.C. Drucker, On the postulate of stability of material in the mechanics of continua. *J. Méc.* 3, 235-249, 1964.
70. R.N. Dubey and S.T. Ariaratnam, Bifurcation in elastic-plastic solids in plane stress. *Q. appl. Math.* 3, 381-390, 1969.
71. M.K.Duszek and P.Perzyna, Influence of kinematic hardening on plastic flow localization in damaged solids. *Arch. Mech.* 40, 595-609, 1988.
72. B. Fedelich and A. Ehrlacher, Sur un principe de minimum concernant des matériaux à comportement indépendant du temps physique. *C. R. Acad. Sci. Paris, Série II, t.* 308, 1391-1394, 1989.
73. J. Fish and T. Belytschko, Elements with embedded localization zones for large deformation problems. *Comput. Struct.* 30, 247-256, 1988.
74. N.A. Fleck and J.W. Hutchinson, A phenomenological theory for strain gradient effects in plasticity. *J. Mech. Phys. Solids* 41, 1825-1857, 1993.
75. M. Gotoh, A class of plastic constitutive equation with vertex effect - I, II. *Int. J. Solids Structures* 21, 1101-1116, 1117-1129, 1985.
76. M. Gotoh, A class of plastic constitutive equation with vertex effect - III. Applications to calculation of FLD of metal sheets. *Int. J. Solids Structures* 21, 1131-1145, 1985.
77. M. Gotoh, A class of plastic constitutive equation with vertex effect - IV. Applications to prediction of forming limit strains of metal sheets under nonproportional loadings. *Int. J. Solids Structures* 21, 1149-1163, 1985.
78. G. Gudehus, J. Tejchman, Some mechanisms of a granular mass in a silo-model tests and a numerical Cosserat approach, *Advances in Continuum Mechanics*, pp. 178-193, Springer, Berlin 1991.
79. A.L. Gurson, Continuum theory of ductile rupture by void nucleation and growth, Part I: Yield criteria and flow rules for porous ductile materials. *ASME J. Eng. Mat. Techn.* 99, 2-15, 1977.
80. J. Hadamard, *Leçons sur la Propagation des Ondes et les Equations de l'Hydro-dynamique.* Hermann, Paris 1903.
81. R. Hill, On discontinuous plastic states, with special reference to localized necking in thin sheets. *J. Mech. Phys. Solids* 1, 19-30, 1952.
82. R. Hill, New horizons in the mechanics of solids. *J. Mech. Phys. Solids*, 5, 66-74, 1956.
83. R. Hill, On the problem of uniqueness in the theory of a rigid-plastic solid. *J. Mech. Phys. Solids* 5, 153-161, 302-307, 1957.
84. R. Hill, Stability of rigid-plastic solids. *J. Mech. Phys. Solids* 6, 1-8, 1957.
85. R. Hill, A general theory of uniqueness and stability in elastic-plastic solids. *J. Mech. Phys. Solids* 6, 236-249, 1958.
86. R. Hill, Some basic principles in the mechanics of solids without a natural time. *J. Mech. Phys. Solids*, 7 , 209-225, 1959.
87. R. Hill, Bifurcation and uniqueness in non-linear mechanics of continua. In: *Problems of Continuum Mechanics, N. I. Muskhelishvili Anniversary Volume*, pp. 155-164, SIAM, Philadelphia 1961.
88. R. Hill, Acceleration waves in solids. *J. Mech. Phys. Solids* 10, 1-16, 1962.
89. R. Hill, Uniqueness criteria and extremum principles in self-adjoint problems of continuum mechanics, *J. Mech. Phys. Solids* 10, 185-194, 1962.
90. R. Hill, Generalized constitutive relations for incremental deformation of metal crystals by multislip. *J. Mech. Phys. Solids* 14, 95-102, 1966.
91. R. Hill, On the classical constitutive laws for elastic//plastic solids. In: *Recent Progress in Applied Mechanics, The Folke Odkvist Volume* (B. Broberg *at. al.*, Eds.), pp. 241-249, Almqvist & Wiksell, Stockholm 1967.

92. R. Hill, The essential structure of constitutive laws for metal composites and polycrystals. *J. Mech. Phys. Solids* 15, 79-95, 1967.
93. R. Hill, Eigenmodal deformations in elastic//plastic continua. *J. Mech. Phys. Soids* 15, 371-386, 1967.
94. R. Hill, On consitutive inequalities for simple materials. *J. Mech. Phys. Solids* 16, 229-242, 315-322, 1968.
95. R. Hill, Constitutive inequalities for isotropic elastic solids under finite strain. *Proc. Roy. Soc. Lond.* A 314, 457-472, 1970.
96. R. Hill, Aspects of invariance in solids mechanics. *Advances in Applied Mechanics, Vol.* 18, pp. 1-75, Acad. Press, New York 1978.
97. R. Hill, On the theory of plane strain in finitely deformed compressible materials. *Math. Proc. Camb. Phil. Soc.* 86, 161-178, 1979.
98. R. Hill, On intrinsic eigenstates in plasticity with generalized variables. *Math. Proc. Camb. Phil. Soc.* 93, 177-189, 1983.
99. R. Hill, Path sensitivity of material response at intrinsic eigenstates in classical plasticity. *Math. Proc. Camb. Phil. Soc.* 103, 371-381, 1988.
100. R. Hill, A theoretical perspective on in-plane forming of sheet metal. *J. Mech. Phys. Solids* 39, 295-307, 1991.
101. R. Hill and J.W. Hutchinson, Bifurcation phenomena in the plane tension test. *J. Mech. Phys. Solids* 23, 239-264, 1975.
102. R. Hill and J.W. Hutchinson, Differential hardening in sheet metal under biaxial loading: a theoretical framework. *ASME J. Appl. Mech.* 59, S1-S9, 1992.
103. R. Hill and J.R. Rice, Constitutive analysis of elastic- plastic crystals at arbitrary strain. *J. Mech. Phys. Solids* 20, 401-413, 1972.
104. R. Hill and M.J. Sewell, A general theory of inelastic column failure - I. *J. Mech. Phys. Solids* 8, 105-111, 1960.
105. R. Hill and M.J. Sewell, A general theory of inelastic column failure - II. *J. Mech. Phys. Solids* 8, 112-118, 1960.
106. H. Horii and S. Nemat-Nasser, Instability of a half-space with frictional materials, ZAMP 33, 1-16, 1982.
107. L. van Hove, Sur l'extension de la condition de Legendre du calcul des variations aux intégrales multiples à plusieurs fonctions inconnues. *Proc. Kon. Ned. Acad. Wet.* 50, 18-23, 1947.
108. J.W. Hutchinson, Elastic-plastic behaviour of polycrystalline metals and composites. *Proc. Roy. Soc. Lond.* A 319, 247-272, 1970.
109. J.W. Hutchinson, Post-bifurcation behavior in the plastic range. *J. Mech. Phys. Solids* 21, 163-190, 1973.
110. J.W. Hutchinson, Imperfection sensitivity in the plastic range *J. Mech. Phys. Solids* 21, 191-204, 1973.
111. J.W. Hutchinson, Plastic buckling. *Advances in Applied Mechanics, Vol.* 14, pp. 67-144, Acad. Press, New York 1974.
112. J.W. Hutchinson and B. Budiansky, Analytical and numerical study of the effects of initial imperfections on the inelastic buckling of a cruciform column. In: *Buckling of Structures* (B. Budiansky, Ed.), pp.98-105, Springer, Berlin 1976.
113. J.W. Hutchinson and J.P. Miles, Bifurcation analysis of the onset of necking in an elastic//plastic cylinder under uniaxial tension. *J. Mech. Phys. Solids* 22, 61-71, 1974.
114. J.W. Hutchinson and K.W. Neale, Sheet necking - II: Time-independent behavior. *Mechanics of Sheet Metal Forming* (D.P. Koistinen and N.M. Wang, Eds.), pp. 127-150. Plenum, New York - London 1978.
115. J.W. Hutchinson, K.W. Neale and A. Needleman, Sheet necking - I. Validity of plane stress assumptions of the long-wavelength approximation. *Mechanics of Sheet Metal Forming* (D.P. Koistinen and N.M. Wang, Eds.), pp. 111-126. Plenum, New York - London 1978.
116. J.W. Hutchinson and V. Tvergaard, Surface instabilities on statically strained plastic solids. *Int. J. Mech. Sci.* 22, 339-354, 1980.

117. J.W. Hutchinson and V. Tvergaard, Shear band formation in plane strain. *Int. J. Solids Structures* 17, 451-470, 1981.
118. T. Iwakuma and S. Nemat-Nasser, An analytical estimate of shear band initiation in a necked bar. *Int. J. Solids Structures* 18, 69-83, 1982.
119. H. Kitagawa and H. Matsushita, Flow localization in elastic-plastic material developing from stress-free surface. *Int. J. Solids Structures* 23, 351-368, 1987.
120. M. Kleiber, Numerical study of necking-type bifurcations in void-containing elastic-plastic materials. *Int. J. Solids Structures* 20, 191-210, 1984.
121. M. Kleiber, On plastic localization and failure in plane strain and round void containing tensile bars. *Int. J. Plasticity* 2, 205-221, 1986.
122. M. Kleiber, Shape and non-shape structural sensitivity analysis for problems with any material and kinematic non-linearity. *Comp. Meth. Appl. Mech. Engng.* 108, 73-97, 1993.
123. V.D. Klushnikov, On a possible way of constructing the plastic deformation laws (in Russian). *Prikl. Mat. Mekh.* 23, 282-291, 1959.
124. V.D. Klushnikov, New concepts in plasticity and the deformation theory (in Russian). *Prikl. Mat. Mekh.* 23, 722-731, 1959.
125. V.D. Klushnikov, *Stability of Elastic-Plastic Systems* (in Russian), Nauka, Moscow 1980.
126. W.T. Koiter, Stress-strain relations, uniqueness and variational theorems for elastic-plastic materials with a singular yield surface. *Quart. Appl. Math.* 11, 350-353, 1953.
127. D. Kolymbas, Bifurcation analysis for sand samples with a non-linear constitutive equation. *Ing.-Archiv* 50, 131-140, 1981.
128. D. Kolymbas, A novel constitutive law for soils, In: *Constitutive Laws for Engineering Materials* (S.C. Desai *et. al.*, Eds.) pp. 319-326, Elsevier, New York 1987.
129. D. Kolymbas, An outline of hypoplasticity. *Arch. Appl. Mech.* 61, 143-151, 1991.
130. D. Kolymbas and G. Rombach, Shear band formation in generalized hypoplasticity. Ing.-Archiv 59, 177-186, 1989.
131. M. Larsson, A. Needleman, V. Tvergaard, B. Storakers, Instability and failure of internally pressurized ductile metal cylinders. *J. Mech. Phys. Solids* 30, 121-154, 1982.
132. R. Larsson, K. Runesson and M. Akesson, Embedded localization band based on regularized strong discontinuity. In: *Computational Plasticity: Fundamentals and Applications* (D.R.J. Owen and E. Oñate, Eds.) Vol. I, pp. 599-610, Pineridge Press, Swansea 1995.
133. R. Larsson, K. Runesson and N.S. Ottosen, Discontinuous displacement approximation for capturing plastic localization. *Int. J. Num. Meth. Engng.* 36, 2087-2105, 1993.
134. R. Larsson, K. Runesson and S. Sture, Numerical simulation of localized plastic deformation. *Arch. Appl. Mech.* 66, 305-317, 1991.
135. J.B. Leblond, G. Perrin and J. Devaux, Bifurcation effects in ductile metals with nonlocal damage. *ASME J. Appl. Mech.* 61, 236-242, 1994.
136. A. Leger and M. Potier-Ferry, Elastic-plastic post-buckling from a heterogeneous state. *J. Mech. Phys. Solids* 41, 783-807, 1993.
137. Y.M. Leroy and A. Molinari, Spatial patterns and size effects in shear zones: a hyperelastic model with higher-order gradients. *J. Mech. Phys. Solids* 41, 631-663, 1993.
138. Y. Leroy and M. Ortiz, Finite element analysis of strain localization in frictional solids. *Int. J. Numer Anal. Methods Geomech.* 14, 93-124, 1990.
139. J. Lian, F. Barlat and B. Baudelet, Plastic behaviour and stretchability of sheet metals. Part II: Effect of yield surface shape on sheet forming limit. *Int. J. Plasticity* 5, 131-147, 1989.
140. T.H. Lin, Physical theory of plasticity. *Advances in Applied Mechanics, Vol.* 11, pp. 255-311, Acad. Press, New York 1971.
141. B. Loret, Non-linéarité incrémentale et localisation des déformations: quelque remarques. *J. Méc. Théor. Appl.* 6, 423-459, 1987.
142. B. Loret, Does deviation from deviatoric associativity lead to the onset of flutter instability? *J. Mech. Phys. Solids* 40, 1363-1375, 1992.
143. B. Loret, O. Harireche, Acceleration waves, flutter instabilities and stationary discontinuities in inelastic porous media. *J. Mech. Phys. Solids* 39, 569-606, 1991.

144. B. Loret, J.A.C. Martins and F.M.F. Simões, Surface boundary conditions trigger flutter instability in non-associative elastic-plastic solids. *Int. J. Solids Structures* 32, 2155-2190, 1995.
145. B. Loret, J.H. Prevost and O. Harireche, Loss of hyperbolicity in elastic-plastic solids with deviatoric associativity. *Eur. J. Mech. A//Solids* 9, 225-231, 1990.
146. G. Maier and T. Hueckel, Non-associated and coupled flow rules of elastoplasticity for rock-like materials. *Int. J. Rock Mech. Min. Sci. & Geomech. Abstr.* 16, 77-92, 1979.
147. J. Mandel, Generalisation de la théorie de plasticité de W.T. Koiter. *Int. J. Solids Structures* 1, 273-295, 1965.
148. J. Mandel, Conditions de stabilité et postulat de Drucker. *Rheology and Soil Mechanics* (J. Kravtchenko, Ed.), pp.58-67, Springer, Berlin 1966.
149. Z. Marciniak, Stability of plastic shells under tension with kinematic boundary conditions. *Arch. Mech. Stos.* 17, 577-592, 1965.
150. Z. Marciniak and K. Kuczyński, Limit strains in the processes of stretch-forming sheet metal. *Int. J. Mech. Sci* 9, 609-620, 1967.
151. Z. Marciniak, K. Kuczyński, T. Pokora, Influence of the plastic properties of a material on the forming limit diagram for sheet metal in tension. *Int. J. Mech. Sci.* 15, 789-805, 1973.
152. M.E. Mear and J.W. Hutchinson, Influence of yield surface curvature on flow localization in dilatant plasticity. *Mech. Mater.* 4, 395-407, 1985.
153. R.M. McMeeking, J.R. Rice, Finite-element formulations for problems of large elastic-plastic deformation. *Int. J. Solids Structures* 11, 601-616, 1975.
154. J.P.Miles, Bifurcation in rigid-plastic materials under spherically symmetric loading conditions. *J. Mech. Phys. Solids* 17, 303-313, 1969.
155. J.P. Miles, Fluid-pressure eigenstates and bifurcation in tension specimens under lateral pressure, *J. Mech. Phys. Solids* 21, 145-162, 1973.
156. J.P. Miles, The initiation of necking in rectangular elastic//plastic specimens under uniaxial and biaxial tension. *J. Mech. Phys. Solids* 23, 197-213, 1975.
157. Z. Mróz, Non-associated flow laws in plasticity. *J. Mécanique* 2, 21-42, 1963.
158. Z. Mróz, On forms of constitutive laws for elastic-plastic solids. *Arch. Mech. Stos.* 18, 1-35, 1966.
159. H.B. Mühlhaus, Sherfugenanalyse bei granularem Material im Rahmen der Cosserat-Theorie. *Ing.-Archiv* 56, 389-399, 1986.
160. H.B. Mühlhaus, Application of Cosserat theory in numerical simulations of limit load problems. *Ing.-Archiv* 59, 124-137, 1989.
161. H.B. Mühlhaus and E.C. Aifantis, A variational principle for gradient plasticity. *Int. J. Solids Structures* 28, 845 857, 1991.
162. H.B. Mühlhaus and I. Vardoulakis, The thickness of shear bands in granular materials. *Géotechnique*, 37, 271-283, 1987.
163. A. Nacar, A. Needleman and M. Ortiz, A finite element method for analyzing localization in rate dependent solids at finite strains. *Comp. Meth. Appl. Mech. Engng.* 73, 235-258, 1989.
164. A. Needleman, A numerical study of necking in circular cylindrical bars. *J. Mech. Phys. Solids* 20, 111-127, 1972.
165. A. Needleman, A numerical study of uniaxial compression in circular elastic-plastic columns. *Int. J. Solids Structures* 9, 981-998, 1973.
166. A. Needleman, Bifurcation of elastic-plastic spherical shells subject to internal pressure. *J. Mech. Phys. Solids*, 23, 357-367, 1975.
167. A. Needleman, Postbifurcation behavior and imperfection sensitivity of elastic-plastic circular plates. *Int. J. Mech. Sci.* 17, 1-13, 1975.
168. A. Needleman, Necking of pressurized spherical membranes. *J. Mech. Phys. Solids* 24, 339-359, 1976
169. A. Needleman, Non-normality and bifurcation in plane strain tension and compression. *J. Mech. Phys. Solids* 27, 231-254, 1979.
170. A. Needleman and M. Ortiz, Effect of boundaries and interfaces on shear band localization. *Int. J. Solids Structures* 28, 859-877, 1991.

171. A. Needleman and J.R. Rice, Limits to ductility set by plastic flow localization. *Mechanics of Sheet Metal Forming* (D.P. Koistinen, N.-M. Wang, Eds.), pp. 237-264, Plenum, New York 1978.
172. A. Needleman and V. Tvergaard, An analysis of the imperfection sensitivity of square elastic-plastic plates under axial compression. *Int. J. Solids Structures* 12, 185-201, 1976.
173. Needleman, A. and Tvergaard, V. Necking of biaxially stretched elastic-plastic circular plates. *J. Mech. Phys. Solids*, 25, 159-183, 1977.
174. A. Needleman and V. Tvergaard, Aspects of plastic postbuckling behavior. In: *Mechanics of Solids, The Rodney Hill 60-th Anniversary Volume* (H.G. Hopkins and M.J. Sewell, Eds.), pp. 453-498, Pergamon Press, Oxford 1982.
175. A. Needleman and V. Tvergaard, Finite element analysis of localization in plasticity. In: *Finite Elements: Special Problems in Solid Mechanics* (J.T. Oden and G.F.Carey, Eds.), Vol. V, pp. 94-157, Prentice-Hall, Englewood Cliffs, 1984.
176. A. Needleman and V. Tvergaard, Analyses of plastic flow localization in metals. *Appl. Mech. Rev.* 45, no 3, part 2, S3-S18, 1992.
177. Q.S. Nguyen, Bifurcation et stabilité des systèmes irréversibles obéissant au principe de dissipation maximale. J. Méc. theor. appl. 3, 41-61, 1984.
178. Q.S. Nguyen, Bifurcation and post-bifurcation analysis in plasticity and brittle fracture. *J. Mech. Phys. Solids* 35, 303-324, 1987.
179. Q.S. Nguyen, Stabilité et bifurcation des systèmes dissipatifs standarts à comportement indépendant du temps physique, *C. R. Acad. Sci. Paris, t.* 310, *Série II*, 1375-1380, 1990.
180. Q.S. Nguyen, Bifurcation and stability of time-independent standard dissipative systems. In: *CISM Lecture Notes No.* 327 (Q.S. Nguyen, Ed.), pp. 45-94, Springer, Wien - New York, 1993.
181. Q.S. Nguyen, Bifurcation and stability in dissipative media (plasticity, friction, fracture). *Appl. Mech. Rev.* 47, no 1, Part 1, 1-31, 1994.
182. Q.S. Nguyen and D. Radenkovic, Stability of equilibrium in elastic plastic solids. *Lecture Notes in Mathematics, Vol.* 503, pp. 403-414, Springer, Berlin 1975.
183. Q.S. Nguyen and N. Triantafyllidis, Plastic bifurcation and postbifurcation analysis for generalized standard continua. *J. Mech. Phys. Solids* 37, 545-566, 1989.
184. D.M. Norris, B. Moran, J.K. Soudder, D.F. Quiñones, A computer simulation of the tension test. *J. Mech. Phys. Solids* 26, 1-19, 1978.
185. J. Olivier, Continuum modelling of strong discontinuities in solid mechanics. In: *Computational Plasticity: Fundamentals and Applications* (D.R.J. Owen and E. Oñate, Eds.) Vol. I, pp. 455-479, Pineridge Press, Swansea 1995.
186. E.T. Onat and D.C. Drucker, Inelastic instability and incremental theories of plasticity. *J. Aero. Sci.* 20, 181-186, 1953.
187. M. Ortiz, Y. Leroy and A. Needleman, A finite element method for localized failure analysis, *Comp. Meth. Appl. Mech. Engng.* 61, 189-214, 1987.
188. M. Ortiz and J.J. Quigley, Adaptive mesh refinement in strain localization problems. *Comp. Meth. Appl. Mech. Engng.* 90, 781-804, 1991.
189. N.S. Ottosen and K. Runesson, Properties of discontinuous bifurcation solutions in elasto-plasticity. *Int. J. Solids Structures* 27, 401-421, 1991.
190. N.S. Ottosen and K. Runesson, Acceleration waves in elasto-plasticity. *Int. J. Solids Structures* 28, 135-159, 1991.
191. P.C. Papanastasiou and I.G. Vardoulakis, Numerical treatment of progressive localization in relation to borehole stability. *Int. J. Numer Anal. Methods Geomech.* 16, 389-424, 1992.
192. A. Parmar and P.B. Mellor, Predictions of limit strains in sheet metal using a more general yield criterion. *Int. J. Mech. Sci.* 20, 385-391, 1978.
193. M. Pastor, J. Peraire and O.C. Zienkiewicz, Adaptive remeshing for shear band computations. *Arch. Appl. Mech.* 61, 30-39, 1991.
194. D. Peirce, Shear band bifurcations in ductile single crystals. *J. Mech. Phys. Solids* 31, 133-153, 1983.
195. D. Peirce, R.J. Asaro and A. Needleman, An analysis of nonuniform and localized deformation in ductile single crystals. *Acta Metall.* 30, 1087-1119, 1982.

196. D. Perić, J. Yu and D.R.J. Owen, On error estimates and adaptivity in elastoplastic solids: applications to the numerical simulation of strain localization in classical and cosserat continua. *Int. J. Num. Meth. Engng.* 37, 1351-1379, 1994.
197. G. Perrin and J.B. Leblond, Rudnicki and Rice's analysis of strain localization revisited. *ASME J. Appl. Mech.* 60, 842-846, 1993.
198. H. Petryk, A consistent energy approach to defining stability of plastic deformation processes. In: *Stability in the Mechanics of Continua, Proc. IUTAM Symp. Nümbrecht* 1981 (F.H. Schroeder, Ed.), pp. 262-272, Springer, Berlin - Heidelberg 1982.
199. H. Petryk, On stability and symmetry conditions in time-independent plasticity. *Arch. Mech.* 37, 503-520, 1985.
200. H. Petryk, On energy criteria of plastic instability. In: *Plastic Instability, Proc. Considère Memorial,* pp. 215-226, Ecole Nat. Ponts Chauss., Paris 1985.
201. H. Petryk, On constitutive inequalities and bifurcation in elastic-plastic solids with a yield-surface vertex. *J. Mech. Phys. Solids* 37, 265-291, 1989.
202. H. Petryk, The energy criteria of instability in time-independent inelastic solids. *Arch. Mech.* 43, 519-545, 1991.
203. H.Petryk. On stability of time-independent materials at finite strain. in: *Finite Inelastic Deformations - Theory and Applications* (D.Besdo and E.Stein, Eds.), pp. 189-196, Springer-Verlag, Berlin Heidelberg 1992.
204. H. Petryk, Material instability and strain-rate discontinuities in incrementally nonlinear continua. *J. Mech. Phys. Solids* 40, 1227-1250, 1992.
205. H. Petryk, Theory of bifurcation and instability in time-independent plasticity. In: *CISM Lecture Notes No.* 327 (Q.S. Nguyen, Ed.), pp. 95-152, Springer, Wien - New York, 1993.
206. H.Petryk, Stability and constitutive inequalities in plasticity. *CISM Lecture Notes No.* 336 (W. Muschik, Ed.), pp. 259-329, Springer, Wien - New York 1993.
207. H.Petryk, Thermodynamic stability of equilibrium in plasticity, *J. Non-Equil. Thermodyn.* 20, 132-149, 1995.
208. H. Petryk and K. Thermann, Second-order bifurcation in elastic-plastic solids. *J. Mech. Phys. Solids* 33, 577-593, 1985.
209. H. Petryk and K. Thermann, On discretized plasticity problems with bifurcations. *Int. J. Solids Structures* 29, 745-765, 1992.
210. H. Petryk and K. Thermann, On post-bifurcation paths in plastic deformation processes. In: *Computational Plasticity: Fundamentals and applications* (D.R.J. Owen *et al*, Eds.) Vol. I, 593-604, Pineridge Press, Swansea 1992.
211. H. Petryk and K. Thermann, On plastic strain localisation in the non-elliptic range under plane stress. In: *Computational Plasticity: Fundamentals and Applications* (D.R.J. Owen and E. Oñate, Eds.) Vol. I, pp. 647-658, Pineridge Press, Swansea 1995.
212. H. Petryk and K. Thermann, Post-critical plastic deformation of biaxially stretched sheets. *Int. J. Solids Structures* (in press)
213. St. Pietruszczak and Z. Mróz, Finite element analysis of deformation of strain-softening materials. *Int. J. Num. Meth. Engng.* 17, 327-334, 1981.
214. G. Pijaudier-Cabot and Z.P. Bažant, Nonlocal damage theory. *ASCE J. Eng. Mech.* 113, 1512-1533, 1987.
215. G. Pijaudier-Cabot and A. Benallal, Strain localization and bifurcation in a nonlocal continuum. *Int. J. Solids Structures* 30, 1761-1775, 1993.
216. M. Potier-Ferry, Towards a catastrophe theory for the mechanics of plasticity and fracture. *Int. J. Engng. Sci.* 23, 821-837, 1985.
217. W. Prager, Three dimensional plastic flow under uniform stress, *Rev. Fac. Sci. Univ. Istanbul* 19, 23-26, 1954.
218. B. Raniecki, Uniqueness criteria in solids with non-associated plastic flow laws at finite deformations. *Bull. Acad. Polon. Sci., Sér. sci. techn.* 27, 391-399, 1979.
219. B. Raniecki and O.T. Bruhns, Bounds to bifurcation stresses in solids with non-associated plastic flow law at finite strain. *J. Mech. Phys. Solids* 29, 153-172, 1981.
220. S.N. Rasmussen, Effect of rounded vertices in the yield surface on localized necking in biaxially stretched sheets. *Int. J. Mech. Sci.* 12, 729-739, 1982.

221. J.R. Rice, The localization of plastic deformation. In: *Theoretical and Applied Mechanics* (W.T. Koiter, Ed.), pp. 207-220, North- Holland, Amsterdam 1977.
222. J.R. Rice and J.W. Rudnicki, A note on some features of the theory of localization of deformation. *Int. J. Solids Structures* 16, 597-605, 1980.
223. E. Riks, An incremental approach to the solution of snapping and buckling problems. *Int. J. Solids Structures* 15, 529-551, 1979.
224. A. Rosochowski, Shear band localization in ductile porous materials. *Plastic Instability, Proc. Considère Memorial,* pp. 135-146, Ecole Nat. Ponts Chauss. Press, Paris 1985.
225. J.W. Rudnicki and J.R. Rice, Conditions for the localization of deformation in pressure-sensitive dilatant materials. *J. Mech. Phys. Solids* 23, 371-394, 1975.
226. E.I. Ryzhak, Necessity of Hadamard conditions for stability of elastic-plastic solids (in Russian). *Mekh. Tv. Tela No.* 4, 101-104, 1987.
227. E.I. Ryzhak, On stable deformation of 'unstable' materials in a rigid triaxial testing machine. *J. Mech. Phys. Solids* 41, 1345-1356, 1993.
228. M. Saje, J. Pan and A. Needleman, Void nucleation effects on shear localization in porous plastic solids. *Int. J. Fracture* 19, 163-182, 1982.
229. J.L. Sanders, Plastic stress-strain relations based on linear loading functions. *Proc. 2nd U.S. Nat. Congr. Appl. Mech.*, pp. 455-460, Univ. of Michigan, Ann Arbor, 1954.
230. I.S. Sandler and D. Rubin, The consequences of non-associated plasticity in dynamic problems. In: *Constitutive Laws for Engineering Materials* (S.C. Desai *et. al.*, Eds.) pp. 345-352, Elsevier, New York 1987.
231. H.L. Schreyer, Analytical solutions for nonlinear strain-gradient softening and localization. *ASME J. Appl. Mech.* 57, 522-528, 1990.
232. H.L. Schreyer and Z. Chen, One-dimensional softening with localization. *ASME J. Appl. Mech.* 53, 791-797, 1986.
233. M.J. Sewell, A general theory of elastic and inelastic plate failure - I. *J. Mech. Phys. Solids* 11, 377-393, 1963.
234. M.J. Sewell, On configuration-dependent loading. *Arch. Rat. Mech. Anal.* 23, 327-351, 1967.
235. M.J. Sewell, A survey of plastic buckling. In: *Stability* (H.H.E. Leipholz, Ed.), pp. 85-197, Univ. of Waterloo Press, Ontario 1972.
236. M.J. Sewell, A yield surface corner lowers the buckling stress of an elastic-plastic plate under compression. *J. Mech. Phys. Solids* 21, 19-45, 1973
237. F.R. Shanley, The column paradox. *J. Aero. Sci.* 13, 678, 1946.
238. F.R. Shanley, Inelastic column theory. *J. Aero. Sci.* 14, 261-267, 1947.
239. J.C. Simo, Nonlinear stability of the time-discrete variational problem of evolution in nonlinear heat conduction, plasticity and viscoplasticity. *Comp. Meth. Appl. Mech. Engng.* 88, 111-131, 1991.
240. J.C. Simo, J. Olivier and F. Armero, An analysis of strong discontinuities induced by strain softening in rate-independent inelastic solids. *Comput. Mech.* 12, 277-296, 1993.
241. R. Sowerby and J.L. Duncan, Failure in sheet metal in biaxial tension. *Int. J. Mech. Sci.* 13, 217-229, 1971.
242. P.S. Steif, Growth of interface imperfections in layered plastic solids. In: *Plastic Instability, Proc. Considère Memorial,* pp. 175-183, Ecole Nat. Ponts Chauss., Paris 1985.
243. E. Stein, W. Wagner and P. Wriggers, Numerical procedures for the computation of singular points and postcritical branches, in: *Inelastic Solids and Structures, Antoni Sawczuk Memorial Volume* (M.Kleiber and J.A.König, Eds.), pp. 145-164, Pineridge Press, Swansea 1990.
244. P. Steinmann, C. Miehe, E. Stein, On the localization analysis of orthotropic Hill type elasto-plastic solids. *J. Mech. Phys. Solids* 42, 1969-1994, 1994.
245. B. Storakers, On uniqueness and stability of elastic-plastic deformation. *Arch. Mech.* 27, 821-839, 1975.
246. B. Storakers, On uniqueness and stability under configuration-dependent loading of solids with or without a natural time. *J. Mech. Phys. Solids* 25, 269-287, 1977.
247. S. Stören and J.R. Rice, Localized necking in thin sheets. *J. Mech. Phys. Solids* 23, 421-441, 1975.

248. H. Strifords and B. Storakers, Uniqueness and stability of rigid-plastic spherical shells subjected to finite deformation under hydrostatic pressure. *J. Mech. Phys. Solids* 21, 125-134, 1973.
249. Z. Suo, M. Ortiz and A. Needleman, Stability of solids with interfaces. *J. Mech. Phys. Solids* 40, 613-640, 1992.
250. J. Tejchman and W. Wu, Numerical study on patterning of shear bands in a Cosserat continuum. *Acta Mech.* 99, 61-74, 1993.
251. T.Y. Thomas, *Plastic Flow and Fracture in Solids*, Acad. Press, New York 1961.
252. Y. Tomita, Simulation of plastic instabilities in solid mechanics. *Appl. Mech. Rev.* 47, no 6, Part 1, 171-205, 1994.
253. Y. Tomita and A. Shindo, Bifurcation behaviour of thin square elastic-plastic orthotropic plates subjected to diagonal tension. In: *Plastic Instability, Proc. Considère Memorial,* pp. 203-213, Ecole Nat. Ponts Chauss., Paris 1985.
254. Y. Tomita and A. Shindo, Onset and growth of wrinkles in thin square plates subjected to diagonal tension. *Int. J. Mech. Sci.* 30, 921-931, 1988.
255. Y. Tomita, A. Shindo, Y.S. Kim, K. Michiura, Deformation behaviour of elastic-plastic tubes under external pressure and axial load. *Int. J. Mech. Sci.* 28, 263-274, 1986.
256. Y. Tomita, A. Shindo and H. Kitagawa, Bifurcation and post-bifurcation behaviour of internally pressurized elastic-plastic circular tubes under plane strain conditions. *Int. J. Mech. Sci.* 23, 723-732, 1981.
257. Y. Tomita, A. Shindo and M. Nagai, Axisymmetric deformation of circular elastic-plastic tubes under axial tension and internal pressure. *Int. J. Mech. Sci.* 26, 437-444, 1984.
258. N. Triantafyllidis, Bifurcation phenomena in pure bending. *J. Mech. Phys. Solids* 28, 221-245, 1980.
259. N. Triantafyllidis, On the bifurcation and postbifurcation analysis of elastic-plastic solids under general prebifurcation conditions. *J. Mech. Phys. Solids* 31, 499-510, 1983.
260. N. Triantafyllidis, The localization of deformation in finitely strained shells. In: *Plastic Instability, Proc. Considère Memorial,* pp. 115-124, Ecole Nat. Ponts Chauss., Paris 1985.
261. N. Triantafyllidis and E.C. Aifantis, A gradient approach to localization of deformation - I. Hyperelastic materials. *J. Elasticity*, 16, 225-238, 1986.
262. N. Triantafyllidis, A. Needleman and V. Tvergaard, On the development of shear bands in pure bending. *Int. J. Solids Structures* 18, 121-138, 1982.
263. J.J. Tsay and J.S. Arora, Nonlinear structural design sensitivity analysis in nonlinear structural mechanics. Part I: General theory. *Comp. Meth. Appl. Mech. Engng.* 81, 183-208, 1990.
264. V. Tvergaard, Effect of thickness inhomogeneities in internally pressurized elastic-plastic spherical shells. *J. Mech. Phys. Solids* 24, 291-304, 1976.
265. V. Tvergaard, Buckling behavior of plate and shell structures. *Theoretical and Applied Mechanics* (W.T. Koiter, Ed.), pp. 233-247, North-Holland, Amsterdam 1977.
266. V. Tvergaard, Buckling of elastic-plastic cylindrical panel under axial compression. *Int. J. Solids Structures* 13, 957-970, 1977.
267. V. Tvergaard, On the burst strength and necking behaviour of rotating disks. *Int. J. Mech. Sci.* 20, 109-120, 1978.
268. V. Tvergaard, Effect of kinematic hardening on localized necking in biaxially stretched sheets. *Int. J. Mech. Sci.* 20, 651-658, 1978.
269. V. Tvergaard, Bifurcation and imperfection-sensitivity at necking instabilities. *ZAMM* 60, T26-T34, 1980.
270. V. Tvergaard, Ductile fracture by cavity nucleation between larger voids. *J. Mech. Phys. Solids*, 30, 265-286, 1982.
271. V. Tvergaard, Influence of void nucleation on ductile shear fracture at a free surface. *J. Mech. Phys. Solids* 30, 399-425, 1982.
272. V. Tvergaard, Effect of yield surface curvature and void nucleation on plastic flow localization, *J. Mech. Phys. Solids* 35 43-60, 1987.
273. V. Tvergaard, Bifurcation in elastic-plastic tubes under internal pressure. *Eur. J. Mech. A//Solids* 9, 21-35, 1990.

274. V. Tvergaard, Necking in tensile bars with rectangular cross-section. *Comp. Meth. Appl. Mech. Engng*, 103, 273-290, 1993.
275. V.Tvergaard and E. van der Giessen, Effect of plastic spin on localization predictions for a porous ductile material. *J. Mech. Phys. Solids* 39, 763-781, 1991.
276. V. Tvergaard and A. Needleman, On the buckling of elastic-plastic columns with asymmetric cross-sections. *Int. J. Mech. Sci.* 17, 419-424, 1975.
277. V. Tvergaard, A. Needleman and K.K. Lo, Flow localization in plane strain tensile test. *J. Mech. Phys. Solids* 29, 115-142, 1981.
278. I. Vardoulakis, Shear band inclination and shear modulus of sand in biaxial tests. *Int. J. Num. Anal. Methods Geomech.* 4, 103-119, 1980.
279. I. Vardoulakis, Shear banding and liquefaction in granular materials on the basis of a Cosserat continuum theory. *Ing.-Archiv* 59, 106-113, 1989
280. I. Vardoulakis and E.C. Aifantis, A gradient flow theory of plasticity for granular materials. *Acta Mech.* 87, 197-217, 1991.
281. I. Vardoulakis and G. Frantziskonis, Micro-structure in kinematic-hardening plasticity. *Eur. J. Mech. A//Solids* 11, 467-486, 1992.
282. I. Vardoulakis and B. Graf, Calibration of constitutive models for granular materials using data from biaxial experiments. *Géotechnique* 35, 299-317, 1985.
283. P.A. Vermeer and R.B.J. Brinkgreve, A new effective non-local strain-measure for softening plasticity. In: *Localisation and Bifurcation Theory for Soils and Rocks* (R. Chambon, J. Desrues and I. Vardoulakis, Eds.), pp. 89-100, Balkema, Rotterdam 1994.
284. A.A. Vidal and R.B. Haber, Design sensitivity analysis for rate-independent elastoplasticity. *Comp. Meth. Appl. Mech. Engng.* 107, 393-431, 1993.
285. P. Wriggers and J.C. Simo, A general procedure for the direct computation of turning and bifurcation points. *Int. J. Num. Meth. Engng.* 30, 155-176, 1990
286. H. Yamamoto, Conditions for shear localization in the ductile fracture of void-containing materials. *Int. J. Fracture* 14, 347-365, 1978.
287. N.J.B. Young, Bifurcation phenomena in the plane compression test. *J. Mech. Phys. Solids* 24, 77-91, 1976.
288. Zbib, H.M. and Aifantis, E.C., On the localization and postlocalization behavior of plastic deformation, I, II and III. *Res. Mechanica* 23, 261-305, 1988.
289. Z. Zhou and D.W. Murray, An incremental solution technique for unstable equilibrium paths of shell structures. *Computers & Structures* 55, 749-759, 1994.
290. A. Zubelewicz and Z. Mróz, Numerical simulation of rock burst process treated as problems of dynamic stability, Rock Mechanics and Rock Engineering, 16, 253-274, 1983.

Chapter 10
On the Micro-Macro Transition and Hardening Moduli in Plasticity

10.1 Introduction

This paper focuses upon the theoretical foundations of the modeling of elastoplastic polycrystalline aggregates. As a starting point, the known constitutive framework for a time-independent elastic-plastic material with a finite number of internal mechanisms of plastic deformation obeying the normality flow rule is adopted at a mesoscopic level (Subsection 10.2.1). Implications of that assumption for the macroscopic properties of a heterogeneous material have been studied theoretically and numerically in a number of papers. No survey of previous results is attempted here, rather, the aim of this article is to present nowel qualitative conclusions. In Subsection 10.2.2 the transmissibility of the structure of incremental constitutive relationships from the meso- to macro-level is proved for a spatially discretized aggregate, and expressions are derived for macroscopic effective hardening moduli. This goes beyond earlier results concerning the transition rules expressed in terms of the strain or stress rates and of the fourth-order tensors of instantaneous moduli or compliances.

A known difficulty in the modelling of time-independent elastic-plastic crystalline solids is that the set of incremental constitutive relationships need not have a unique solution. This problem is addressed here in relation to the question of stability of an equilibrium state and of a solution path for a material element. With the thermodynamic framework presented in Subsection 10.2.3 as a basis, in Sect. 10.3 the stability conditions are formulated in terms of the energy function and of the hardening moduli. The requirement of the stability of a deformation process within the material element is introduced in Subsection 10.3.2 and used to generate criteria for selection of a solution when the material response is non-uniquely defined.

Throughout the paper, bold-face small letters (Roman or Greek) denote second-order symmetric tensors, and bold-face capital letters denote fourth-order tensors

[0] Reprinted from: Proc. IUTAM Symposium on Micro- and Macrostructural Aspects of Thermoplasticity, Bochum 1997 (Ed. O.T. Bruhns and E. Stein), pp. 219-230, Kluwer, Dordrecht 1999. H.Petryk, On the micro-macro transition and hardening moduli in plasticity. Copyright ©1999 Kluwer, with permission from Springer.

H. Petryk, *Theory of Plastic Instability*,
https://doi.org/10.1007/978-3-031-98223-1_10

with the minor symmetries, with $\mathbf{I}$ as the respective unit tensor and a superscript T indicating a transpose. Juxtaposition of two tensor symbols denotes double contraction, and a superimposed dot denotes the forward rate with respect to a time-like parameter t.

10.2 Structure of Constitutive Equations in Elastoplasticity

10.2.1 Constitutive Assumptions

An elastic-plastic material element described at a mesoscopic level, e.g. a single crystal in a polycrystalline aggregate, is assumed to have a finite number of internal mechanisms of time-independent plastic deformation, each obeying the normality flow rule. Only isothermal processes are considered here. The constitutive framework for such materials at small strain [11][5] and at finite strain [8][20] is well known and can be summarized as

$$\dot{\mathbf{e}} = \mathbf{M}\dot{\mathbf{t}} + \sum_K \boldsymbol{\mu}_K \dot{\gamma}_K \,, \qquad \dot{\gamma}_K \geq 0, \qquad \dot{f}_K = \boldsymbol{\mu}_K \dot{\mathbf{t}} - \sum_L h_{KL} \dot{\gamma}_L \,, \tag{10.1}$$

$$f_K \leq 0, \qquad f_K \, \dot{\gamma}_K = 0 \,. \tag{10.2}$$

A scalar $\mathring{\gamma}_K$ measures the rate of activity of the K-th mechanism whose individual yield function $f_K(\mathbf{t}, \mathcal{H})$ is smooth with the gradient $\boldsymbol{\mu}_K = \partial f_K / \partial \mathbf{t}$ in the stress space. $(\mathbf{t}, \mathbf{e})$ stands for any pair of *work-conjugate* measures of stress and strain [7]. $\mathcal{H}$ symbolizes the pattern of internal rearrangement or the prior history of the $\mathring{\gamma}_K$'s [19], and is not necessarily represented by internal state variables. Formally, $\dot{\mathcal{H}}$ is regarded as a positively homogeneous function of degree one in the $\mathring{\gamma}_K$'s. The elastic compliance tensor $\mathbf{M}$ and the matrix (h_{KL}) of *effective* hardening moduli may depend continuously on $(\mathbf{t}, \mathcal{H})$. The usual assumption is adopted that $\mathbf{M}$ is diagonally symmetric ($\mathbf{M} = \mathbf{M}^{\mathrm{T}}$) and positive definite, noting that the latter property need not hold for all pairs $(\mathbf{t}, \mathbf{e})$. The matrix (h_{KL}) at this stage of considerations is arbitrary.

The incremental problem to be solved in a given state of the material follows by differentiating (10.2) and involves only potentially active mechanisms, viz.

$$\dot{f}_K \leq 0, \quad \dot{f}_K \, \dot{\gamma}_K = 0 \qquad \text{for } K \in \mathcal{P} \equiv \{K \,:\, f_K = 0\} \,, \tag{10.3}$$

while $\mathring{\gamma}_K = 0$ for $K \notin \mathcal{P}$. A vector of $\mathring{\gamma}_K$'s with $K \in \mathcal{P}$ will be denoted by $[\mathring{\gamma}]$.

In the finite-strain theory of single crystals deformed plastically by multislip, $\mathring{\gamma}_K$ is identified as the *invariant* rate of shearing on K-th slip system, and $f_K = \tau_K - \tau_K^{\mathrm{c}}$, where τ_K is the *generalized* resolved shear stress on that system and τ_K^{c} is its critical value; cf. [8][4].

10.2.2 Macroscopic Constitutive Law

Consider a representative volume element (macro-element) of a heterogeneous material that behaves at the macroscopic level as if it were homogeneous. At the mesoscopic level where the continuum mechanics is applicable, the material is assumed to fit into the constitutive framework described in Subsection 10.2.1, with the material functions and parameters dependent on place. For simplicity, throughout this subsection the small strain theory is employed, so that the Cauchy stress $\boldsymbol{\sigma}$ and small strain $\boldsymbol{\epsilon}$ are used in place of $(\mathbf{t}, \mathbf{e})$. In comparison with the general treatment by Hill [6], an additional assumption is introduced here that the macro-element is divided into an arbitrarily large but finite number of cells within which the material state, including stress and strain, is uniform at every instant. That spatial (finite element) discretization may be arbitrarily fine, e.g. also within a single crystal.

A Greek subscript, typically η, will be used to denote quantities in the η-th cell of a volume fraction v_η. Angular brackets will denote unweighted volume averaging over the macroelement, i.e. $\langle\psi\rangle = \sum_\eta v_\eta \psi_\eta$ for any quantity ψ. The stresses $\boldsymbol{\sigma}_\eta$ are assumed below to be in 'weak' equilibrium in the sense that the principle of virtual power at zero body forces is satisfied for any admissible (i.e. continuous and piecewise linear) velocity field. Macroscopic (overall) quantities will be distinguished by a bar, and in particular $\overline{\boldsymbol{\sigma}} = \langle\boldsymbol{\sigma}\rangle$, $\overline{\boldsymbol{\epsilon}} = \langle\boldsymbol{\epsilon}\rangle$.

Validity of Hill's formula for product averaging [6] is assumed for the discretized macro-element, viz.

$$\langle\boldsymbol{\sigma}\rangle\langle\dot{\boldsymbol{\epsilon}}\rangle = \langle\boldsymbol{\sigma}\,\dot{\boldsymbol{\epsilon}}\rangle, \qquad \langle\delta\boldsymbol{\sigma}\rangle\langle\dot{\boldsymbol{\epsilon}}\rangle = \langle\delta\boldsymbol{\sigma}\,\dot{\boldsymbol{\epsilon}}\rangle, \tag{10.4}$$

where $\delta\boldsymbol{\sigma}_\eta$ is any variation in the piecewise-uniform stresses subject to the condition of 'weak' equilibrium, and $\dot{\boldsymbol{\epsilon}}_\eta$ is any piecewise-uniform and compatible field of strain-rates, in general unrelated to $\delta\boldsymbol{\sigma}_\eta$. We will also assume that the linear elasticity problem for the *representative* macro-element is uniquely solvable for a prescribed variation $\delta\langle\boldsymbol{\sigma}\rangle$ of the average stress at a fixed plastic strain field, or for a prescribed variation $\delta\boldsymbol{\epsilon}^{\mathrm{p}}_\eta = (\delta\boldsymbol{\epsilon} - \mathbf{M}\,\delta\boldsymbol{\sigma})_\eta$ of a local plastic strain at a fixed average stress. It is recalled that elastic compliance tensors $\mathbf{M}_\eta$ are assumed to be diagonally symmetric and positive definite. A general solution is written as

$$\delta\boldsymbol{\sigma}_\eta = \mathbf{B}_\eta\,\delta\langle\boldsymbol{\sigma}\rangle - \sum_\iota v_\iota \mathbf{Q}_{\eta\iota}\,\delta\boldsymbol{\epsilon}^{\mathrm{p}}_\iota, \qquad \langle\mathbf{B}\rangle = \mathbf{I}, \qquad \langle\mathbf{Q}_{\cdot\iota}\rangle = \mathbf{0}, \tag{10.5}$$

where the place-dependent fourth-order tensors $\mathbf{B}_\eta$ and $\mathbf{Q}_{\eta\iota}$ represent the *elastic* influence functions, expressible through the Green function for the macro-element, in analogy to Hill [6] and Berveiller *at al.* [1], respectively.

We shall prove the following statement:

Theorem 1 *If the constitutive law in each η-th cell in the discretized macro-element has the form (10.1), (10.2)* (with $(\mathbf{t}, \mathbf{e})$ replaced by $(\boldsymbol{\sigma}, \boldsymbol{\epsilon})$ and with the state parameters dependent on η) *then from (10.4) and (10.5) it follows that the macroscopic constitutive law has an analogous structure* (with a double index $(K\eta)$

in place of a single index (K)), *viz.*

$$\dot{\bar{\epsilon}} = \overline{\mathbf{M}}\dot{\bar{\sigma}} + \sum_{\eta,K} \bar{\mu}_{K\eta}\bar{\dot{\gamma}}_{K\eta}, \qquad \bar{\dot{\gamma}}_{K\eta} \geq 0, \qquad \dot{\bar{f}}_{K\eta} = \bar{\mu}_{K\eta}\dot{\bar{\sigma}} - \sum_{\iota,L} \bar{h}_{K\eta L\iota}\bar{\dot{\gamma}}_{L\iota}, \tag{10.6}$$

$$\bar{f}_{K\eta} \leq 0, \qquad \dot{\bar{f}}_{K\eta}\bar{\dot{\gamma}}_{K\eta} = 0, \tag{10.7}$$

where

$$\dot{\bar{\epsilon}} = \langle\dot{\epsilon}\rangle, \quad \dot{\bar{\sigma}} = \langle\dot{\sigma}\rangle, \quad \bar{f}_{K\eta} = f_{K\eta}, \quad \bar{\dot{\gamma}}_{K\eta} = \nu_\eta\dot{\gamma}_{K\eta}, \quad \overline{\mathbf{M}} = \langle\mathbf{B}^{\mathrm{T}}\mathbf{M}\mathbf{B}\rangle, \tag{10.8}$$

$$\bar{\mu}_{K\eta} = \mu_{K\eta}\mathbf{B}_\eta = \partial\bar{f}_{K\eta}/\partial\bar{\sigma}, \quad \bar{h}_{K\eta L\iota} = \mu_{K\eta}\mathbf{Q}_{\eta\iota}\mu_{L\iota} + \delta_{\eta\iota}h_{KL\eta}/\nu_\iota. \tag{10.9}$$

Proof. We have to show that the equalities in (10.6) hold when the macroscopic quantities are defined by (10.8) and (10.9) (with $\delta_{\eta\iota}$ as the Kronecker delta). The substitution of (10.5) and $(10.1)_1$ into (10.4) shows that for purely elastic changes

$$\delta\bar{\sigma}\,\dot{\bar{\epsilon}} = \langle\delta\sigma\,\dot{\epsilon}\rangle = \delta\bar{\sigma}\langle\mathbf{B}^{\mathrm{T}}\mathbf{M}\mathbf{B}\rangle\dot{\bar{\sigma}} \qquad (\dot{\gamma}_{K\eta} \equiv 0,\ \delta\epsilon^{\mathrm{p}}_\eta \equiv \mathbf{0}) \tag{10.10}$$

for arbitrary $\delta\bar{\sigma}$, $\dot{\bar{\sigma}}$, so that the elasticity law in $(10.6)_1$ agrees with the definition $(10.8)_5$.[1] From $\mathbf{M} = \mathbf{M}^{\mathrm{T}}$ and $(10.8)_5$ it follows that $\overline{\mathbf{M}} = \overline{\mathbf{M}}^{\mathrm{T}}$, as expected. When $\dot{\gamma}_{K\eta} \not\equiv 0$ while δ-variations are still purely elastic then from (10.4) by using the diagonal symmetry of $\mathbf{M}$ and $\overline{\mathbf{M}}$, we obtain

$$\delta\bar{\sigma}(\dot{\bar{\epsilon}} - \overline{\mathbf{M}}\dot{\bar{\sigma}}) = \delta\bar{\sigma}\,\dot{\bar{\epsilon}} - \delta\bar{\epsilon}\,\dot{\bar{\sigma}} = \langle\delta\sigma\,\dot{\epsilon} - \delta\epsilon\,\dot{\sigma}\rangle = \delta\bar{\sigma}\langle\mathbf{B}^{\mathrm{T}}(\dot{\epsilon} - \mathbf{M}\dot{\sigma})\rangle \qquad (\delta\epsilon^{\mathrm{p}}_\eta \equiv \mathbf{0}) \tag{10.11}$$

for arbitrary $\delta\bar{\sigma}$. It follows that

$$\dot{\bar{\epsilon}} - \overline{\mathbf{M}}\dot{\bar{\sigma}} = \langle(\dot{\epsilon} - \mathbf{M}\dot{\sigma})\mathbf{B}\rangle = \langle\sum_K(\mu_K\mathbf{B})\dot{\gamma}_K\rangle \tag{10.12}$$

by $(10.1)_1$, so that $(10.6)_1$ has been proved. The substitution of (10.5), with the variations replaced by rates, into $(10.1)_3$ shows that $(10.6)_3$ is also satisfied, which completes the proof. □

The above statement on the transmissibility of the structure of incremental constitutive relationships from a meso-level to the level of a discretized macro-element appears to be new in the present form, to the author's knowledge, although its validity could be conjectured on the basis of similar theorems established for elastic-plastic structures by Maier [10]. A significant part of Theorem 1 and the main line of the proof, with the exception of the use of the influence tensors $\mathbf{Q}_{\eta\iota}$, are analogous to those for a non-discretized macroelement, cf. [6].

The explicit formula $(10.9)_3$ for effective hardening moduli for a discretized macro-element may be useful in a study of a hardening law for a polycrystal. It can be shown, e.g. by comparing the elastic energy of the aggregate after imposing

[1] A direct use of (10.5) and $(10.1)_1$ shows that $\overline{\mathbf{M}} = \langle\mathbf{M}\mathbf{B}\rangle$ as well.

arbitrary variations in plastic strain in two different cells in a different order, that $\mathbf{Q}_{\eta\iota} = \mathbf{Q}^{\mathrm{T}}_{\iota\eta}$ on account of $\mathbf{M} = \mathbf{M}^{\mathrm{T}}$. Consequently, if the hardening moduli matrix at the meso-level is symmetric, $h_{KL} = h_{LK}$, then this property is transmitted to the macro-level, $\overline{h}_{K\eta L\iota} = \overline{h}_{L\iota K\eta}$. A quantitative study could be done by using $\mathbf{Q}_{\eta\iota}$ determined from the analysis of interaction between two ellipsoidal inclusions, cf. [1]. Here, we restrict ourselves to identifying $\overline{h}_{K\eta L\iota}$ for three well-known simple models of a polycrystal (details of the derivation are omitted):

Uniform stress approximation:

$$\dot{\sigma}_\eta = \dot{\overline{\sigma}}, \qquad \mathbf{Q}_{\eta\iota} = \mathbf{0}, \qquad \overline{h}_{K\eta L\iota} = \delta_{\eta\iota} h_{KL\eta}/\nu_\iota\,. \tag{10.13}$$

Uniform strain approximation:

$$\begin{gathered}\dot{\epsilon}_\eta = \dot{\overline{\epsilon}}, \qquad \mathbf{Q}_{\eta\iota} = -\mathbf{L}_\eta \langle \mathbf{L}\rangle^{-1}\mathbf{L}_\iota + \delta_{\eta\iota}\mathbf{L}_\eta/\nu_\iota\,, \qquad \mathbf{L} = \mathbf{M}^{-1}, \\ \overline{h}_{K\eta L\iota} = -\boldsymbol{\mu}_{K\eta}\mathbf{L}_\eta\langle \mathbf{L}\rangle^{-1}\mathbf{L}_\iota\boldsymbol{\mu}_{L\iota} + \delta_{\eta\iota}(\boldsymbol{\mu}_K\mathbf{L}\boldsymbol{\mu}_L + h_{KL})_\eta/\nu_\iota\,.\end{gathered} \tag{10.14}$$

KBW self-consistent scheme:

$$\begin{gathered}\mathbf{Q}_{\eta\iota} = \mathbf{Q}_{\varrho\iota} \quad \text{for } \eta \neq \iota \neq \varrho, \qquad \mathbf{Q}'_{\eta\iota} = -2G(1-b)(1-\delta_{\eta\iota}/\nu_\iota)\mathbf{I}', \\ \overline{h}_{K\eta L\iota} = -2G(1-b)\boldsymbol{\mu}_{K\eta}\boldsymbol{\mu}_{L\iota} + \delta_{\eta\iota}(2G(1-b)\boldsymbol{\mu}_K\boldsymbol{\mu}_L + h_{KL})_\eta/\nu_\iota\,.\end{gathered} \tag{10.15}$$

In the last example, uniform isotropic elasticity and Eshelby's solution for a spherical inclusion are used along with plastic incompressibility, a prime denotes a deviatoric part, G is the elastic shear modulus, and $b = 0.4 \div 0.5$ depends on Poisson's ratio.

It can be seen that $\overline{h}_{K\eta L\iota}$ can generally be decomposed into three parts, due to: self-hardening and constraint-hardening within a grain, and interaction hardening (or rather softening in the two last models) between different grains. The interaction softening moduli between different grains are here much smaller in magnitude than the constraint-hardening term with a large factor $(1/\nu_\iota)$, but their influence on the overall hardening need not be negligible unless the volume fraction of the plastically deforming grains is small. If this fraction is small and the plastic deformation in each grain is due to single slip, e.g. when the loading direction approaches a generator of the current elastic unloading cone for a polycrystal, then the matrix $(\overline{h}_{K\eta L\iota})$ becomes essentially a diagonal matrix, as in the Koiter's theory [9]. It can be added that the constraint-hardening term in (10.14) or (10.15) is overestimated when the volume fraction of the plastically deforming grains is closer to unity.

10.2.3 The Energy Function for a Material Element

Henceforth, in view of the analogy between (10.6), (10.7) and (10.1), (10.2), the notation from Subsection 10.2.1 will be used in the constitutive description of the material either at the meso- or macro-level. The yield functions f_K at a fixed temperature can be given the following thermodynamic interpretation [19] [2]

$$f_K = \pi_K - \pi_K^{\mathrm{c}}, \qquad \dot{\phi} = \mathbf{t}\dot{\mathbf{e}} - \sum_K \pi_K \overset{\circ}{\gamma}_K, \qquad D = \sum_K \pi_K^{\mathrm{c}} \overset{\circ}{\gamma}_K \geq 0, \tag{10.16}$$

where $\phi = \phi(\mathbf{e}, \mathcal{H})$ is the isothermal free energy per unit reference volume, and $D = D(\dot{\mathcal{H}}, \mathcal{H})$ is the dissipation function that defines the rate of entropy production due to *virtual* $\overset{\circ}{\gamma}_K$'s. The first condition in (10.2) defines now an admissible (elastic) domain in the space of thermodynamic forces π_K which are limited by their threshold values π_K^{c}, and the second means that the actual plastic dissipation rate is equal to D. In the stability analysis below, we will consider infinitesimal perturbations of π_K such that in a *given* state the conditions in (10.2) are still satisfied but their rate-counterparts in (10.3) may be violated. The constraints (10.1) are imposed in every case.

In order to deal with possibly simple formulae, the attention will be restricted to a material element at the surface data such that the chosen measure of either external stress $\mathbf{t}^{\mathrm{ext}}$ or overall strain $\mathbf{e}$ is controlled, cf. Hill [7]. In analogy to the rates of internal thermodynamic forces, the rates of $\mathbf{t}$ and $\mathbf{t}^{\mathrm{ext}}$ need not coincide *a priori* in the stability analysis, while kinematic constrains, if present, are assumed to be always satisfied. The tractions on the element surface are associated here with a potential energy ω (per unit reference volume) equal to $-\mathbf{t}^{\mathrm{ext}}\mathbf{e}$ for controlled $\mathbf{t}^{\mathrm{ext}}$ or to zero for controlled $\mathbf{e}$. The dissipated energy is imagined to be transported to a heat reservoir. At constant loading (here at fixed $\mathbf{t}^{\mathrm{ext}}$ or $\mathbf{e}$) and in the absence of disturbances, the compound thermodynamic system: {material element + loading environment + heat reservoir} can be treated as isolated. Consider an amount of energy E which has to be supplied to the *system* in a quasi-static isothermal process of plastic deformation, in general at varying loading and in the presence of additional perturbing forces [14, 17]. If perturbations are allowed to enter the level of thermodynamic forces within a material element then the rate of E (i.e. of the internal energy of the system) can be specified in the present context as

$$\begin{aligned} \dot{E} &= \dot{\phi} + D + \dot{\omega} \\ &= \mathbf{t}\dot{\mathbf{e}} - \sum_K f_K \overset{\circ}{\gamma}_K - \dot{\mathbf{t}}^{\mathrm{ext}}\mathbf{e} - \mathbf{t}^{\mathrm{ext}}\dot{\mathbf{e}} \quad \text{for controlled } \mathbf{t}^{\mathrm{ext}}, \\ &= \mathbf{t}\dot{\mathbf{e}} - \sum_K f_K \overset{\circ}{\gamma}_K \quad \text{for controlled } \mathbf{e}, \end{aligned} \tag{10.17}$$

per unit reference volume of the material.

An increment ΔE in the value of E from a given state, augmented by a kinetic energy term, represents a natural Lyapunov function for examining isothermal stability of *equilibrium* [12][17]. A time integral of D is generally path-dependent, however, in the first instance the stability conditions can be formulated in terms of

[2] There is a distinction from Rice's theory [19] in that the concept of a dissipation function D is explicitly used here. In turn, a distinction from the theory of 'standard' materials based on two potentials (cf. [13]) is that here no explicit reference to internal *state* variables is needed.

the increment in E taken along a 'direct' path of departure from an equilibrium state. That increment evaluated to *second* order is denoted by $\Delta^{\mathrm{d}}E \equiv \dot{E}\Delta t + \frac{1}{2}\ddot{E}(\Delta t)^2$, where the (forward) rates are taken in a given state of mechanical equilibrium in any admissible direction. If direct perturbations of thermodynamic forces are not considered then $\dot{\phi} + D = \mathbf{t}\dot{\mathbf{e}}$ at every instant, and the condition of stability of equilibrium for a material element can be formulated in terms of the second-order work, cf. [7]. If the prescribed values of $\mathbf{t}^{\mathrm{ext}}$ or $\mathbf{e}$ are varying in time then the *difference* between $\Delta^{\mathrm{d}}E$ for two (perturbed and fundamental) processes can be used to examine stability of an isothermal quasi-static *process* [14, 16, 17][2].

10.3 Non-Uniqueness and Stability Within a Material Element

10.3.1 Stability of Equilibrium

By specifying the general concepts of thermodynamic stability of an equilibrium state in plasticity [18] for the material element at a constant temperature, the following condition for stability *in the first approximation* of a given state $\mathcal{S}$ of mechanical equilibrium of the element is obtained:

$$\ddot{E} > 0 \quad \text{for every } [\mathring{\gamma}] \geq [0],\ [\mathring{\gamma}] \neq [0] \text{ subject to (10.1) and (10.2) in } \mathcal{S} \tag{10.18}$$

at fixed either $\mathbf{t}^{\mathrm{ext}}$ or $\mathbf{e}$.[3] This means that $\Delta^{\mathrm{d}}E > 0$ when $\dot{E} = 0$ in $\mathcal{S}$. There is a close analogy between (10.18) and the stability condition discussed by Nguyen [12], with the distinction that here (10.18) can be applied without the need of introducing internal *state* variables. The condition (10.18) refers to the stability against internal rearrangements within a material element at the prescribed macroscopically uniform boundary data over the element surface.

By taking the rate of the expressions in (10.17) and rearranging with the help of (10.1), we obtain

$$\begin{aligned} \ddot{E} &= \dot{\mathbf{t}}\mathbf{M}\dot{\mathbf{t}} + \sum_{K,L\in\mathcal{P}} h_{KL}\mathring{\gamma}_K\mathring{\gamma}_L - \sum_{K\notin\mathcal{P}} f_K\ddot{\gamma}_K \qquad \text{at fixed } \mathbf{t}^{\mathrm{ext}}, \\ &= \sum_{K,L\in\mathcal{P}} g_{KL}\mathring{\gamma}_K\mathring{\gamma}_L - \sum_{K\notin\mathcal{P}} f_K\ddot{\gamma}_K \qquad \text{at fixed } \mathbf{e}, \end{aligned} \tag{10.19}$$

where the equilibrium condition $\mathbf{t} = \mathbf{t}^{\mathrm{ext}}$ has been substituted in the former expression, and g_{KL} are the interaction moduli that appear in a dual formulation of (10.1) [8, 20],

$$g_{KL} = h_{KL} + \boldsymbol{\mu}_K\mathbf{L}\boldsymbol{\mu}_L\,, \qquad \mathbf{L} = \mathbf{M}^{-1}\,. \tag{10.20}$$

Since $\mathbf{M}$ has been assumed positive definite, and $f_K\ddot{\gamma}_K \leq 0$ by (10.2) and by the non-negativeness of $\mathring{\gamma}_K$'s at every instant, the stability condition (10.18) is satisfied

[3] The relations for vectors $[\mathring{\gamma}]$ should be read componentwise for every $K \in \mathcal{P}$.

if

$$\left.\begin{aligned}&\sum_{K,L\in\mathcal{P}} h_{KL}\mathring{\gamma}_K\mathring{\gamma}_L > 0 && \text{at fixed } \mathbf{t}^{\text{ext}},\\ & && \text{for every } [\mathring{\gamma}] \geq [0],\ [\mathring{\gamma}] \neq [0]\\ &\sum_{K,L\in\mathcal{P}} g_{KL}\mathring{\gamma}_K\mathring{\gamma}_L > 0 && \text{at fixed } \mathbf{e}\,.\end{aligned}\right. \tag{10.21}$$

Conversely, the fulfillment of (10.18) implies (10.21) since the contribution of the remaining terms in (10.19) can be made zero independently of $[\mathring{\gamma}]$.

Although the present derivation of (10.21) is not typical, [4] the restrictions in (10.21) on the hardening moduli matrix are themselves not new, especially when viewed in connection with the familiar conditions

$$\dot{\mathbf{t}}\dot{\mathbf{e}}^{\mathrm{p}} > 0, \quad \dot{\mathbf{e}}^{\mathrm{p}} \equiv \dot{\mathbf{e}} - \mathbf{M}\dot{\mathbf{t}} \qquad \text{or} \qquad -\dot{\mathbf{t}}^{\mathrm{p}}\dot{\mathbf{e}} > 0, \quad \dot{\mathbf{t}}^{\mathrm{p}} \equiv \dot{\mathbf{t}} - \mathbf{L}\dot{\mathbf{e}}\,. \tag{10.22}$$

The first inequality in (10.21) reduces to the first inequality in (10.22) if $[\mathring{\gamma}]$ is additionally restricted by (10.3), and the same concerns the second inequalities. It must be noted that while the second inequalities in (10.22) and (10.21) (and in (10.28) below) hold either for all or for none of work-conjugate pairs $(\mathbf{t}, \mathbf{e})$ of stress and strain measures, this is not so for the former inequalities, cf. [7][4]. The lack of invariance is related to the dependence of stability on *which* stress measure is kept constant. In spite of this, for a reasonable choice of the measures, the first condition in (10.21) is acceptable for effectively hardening solids. Observe that (10.21) holds automatically if all h_{KL} are positive, contrary to the stronger condition of positive definiteness of the matrix (h_{KL}) which fails in crystals when effective latent hardening of the slip systems is greater than their self-hardening. The second (invariant) condition imposed on g_{KL} in (10.21) is still less restrictive (for $\mathbf{L}$ positive definite) and refers to the most rigid boundary constraints. It may thus be regarded as a requirement of the intrinsic thermodynamic stability of the material, either at the meso- or macro-level. This condition appeared previously in the literature on crystal plasticity in a different context, cf. Havner [4].

10.3.2 Non-Uniqueness and Stability of a Deformation Process

Throughout this subsection we shall assume that

$$h_{KL} = h_{LK}\,. \tag{10.23}$$

This symmetry assumption was sometimes criticized as possessing apparently no physical justification. On the other hand, (10.23) can be supported by a thermodynamic argument. Here, (10.23) is adopted as a working hypothesis.

[4] A somewhat different derivation of the condition for g_{KL} was given in [17].

Consider the incremental problem from Subsection 10.2.1 formulated for a single crystal or, by appealing to the structural transmissibility property from Subsection 10.2.2, for a polycrystal. As in Subsection 10.2.3, we assume that either an external stress or strain is controlled. The following problem is to be solved in a given equilibrium state $\mathcal{S}$ where $\mathbf{M}$, $\boldsymbol{\mu}_K$, f_K, h_{KL} and either $\dot{\mathbf{t}} = \dot{\mathbf{t}}^{\mathrm{ext}}$ or $\dot{\mathbf{e}}$ are given:

(P1) *Find* $[\mathring{\gamma}]$ *that satisfies (10.3) subject to (10.1) and (10.2) in* $\mathcal{S}$.

It is well known that uniqueness of a solution to (P1) is not guaranteed unless (h_{KL}) is positive definite. The lack of a physical support to the latter property, and in consequence the non-uniqueness of a solution, constitute a known difficulty in the modeling of time-independent crystalline solids.

The non-uniqueness problem is addressed here (for both crystals and polycrystals) by imposing an additional requirement of the energy-type stability of a *process* of plastic deformation within the material element, under the symmetry restriction on (h_{KL}). The following stability criterion is postulated (cf. [14, 17]): *In a stable quasi-static process, the increment* $\Delta^{\mathrm{d}}E$ *of the energy function* E *is minimized within the class of all nonnegative increments in* γ_K*'s accompanied by kinematically admissible deformation increments.* From (10.17) it follows that a minimum of $\dot{E}$ on the set of all $\mathring{\gamma}_K \geq 0$ in an admissible (i.e. with all $f_K \leq 0$) equilibrium state is attained for any $[\mathring{\gamma}]$ provided $\mathring{\gamma}_K = 0$ for $K \notin \mathcal{P}$. The stability criterion reduces to

$$\ddot{E}([\mathring{\gamma}]) \geq \ddot{E}([\mathring{\gamma}]^0) \qquad \text{for every } [\mathring{\gamma}] \text{ subject to (10.1) and (10.2) in } \mathcal{S} \tag{10.24}$$

if $[\mathring{\gamma}]^0$ corresponds to a stable process. Similarly to (10.18), the condition (10.24) refers to the stability against internal rearrangements within a material element at the macroscopically uniform boundary data which now are *varying* in time.

The interpretation of (10.24) as a condition *necessary* for stability of a process is supported by the following result, which appears to be new.

Theorem 2 *Suppose that along a segment of a solution path* $(\dot{\mathbf{e}}^0, \mathring{\gamma}_K^0)(t)$ *the condition (10.24) is not satisfied while the condition (10.18) of stability of equilibrium holds.*

Then at every point of that segment there exists another solution to (P1) that is energetically preferable to $[\mathring{\gamma}]^0$ *in the sense of a smaller value of* $\Delta^{\mathrm{d}}E$.

Proof of the theorem follows as a corollary of Lemmas 2 and 3 given below. □

The conclusion of the above theorem can be compared with that obtained in [16]. From Theorem 2 it follows, in particular, that if (10.18) holds then the fulfillment of (10.24) is *necessary* for the uniqueness of a solution to (P1). In the case of non-uniqueness, solutions which are unstable in the sense of Theorem 2 can be eliminated by replacing (P1) by the following problem to be solved:

(P2) *Find* $[\mathring{\gamma}]$ *that minimizes* $\ddot{E}$ *subject to (10.1) and (10.2) in* $\mathcal{S}$.

Some needed auxiliary statements are now provided.

Lemma 1 (P2) *is equivalent to*:

(P3) *Find* $[\mathring{\gamma}] \geq [0]$ *that minimizes* $J([\mathring{\gamma}])$ *in* $\mathcal{S}$, *where*

$$J = \begin{cases} \frac{1}{2} \sum_{K,L\in\mathcal{P}} h_{KL}\mathring{\gamma}_K\mathring{\gamma}_L - \dot{\mathbf{t}} \sum_{K\in\mathcal{P}} \mu_K\mathring{\gamma}_K & \text{for given } \dot{\mathbf{t}} = \dot{\mathbf{t}}^{\text{ext}}, \\ \frac{1}{2} \sum_{K,L\in\mathcal{P}} g_{KL}\mathring{\gamma}_K\mathring{\gamma}_L - \dot{\mathbf{e}}\mathbf{L} \sum_{K\in\mathcal{P}} \mu_K\mathring{\gamma}_K & \text{for given } \dot{\mathbf{e}}. \end{cases} \tag{10.25}$$

Proof. In analogy to $(10.19)_1$ but for $\ddot{\mathbf{t}}^{\text{ext}} \neq \mathbf{0}$ at controlled $\mathbf{t}^{\text{ext}}$, from (10.17) and (10.1) in an equilibrium state we have

$$\ddot{E} = \dot{\mathbf{t}}\mathbf{M}\dot{\mathbf{t}} - 2\dot{\mathbf{t}}^{\text{ext}}\mathbf{M}\dot{\mathbf{t}} + \sum_{K,L\in\mathcal{P}} h_{KL}\mathring{\gamma}_K\mathring{\gamma}_L - 2\dot{\mathbf{t}}^{\text{ext}} \sum_{K\in\mathcal{P}} \mu_K\mathring{\gamma}_K - \sum_{K\notin\mathcal{P}} f_K\ddot{\gamma}_K - \ddot{\mathbf{t}}^{\text{ext}}\mathbf{e}\,. \tag{10.26}$$

For $\mathbf{M}$ symmetric and positive definite, the first two terms in this expression give the smallest contribution when $\dot{\mathbf{t}} = \dot{\mathbf{t}}^{\text{ext}}$; this condition of continuing equilibrium is thus a prerequisite for $[\mathring{\gamma}]$ to be a solution to (P2). The smallest value of the term with $\ddot{\gamma}_K$'s is zero, as in (10.19), and the last term in (10.26) has a given value. The remaining two terms constitute the quadratic form $J([\mathring{\gamma}])$ minimized in the problem (P3), with a multiplier two. This proves validity of the lemma for controlled $\mathbf{t}^{\text{ext}}$.

The proof for controlled $\mathbf{e}$ follows in an analogous (somewhat simpler) way from the expression

$$\ddot{E} = \dot{\mathbf{e}}\mathbf{L}\dot{\mathbf{e}} + \sum_{K,L\in\mathcal{P}} g_{KL}\mathring{\gamma}_K\mathring{\gamma}_L - 2\dot{\mathbf{e}}\,\mathbf{L} \sum_{K\in\mathcal{P}} \mu_K\mathring{\gamma}_K - \sum_{K\notin\mathcal{P}} f_K\ddot{\gamma}_K + \ddot{\mathbf{e}}\mathbf{t}\,. \tag{10.27}$$

□

The minimization problem in (P2) reduces thus to a quadratic programming problem (P3), while (P1) is the associated linear complementarity problem. It is well known that (P1) is equivalent to (P3) under the additional assumption that (h_{KL}) is positive semidefinite, since then (P3) becomes a problem of *convex* programming. We do *not* make this additional assumption, so that our problem (P3) (and hence (P2)) is in general a *nonconvex* programming problem and is no longer equivalent to (P1).

The statements given below (with shortened proofs), which are essentially standard in relation to a mathematical problem (P3), become of more interest for the problem (P2) which has a physical meaning. In spite of certain similarities, the approach presented here differs from that by Sewell [21].

Lemma 2 *Any solution to (P2) is also a solution to (P1).*

Proof. Let $[\mathring{\gamma}]^0$ be a solution to (P2); by Lemma 1 $[\mathring{\gamma}]^0$ is also a solution to (P3). The conditions (10.3) are satisfied for $[\mathring{\gamma}] = [\mathring{\gamma}]^0$ as the Kuhn-Tucker conditions for the minimization problem in (P3), which completes the proof. □

Lemma 3 *If (10.21) holds then a solution to (P2) exists.*

Proof. (10.21) implies that $J([\mathring{\gamma}]) \to +\infty$ as $\mathring{\gamma}_K \to +\infty$ for some $K \in \mathcal{P}$ while $[\mathring{\gamma}] \geq [0]$. It follows that $J([\mathring{\gamma}])$ is bounded below on the set of $[\mathring{\gamma}] \geq [0]$ and, by the Frank-Wolf theorem, attains its minimum on this set. The use of Lemma 1 completes the proof. □

Lemma 4 *Among all solutions to (P1), the solution to (P2)*
maximizes $\dot{\mathbf{t}}\dot{\mathbf{e}}$ *(and* $\dot{\mathbf{t}}\dot{\mathbf{e}}^{\mathrm{p}}$*) for controlled* $\mathbf{t}^{\mathrm{ext}}$,
minimizes $\dot{\mathbf{t}}\dot{\mathbf{e}}$ *(and* $\dot{\mathbf{t}}^{\mathrm{p}}\dot{\mathbf{e}}$*) for controlled* $\mathbf{e}$.

Proof follows by rearranging the minimized expression for $\ddot{E}$ with the help of (10.1) and $(10.3)_2$. □

The conclusions of Lemma 4 are analogous to the criterion of selection of an 'optimal' solution proposed for single crystals by Franciosi and Zaoui [3] in a somewhat different form for a wider class of constitutive laws without (10.23). Their proposal was based on the concept of minimizing the energy supplied to the mechanical system as in [14], while in this paper the argumentation is more in the spirit of thermodynamics.

Lemma 5 *A necessary condition for* $[\mathring{\gamma}]^0$ *to be a solution to (P2) is that*

$$\begin{aligned} &\sum_{K,L\in\mathcal{A}^0} h_{KL} w_K w_L \geq 0 && \textit{for controlled } \mathbf{t}^{\mathrm{ext}}, \\ &\qquad\qquad \textit{for every } [w],\ \mathcal{A}^0 = \{K : \mathring{\gamma}_K^0 > 0\} && \qquad (10.28) \\ &\sum_{K,L\in\mathcal{A}^0} g_{KL} w_K w_L \geq 0 && \textit{for controlled } \mathbf{e}\,. \end{aligned}$$

Proof. If $[\mathring{\gamma}]^0$ is a solution to (P2) and hence to (P3) then it must minimize $J([\mathring{\gamma}])$ on the set of $[\mathring{\gamma}]$ having the same set $\mathcal{A}^0$ of the indices of active mechanisms. (10.28) follows as a classical condition necessary for the minimum. □

The conditions (10.28) or (10.24) need not be understood as a restriction on the assumed constitutive law. Rather, they provide a criterion of choice between non-uniquely defined plastic deformation modes of the material, in analogy to a similar implication of the general energy criterion of plastic instability in finite bodies [16].

On the above theoretical basis, the following procedures can be suggested for determining the material response in the case of constitutive non-uniqueness met for the problem (P1):

(i) Replace the problem (P1) by the non-convex quadratic programming problem (P3).

(ii) Complement the problem (P1) by selecting the solution according to the conclusion of Lemma 4.

(iii) Complement the problem (P1) by selecting the solution that satisfies (10.28).

A more detailed discussion of the procedures, and of their relation to other proposals and to experimental data, is beyond the scope of this article. We restrict ourselves to a simple illustrative example of double slipping in a single crystal with the effective hardening matrix for a selected pair of stress and strain measures given by

$$(h_{KL}) = h \cdot \begin{pmatrix} 1 & 1+q \\ 1+q & 1 \end{pmatrix}, \qquad K, L = 1, 2, \qquad h > 0, \qquad q > 0. \qquad (10.29)$$

It is clear that the condition (10.18) of stability of equilibrium is satisfied under both strain and stress control, although the matrix (h_{KL}) is indefinite. The process of uniform double slipping is unstable under stress control according to the criterion (10.24) and Lemma 5, and may be stable under strain control. According to Theorem 2 or Lemma 3, a solution under stress control exists which satisfies (10.24), and in the loading case this will be single slipping $\mathring{\gamma}_1^0 > 0$, $\mathring{\gamma}_2^0 = 0$ or $\mathring{\gamma}_1^0 = 0$, $\mathring{\gamma}_2^0 > 0$.

For polycrystals with a large number of internal plastic deformation mechanisms, a reformulation of the path stability conditions in terms of the second-order work expression for the material element may be found useful. This can be done with the help of the second-order work formula derived in [15] and valid for *indirect* deformation paths whose complexity is preserved as their length tends to zero. The comparison of that formula with the expression in (10.27) shows that the second-order work along a solution path found from (P3) or (P2) under strain control is minimized within the class of arbitrary circuitous short paths that lead to the same final strain increment.

Acknowledgements

This work was supported by the State Committee of Research (KBN) in Poland under the project No. 3 P404 035 07.

References

1. Berveiller, M., Fashi-Fehri, 0. and Hihi, A. (1987) The problem of two plastic and heterogeneous inclusions in an anisotropic medium. *Int. J. Engng Sci.* **25**, 691-709.
2. Fedelich, B. and Ehrlaher, A. (1997) An analysis of stability of equilibrium and of quasi-static transformations on the basis of the dissipation function, *Eur. J. Mech. A/Solids* **16**, 833-855.
3. Franciosi, P. and Zaoui, A. (1991) Crystal hardening and the issue of uniqueness, *Int. J. Plasticity* **7**, 295-311.
4. Havner, K.S. (1992) *Finite Plastic Deformation of Crystalline Solids*, University Press, Cambridge.
5. Hill, R. (1966) Generalized constitutive relations for incremental deformation of metal crystals by multislip, *J. Mech. Phys. Solids* **14**, 95-102.
6. Hill, R. (1967) The essential structure of constitutive laws for metal composites and polycrystals, *J. Mech. Phys. Solids* **15**, 79-95.

7. Hill, R. (1978) Aspects of invariance in solids mechanics, *Advances in Applied Mechanics, Vol.* **18**, Acad. Press, New York, pp. 1-75.
8. Hill, R. and Rice, J.R. (1972) Constitutive analysis of elastic- plastic crystals at arbitrary strain, *J. Mech. Phys. Solids* **20**, 401-413.
9. Koiter, W.T. (1953) Stress-strain relations, uniqueness and variational theorems for elastic-plastic materials with a singular yield surface, *Quart. Appl. Math.* **11**, 350-353.
10. Maier, G. (1969) "Linear" flow-laws of elastoplasticity: a unified general approach, *Rend. Acc. Naz. Lincei, Ser. VIII*, Vol. XLVII, 266-276.
11. Mandel, J. (1965) Generalisation de la théorie de plasticité de W.T. Koiter, *Int. J. Solids Structures* **1**, 273-295.
12. Nguyen, Q.S. (1993) Bifurcation and stability of time-independent standard dissipative systems, in Q.S. Nguyen (ed.), *CISM Lecture Notes No.* 327, Springer, Wien - New York, pp. 45-94.
13. Nguyen, Q.S. (1994) Bifurcation and stability in dissipative media (plasticity, friction, fracture), *Appl. Mech. Rev.* **47**, no 1, Part 1, 1-31.
14. Petryk, H. (1982) A consistent energy approach to defining stability of plastic deformation processes, in F.H. Schroeder (ed.), *Stability in the Mechanics of Continua, Proc. IUTAM Symp. Nümbrecht* 1981, Springer, Berlin - Heidelberg, pp. 262-272.
15. Petryk, H. (1991a) On the second-order work in plasticity, *Arch. Mech.* **43**, 377-397.
16. Petryk, H. (1991b) The energy criteria of instability in time-independent inelastic solids, *Arch. Mech.* **43**, 519-545.
17. Petryk, H. (1993) Stability and constitutive inequalities in plasticity, in W. Muschik (ed.), *CISM Lecture Notes No.* 336, Springer, Wien - New York, pp. 259-329.
18. Petryk, H. (1995) Thermodynamic stability of equilibrium in plasticity, *J. Non-Equil. Thermodyn.* **20**, 132-149.
19. Rice, J.R. (1971) Inelastic constitutive relations for solids: an internal-variable theory and its application to metal plasticity. *J. Mech. Phys. Solids*, **19**, 433-455.
20. Sewell, M.J. (1972) A survey of plastic buckling, in H.H.E. Leipholz (ed.), *Stability*, Univ. of Waterloo Press, Ontario, pp. 85-197.
21. Sewell, M.J. (1987) *Maximum and minimum principles*, University Press, Cambridge.

Chapter 11
General Conditions for Uniqueness in Materials With Multiple Mechanisms of Inelastic Deformation

Abstract This study is concerned with multi-mode inelastic behaviour at macroscopically uniform deformation. The material is assumed to be time-independent; the physical origin of inelasticity may be otherwise arbitrary, including plasticity of crystals and polycrystals, micro-cracking, phase transformation, etc. A non-linear rate-problem of continuing mechanical equilibrium at finite strain is examined for a material element subject to deformation-sensitive loading under partial kinematic constraints. General conditions for uniqueness of the material response are established. As an application to predicting the onset of strain localization or failure, the condition is derived that excludes the bifurcation in a band from homogeneous deformation. In contrast to the usual requirement of ellipticity of the tangent stiffness moduli, the present condition for uniqueness takes into account any possible unloading and is directly imposed on the matrix of interaction moduli of internal mechanisms. Lower and upper bounds are established for the primary shear-band bifurcation along a smooth straining path.

Keywords: Microstructures, Multi-mode inelasticity, Constitutive behaviour, Finite strain, Bifurcation

11.1 Introduction

The aim of this paper is to study the incremental response of time-independent materials whose inelastic deformation is due to activity of a finite number of internal mechanisms of arbitrary physical nature. The attention is focused on a representative volume element of the material, in general heterogeneous, which can be treated at a macroscopic level as if it were homogeneous and uniformly deformed. The rates of activity of the internal mechanisms (inelastic multipliers) are related to the

---

[0] Reprinted from: Journal of the Mechanics and Physics of Solids 48, 367-396, 2000. H.Petryk, General conditions for uniqueness in materials with multiple mechanisms of inelastic deformation. Copyright © 2000, with permission from Elsevier.

H. Petryk, *Theory of Plastic Instability*,
https://doi.org/10.1007/978-3-031-98223-1_11

rates of macroscopic strain or stress by a set of linear constitutive equations and inequalities. The existence of an elastic domain in a stress or strain space is not required. The set of active mechanisms is not given in advance, which results in the incremental nonlinearity (in fact, piecewise linearity) of the entire constitutive law. The rate-problem of continuing mechanical equilibrium of the material element will be examined in a possibly general setting, with the particular emphasis on the question of uniqueness of the material response under a broad class of external loadings. This encompasses the uniqueness of a constitutive law itself, as well as the problem of bifurcation in a band from homogeneous deformation, related to the onset of strain localization or failure.

The assumption of multiple internal mechanisms of time-independent inelastic deformation may be relevant to various physical phenomena. It may correspond to the interactive plastic shearing on multiple slip systems in metal crystals [15, 40, 21, 1, 12], or more generally, to plasticity with many intersecting yield surfaces [24, 27, 26, 46], to single-mode plasticity coupled with damage [8, 3, 50, 11], to propagation of a number of microcracks [41, 28, 30, 5, 38], to multi-variant martensitic phase transformation in shape memory alloys [32, 48, 33, 22], to the transformation induced plasticity (TRIP) in steels [25, 10, 6], and others. A general conclusion can be drawn that the structure of the incremental constitutive relations, typical for plasticity with multiple yield mechanisms, is relevant to a broad class of inelastic materials, not necessarily characterized by a permanent strain after unloading to zero stress.

The restriction to macroscopically uniform deformation means that the question of uniqueness will primarily concern the set and magnitudes of non-zero inelastic multipliers corresponding to active mechanisms of inelastic deformation within the material element. The lack of such uniqueness in the incremental constitutive law represents the well-known difficulty in the theory of plasticity of single crystals deformed by multislip, which might also refer to other inelastic phenomena. A sufficient condition for uniqueness of plastic multipliers for a given stress-rate in the small strain framework was given by Mandel [27] and Hill [15] in the form of positive definiteness of the hardening moduli matrix. Maier [26] sharpened that result and gave necessary and sufficient conditions for the uniqueness to hold for every given stress-rate or strain-rate.[1] The principal requirement was that the matrix of the hardening or interaction moduli had all principal minors positive. Maier used results of the mathematical theory of the 'linear complementarity problem' (for a more recent account, see Cottle et al. [7]), emphasizing the fundamental analogy between that problem and the incremental constitutive relations in plasticity. Without taking advantages of that analogy, it may be not easy to arrive at precise general results, cf. [23]. An alternative approach, not used here, is to work with the related variational inequality problem, obtained under certain assumptions within a thermodynamic formulation, cf. [29].

Nowadays, it is well known that the 'small strain' theory of the incremental inelastic behaviour is frequently insufficient and that the constitutive rate-equations

[1] Maier's analysis [26] was extended to generalized stresses and strain, while the problems connected with finite deformations were not addressed.

should be analysed in a geometrically exact formulation. The reason is that the changes in stress-rate due to the geometric effect can be of the same order of magnitude as the stress-rate itself when tangent moduli are of the order of the stress, no matter whether the current strain is 'small' or not [16]. The transition to a geometrically exact theory influences the conditions of constitutive uniqueness. The problem is not so important in the case of strain control where the uniqueness condition, given at unrestricted strain by Hill and Rice [21] in the form (which will be sharpened below) of positive definiteness of the matrix of interaction moduli, is invariant for all finite strain measures. The situation changes when a stress control is considered. As pointed out in [16, 21], the sign of a hardening modulus and the uniqueness of the material response for a prescribed stress-rate can depend on the stress-rate measure used.

That conclusion is not in conflict with the objectivity requirement that the behaviour of a *given* material element under *given* loading must not depend on a subjective choice of the measures of stress and strain or their rates. The key observation is that if one stress measure is fully controlled then another stress measure becomes deformation-sensitive [19]. It follows that a measure-invariant condition for the uniqueness of the inelastic multipliers under stress control must be derived by studying a more general case of deformation-sensitive incremental loading. This is done in this paper for a general class of linear inhomogeneous (affine) constitutive rate-equations for the external loading, formulated independently of the constitutive law for the material. That approach offers, moreover, the possibility of analysing the behaviour of a material element, and the uniqueness problem in particular, taking into account the flexibility of a surrounding material.

To enhance the generality of the analysis and results, an extension to arbitrary linear constraints on the velocity gradient will be presented. This includes typical partial kinematic constraints due to plane strain, mixed traction/displacement boundary data or a prescribed material spin. The kinematic constraints can also be chosen to coincide with those for a material element within a bifurcation band in an infinite homogeneous solid, cf. [42]. The respective uniqueness condition, to be established below, represents the exclusion condition for the bifurcation in a band in an incrementally piecewise-linear material with multiple mechanisms of inelastic deformation, with any possible unloading taken into account. That result goes beyond the ellipticity condition for the tangent stiffness moduli [13, 42] and has not been found in related general studies of the incipient strain localization in incrementally nonlinear materials with more than two linear branches, cf. [34, 35, 36, 44, 47] and a survey [37].

In the tensor notation employed, bold-face letters denote vectors (in R^3) or second-order tensors, and 'sans serif' bold capitals (like A) stand for fourth-order tensors. Direct juxtaposition of two vectors or tensors denotes simple contraction, separation by a dot - double contraction, and the symbol ⊗ - the tensor product. In components on a fixed orthonormal basis, which are denoted by lower case Latin subscripts with the summation convention for repeated indices, it means that $s_j n_j = \mathbf{sn}$, $F_{ij} n_j = (\mathbf{Fn})_i$, $\sigma_{ij}\omega_{jk} = (\boldsymbol{\sigma}\mathbf{W})_{ik}$, $m_{ij}\omega_{ij} = \mathbf{m}\cdot\mathbf{W}$, $C_{ijkl}F_{kl} = (\mathsf{C}\cdot\mathbf{F})_{ij}$, $C_{ijkl}P_{klpq} = (\mathsf{C}\cdot\mathsf{P})_{ijpq}$, $s_i n_j = (\mathbf{s}\otimes\mathbf{n})_{ij}$, $\lambda_{ij}\mu_{kl} = (\boldsymbol{\lambda}\otimes\boldsymbol{\mu})_{ijkl}$. An upper index T denotes a transpose,

$(\mathbf{F}^{\mathrm{T}})_{ij} = F_{ji}$, $(\mathbf{A}^{\mathrm{T}})_{ijkl} = A_{klij}$, and a superimposed dot over a symbol denotes the *forward* rate, i.e. the right-hand material time derivative with respect to a time-like variable t.

11.2 Constitutive Framework

Consider a representative volume element of a time-independent material, in general heterogeneous, which at a macroscopic level is treated as uniformly deformed. It is assumed that the temperature within the element is uniform and equal to a constant temperature of the environment, and that body forces within the element are negligible. In the absence of disturbances, the element is assumed to be in mechanical equilibrium under surface tractions that result from interaction between the element and a loading device or surrounding material. Incremental mechanical properties of the material are described at the macroscopic level at deformations of unrestricted magnitude. Assuming that the material is free of couple stresses, symmetric second-order tensors objective in the Lagrangean sense [19, 31] can be used as macroscopic measures of finite strain $\mathbf{e}$ and stress $\mathbf{t}$. Their material right-hand time derivatives $(\dot{\mathbf{e}}, \dot{\mathbf{t}})$ can serve as objective measures of strain-rate and stress-rate, respectively, insensitive to the material spin. To enhance the generality of considerations, we will work with a broader class of spin-insensitive measures of strain-rate $\mathring{\mathbf{e}}$ and stress-rate $\mathring{\mathbf{t}}$ which need not be material time derivatives of any strain or stress measures. The choice of a pair $(\mathring{\mathbf{e}}, \mathring{\mathbf{t}})$ will be left arbitrary in order to examine the essential question of invariance of basic conditions with respect to that choice.

The material element is assumed to undergo time-independent inelastic deformation due to activity of a finite number of internal mechanisms. The physical origin of the inelasticity of deformation is arbitrary here; as mentioned in Introduction, it can correspond to typical plasticity sources in a variety of materials, but also to nucleation and growth of voids, to propagation of internal micro-cracks or diffusionless phase transformation fronts, etc. The class of material models examined here is restricted by the requirement, familiar from the plasticity theory, that for a particular mechanism of inelastic deformation to be active, it is necessary that a critical stress or strain level for that mechanism, dependent on the current material state but independent of the loading rate, be reached and maintained. Mathematically, the rate of activity of K-th mechanism, $K = 1, \dots, N$, is measured by a scalar rate-variable $\mathring{\gamma}_K$, taken nonnegative by convention. It can be nonzero only if the respective loading variable f_K has reached from below the threshold value, without loss of generality set equal to zero, viz.

$$\mathring{\gamma}_K \geq 0, \qquad f_K \leq 0, \qquad f_K\, \mathring{\gamma}_K = 0\,. \tag{11.1}$$

No implicit summation convention is adopted for the upper case Latin indices of the mechanisms: the summation over an index K will always be indicated explicitly. If all the $\mathring{\gamma}_K$'s are zero then the material undergoes purely elastic deformation. A circle rather than a dot over γ_K is used to emphasize that $\mathring{\gamma}_K$'s need not represent rates of any internal state variables, cf. [40].

Current values of the strain, stress and f_K's are obtained by the integration of their rates along a straining path with given initial conditions. The basic constitutive expressions for the rate-variables $\mathring{\mathbf{t}}$ and $\dot{f}_K$ in a given state of the material are assumed as linear forms in $(\mathring{\mathbf{e}}, \mathring{\gamma}_K)$:

$$\mathring{\mathbf{t}} = \mathbf{L}^{\mathrm{e}} \cdot \mathring{\mathbf{e}} - \sum_K \boldsymbol{\lambda}^{\mathrm{p}}_K \mathring{\gamma}_K, \qquad \dot{f}_K = \boldsymbol{\lambda}_K \cdot \mathring{\mathbf{e}} - \sum_L g_{KL} \mathring{\gamma}_L. \tag{11.2}$$

A dual form, fully equivalent to (11.2) under the sole additional assumption that $\mathbf{L}^{\mathrm{e}}$ represents an invertible linear operator acting in the space $\mathcal{S}$ of symmetric second-order tensors, reads

$$\mathring{\mathbf{e}} = \mathbf{M}^{\mathrm{e}} \cdot \mathring{\mathbf{t}} + \sum_K \boldsymbol{\mu}^{\mathrm{p}}_K \mathring{\gamma}_K, \qquad \dot{f}_K = \boldsymbol{\mu}_K \cdot \mathring{\mathbf{t}} - \sum_L h_{KL} \mathring{\gamma}_L. \tag{11.3}$$

By definition, all the second-order tensors in these equations are symmetric, and $\mathbf{L}^{\mathrm{e}}$, $\mathbf{M}^{\mathrm{e}}$ possess the corresponding minor symmetries, $L^{\mathrm{e}}_{ijkl} = L^{\mathrm{e}}_{jikl} = L^{\mathrm{e}}_{ijlk}$ (but generally $\mathbf{L}^{\mathrm{e}} \neq \mathbf{L}^{\mathrm{eT}}$ and $\mathbf{M}^{\mathrm{e}} \neq \mathbf{M}^{\mathrm{eT}}$).

The nonnegative $\mathring{\gamma}_K$'s for potentially active mechanisms are to be determined from the consistency conditions that follow from (11.1) by differentiation, viz.

$$\dot{f}_K \leq 0, \quad \dot{f}_K \mathring{\gamma}_K = 0 \qquad \text{for } K \in \mathcal{P} \equiv \{K \, : \, f_K = 0\}, \tag{11.4}$$

while $\mathring{\gamma}_K = 0$ for $K \notin \mathcal{P}$. The elastic moduli tensor $\mathbf{L}^{\mathrm{e}}$, direction tensors $\boldsymbol{\lambda}^{\mathrm{p}}_K$ of inelastic stress-rate, normals $\boldsymbol{\lambda}_K$ to loading hyperplanes in $\mathring{\mathbf{e}}$-space, and interaction moduli g_{KL}, depend on the material state and not on values of any rate-variables. In the dual formulation, the elastic compliance tensor $\mathbf{M}^{\mathrm{e}}$, direction tensors $\boldsymbol{\mu}^{\mathrm{p}}_K$ of inelastic strain-rate, normals $\boldsymbol{\mu}_K$ to loading hyperplanes in $\mathring{\mathbf{t}}$-space, and *effective* hardening (or softening) moduli h_{KL}, are likewise state-dependent, with the mutual relationships:

$$\mathbf{M}^{\mathrm{e}} = (\mathbf{L}^{\mathrm{e}})^{-1}, \quad \boldsymbol{\mu}^{\mathrm{p}}_K = \mathbf{M}^{\mathrm{e}} \cdot \boldsymbol{\lambda}^{\mathrm{p}}_K, \quad \boldsymbol{\mu}_K = \boldsymbol{\lambda}_K \cdot \mathbf{M}^{\mathrm{e}}, \quad h_{KL} = g_{KL} - \boldsymbol{\lambda}_K \cdot \mathbf{M}^{\mathrm{e}} \cdot \boldsymbol{\lambda}^{\mathrm{p}}_L. \tag{11.5}$$

These relationships can easily be verified by multiplying from the left the first equation in (11.2) by the inverse[2] of $\mathbf{L}^{\mathrm{e}}$, substituting the resulting expression into $(11.2)_2$ and comparing with (11.3). The formulation (11.2), taken here as primary, remains meaningful also for $\mathbf{L}^{\mathrm{e}}$ singular, and only in this respect it is more general than (11.3). It should be noted that the choice between (11.2) and (11.3) imposes no restriction on the selection of control variables, to be discussed later.

The above constitutive framework may be nowadays regarded as typical for time-independent materials with multiple internal mechanisms of plastic deformation, although it was adopted usually in a less general form, cf. the references listed in

[2] The inverse of a fourth-order tensor with assumed minor symmetries is meant in the sense appropriate for the space $\mathcal{S} \otimes \mathcal{S}$.

Introduction. In the present formulation no restriction (except those that follow from (11.1)) is imposed on the quantities used in (11.2). In particular, neither symmetry of any moduli matrix nor any normality law are assumed, and the existence of an elastic domain in strain or stress space is not required. The framework is not restricted to plasticity since other sources of time-independent inelastic behaviour, mentioned above, are also included. The formalism is directly applicable to micromechanical models when the strain-rate and stress-rate are understood as macroscopic variables, while $\mathring{\gamma}_K$'s may represent local (possibly strongly nonuniform) inelastic phenomena specified at a more refined scale.

11.3 Transformation Rules and Measure-Invariance

The question of transformation of the quantities involved in the constitutive rate equations (11.2) and (11.3) under a change of the measures of strain-rate and stress-rate is now addressed. This is an essential ingredient of a constitutive theory since it establishes the link between relationships formulated with different preferences concerning those measures. The analysis below provides an extension of the results due to Hill [16, 19, 20].

We shall assume that

$$f_K \text{ and } \mathring{\gamma}_K \text{ are measure-invariant,} \tag{11.6}$$

that is, the values of these quantities do not change under transformation to any other measures of the strain-rate and stress-rate.

Denote by $\mathbf{D}$ the Eulerian strain-rate and by $\overset{\nabla}{\sigma}$ the Zaremba-Jaumann flux of the Cauchy stress $\sigma = \sigma^{\mathrm{T}}$, corotational with the material spin $\mathbf{W}$. We take the pair $(\mathbf{D}, \overset{\nabla}{\sigma})$ as the basic spin-insensitive measures of the strain-rate and stress-rate. Recall the standard formulae

$$\mathbf{D} = (\nabla \mathbf{v})_{\mathrm{s}}, \qquad \mathbf{W} = (\nabla \mathbf{v})_{\mathrm{a}}, \qquad \overset{\nabla}{\sigma} = \dot{\sigma} + \sigma \mathbf{W} - \mathbf{W} \sigma, \tag{11.7}$$

where $\nabla \mathbf{v}$ is the spatial gradient (in the current configuration) of a material velocity $\mathbf{v}$. Henceforth, subscripts 's' and 'a' denote the symmetric and antisymmetric part, $\frac{1}{2}(\mathbf{a} + \mathbf{a}^{\mathrm{T}})$ and $\frac{1}{2}(\mathbf{a} - \mathbf{a}^{\mathrm{T}})$, respectively, of a second-order tensor $\mathbf{a}$.

Consider a class of the symmetric measures of strain-rate and stress-rate, in a given state of the deformation and stress in the material element defined by

$$\mathring{\mathbf{e}} = \mathsf{A} \cdot \mathbf{D} \qquad \text{and} \qquad \mathring{\mathbf{t}} = \mathsf{B} \cdot \overset{\nabla}{\sigma} + \mathsf{Z} \cdot \mathbf{D}, \tag{11.8}$$

respectively. A, B, Z are fourth order tensors with minor symmetries, i.e. are elements of $\mathcal{S} \otimes \mathcal{S}$, independent of $\mathbf{D}$ and $\overset{\nabla}{\sigma}$ and dependent in general on the material state[3]. A and B are assumed to be invertible when regarded as linear operators acting

[3] Typically, Z depends linearly on σ, and A, B and Z may be dependent of the deformation gradient.

in the space $\mathcal{S}$ of symmetric second-order tensors. That is, their unique inverses $\mathbf{A}^{-1}$, $\mathbf{B}^{-1} \in \mathcal{S}\otimes\mathcal{S}$ satisfy $\mathbf{A}\cdot\mathbf{A}^{-1} = \mathbf{A}^{-1}\cdot\mathbf{A} = \mathbf{I}_s$ and $\mathbf{B}\cdot\mathbf{B}^{-1} = \mathbf{B}^{-1}\cdot\mathbf{B} = \mathbf{I}_s$, where $\mathbf{I}_s$ is the unit tensor in the space $\mathcal{S}\otimes\mathcal{S}$, of components $(\mathbf{I}_s)_{ijkl} = \frac{1}{2}(\delta_{ik}\delta_{jl}+\delta_{il}\delta_{jk})$ on a fixed orthonormal basis, and δ_{ij} denotes the Kronecker delta. These are the minimal mathematical restrictions imposed on the rate-measures $\overset{\circ}{\mathbf{e}}$ and $\overset{\circ}{\mathbf{t}}$. A superimposed circle instead of a dot is used in order to indicate that a strain-rate or stress-rate defined by (11.8) need not be a material time derivative of any specified measure of strain or stress, respectively. A broad subclass of (11.8) was recently investigated by Xiao *et al.* [49].

Take $(\overset{\circ}{\mathbf{e}}^{-}, \overset{\circ}{\mathbf{t}}^{-})$ as an arbitrarily selected pair from the class (11.8), and distinguish the related (non-invariant) quantities by a bar. To fix attention, we choose the pair $(\overset{\circ}{\mathbf{e}}^{-}, \overset{\circ}{\mathbf{t}}^{-})$ to coincide with $(\mathbf{D}, \overset{\nabla}{\sigma})$. Taking into account the invariance property (11.6), the constitutive expressions (11.2) are now rewritten in terms of $(\mathbf{D}, \overset{\nabla}{\sigma})$ as

$$\overset{\nabla}{\sigma} = \bar{\mathbf{L}}^{e}\cdot\mathbf{D} - \sum_{K} \bar{\lambda}^{p}_{K}\overset{\circ}{\gamma}_{K}, \qquad \dot{f}_{K} = \bar{\lambda}_{K}\cdot\mathbf{D} - \sum_{L} \bar{g}_{KL}\overset{\circ}{\gamma}_{L}, \tag{11.9}$$

and the dual expressions (11.3) as

$$\mathbf{D} = \bar{\mathbf{M}}^{e}\cdot\overset{\nabla}{\sigma} + \sum_{K} \bar{\mu}^{p}_{K}\overset{\circ}{\gamma}_{K}, \qquad \dot{f}_{K} = \bar{\mu}_{K}\cdot\overset{\nabla}{\sigma} - \sum_{L} \bar{h}_{KL}\overset{\circ}{\gamma}_{L}. \tag{11.10}$$

In order for the material behaviour to be not influenced by the transition from the measures $(\mathbf{D}, \overset{\nabla}{\sigma})$ to any $(\overset{\circ}{\mathbf{e}}, \overset{\circ}{\mathbf{t}})$ or conversely, the respective quantities with a bar in (11.9) and without a bar in (11.2) must by mutually related by the following transformation formulae:

$$\mathbf{L}^{e} = (\mathbf{B}\cdot\bar{\mathbf{L}}^{e} + \mathbf{Z})\cdot\mathbf{A}^{-1}, \qquad \lambda^{p}_{K} = \mathbf{B}\cdot\bar{\lambda}^{p}_{K}, \qquad \lambda_{K} = \bar{\lambda}_{K}\cdot\mathbf{A}^{-1}, \qquad g_{KL} = \bar{g}_{KL} \tag{11.11}$$

derived in Appendix. The proof remains unchanged and the transformation rules remain valid if the pair $(\mathbf{D}, \overset{\nabla}{\sigma})$ is replaced in (11.8) and (11.9) by another specified pair $(\overset{\circ}{\mathbf{e}}^{-}, \overset{\circ}{\mathbf{t}}^{-})$ from the class (11.8). Note the measure-invariance of interaction moduli g_{KL}.

Transformation rules that relate the quantities involved in (11.3) and (11.10) are derived in Appendix (ii). Contrary to the interaction moduli g_{KL}, the hardening moduli

$$h_{KL} = \bar{h}_{KL} + \mu_{K}\cdot\mathbf{Z}\cdot\bar{\mu}^{p}_{L} \tag{11.12}$$

are not measure-invariant in value and may change their sign under the transformation (11.8). An important consequence is that the distinction between hardening and softening is not a measure-invariant concept [16]. In measure-invariance aspects, the formulation (11.2) of constitutive rate equations is advantageous to (11.3), although the structure of the constitutive relations itself is preserved under the transformation (11.8) in both formulations (11.2) and (11.3).

In investigations of boundary-value problems, it is of basic interest to reformulate the rate equations (11.2) in terms of the rates of the deformation gradient $\mathbf{F}$ (restricted by $J \equiv \det\mathbf{F} > 0$) and nominal stress $\mathbf{N}$ or its transpose $\mathbf{S} = \mathbf{N}^{\mathrm{T}}$ that satisfies

$\mathbf{SF}^{\mathrm{T}} = J\boldsymbol{\sigma}$. Below the tensor $\mathbf{S}$ will be used and for brevity referred to as the Piola stress (rather than the first Piola-Kirchhoff stress tensor). Under the symmetry restriction on $\dot{\boldsymbol{\sigma}}$, the equations (11.2) are equivalent to

$$\dot{\mathbf{S}} = \mathbf{C}^{\mathrm{e}} \cdot \dot{\mathbf{F}} - \sum_{K} \boldsymbol{\Lambda}_K^{\mathrm{p}} \mathring{\gamma}_K , \qquad \dot{f}_K = \boldsymbol{\Lambda}_K \cdot \dot{\mathbf{F}} - \sum_{L} g_{KL} \mathring{\gamma}_L , \tag{11.13}$$

with the relationships

$$\boldsymbol{\Lambda}_K^{\mathrm{p}} = J \bar{\boldsymbol{\lambda}}_K^{\mathrm{p}} \mathbf{F}^{-1^{\mathrm{T}}} , \qquad \boldsymbol{\Lambda}_K = \bar{\boldsymbol{\lambda}}_K \mathbf{F}^{-1^{\mathrm{T}}} , \tag{11.14}$$

which are easily obtained by substituting $\dot{\mathbf{F}} = \mathbf{0}$ or $\dot{\mathbf{F}} = (\nabla \mathbf{v})\mathbf{F}$ with $\mathring{\gamma}_K = 0$, respectively (cf. also Appendix (i)). The transformation is completed by the known formula (cf. [19]) for the components of $\mathbf{C}^{\mathrm{e}}$:

$$J^{-1} F_{ip} F_{kq} C^{\mathrm{e}}_{jplq} = \bar{C}^{\mathrm{e}}_{jilk} = \bar{L}^{\mathrm{e}}_{ijkl} + \sigma_{ij}\delta_{kl} - \frac{1}{2}(\sigma_{jl}\delta_{ik} + \sigma_{jk}\delta_{il} + \sigma_{il}\delta_{jk} - \sigma_{ik}\delta_{jl}). \tag{11.15}$$

11.4 Deformation-Sensitive Loading of a Material Element

11.4.1 No Kinematic Constraint

In the mechanics of continua it is common to consider external surface tractions which are prescribed functions of time. However, in many cases the applied external tractions are coupled with surface deformations, i.e. the loading is deformation-sensitive (or configuration-dependent). Loading by fluid pressure or by contact with an elastic tool are among the well-known examples; cf. Sewell [45] for a general discussion. A material element surrounded by similar elements is subject to surface tractions dependent on the element configuration, also if a larger sample of the material is tested under the control of stress or strain. It is thus of interest to examine uniqueness of the material response for a possibly general class of deformation-sensitive loadings. A strong motivation is also provided by the fact, already mentioned in Introduction, that a direct control of one stress measure makes another stress measure deformation-sensitive.

To retain consistency with the assumption of macroscopically uniform deformation of a material element, we shall assume that the rate of work done by surface tractions applied to the element boundary from the outside is expressible as a linear form in $\dot{\mathbf{F}}$,

$$\dot{w}^{\mathrm{ext}} = \mathbf{S}^{\mathrm{ext}} \cdot \dot{\mathbf{F}} , \tag{11.16}$$

per unit reference volume of the material. In analogy to Hill's concept of a (symmetric) work-conjugate stress [18, 19], the rate of work is understood here as a basic quantity, and (11.16) serves as a definition of the *external* Piola stress $\mathbf{S}^{\mathrm{ext}}$. The product in (11.16) can be decomposed as

$$\begin{aligned} \mathbf{S}^{\text{ext}} \cdot \dot{\mathbf{F}} &= J(\boldsymbol{\sigma}^{\text{ext}} \cdot \mathbf{D} + \mathbf{m}^{\text{ext}} \cdot \mathbf{W})\,, \\ \boldsymbol{\sigma}^{\text{ext}} = \boldsymbol{\sigma}^{\text{ext}\mathrm{T}} = J^{-1}(\mathbf{S}^{\text{ext}}\mathbf{F}^{\mathrm{T}})_{\mathrm{s}}\,, &\qquad \mathbf{m}^{\text{ext}} = -\mathbf{m}^{\text{ext}\mathrm{T}} = J^{-1}(\mathbf{S}^{\text{ext}}\mathbf{F}^{\mathrm{T}})_{\mathrm{a}}\,. \end{aligned} \tag{11.17}$$

The need for distinguishing between the external ($\mathbf{S}^{\text{ext}}$) and internal ($\mathbf{S}$) Piola stress can be seen from the absence of a counterpart to $\mathbf{m}^{\text{ext}}$ in the constitutive framework for the material without couple stresses. The classical requirement $\mathbf{S}\mathbf{F}^{\mathrm{T}} = \mathbf{F}\mathbf{S}^{\mathrm{T}}$ of rotational equilibrium need not apply *in advance* to $\mathbf{S}^{\text{ext}}$, so that $\mathbf{m}^{\text{ext}} \neq \mathbf{0}$ in general. A simple example is the case of dead loading, i.e. of prescribed $\mathbf{S}^{\text{ext}}$, when $\mathbf{S}^{\text{ext}}\mathbf{F}^{\mathrm{T}}$ need not be equal to $\mathbf{F}\mathbf{S}^{\text{ext}\mathrm{T}}$ for an arbitrary deformation gradient $\mathbf{F}$.

The actual history of $\mathbf{F}$ in a quasi-static motion must be calculated, of course, from an equilibrium condition. If $\mathbf{F}$ is kinematically unconstrained then in a state of mechanical equilibrium we must have

$$\mathbf{S} = \mathbf{S}^{\text{ext}} \qquad \Leftrightarrow \qquad (\,\boldsymbol{\sigma} = \boldsymbol{\sigma}^{\text{ext}} \quad \text{and} \quad \mathbf{m}^{\text{ext}} = \mathbf{0}\,), \tag{11.18}$$

by the principle of virtual power. In view of the incremental character of the constitutive law for the material, the basic condition of continuing mechanical equilibrium of a material element is formulated for the stress rates:

$$\dot{\mathbf{S}} = \dot{\mathbf{S}}^{\text{ext}}\,. \tag{11.19}$$

Clearly, a constitutive expression (different from that for $\dot{\mathbf{S}}$) is needed for $\dot{\mathbf{S}}^{\text{ext}}$ in order to specify the loading conditions. In an equilibrium state regarded as known, we shall consider a class of deformation-sensitive incremental loadings of a material element, defined by

$$\dot{\mathbf{S}}^{\text{ext}} = \dot{\mathbf{S}}^{*} - \mathbf{C}^{*} \cdot \dot{\mathbf{F}}\,. \tag{11.20}$$

Here, a fourth-order tensor $\mathbf{C}^{*}$ of external stiffness moduli and an applied stress-rate $\dot{\mathbf{S}}^{*}$ are given quantities in a given state of the material element. $\dot{\mathbf{S}}^{*}$ is a controllable part and $(-\mathbf{C}^{*} \cdot \dot{\mathbf{F}})$ is a deformation-sensitive part of $\dot{\mathbf{S}}^{\text{ext}}$. For example, for the loading by all-around hydrostatic pressure ($\boldsymbol{\sigma}^{\text{ext}} = -p\mathbf{I}$, $\mathbf{I}$ = unit tensor) and in the current configuration taken as reference ($\mathbf{F} = \mathbf{I}$ instantaneously), we have $\dot{\mathbf{S}}^{*} = -\dot{p}\mathbf{I}$ and $C^{*}_{ijkl} = (\delta_{il}\delta_{jk} - \delta_{ij}\delta_{kl})p$.

In a given equilibrium state where $\mathbf{S}^{\text{ext}} = \mathbf{S} = J\boldsymbol{\sigma}\mathbf{F}^{-1\mathrm{T}}$, we have

$$\begin{aligned} \dot{\mathbf{m}}^{\text{ext}} &= J^{-1}(\mathbf{S}^{\text{ext}}\dot{\mathbf{F}}^{\mathrm{T}} + \dot{\mathbf{S}}^{\text{ext}}\mathbf{F}^{\mathrm{T}})_{\mathrm{a}} \\ &= \left(\boldsymbol{\sigma}(\mathbf{D} - \mathbf{W}) + J^{-1}(\dot{\mathbf{S}}^{*} - \mathbf{C}^{*} \cdot \dot{\mathbf{F}})\mathbf{F}^{\mathrm{T}}\right)_{\mathrm{a}} \\ &= -(\boldsymbol{\sigma}\mathbf{W} + \bar{\mathbf{C}}^{*} \cdot \mathbf{W})_{\mathrm{a}} + (\boldsymbol{\sigma}\mathbf{D} - \bar{\mathbf{C}}^{*} \cdot \mathbf{D} + J^{-1}\dot{\mathbf{S}}^{*}\mathbf{F}^{\mathrm{T}})_{\mathrm{a}}\,, \end{aligned} \tag{11.21}$$

where, in analogy to $(11.15)_1$,

$$\bar{C}^{*}_{jilk} \equiv J^{-1}F_{ip}F_{kq}C^{*}_{jplq}\,. \tag{11.22}$$

The condition (11.19) of continuing equilibrium implies $\dot{\mathbf{m}}^{\text{ext}} = \mathbf{0}$, that is,

$$(\bar{\mathbf{C}}^* \cdot \mathbf{W} + \sigma\mathbf{W})_{\text{a}} = (-\bar{\mathbf{C}}^* \cdot \mathbf{D} + \sigma\mathbf{D} + J^{-1}\dot{\mathbf{S}}^*\mathbf{F}^{\text{T}})_{\text{a}} . \tag{11.23}$$

This is a system of three linear algebraic equations for three independent components of $\mathbf{W}$ if $\mathbf{D}$ is regarded as given. In general, a solution to the system need not be uniquely defined, cf. the discussion by Hill [17] on the special case of dead loading. The solution is unique, i.e. the material spin $\mathbf{W}$ is expressible through (11.23) as a function of $\mathbf{D}$, if and only if

$$\det(\hat{L}_{ij}) \neq 0 , \tag{11.24}$$

where

$$(\hat{L}_{ij}) = \begin{pmatrix} \hat{L}_{11} & \hat{L}_{12} & \hat{L}_{13} \\ \hat{L}_{21} & \hat{L}_{22} & \hat{L}_{23} \\ \hat{L}_{31} & \hat{L}_{32} & \hat{L}_{33} \end{pmatrix} = \begin{pmatrix} \bar{L}^{\text{a}}_{2323} & \bar{L}^{\text{a}}_{2331} & \bar{L}^{\text{a}}_{2312} \\ \bar{L}^{\text{a}}_{3123} & \bar{L}^{\text{a}}_{3131} & \bar{L}^{\text{a}}_{3112} \\ \bar{L}^{\text{a}}_{1223} & \bar{L}^{\text{a}}_{1231} & \bar{L}^{\text{a}}_{1212} \end{pmatrix} , \tag{11.25}$$

$$\bar{L}^{\text{a}}_{ijkl} = \frac{1}{4}(\bar{C}^*_{ijkl} - \bar{C}^*_{ijlk} - \bar{C}^*_{jikl} + \bar{C}^*_{jilk} + \sigma_{ik}\delta_{jl} - \sigma_{il}\delta_{jk} - \sigma_{jk}\delta_{il} + \sigma_{jl}\delta_{ik}) . \tag{11.26}$$

This can be verified by rewriting (11.23) in the vector form, where the term dependent on $\mathbf{W}$ becomes $2\hat{\mathbf{L}}\hat{\mathbf{W}}$ with $\hat{\mathbf{W}} = (\omega_{23},\ \omega_{31},\ \omega_{12})^{\text{T}}$.

Suppose that (11.24) holds, so that the material spin $\mathbf{W}$ becomes a function of $\mathbf{D}$ determined from (11.23). Then $\dot{\mathbf{F}}$ becomes a function of $\mathbf{D}$, or alternatively, of $\mathring{\mathbf{e}}$ for any selected strain-rate measure from the class (11.8). In effect, $\dot{\mathbf{S}}^{\text{ext}}$ and $\overset{\nabla}{\sigma}{}^{\text{ext}}$ become linear inhomogeneous functions of $\mathring{\mathbf{e}}$. It follows that the equation (11.20) can be reduced to

$$\mathring{\mathbf{t}}^{\text{ext}} = \mathring{\mathbf{t}}^* - \mathbf{L}^* \cdot \mathring{\mathbf{e}} , \tag{11.27}$$

where the external stress-rate measure $\mathring{\mathbf{t}}^{\text{ext}}$ is defined by $\mathring{\mathbf{t}}^{\text{ext}} = \mathbf{B} \cdot \overset{\nabla}{\sigma}{}^{\text{ext}} + \mathbf{Z} \cdot \mathbf{D}$ with the same $\mathbf{B}$ and $\mathbf{Z}$ as in (11.8). Here, $\mathring{\mathbf{t}}^*$ and $\mathbf{L}^*$ are known quantities, $\mathring{\mathbf{t}}^* \in \mathcal{S}$, $\mathbf{L}^* \in \mathcal{S} \otimes \mathcal{S}$. The following transformation rules derived in Appendix (iii) apply:

$$\mathring{\mathbf{t}}^* = \mathbf{B} \cdot \overset{\nabla}{\sigma}{}^*, \qquad \mathbf{L}^* = (\mathbf{B} \cdot \bar{\mathbf{L}}^* - \mathbf{Z}) \cdot \mathbf{A}^{-1} \qquad \text{if } \overset{\nabla}{\sigma}{}^{\text{ext}} = \overset{\nabla}{\sigma}{}^* - \bar{\mathbf{L}}^* \cdot \mathbf{D} . \tag{11.28}$$

Note the opposite signs at the second terms in the otherwise similar expressions for $\mathbf{L}^*$ and $\mathbf{L}^{\text{e}}$. Note also that if $\mathbf{L}^* = \mathbf{0}$ for one choice of a measure $\mathring{\mathbf{t}}^{\text{ext}}$ then generally $\mathbf{L}^* \neq \mathbf{0}$ for another choice. In words, if one stress-rate measure is fully controlled then another stress-rate measure becomes strain-rate dependent in general, cf. [16, 19].

It may sometimes be convenient to take (11.27), rather than (11.20), as a definition of the material loading insensitive from the outset to the material spin, with given $\mathring{\mathbf{t}}^*$ and $\mathbf{L}^*$. In particular, this may be adopted in the 'small strain' theory where all effects of changes in geometry of a material element, including the effects of material rotation, are disregarded in advance. A special case of interest arises when the material element is imagined to be of ellipsoidal shape and embedded in an infinite homogeneous mass of an incrementally linear material, which is consistent with uniform straining of the ellipsoid (Eshelby [9]). Then the meaning of $\mathbf{L}^*$ above corresponds to the 'overall constraint tensor' introduced by Hill [14], with a similar notation.

11.4.2 Partial Kinematic Constraints

Suppose that the forward rate of the deformation gradient $\mathbf{F}$ is restricted by a set of linearly independent kinematic constraints

$$\mathbf{K}_\alpha \cdot (\dot{\mathbf{F}} - \dot{\mathbf{F}}^*) = 0, \qquad \alpha = 1, \ldots, n^*, \quad 0 \le n^* \le 9, \tag{11.29}$$

where $\mathbf{K}_\alpha \neq \mathbf{0}$ and $\dot{\mathbf{F}}^*$ are given second-order tensors, state dependent in general, and $(\dot{\mathbf{F}} - \dot{\mathbf{F}}^*)$ orthogonal to every $\mathbf{K}_\alpha$ is an unconstrained part of $\dot{\mathbf{F}}$. For example, if a component $\dot{F}_{pq} = \dot{F}^*_{pq}$ of $\dot{\mathbf{F}}$ is prescribed for some given p, q then the respective $(\mathbf{K}_\alpha)_{ij} = \delta_{pi}\delta_{qj}$. Similarly, if a component d_{pq} or ω_{pq} is prescribed then we can substitute $(\mathbf{K}_\alpha)_{ij} = \delta_{pi}F^{-1}_{jq} \pm \delta_{qi}F^{-1}_{jp}$, respectively with the (+) or (−) sign. It is convenient to rewrite the kinematic restriction (11.29) as

$$\dot{\mathbf{F}} - \dot{\mathbf{F}}^* \in \mathcal{K}, \tag{11.30}$$

where $\mathcal{K}$ is the space of the second order tensors orthogonal to all $\mathbf{K}_\alpha$ in the sense of (11.29).

The principle of virtual power under the constraint (11.30) reads

$$\delta\dot{w} - \delta\dot{w}^{\mathrm{ext}} = (\mathbf{S} - \mathbf{S}^{\mathrm{ext}}) \cdot \delta\dot{\mathbf{F}} = 0 \qquad \text{for every } \delta\dot{\mathbf{F}} \in \mathcal{K}, \tag{11.31}$$

where a prefix δ denotes a kinematically admissible variation. A non-zero part of $(\mathbf{S} - \mathbf{S}^{\mathrm{ext}})$ not restricted by the equilibrium condition (11.31) represents reaction forces. Without loss of generality, in a given equilibrium state we can include the (known) reaction part when defining the current external stress. This means that we can *choose* $\mathbf{S}^{\mathrm{ext}}$ to coincide *momentarily* with $\mathbf{S}$. With that convention, the rate form of (11.31) yields the following condition of continuing mechanical equilibrium of the material element with partial kinematic constraints:

$$(\dot{\mathbf{S}} - \dot{\mathbf{S}}^{\mathrm{ext}}) \cdot \delta\dot{\mathbf{F}} = 0 \qquad \text{for every } \delta\dot{\mathbf{F}} \in \mathcal{K}. \tag{11.32}$$

This encompasses (11.19) when $n^* = 0$, i.e. when $\dot{\mathbf{F}}$ is unconstrained.

The needed constitutive expression for $\dot{\mathbf{S}}^{\mathrm{ext}}$ in the presence of kinematic constraints is still assumed in the form (11.20). The part of that expression which is orthogonal to $\mathcal{K}$ has no influence on the fulfillment of (11.32) and appears thus superfluous. However, in general considerations it is advantageous to operate an external stress-rate defined independently of kinematic constraints.

On decomposing (11.32) similarly to (11.17), it follows that (11.32) implies

$$\dot{\mathbf{m}}^{\mathrm{ext}} \cdot \delta\mathbf{W} = 0 \qquad \text{for every } \delta\mathbf{W} = (\delta\dot{\mathbf{F}}\mathbf{F}^{-1})_{\mathrm{a}}, \quad \delta\dot{\mathbf{F}} \in \mathcal{K}. \tag{11.33}$$

This includes (11.23) if $\mathbf{W}$ is kinematically unconstrained. If the condition (11.29) imposes one or two linearly independent equality constraints on $\mathbf{W}$ then (11.33) yields two or one, respectively, complementary linear algebraic equations for components of $\mathbf{W}$. These are uniquely solvable for ω_{ij} in terms of $\mathbf{D}$ if an analog of (11.24) of the lower dimensionality is satisfied. Finally, if $\mathbf{W}$ is fully prescribed by (11.29) then $\dot{\mathbf{m}}^{\mathrm{ext}}$ is left arbitrary.

If $\mathbf{W}$ restricted by (11.29) and (11.33) becomes a single-valued function of $\mathbf{D}$ then the equation (11.20) can be reduced to a form analogous to eq. (11.27) examined in the preceding subsection.

11.5 The Rate-Problem of Continuing Equilibrium

11.5.1 Unconstrained Strain-Rate

In this subsection, the rate-problem of continuing mechanical equilibrium of the material element is addressed in the absence of kinematic constraints imposed on the strain-rate and for the material spin eliminated in the manner discussed above. Accordingly, the spin-insensitive form (11.27) of the loading condition is taken with L^* and $\mathring{\mathbf{t}}^*$ regarded as known. If (11.29) imposes no constraint on $\mathbf{D}$ then the rate-problem is defined as follows: in a given material state, find $(\mathring{\mathbf{e}}, \mathring{\gamma}_K)$ such that

$$\mathring{\mathbf{t}} = \mathring{\mathbf{t}}^{\mathrm{ext}} \qquad \text{at unconstrained strain-rate}\,, \tag{11.34}$$

where $\mathring{\mathbf{t}}$ and $\mathring{\mathbf{t}}^{\mathrm{ext}}$ are defined by $(11.2)_1$ and (11.27), respectively, and the $\mathring{\gamma}_K$'s are subject to the constraints (11.1) and (11.4).

On eliminating $\mathring{\mathbf{t}}$ between (11.34) and $(11.2)_1$ and substituting (11.27), we obtain

$$(\mathsf{L}^{\mathrm{e}} + \mathsf{L}^*) \cdot \mathring{\mathbf{e}} = \mathring{\mathbf{t}}^* + \sum_K \boldsymbol{\lambda}_K^{\mathrm{p}} \mathring{\gamma}_K\,. \tag{11.35}$$

It is worth noting that an identical equation is obtained when $(11.3)_1$ (multiplied by L^{e}) is used instead of $(11.2)_1$. The tensor $(\mathsf{L}^{\mathrm{e}} + \mathsf{L}^*)$ is related to the moduli tensors $\bar{\mathsf{L}}^{\mathrm{e}}$ and $\bar{\mathsf{L}}^*$ by the formula

$$\mathsf{L}^{\mathrm{e}} + \mathsf{L}^* = \mathsf{B} \cdot (\bar{\mathsf{L}}^{\mathrm{e}} + \bar{\mathsf{L}}^*) \cdot \mathsf{A}^{-1} \tag{11.36}$$

that follows from (11.77) and (11.92) in Appendix.

Although L^{e} or L^* taken separately may be singular, we shall assume that their sum $(\mathsf{L}^{\mathrm{e}} + \mathsf{L}^*)$ possesses a unique inverse $(\mathsf{L}^{\mathrm{e}} + \mathsf{L}^*)^{-1}$ when regarded as a linear operator acting in $\mathcal{S}$, cf. the definition of A^{-1}. It is evident that this condition is necessary and sufficient for the uniqueness of a solution to the purely elastic rate-problem (11.35) with all $\mathring{\gamma}_K$ given in advance (e.g. set equal to zero), the consistency conditions (11.4) being then disregarded. An essential conclusion drawn from (11.36) is that the invertibility of $(\mathsf{L}^{\mathrm{e}} + \mathsf{L}^*)$ is a measure-invariant property.

The elimination of $\mathring{\mathbf{e}}$ between (11.35) and $(11.2)_2$ now yields

$$\dot{f}_K = \boldsymbol{\lambda}_K \cdot (\mathsf{L}^{\mathrm{e}} + \mathsf{L}^*)^{-1} \cdot \mathring{\mathbf{t}}^* - \sum_{L \in \mathcal{P}} (g_{KL} - \boldsymbol{\lambda}_K \cdot (\mathsf{L}^{\mathrm{e}} + \mathsf{L}^*)^{-1} \cdot \boldsymbol{\lambda}_L^{\mathrm{p}}) \mathring{\gamma}_L\,. \tag{11.37}$$

Introduce the notation

$$\mathbf{G}^* = (\mathbf{L}^e + \mathbf{L}^*)^{-1}, \qquad \mathbf{H}^* = \mathbf{L}^* \cdot (\mathbf{L}^e + \mathbf{L}^*)^{-1} \cdot \mathbf{L}^e \tag{11.38}$$

and denote the expressions in (11.37) by

$$\begin{aligned} b^*_K &= \lambda_K \cdot \mathbf{G}^* \cdot \overset{\circ}{\mathbf{t}}{}^* & \qquad g^*_{KL} &= g_{KL} - \lambda_K \cdot \mathbf{G}^* \cdot \lambda^{\mathrm{p}}_L \\ &= \mu_K \cdot \mathbf{L}^e \cdot \mathbf{G}^* \cdot \overset{\circ}{\mathbf{t}}{}^*, & &= h_{KL} + \mu_K \cdot \mathbf{H}^* \cdot \mu^{\mathrm{p}}_L. \end{aligned} \tag{11.39}$$

The dual expressions in (11.39) involving μ_K follow from (11.5) and require the assumption of the invertibility of $\mathbf{L}^e$. The equivalence of the two expressions for b^*_K is obvious, and the equivalence of the second pair can be verified as follows:

$$\begin{aligned} g^*_{KL} &= h_{KL} + \mu_K \cdot \mathbf{L}^e \cdot \mu^{\mathrm{p}}_L - \mu_K \cdot \mathbf{L}^e \cdot \mathbf{G}^* \cdot \mathbf{L}^e \cdot \mu^{\mathrm{p}}_L \\ &= h_{KL} + \mu_K \cdot ((\mathbf{L}^e + \mathbf{L}^*) \cdot \mathbf{G}^* - \mathbf{L}^e \cdot \mathbf{G}^*) \cdot \mathbf{L}^e \cdot \mu^{\mathrm{p}}_L \\ &= h_{KL} + \mu_K \cdot \mathbf{H}^* \cdot \mu^{\mathrm{p}}_L. \end{aligned} \tag{11.40}$$

The same expressions appear in the formula for $\dot{f}_K$ obtained by starting from $(11.3)_2$ instead of $(11.2)_2$.

Further transformations yield

$$\begin{aligned} \mathbf{H}^* &= \mathbf{L}^* \cdot (\mathbf{I}_s + \mathbf{M}^e \cdot \mathbf{L}^*)^{-1} & \qquad &\text{if } \mathbf{M}^e = (\mathbf{L}^e)^{-1} \text{ exists,} \\ &= (\mathbf{M}^e + \mathbf{M}^*)^{-1} & &\text{if } \mathbf{M}^* = (\mathbf{L}^*)^{-1} \text{ exists.} \end{aligned} \tag{11.41}$$

From (11.4), (11.37) and (11.39) we arrive at the following rate problem to be solved with respect to $\overset{\circ}{\gamma}_K$'s in a given material state :

$$\dot{f}_K = b^*_K - \sum_{L\in\mathcal{P}} g^*_{KL}\overset{\circ}{\gamma}_L, \quad \overset{\circ}{\gamma}_K \geq 0, \quad \dot{f}_K \leq 0, \quad \dot{f}_K\,\overset{\circ}{\gamma}_K = 0 \qquad \text{for } K \in \mathcal{P}. \tag{11.42}$$

If this is done then a full solution to (11.34) is completed by using the expression for $\overset{\circ}{\mathbf{e}}$:

$$\overset{\circ}{\mathbf{e}} = \mathbf{G}^* \cdot \overset{\circ}{\mathbf{t}}{}^* + \mathbf{G}^* \cdot \sum_K \lambda^{\mathrm{p}}_K \overset{\circ}{\gamma}_K \tag{11.43}$$

obtained from (11.35).

In a given state of the material element and for given incremental constitutive laws (11.2) or (11.3) for the material and (11.27) for the loading, the quantities b^*_K and g^*_{KL} are known. The substitution of the transformation formulae (11.11) and (11.28) into (11.38) and (11.39) shows that both b^*_K and g^*_{KL} are measure-invariant. Hence, precisely the same system of equations and inequalities for a solution $(\overset{\circ}{\mathbf{e}}, \overset{\circ}{\gamma}_K)$ to the rate-problem of continuing equilibrium for a given material and loading is obtained irrespectively of the choice of the measures for strain-rate and stress-rate, and no matter whether (11.2) or (11.3) is taken as the basic constitutive framework. Obviously, this conclusion is in direct connection with the assumed invariance of the f_K's and $\overset{\circ}{\gamma}_K$'s.

If $\mathbf{L}^* = \mathbf{0}$ for *some* specific measure $\overset{\circ}{\mathbf{t}}$ then the respective $\mathbf{H}^*$ vanishes, $\mathbf{G}^* = \mathbf{M}^{\mathrm{e}}$ and $g^*_{KL} = h_{KL}$ for *this* measure $\overset{\circ}{\mathbf{t}}$. This is, of course, in agreement with the problem defined by (11.3) and (11.4) for prescribed $\overset{\circ}{\mathbf{t}}$. For the same problem expressed in another measure $\overset{\circ}{\mathbf{t}}$, as a rule we obtain $\mathbf{L}^* \neq \mathbf{0}$ by (11.92) and must return to the more general formulation given above that takes the deformation sensitivity of the loading into account.

In turn, if a norm of $\mathbf{L}^*$ increases unboundedly (which corresponds to an infinitely stiff environment of the element) for some measures $(\overset{\circ}{\mathbf{e}}, \overset{\circ}{\mathbf{t}})$ then the same happens, by (11.92) and the non-singularity of $\mathbf{B}$, for *each* choice of the measures from the class (11.8). Then $\mathbf{G}^*$ tends in the limit to $\mathbf{0}$, so that $g^*_{KL} \to g_{KL}$ for any measures. The agreement with the problem defined by (11.2) and (11.4) for prescribed $\overset{\circ}{\mathbf{e}}$ is obtained if we let $(\mathbf{L}^*)^{-1} \cdot \overset{\circ}{\mathbf{t}}{}^*$ tend in the limit to the prescribed $\overset{\circ}{\mathbf{e}}$. A more direct transition to the case of a prescribed strain-rate will be given below.

11.5.2 General Formulation

Consider now the class of deformation-sensitive incremental loadings of the material element in the presence of partial kinematic constraints, defined by the equations (11.20) and (11.29). The corresponding condition (11.32) of continuing equilibrium has the form of a variational equality. To express it in a more convenient form of a tensorial equation, we split the space of second-order tensors into an orthogonal sum $\mathcal{K} \oplus \mathcal{K}^{\perp}$ of two subspaces, where $\mathcal{K}^{\perp}$ is the orthogonal complement of $\mathcal{K}$, and the basis in $\mathcal{K}^{\perp}$ is defined by the $\mathbf{K}_\alpha$'s. Let the operator of orthogonal projection of a second-order tensor on $\mathcal{K}$ be represented by a fourth-order tensor $\mathbf{P}$ which by definition has the properties $\mathbf{P} = \mathbf{P}^{\mathrm{T}}$, $\mathbf{P} \cdot \mathbf{P} = \mathbf{P}$, $\mathbf{P} \cdot \mathbf{K}_\alpha = \mathbf{0}$ and $\mathbf{P} \cdot \tilde{\mathbf{F}} = \tilde{\mathbf{F}}$ for every $\tilde{\mathbf{F}} \in \mathcal{K}$. Analogously, the operator of orthogonal projection on $\mathcal{K}^{\perp}$ is represented by a fourth-order tensor $\mathbf{P}^{\perp} = \mathbf{I} - \mathbf{P}$ (where $\mathbf{I}$ is the unit fourth-order tensor) such that $\mathbf{P}^{\perp} = (\mathbf{P}^{\perp})^{\mathrm{T}}$, $\mathbf{P}^{\perp} \cdot \mathbf{P}^{\perp} = \mathbf{P}^{\perp}$, $\mathbf{P}^{\perp} \cdot \mathbf{K}_\alpha = \mathbf{K}_\alpha$ and $\mathbf{P}^{\perp} \cdot \tilde{\mathbf{F}} = \mathbf{0}$ for every $\tilde{\mathbf{F}} \in \mathcal{K}$. In terms of the reciprocal basis $\mathbf{K}^\alpha$ in $\mathcal{K}^{\perp}$, $\mathbf{P}$ and $\mathbf{P}^{\perp}$ are given by

$$\mathbf{P} = \mathbf{I} - \mathbf{P}^{\perp}, \qquad \mathbf{P}^{\perp} = \sum_\alpha \mathbf{K}_\alpha \otimes \mathbf{K}^\alpha, \qquad \mathbf{K}_\alpha \cdot \mathbf{K}^\beta \equiv \delta_{\alpha\beta}, \qquad \mathbf{K}^\beta \in \mathcal{K}^{\perp}. \tag{11.44}$$

The basic rate-problem of continuing mechanical equilibrium of the material element with partial kinematic constraints, that results from (11.29) and (11.32), is now formulated as follows: in a given material state, find $(\dot{\mathbf{F}}, \overset{\circ}{\gamma}_K)$ such that

$$\mathbf{P} \cdot (\dot{\mathbf{S}} - \dot{\mathbf{S}}^{\mathrm{ext}}) = \mathbf{0} \qquad \text{and} \qquad \mathbf{P}^{\perp} \cdot (\dot{\mathbf{F}} - \dot{\mathbf{F}}^*) = \mathbf{0}, \tag{11.45}$$

where $\dot{\mathbf{S}}$ and $\dot{\mathbf{S}}^{\mathrm{ext}}$ are defined by $(11.13)_1$ and (11.20), respectively, and the $\overset{\circ}{\gamma}_K$'s are subject to the constraints (11.1) and (11.4). It is clear that the conditions in (11.45) provide jointly nine equations in components, and that the data $\mathbf{P}^{\perp} \cdot \dot{\mathbf{S}}^*$ and $\mathbf{P} \cdot \dot{\mathbf{F}}^*$ are irrelevant to the solution.

The elimination of the internal and external Piola stress-rate between $(11.13)_1$, (11.20) and (11.45) yields

$$\mathbf{P}\cdot\Big((\mathbf{C}^{\mathrm{e}}+\mathbf{C}^{*})\cdot\dot{\mathbf{F}}-\dot{\mathbf{S}}^{*}-\sum_{K}\boldsymbol{\Lambda}_K^{\mathrm{p}}\mathring{\gamma}_K\Big)=\mathbf{0}\,. \tag{11.46}$$

Clearly, it is the part $\mathbf{P}^{\perp}\cdot(\mathbf{C}^{\mathrm{e}}+\mathbf{C}^{*})$ of the stiffness moduli which is irrelevant to the solution searched. On substituting the identity $\dot{\mathbf{F}}=\mathbf{P}\cdot\mathbf{P}\cdot\dot{\mathbf{F}}+\mathbf{P}^{\perp}\cdot\dot{\mathbf{F}}$ into (11.46) and rearranging with the help of $(11.45)_2$, we obtain

$$\Big(\mathbf{P}\cdot(\mathbf{C}^{\mathrm{e}}+\mathbf{C}^{*})\cdot\mathbf{P}\Big)\cdot\Big(\mathbf{P}\cdot\dot{\mathbf{F}}\Big)=\mathbf{P}\cdot\Big(\dot{\mathbf{S}}^{*}-(\mathbf{C}^{\mathrm{e}}+\mathbf{C}^{*})\cdot\mathbf{P}^{\perp}\cdot\dot{\mathbf{F}}^{*}+\sum_{K}\boldsymbol{\Lambda}_K^{\mathrm{p}}\mathring{\gamma}_K\Big)\,. \tag{11.47}$$

The fourth-order tensor $\Big(\mathbf{P}\cdot(\mathbf{C}^{\mathrm{e}}+\mathbf{C}^{*})\cdot\mathbf{P}\Big)$ is an orthogonal projection of $(\mathbf{C}^{\mathrm{e}}+\mathbf{C}^{*})$ on the subspace $\mathcal{K}\otimes\mathcal{K}$ that contains all finite sums $\sum_{\beta}\mathbf{k}_{\beta}\otimes\mathbf{m}_{\beta}$ with $\mathbf{k}_{\beta}\in\mathcal{K}$, $\mathbf{m}_{\beta}\in\mathcal{K}$. That tensor defines a linear operator acting in $\mathcal{K}$, whose unique inverse, represented by a fourth-order tensor $\mathbf{G}\in\mathcal{K}\otimes\mathcal{K}$ that satisfies

$$\mathbf{G}\cdot\Big(\mathbf{P}\cdot(\mathbf{C}^{\mathrm{e}}+\mathbf{C}^{*})\cdot\mathbf{P}\Big)=\Big(\mathbf{P}\cdot(\mathbf{C}^{\mathrm{e}}+\mathbf{C}^{*})\cdot\mathbf{P}\Big)\cdot\mathbf{G}=\mathbf{P}\,, \tag{11.48}$$

is assumed to exist. It is evident that this assumption is necessary and sufficient for the uniqueness of a solution to the purely elastic, linear rate problem defined by (11.47) and $(11.45)_2$ with all $\mathring{\gamma}_K$ given in advance (e.g. set equal to zero), the consistency conditions (11.4) being then disregarded. It can be shown that $\mathbf{G}$, if exists, is unique, and $\mathbf{G}\cdot\mathbf{P}=\mathbf{P}\cdot\mathbf{G}=\mathbf{G}$, $\mathbf{G}\cdot\mathbf{P}^{\perp}=\mathbf{P}^{\perp}\cdot\mathbf{G}=\mathbf{0}$. In the degenerate case of a full kinematic constraint we have $\mathbf{P}^{\perp}=\mathbf{I}$, $\mathbf{P}=\mathbf{0}$, $\mathcal{K}=\{\mathbf{0}\}$ and $\mathbf{G}=\mathbf{0}$ as the only element of $\mathcal{K}\otimes\mathcal{K}$.

On determining $\mathbf{P}\cdot\dot{\mathbf{F}}$ from (11.47) and adding to $\mathbf{P}^{\perp}\cdot\dot{\mathbf{F}}$ given by $(11.45)_2$, the solution $\dot{\mathbf{F}}$ to the system of equations (11.45) is expressed in terms of the $\mathring{\gamma}_K$'s as

$$\begin{aligned}\dot{\mathbf{F}}&=\Big(\mathbf{I}-\mathbf{G}\cdot\mathbf{P}\cdot(\mathbf{C}^{\mathrm{e}}+\mathbf{C}^{*})\Big)\cdot\mathbf{P}^{\perp}\cdot\dot{\mathbf{F}}^{*}+\mathbf{G}\cdot\mathbf{P}\cdot\Big(\dot{\mathbf{S}}^{*}+\sum_{K\in\mathcal{P}}\boldsymbol{\Lambda}_K^{\mathrm{p}}\mathring{\gamma}_K\Big)\\&=\Big(\mathbf{I}-\mathbf{G}\cdot(\mathbf{C}^{\mathrm{e}}+\mathbf{C}^{*})\Big)\cdot\dot{\mathbf{F}}^{*}+\mathbf{G}\cdot\dot{\mathbf{S}}^{*}+\mathbf{G}\cdot\sum_{K\in\mathcal{P}}\boldsymbol{\Lambda}_K^{\mathrm{p}}\mathring{\gamma}_K\,.\end{aligned} \tag{11.49}$$

The possibility of omitting the projection operators in the latter expression follows from the properties of $\mathbf{PG}$ and $\mathbf{P}^{\perp}=\mathbf{I}-\mathbf{P}$.

Finally, on eliminating $\dot{\mathbf{F}}$ between (11.49) and $(11.13)_2$ and comparing (11.4) with (11.42), we arrive at the following

Conclusion 1 *If the inverse tensor* $\mathbf{G}$ *defined by (11.48) exists then the basic rate-problem (11.45) reduces to the problem (11.42) to be solved with respect to* $\mathring{\gamma}_K$*'s, where*

$$\begin{aligned}b_K^{*}&=\boldsymbol{\Lambda}_K\cdot\Big((\mathbf{I}-\mathbf{G}\cdot(\mathbf{C}^{\mathrm{e}}+\mathbf{C}^{*}))\cdot\dot{\mathbf{F}}^{*}+\mathbf{G}\cdot\dot{\mathbf{S}}^{*}\Big)\,,\\g_{KL}^{*}&=g_{KL}-\boldsymbol{\Lambda}_K\cdot\mathbf{G}\cdot\boldsymbol{\Lambda}_L^{\mathrm{p}}\,.\end{aligned} \tag{11.50}$$

The solution is completed by determining $\dot{\mathbf{F}}$ *from (11.49)* •

11.6 Uniqueness

11.6.1 The Uniqueness Theorem

Denote by M the number of the elements of $\mathcal{P}$, i.e. the number of potentially active mechanisms ($f_K = 0$) in a given material state. Introduce the matrix notation, where vectors from the real M-dimensional space R^M and square $M \times M$ real matrices, both of the indices from $\mathcal{P}$, are denoted by square brackets. In that notation, (11.42) is rewritten as

$$[-\dot{f}] = [g^*][\mathring{\gamma}] + [-b^*], \quad [\mathring{\gamma}] \geq [0], \quad [-\dot{f}] \geq [0], \quad [-\dot{f}]^{\mathrm{T}}[\mathring{\gamma}] = 0, \quad (11.51)$$

where inequalities between vectors should be read componentwise. The mathematical structure of the system (11.51) is precisely that of the linear complementarity problem, cf. [7]. The expressions (11.50) or (11.39) for $[b^*]$ and $[g^*]$ specify how the rate-problem of continuing mechanical equilibrium is reduced to the linear complementarity problem. The fundamental correspondence between the incremental constitutive law of multi-mode plasticity and the linear complementarity problem was emphasized by Maier [26], for a (generalized) strain or stress control and in the small strain context. The correspondence is extended in this paper to finite deformation under deformation-sensitive loading and in the presence of partial kinematic constrains.

The fundamental theorem concerning the existence and uniqueness of a solution to the linear complementarity problem (cf. [7]) in the present notation reads

Theorem 1 *The problem (11.51) has a unique solution for all vectors* $[b^*] \in R^M$ *if and only if*

$$\textit{all principal minors of the matrix } [g^*] \textit{ are positive} \bullet \quad (11.52)$$

Recall that for an index set $\mathcal{A} \subseteq \mathcal{P}$, the determinant of a principal submatrix (a_{KL}), $K, L \in \mathcal{A}$, of a matrix $[a]$ is called a principal minor of $[a]$.

Generally, (11.52) is less restrictive than the positive definiteness of $[g^*]$, i.e. than

$$[x]^{\mathrm{T}}[g^*][x] > 0 \qquad \text{for every } [x] \neq [0]\,. \quad (11.53)$$

It is well known that if the matrix $[g^*]$ is symmetric then the conditions (11.52) and (11.53) are equivalent to each other.

In the monograph by Cottle at al. [7], a number of other useful theorems (e.g. on the existence without uniqueness) can be found which are not used here.

In view of the reduction of the basic rate-problem (11.45) to the linear complementarity problem (11.51) as shown above, the 'if' part of Theorem 1 provides a general condition sufficient for the existence and uniqueness of a solution to the rate-problem of continuing mechanical equilibrium formulated for the inelastic material element subject to the deformation-sensitive loading under partial kinematic constraints. The 'only if' part of Theorem 1 requires some care since it concerns *all* vectors $[b^*] \in R^M$. First, the fulfillment of (11.52) is *not* necessary for the existence of a unique solution for a particular set of vectors $[b^*]$; examples can easily be constructed. Second, the set of $[b^*]$ for which the solution is nonunique if (11.52) is violated might be excluded by constraints imposed on $[b^*]$. In the rate-problem posed in the preceding section, the vector $[b^*]$ is constrained by the equation $(11.50)_1$, which on substituting $(11.14)_2$ and $(11.11)_3$ can be rewritten as

$$b_K^* = \lambda_K \cdot \mathsf{A} \cdot \left((\mathsf{I} - \mathsf{G} \cdot (\mathsf{C}^e + \mathsf{C}^*)) \cdot \dot{\mathbf{F}}^* + \mathsf{G} \cdot \dot{\mathbf{S}}^*\right)\mathbf{F}^{-1}. \tag{11.54}$$

The symmetric second-order tensors λ_K, $K \in \mathcal{P}$, when identified with M six-dimensional transposed vectors, define the rows of a $M \times 6$ matrix (λ). From (11.54) it follows that every vector $[b^*] \in R^M$ can be generated by suitably chosen $\dot{\mathbf{F}}^*$ and $\dot{\mathbf{S}}^*$ if and only if the rank of the matrix (λ) is equal to M. This condition means that $M \leq 6$ and that the λ_K's are linearly independent in $\mathcal{S}$, which is a measure-invariant requirement on account of $(11.11)_3$.

The Conclusion 1 of Sect. 11.5, Theorem 1 and the above statement provide the proof of the following theorem on the uniqueness of a solution to the basic rate-problem (11.45) of continuing mechanical equilibrium at macroscopically uniform deformation.

Theorem 2 *Suppose that the inverse tensor* G *defined by (11.48) exists.*
(i) If (11.52) holds then the problem (11.45) has a unique solution for every $\dot{\mathbf{F}}^*$, $\dot{\mathbf{S}}^*$.
(ii) Let rank$(\lambda) = M$; *then the problem (11.45) has a unique solution for every* $\dot{\mathbf{F}}^*$, $\dot{\mathbf{S}}^*$ *if and only if (11.52) holds* •

It is worth recalling that the existence of the inverse tensor G has a clear mechanical meaning as being equivalent to the existence of a unique solution to the problem (11.45) posed for purely elastic rates. Theorem 2 (i) yields thus the following

Conclusion 2 *The rate-problem (11.45) or (11.34) of continuing mechanical equilibrium of the inelastic material element has a unique solution for all* $\dot{\mathbf{F}}^*$, $\dot{\mathbf{S}}^*$ *if simultaneously:*
(a) the problem is uniquely solvable for purely elastic rates (so that the tensor G *exists),*
(b) all principal minors of the matrix (g^*_{KL}), $K, L \in \mathcal{P}$, *are positive* •

11.6.2 Unique Solution

Suppose that the assumptions of Theorem 2 (i), or equivalently, of Conclusion 2 above, are satisfied. Denote by $\mathcal{A} \subseteq \mathcal{P}$ the set of indices of the active mechanisms ($\mathring{\gamma}_K > 0$) in the guaranteed unique solution to (11.45) for some given $\dot{\mathbf{F}}^*$ and $\dot{\mathbf{S}}^*$.

The set $\mathcal{A}$ is, of course, uniquely determined by $\dot{\mathbf{F}}^*$ and $\dot{\mathbf{S}}^*$, but in general not given explicitly. Denote by $(g^*_{\mathcal{A}})$ the respective principal submatrix of $[g^*]$, of components

$$(g^*_{\mathcal{A}})_{KL} = g_{KL} - \mathbf{\Lambda}_K \cdot \mathsf{G} \cdot \mathbf{\Lambda}^{\mathrm{p}}_L\,, \qquad K, L \in \mathcal{A}\,. \tag{11.55}$$

Introduce also the notation

$$\overset{\circ}{\mathbf{Y}} = (\mathsf{I} - \mathsf{G} \cdot (\mathsf{C}^{\mathrm{e}} + \mathsf{C}^*)) \cdot \dot{\mathbf{F}}^* + \mathsf{G} \cdot \dot{\mathbf{S}}^* \tag{11.56}$$

for the known second-order tensor that appears in (11.49) and $(11.50)_1$. From (11.42), $(11.50)_1$ and (11.49), the solution to (11.45) is given by

$$\overset{\circ}{\gamma}_K = \sum_{L\in\mathcal{A}} (g^*_{\mathcal{A}})^{-1}_{KL} \mathbf{\Lambda}_L \cdot \overset{\circ}{\mathbf{Y}} \quad \text{for } K \in \mathcal{A}, \qquad \overset{\circ}{\gamma}_K = 0 \quad \text{for } K \notin \mathcal{A}, \tag{11.57}$$

$$\dot{\mathbf{F}} = \Big(\mathsf{I} + \mathsf{G} \cdot \sum_{K,L\in\mathcal{A}} (g^*_{\mathcal{A}})^{-1}_{KL} \mathbf{\Lambda}^{\mathrm{p}}_K \otimes \mathbf{\Lambda}_L\Big) \cdot \overset{\circ}{\mathbf{Y}}\,. \tag{11.58}$$

On substituting this into (11.13), we obtain

$$\dot{\mathbf{S}} = \mathsf{C}^{\mathrm{Y}} \cdot \overset{\circ}{\mathbf{Y}}, \qquad \mathsf{C}^{\mathrm{Y}} = \mathsf{C}^{\mathrm{e}} - (\mathsf{I} - \mathsf{C}^{\mathrm{e}} \cdot \mathsf{G}) \cdot \sum_{K,L\in\mathcal{A}} (g^*_{\mathcal{A}})^{-1}_{KL} \mathbf{\Lambda}^{\mathrm{p}}_K \otimes \mathbf{\Lambda}_L\,. \tag{11.59}$$

The set $\mathcal{A}$ and matrix $(g^*_{\mathcal{A}})^{-1}$, and thus also the pseudo-moduli tensor $\mathsf{C}^{\mathrm{Y}} = \mathsf{C}^{\mathrm{Y}}(\mathcal{A})$, are dependent on $\dot{\mathbf{F}}^*$ and $\dot{\mathbf{S}}^*$, which reflects the incremental nonlinearity of the material response.

In the special case of full kinematic control when $\dot{\mathbf{F}} = \dot{\mathbf{F}}^*$ is given, we have $\mathsf{G} = \mathbf{0}$, $[g^*] = [g]$, $\overset{\circ}{\mathbf{Y}} = \dot{\mathbf{F}}$ and recover the familiar form of the tangent stiffness moduli tensor

$$\dot{\mathbf{S}} = \mathsf{C} \cdot \dot{\mathbf{F}}, \qquad \mathsf{C} = \mathsf{C}^{\mathrm{e}} - \sum_{K,L\in\mathcal{A}} (g_{\mathcal{A}})^{-1}_{KL} \mathbf{\Lambda}^{\mathrm{p}}_K \otimes \mathbf{\Lambda}_L\,, \tag{11.60}$$

where $\mathsf{C} = \mathsf{C}(\mathcal{A})$ depends on $\dot{\mathbf{F}}$ only through $\mathcal{A}$ which is not influenced by a given material spin.

11.6.3 Particular Cases

Full kinematic constraints

In the case when $\dot{\mathbf{F}} = \dot{\mathbf{F}}^*$ is fully prescribed ($n^* = 9$), we have $\mathsf{P} = \mathbf{0}$, $\mathsf{G} = \mathbf{0}$ and $b^*_K = \mathbf{\Lambda}_K \cdot \dot{\mathbf{F}}$, $g^*_{KL} = g_{KL}$, in agreement with (11.13). Moreover, since the material spin is given, the rate-problem reduces further to that defined by (11.2) and (11.4) for given $\overset{\circ}{\mathbf{e}}$. The constitutive law (11.20) for the external stress-rate is entirely irrelevant, and the condition (a) in Conclusion 2 imposes no restriction. In effect, the constitutive uniqueness of the $\overset{\circ}{\gamma}_K$'s for given $\overset{\circ}{\mathbf{e}}$ is guaranteed under the single, measure-invariant condition that all principal minors of the matrix (g_{KL}), $K, L \in \mathcal{P}$, are positive. This result is sharper than that given by Hill and Rice [21], in view of the implication (11.52) $\Leftarrow$ (11.53).

No kinematic constraint

In the case when $\dot{\mathbf{F}}$ is unconstrained ($n^* = 0$), we have $\mathbf{P} = \mathbf{I}$ and $\mathbf{G} = (\mathbf{C}^e + \mathbf{C}^*)^{-1}$. Note that $\mathbf{C}^* \neq \mathbf{0}$ unless it is the Piola (or nominal) external stress-rate which directly prescribed. The inverse of $(\mathbf{C}^e + \mathbf{C}^*)$ is assumed to exist in order to satisfy the condition (a) of Conclusion 2. This implies that the material spin is uniquely determined by the condition (11.23), which means that (11.24) holds true. Consequently, the rate-problem (11.45) can be reduced to (11.34), and the expressions (11.50) for b^*_K and g^*_{KL} reduce to those in (11.39). The uniqueness condition (b) in Conclusion 2 is to be applied as it stands; it is a measure-invariant requirement.

If it happens that $\mathbf{L}^* = \mathbf{0}$ for *some* specific measure $\overset{\circ}{\mathbf{t}}$ then the requirement of the existence of $\mathbf{G}^*$ (cf. (11.38)) reduces to that of the invertibility of the respective $\mathbf{L}^e$. As already mentioned, g^*_{KL} reduces then to h_{KL} defined for *that* measure, and the rate-problem (11.34) reduces to that defined by (11.3) and (11.4) for prescribed $\overset{\circ}{\mathbf{t}}$. In effect, the constitutive uniqueness of the $\overset{\circ}{\gamma}_K$'s for given $\overset{\circ}{\mathbf{t}}$ is guaranteed if $\mathbf{M}^e = (\mathbf{L}^e)^{-1}$ exists and all principal minors of the matrix (h_{KL}), $K, L \in \mathcal{P}$, are positive. However, those requirements are not measure-invariant.

Mixed control

Particular cases of the partial kinematic constraints (11.29), frequently met in applications, correspond to prescribed components $\dot{F}_{pq}$, d_{pq} or ω_{pq} for a set of pairs (p, q). As already mentioned, in that cases we can take $(\mathbf{K}_\alpha)_{ij} = \delta_{qi}\delta_{pj}$, $(\mathbf{K}_\alpha)_{ij} = \delta_{pi}F^{-1}_{jq} + \delta_{qi}F^{-1}_{jp}$ or $(\mathbf{K}_\alpha)_{ij} = \delta_{pi}F^{-1}_{jq} - \delta_{qi}F^{-1}_{jp}$, respectively. If the complementary *nominal* surface tractions are prescribed, for instance are equal to zero on a free surface, then we put $\mathbf{C}^* = \mathbf{0}$. The tensor $\mathbf{G}$ and the matrix (g_{KL}) in the uniqueness theorem are to be calculated for a given constitutive law and a given set of the $\mathbf{K}_\alpha$'s as above. The special case of a mixed stress-strain control in the small-strain formulation (cf. [23]) is obtained if $F_{ij} = \delta_{ij}$ and any influence of the material spin $\mathbf{W}$ is disregarded in advance.

Another case which is of particular interest is examined in the next section.

11.7 Application to the Bifurcation in a Band

11.7.1 Exclusion Conditions

The basic rate-problem (11.45) is now specified for a material element within a planar band of possible bifurcation in an infinite homogeneous continuum. A fundamental uniform solution is defined by the rate $\dot{\mathbf{F}}^*$ of the deformation gradient and by inelastic multipliers $\overset{\circ}{\gamma}{}^*_K$, say, with the Piola stress rate $\dot{\mathbf{S}}^*$ known from the constitutive relationship (11.13). An alternative quasi-static solution in the first-order rates in a given uniform state is searched in a bifurcation band of a unit normal $\mathbf{n}$ in the reference configuration. Accordingly, the equations in (11.45) reduce to the well-known form (cf. [42])

$$(\dot{\mathbf{S}} - \dot{\mathbf{S}}^*)\mathbf{n} = \mathbf{0} \qquad \text{and} \qquad \dot{\mathbf{F}} - \dot{\mathbf{F}}^* = \mathbf{g} \otimes \mathbf{n}\,, \tag{11.61}$$

where $\mathbf{g} \in R^3$ is an unknown vector. In that problem $\mathsf{C}^* = \mathbf{0}$ since the fundamental solution outside a bifurcation band is assumed to be not influenced by the bifurcation within a band. Our aim is to specify the general uniqueness conditions (a) and (b) in Conclusion 2 for the problem (11.61) for an arbitrary given $\mathbf{n}$. Of course, the basic solution to (11.61) is trivial and coincides with $(\dot{\mathbf{F}}^*, \mathring{\gamma}_K^*)$. The existence of a non-trivial solution to (11.61) is identified with the bifurcation (in the first-order rates) in a band.

Take two unit vectors $\mathbf{s}_1,\ \mathbf{s}_2 \in R^3$ such that $\mathbf{s}_1\mathbf{n} = 0$, $\mathbf{s}_2\mathbf{n} = 0$, $\mathbf{s}_1\mathbf{s}_2 = 0$. The second condition in (11.61) can be written in the form (11.29) with $n^* = 6$ and

$$\begin{aligned} &\mathbf{K}_1 = \mathbf{s}_1 \otimes \mathbf{s}_1 \qquad \mathbf{K}_2 = \mathbf{s}_2 \otimes \mathbf{s}_1 \qquad \mathbf{K}_3 = \mathbf{n} \otimes \mathbf{s}_1 \\ &\mathbf{K}_4 = \mathbf{s}_1 \otimes \mathbf{s}_2 \qquad \mathbf{K}_5 = \mathbf{s}_2 \otimes \mathbf{s}_2 \qquad \mathbf{K}_6 = \mathbf{n} \otimes \mathbf{s}_2\,. \end{aligned} \tag{11.62}$$

The tensors $\mathbf{K}_\alpha, \alpha = 1, \ldots, 6$, form an orthonormal basis ($\mathbf{K}^\alpha = \mathbf{K}_\alpha$) in the orthogonal complement $\mathcal{K}^\perp$ to the three-dimensional subspace $\mathcal{K} = \{\tilde{\dot{\mathbf{F}}} : \tilde{\dot{\mathbf{F}}} = \mathbf{g} \otimes \mathbf{n}\}$ of the space of second-order tensors, while an orthonormal basis in $\mathcal{K}$ is provided by

$$\mathbf{K}_1^\perp = \mathbf{s}_1 \otimes \mathbf{n} \qquad \mathbf{K}_2^\perp = \mathbf{s}_2 \otimes \mathbf{n} \qquad \mathbf{K}_3^\perp = \mathbf{n} \otimes \mathbf{n}\,. \tag{11.63}$$

Hence

$$\mathsf{P} = \sum_{\beta=1}^{3} \mathbf{K}_\beta^\perp \otimes \mathbf{K}_\beta^\perp\,, \qquad P_{ijkl} = \delta_{ik} n_j n_l\,. \tag{11.64}$$

It follows that

$$\left(\mathsf{P} \cdot (\mathsf{C}^{\mathrm{e}} + \mathsf{C}^*) \cdot \mathsf{P}\right)_{ijkl} = Q^{\mathrm{e}}_{ik} n_j n_l\,, \qquad Q^{\mathrm{e}}_{ik} = C^{\mathrm{e}}_{ipkq} n_p n_q\,. \tag{11.65}$$

The second-order tensor $\mathbf{Q}^{\mathrm{e}}$ is the acoustic tensor (dependent on $\mathbf{n}$) for the purely elastic constitutive branch. Suppose that

$$\det \mathbf{Q}^{\mathrm{e}} \neq 0, \tag{11.66}$$

which is the classical ellipticity condition for elastic moduli that excludes a non-trivial solution to the problem (11.61) posed for purely elastic rates. Then the tensor G defined by (11.48) exists and its components are given by

$$G_{ijkl} = (\mathbf{Q}^{\mathrm{e}})^{-1}_{ik} n_j n_l\,. \tag{11.67}$$

Denote by $\bar{\mathbf{n}}$ the unit normal to the bifurcation band in the current configuration. The acoustic tensor $\mathbf{Q}^{\mathrm{e}}$ can be expressed in terms of the elastic moduli in the current configuration (cf. (11.15)) as follows

$$Q^{\mathrm{e}}_{ik} = J|\mathbf{n}\mathbf{F}^{-1}|^2\,\bar{Q}^{\mathrm{e}}_{ik}\,, \qquad \bar{Q}^{\mathrm{e}}_{ik} = \bar{C}^{\mathrm{e}}_{ipkq}\bar{n}_p\bar{n}_q\,, \qquad \bar{\mathbf{n}} = \frac{\mathbf{n}\mathbf{F}^{-1}}{|\mathbf{n}\mathbf{F}^{-1}|}\,. \tag{11.68}$$

On substituting (11.67) into $(11.50)_2$ and rearranging with the help of the transformation formulae (11.14), (11.68), (11.5) and (11.11), the matrix $[g^{\mathrm{band}}] = [g^*]$ specified for the problem (11.61) in several equivalent forms reads

$$\begin{aligned} g^{\mathrm{band}}_{KL} = g^*_{KL} &= g_{KL} - (\boldsymbol{\Lambda}_K\mathbf{n})(\mathbf{Q}^{\mathrm{e}})^{-1}(\boldsymbol{\Lambda}^{\mathrm{p}}_L\mathbf{n}) \\ &= g_{KL} - \bar{\mathbf{n}}\bar{\boldsymbol{\lambda}}_K(\bar{\mathbf{Q}}^{\mathrm{e}})^{-1}\bar{\boldsymbol{\lambda}}^{\mathrm{p}}_L\bar{\mathbf{n}} \\ &= g_{KL} - (\boldsymbol{\lambda}_K\cdot\mathbf{A}\bar{\mathbf{n}})(\bar{\mathbf{Q}}^{\mathrm{e}})^{-1}(\bar{\mathbf{n}}\mathbf{B}^{-1}\cdot\boldsymbol{\lambda}^{\mathrm{p}}_L) \\ &= \bar{h}_{KL} + \bar{\boldsymbol{\mu}}_K\cdot\bar{\mathbf{L}}^{\mathrm{e}}\cdot\bar{\boldsymbol{\mu}}^{\mathrm{p}}_L - (\bar{\boldsymbol{\mu}}_K\cdot\bar{\mathbf{L}}^{\mathrm{e}}\bar{\mathbf{n}})(\bar{\mathbf{Q}}^{\mathrm{e}})^{-1}(\bar{\mathbf{n}}\bar{\mathbf{L}}^{\mathrm{e}}\cdot\bar{\boldsymbol{\mu}}^{\mathrm{p}}_L) \\ &= h_{KL} + \boldsymbol{\mu}_K\cdot\mathbf{L}^{\mathrm{e}}\cdot\boldsymbol{\mu}^{\mathrm{p}}_L - (\boldsymbol{\mu}_K\cdot\mathbf{L}^{\mathrm{e}}\cdot\mathbf{A}\bar{\mathbf{n}})(\bar{\mathbf{Q}}^{\mathrm{e}})^{-1}(\bar{\mathbf{n}}\mathbf{B}^{-1}\cdot\mathbf{L}^{\mathrm{e}}\cdot\boldsymbol{\mu}^{\mathrm{p}}_L)\,. \end{aligned} \tag{11.69}$$

All the quantities that appear in (11.69), and also the set $\mathcal{P}$ of the indices of potentially active mechanisms, are known in the material state in which the problem (11.61) is posed for a given $\mathbf{n}$ (or $\bar{\mathbf{n}}$). Hence, the matrix $[g^{\mathrm{band}}]$ is known and independent on $(\dot{\mathbf{F}}^*, \overset{\circ}{\gamma}{}^*_K)$. As a corollary of Theorem 2 (i) or Conclusion 2, we obtain the following result.

Exclusion conditions for the bifurcation in a band *The problem (11.61) for a given* $\mathbf{n}$ *admits only the trivial solution if simultaneously :*
(a) the elastic moduli satisfy the ellipticity condition (11.66),
(b) all principal minors of the matrix $[g^{\mathrm{band}}]$ *defined by (11.69) with* $K, L \in \mathcal{P}$ *are positive* •

It is pointed out that (a) and (b) jointly exclude the bifurcation in a band for any possible unloading within or outside the band at the bifurcation instant.

For convenience, the condition (b) is rewritten as

$$\det(g^{\mathrm{band}}_{\mathcal{A}}) > 0 \qquad \text{for all } \mathcal{A} \subseteq \mathcal{P}\,, \tag{11.70}$$

where $(g^{\mathrm{band}}_{\mathcal{A}})$ stands for a principal submatrix of $[g^{\mathrm{band}}]$ with indices from $\mathcal{A} \subseteq \mathcal{P}$.

11.7.2 Lower and Upper Bounds for the Primary Bifurcation

Consider now a (rather typical) homogeneous deformation path $(\dot{\mathbf{F}}^*, \overset{\circ}{\gamma}{}^*_K)(t)$, parameterized by a time-like scalar $t \geq t_0$, along which the following conditions are satisfied:

(i) the condition (a) above holds true at all t,
(ii) the set $\mathcal{P}$ of the indices of potentially active mechanisms is constant,
(iii) $[\overset{\circ}{\gamma}{}^*] > [0]$,

(iv) the matrix $[g^{\mathrm{band}}]$ for any $\mathbf{n}$ varies continuously with t,
(v) the condition (b) above is satisfied for all $\mathbf{n}$ at an initial instant $t = t_0$.

The following bounds to the instant of primary bifurcation in a band are obtained.

Bounds for the primary bifurcation *Along a deformation path that satisfies the conditions (i)-(v) above, the instant t^* of the primary bifurcation (in the first-order rates) in a band is bounded by*

$$t_{\mathcal{A}} \le t^* \le t_{\mathcal{P}}\,, \tag{11.71}$$

where the lower bound is determined from

$$t_{\mathcal{A}} = \min t \ : \quad \det(g^{\mathrm{band}}_{\mathcal{A}})(t) = 0 \quad \textit{for some } \mathcal{A} \subseteq \mathcal{P} \textit{ and some } \mathbf{n}\,, \tag{11.72}$$

and the upper bound from

$$t_{\mathcal{P}} = \min t \ : \quad \det[g^{\mathrm{band}}](t) = 0 \quad \textit{for some } \mathbf{n}\,. \tag{11.73}$$

Proof of the lower bound follows readily from the exclusion condition (11.70) by the continuity argument applied to $\det(g^{\mathrm{band}}_{\mathcal{A}})$. The upper bound is obtained as a particular bifurcation point at which none of the potentially active mechanisms undergoes unloading. To show this explicitly, suppose that $\det[g^{\mathrm{band}}] = 0$ for some $\mathbf{n}$ and $[\mathring{\gamma}^*] > [0]$; then an eigenvector $[w] \neq [0]$ that satisfies $[g^{\mathrm{band}}][w] = [0]$ and is of a sufficiently small magnitude $|w|$ can be added to $[\mathring{\gamma}^*]$ to give $[\mathring{\gamma}] = [\mathring{\gamma}^*] + [w] > 0$. The vector $[\mathring{\gamma}]$ corresponds to $[\dot{f}] = [b^*]-[g^{\mathrm{band}}][\mathring{\gamma}] = [b^*]-[g^{\mathrm{band}}][\mathring{\gamma}^*] = [\dot{f}^*] = [0]$ as required, and jointly with the relationship $\dot{\mathbf{F}} - \dot{\mathbf{F}}^* = \mathbf{G} \cdot \sum_{K\in\mathcal{P}} \mathbf{\Lambda}^{\mathrm{p}}_K(\mathring{\gamma}_K - \mathring{\gamma}^*_K)$ obtained from (11.49) gives a non-trivial solution $(\dot{\mathbf{F}}, \mathring{\gamma}_K)$ to (11.61). This completes the proof of (11.71) under the assumptions (i)-(v). □

The above lower and upper bounds can easily be extended under analogous assumptions to the general case examined in the preceding section. Details are omitted here.

If $M = 1$ (only one mechanism of a given index K is considered) then the lower and upper bounds coincide. If $\mathbf{M}^{\mathrm{e}} = (\mathbf{L}^{\mathrm{e}})^{-1}$ exists for the measures chosen then the uniqueness range $t < t^*$ can be expressed from (11.69) as $h_{KK} > h^{\mathrm{c}}_K$. Here, h_{KK} is the effective hardening modulus for the mechanism, and its critical value h^{c}_K reached at $t = t^*$ reads

$$\begin{aligned} h^{\mathrm{c}}_K &= -\boldsymbol{\lambda}_K \cdot \mathbf{M}^{\mathrm{e}} \cdot \boldsymbol{\lambda}^{\mathrm{p}}_K + \max_{\bar{\mathbf{n}}} \left(\bar{\mathbf{n}} \bar{\boldsymbol{\lambda}}_K (\bar{\mathbf{Q}}^{\mathrm{e}})^{-1} \bar{\boldsymbol{\lambda}}^{\mathrm{p}}_K \bar{\mathbf{n}} \right) \\ &= -\boldsymbol{\mu}_K \cdot \mathbf{L}^{\mathrm{e}} \cdot \boldsymbol{\mu}^{\mathrm{p}}_K + \max_{\bar{\mathbf{n}}} \left((\boldsymbol{\mu}_K \cdot \mathbf{L}^{\mathrm{e}} \cdot \mathbf{A}\bar{\mathbf{n}}) (\bar{\mathbf{Q}}^{\mathrm{e}})^{-1} (\bar{\mathbf{n}} \mathbf{B}^{-1} \cdot \mathbf{L}^{\mathrm{e}} \cdot \boldsymbol{\mu}^{\mathrm{p}}_K) \right), \end{aligned} \tag{11.74}$$

where, from $(11.68)_2$ and (11.15),

$$\bar{\mathbf{Q}}^{\mathrm{e}} = \bar{\mathbf{n}} \bar{\mathbf{L}}^{\mathrm{e}} \bar{\mathbf{n}} + \frac{1}{2} (\bar{\mathbf{n}} \boldsymbol{\sigma} \otimes \bar{\mathbf{n}} - \bar{\mathbf{n}} \otimes \boldsymbol{\sigma} \bar{\mathbf{n}} - \boldsymbol{\sigma} + (\bar{\mathbf{n}} \boldsymbol{\sigma} \bar{\mathbf{n}}) \mathbf{I})\,. \tag{11.75}$$

The formula (11.74) is equivalent to that derived by Rice [43] and Asaro and Rice [2] for the linearized problem without unloading. The validity of the formula (11.74) for the critical hardening modulus was extended to the case of possible unloading by Rice and Rudnicki [43]; cf. also Raniecki and Bruhns [39] and Bigoni and Zaccaria [4]. Hence, the present exclusion condition and the bounds for the bifurcation in a band in the case $M = 1$ yield known results, although with a somewhat more general proof. [4]

If $M \geq 2$ then the upper bound (11.73) can be reduced (see the preceding footnote) to the formulae given by Peirce [34] for finite strain elastoplasticity (of single crystals) with the normality flow rule, and without the restriction to normality by Rizzi et al. [44] and Steinmann [47] in the small strain and finite strain framework, respectively. In those investigations, as a starting point the *linearized* condition for the bifurcation in a band has been taken, in the form of the condition of ellipticity loss of the tangent stiffness moduli [13, 42]. That condition, which in terms of the moduli $\mathbf{C}$ in (11.60) with $\mathcal{A} = \mathcal{P}$ and $\det[g] \neq 0$ reads $\det \mathbf{Q} = 0$, $Q_{ik} = C_{ipkq} n_p n_q$, has not been used in this paper. The resulting coincidence of the instants of $\det[g^{\text{band}}] = 0$ and $\det \mathbf{Q} = 0$ is not unexpected since both conditions, although obtained on different routes, correspond to the same bifurcation in a band without unloading.

The exclusion condition (11.70) and the related lower bound (11.72) for the bifurcation in a band in the incrementally piecewise-linear material with multiple ($M \geq 2$) mechanisms of time-independent inelastic deformation, valid for any possible unloading at the bifurcation instant, have not been found in the literature.

If $M = 2$ and $h_{KK} > h^c_K$ for each mechanism at every instant then (11.70) reduces to $\det[g^{\text{band}}] > 0$, and the bounds (11.72) and (11.73) coincide. It follows that in that case and along a deformation path that satisfies (i)-(v) above, the condition (11.73) defines the exact instant of the primary bifurcation without unloading, any earlier bifurcation with unloading being excluded. This particular result, of interest for certain material models being in use (double slip, classical plasticity coupled with damage or phase transformation, etc.), also appears to be new, to the author's knowledge. That result was conjectured, without providing a proof, by Rizzi et al. [44].

11.8 Conclusions

The principal results of this paper can be summarized as follows:

1. The rate-problem of continuing mechanical equilibrium at macroscopically uniform deformations, formulated for a material element with an incrementally

[4] In the derivation of the uniqueness condition (11.70), the sole assumption concerning the elastic moduli has been the ellipticity of the moduli $\mathbf{C}^{\text{e}}$, and their magnitude with respect to the current stress is arbitrary. The invertibility of $\mathbf{L}^{\text{e}}$ is not required unless the dual form (11.3) of the constitutive rate equations (11.2) is used. The uniqueness condition (b) is expressed in terms of arbitrary constitutive rate-variables at finite deformations, and no restriction has been imposed on the quantities that define the constitutive relationships (11.2) or (11.3).

piecewise-linear constitutive law of the form defined by (11.1), (11.2) and (11.4), has been reduced to the linear complementarity problem (11.42) specified by (11.50) or (11.39). This result is applicable in general circumstances: for an arbitrary number of internal mechanisms of inelastic deformation, for arbitrary measures (11.8) of strain-rate and stress-rate at finite deformations, under no *a priori* restriction on the quantities that may be used to define the constitutive relationships (neither symmetry of moduli nor normality are required), for a class of deformation-sensitive incremental loadings (11.20), and for partial kinematic constraints (11.29).

2. General conditions sufficient, and also necessary to the extent defined, for the existence and uniqueness of a solution to the nonlinear rate-problem of continuing mechanical equilibrium of the material element have been established in Theorem 2. The conditions can be applied to the class of constitutive laws and problems characterized above. The conditions for constitutive uniqueness have been specified as particular cases.

3. As an application to predicting the onset of strain localization or failure, the condition (11.70) has been derived that jointly with (11.66) exclude the bifurcation in a band in an infinite homogeneous continuum. In contrast to other studies based on the linearized condition of ellipticity of the tangent stiffness moduli, any possible unloading is here taken into account. The present uniqueness condition (11.70) is directly imposed on an arbitrary matrix of interaction moduli and concerns all branches of a piecewise-linear incremental constitutive law.

4. Lower and upper bounds (11.72) and (11.73) have been established for the instant of primary bifurcation in a band along a smooth straining path. The upper bound reduces to the instant of ellipticity loss of the tangent stiffness moduli, and the lower bound has not been found in the literature.

The above theoretical conclusions can further be specified for constitutive models of real materials, including plasticity, damage, micro-cracking or phase transformation. There is a variety of possible practical applications, for instance, to deriving restrictions on material models or testing conditions if a unique response is desired, to obtaining lower and upper bounds to a material collapse under given loading conditions, to predicting the onset of strain localization into shear bands in particular materials, etc.

Acknowledgements

This work was partially supported by the State Committee of Research (KBN) in Poland under the project No. 7 T07A 001 14.

Appendix

(i) *Transformation rules* (11.11)

Take any pair $(\overset{\circ}{\mathbf{e}}, \overset{\circ}{\mathbf{t}})$ defined by (11.8). For purely elastic rates we have $\overset{\circ}{\mathbf{t}} = \mathsf{L}^{\mathrm{e}} \cdot \overset{\circ}{\mathbf{e}}$ and $\overset{\nabla}{\sigma} = \bar{\mathsf{L}}^{\mathrm{e}} \cdot \mathbf{D}$. On substituting this into $(11.8)_2$ and eliminating $\mathbf{D}$ between the equations in (11.8), we obtain

$$\mathsf{L}^{\mathrm{e}} \cdot \overset{\circ}{\mathbf{e}} = (\mathsf{B} \cdot \bar{\mathsf{L}}^{\mathrm{e}} + \mathsf{Z}) \cdot \mathsf{A}^{-1} \cdot \overset{\circ}{\mathbf{e}} \qquad \text{if } \overset{\circ}{\gamma}_K \equiv 0 . \tag{11.76}$$

The equality must hold for arbitrary variations in $\overset{\circ}{\mathbf{e}}$. Since the term in parentheses is independent of the values of rate-variables, this implies

$$\mathsf{L}^{\mathrm{e}} = (\mathsf{B} \cdot \bar{\mathsf{L}}^{\mathrm{e}} + \mathsf{Z}) \cdot \mathsf{A}^{-1} . \tag{11.77}$$

In the same way a more general transformation formula is proved for the tangent stiffness moduli tensor in an arbitrary incrementally linear constitutive relationship, viz.

$$\mathsf{L} = (\mathsf{B} \cdot \bar{\mathsf{L}} + \mathsf{Z}) \cdot \mathsf{A}^{-1} \qquad \text{if } \overset{\circ}{\mathbf{t}} = \mathsf{L} \cdot \overset{\circ}{\mathbf{e}}, \ \overset{\nabla}{\sigma} = \bar{\mathsf{L}} \cdot \mathbf{D} . \tag{11.78}$$

From $(11.2)_1$ on applying the transformation formulae (11.8) and (11.77), we obtain

$$\sum_K \lambda^{\mathrm{p}}_K \overset{\circ}{\gamma}_K = \mathsf{L}^{\mathrm{e}} \cdot \overset{\circ}{\mathbf{e}} - \overset{\circ}{\mathbf{t}} = \mathsf{B} \cdot (\bar{\mathsf{L}}^{\mathrm{e}} \cdot \mathbf{D} - \overset{\nabla}{\sigma}) = \mathsf{B} \cdot \sum_K \bar{\lambda}^{\mathrm{p}}_K \overset{\circ}{\gamma}_K . \tag{11.79}$$

This must be true for arbitrarily selected $\overset{\circ}{\gamma}_K$'s, which implies

$$\lambda^{\mathrm{p}}_K = \mathsf{B} \cdot \bar{\lambda}^{\mathrm{p}}_K . \tag{11.80}$$

In turn, from (11.6) and $(11.2)_2$ for purely elastic rates, it follows that

$$\lambda_K \cdot \overset{\circ}{\mathbf{e}} = \dot{f}_K = \bar{\lambda}_K \cdot \mathbf{D} = \bar{\lambda}_K \cdot \mathsf{A}^{-1} \cdot \overset{\circ}{\mathbf{e}} \qquad \text{if } \overset{\circ}{\gamma}_K \equiv 0 . \tag{11.81}$$

The equality must hold for arbitrary variations in $\overset{\circ}{\mathbf{e}}$, so that

$$\lambda_K = \bar{\lambda}_K \cdot \mathsf{A}^{-1} . \tag{11.82}$$

From (11.6) and $(11.2)_2$ for inelastic rates, on using (11.8) and (11.82) we obtain

$$\sum_L g_{KL} \overset{\circ}{\gamma}_L = \lambda_K \cdot \overset{\circ}{\mathbf{e}} - \dot{f}_K = \bar{\lambda}_K \cdot \mathbf{D} - \dot{f}_K = \sum_L \bar{g}_{KL} \overset{\circ}{\gamma}_L . \tag{11.83}$$

Since this must hold for arbitrarily selected $\overset{\circ}{\gamma}_K$'s, it follows that the matrix (g_{KL}) is measure-invariant:

$$g_{KL} = \bar{g}_{KL} . \tag{11.84}$$

This completes the derivation of the transformation rules (11.11).

(ii) *Transformation rules in dual formulation*

From (11.5) and (11.11) it follows that

$$
\begin{aligned}
\mathbf{M}^{\mathrm{e}} &= \left((\mathbf{B}\cdot\bar{\mathbf{L}}^{\mathrm{e}}+\mathbf{Z})\cdot\mathbf{A}^{-1}\right)^{-1},\\
\mu_K^{\mathrm{p}} &= \mathbf{M}^{\mathrm{e}}\cdot\mathbf{B}\cdot\bar{\lambda}_K^{\mathrm{p}},\\
\mu_K &= \bar{\lambda}_K\cdot\mathbf{A}^{-1}\cdot\mathbf{M}^{\mathrm{e}},\\
h_{KL} &= \bar{h}_{KL}-\bar{\lambda}_K\cdot(\mathbf{A}^{-1}\cdot\mathbf{M}^{\mathrm{e}}\cdot\mathbf{B}-\bar{\mathbf{M}}^{\mathrm{e}})\cdot\bar{\lambda}_L^{\mathrm{p}}.
\end{aligned}
\tag{11.85}
$$

On substituting the barred analogs of the relationships (11.5), we obtain

$$
\begin{aligned}
\mathbf{M}^{\mathrm{e}} &= \mathbf{A}\cdot(\mathbf{B}\cdot(\bar{\mathbf{M}}^{\mathrm{e}})^{-1}+\mathbf{Z})^{-1},\\
\mu_K^{\mathrm{p}} &= \mathbf{M}^{\mathrm{e}}\cdot\mathbf{B}\cdot(\bar{\mathbf{M}}^{\mathrm{e}})^{-1}\cdot\bar{\mu}_K^{\mathrm{p}},\\
\mu_K &= \bar{\mu}_K\cdot(\bar{\mathbf{M}}^{\mathrm{e}})^{-1}\cdot\mathbf{A}^{-1}\cdot\mathbf{M}^{\mathrm{e}},\\
h_{KL} &= \bar{h}_{KL}-\bar{\mu}_K\cdot((\mathbf{B}+\mathbf{Z}\bar{\mathbf{M}}^{\mathrm{e}})^{-1}\cdot\mathbf{B}-\mathbf{I}_{\mathrm{s}})\cdot(\bar{\mathbf{M}}^{\mathrm{e}})^{-1}\cdot\bar{\mu}_L^{\mathrm{p}}.
\end{aligned}
\tag{11.86}
$$

After rearranging in a manner similar to that in (11.40) and (11.41), the dual transformation rules can be expressed as follows :

$$
\begin{aligned}
\mathbf{M}^{\mathrm{e}} &= \mathbf{A}\cdot(\mathbf{B}\cdot(\bar{\mathbf{M}}^{\mathrm{e}})^{-1}+\mathbf{Z})^{-1},\\
\mu_K^{\mathrm{p}} &= \mathbf{A}\cdot(\mathbf{I}_{\mathrm{s}}+\bar{\mathbf{M}}^{\mathrm{e}}\cdot\mathbf{B}^{-1}\cdot\mathbf{Z})^{-1}\cdot\bar{\mu}_K^{\mathrm{p}}\\
&= (\mathbf{A}-\mathbf{M}^{\mathrm{e}}\cdot\mathbf{Z})\cdot\bar{\mu}_K^{\mathrm{p}},\\
\mu_K &= \bar{\mu}_K\cdot(\mathbf{B}+\mathbf{Z}\cdot\bar{\mathbf{M}}^{\mathrm{e}})^{-1}\\
&= \bar{\mu}_K\cdot\mathbf{B}^{-1}\cdot(\mathbf{I}_{\mathrm{s}}-\mathbf{Z}\cdot\mathbf{A}^{-1}\cdot\mathbf{M}^{\mathrm{e}}),\\
h_{KL} &= \bar{h}_{KL}+\bar{\mu}_K\cdot(\mathbf{B}+\mathbf{Z}\cdot\bar{\mathbf{M}}^{\mathrm{e}})^{-1}\cdot\mathbf{Z}\cdot\bar{\mu}_L^{\mathrm{p}}\\
&= \bar{h}_{KL}+\mu_K\cdot\mathbf{Z}\cdot\bar{\mu}_L^{\mathrm{p}}.
\end{aligned}
\tag{11.87}
$$

(iii) *Transformation rules* (11.28)

Let

$$
\overset{\circ}{\mathbf{t}}{}^{\mathrm{ext}}=\overset{\circ}{\mathbf{t}}{}^{*}-\mathbf{L}^{*}\cdot\overset{\circ}{\mathbf{e}},\quad \overset{\nabla}{\sigma}{}^{\mathrm{ext}}=\overset{\nabla}{\sigma}{}^{*}-\bar{\mathbf{L}}^{*}\cdot\mathbf{D},\quad \overset{\circ}{\mathbf{e}}=\mathbf{A}\cdot\mathbf{D},\quad \overset{\circ}{\mathbf{t}}{}^{\mathrm{ext}}=\mathbf{B}\cdot\overset{\nabla}{\sigma}{}^{\mathrm{ext}}+\mathbf{Z}\cdot\mathbf{D}. \tag{11.88}
$$

It follows that

$$
\overset{\circ}{\mathbf{t}}{}^{*}-\mathbf{L}^{*}\cdot\overset{\circ}{\mathbf{e}}=\mathbf{B}\cdot\overset{\nabla}{\sigma}{}^{\mathrm{ext}}+\mathbf{Z}\cdot\mathbf{D}, \tag{11.89}
$$

which for $\mathbf{D}=\mathbf{0}$ implies

$$
\overset{\circ}{\mathbf{t}}{}^{*}=\mathbf{B}\cdot\overset{\nabla}{\sigma}{}^{*}. \tag{11.90}
$$

On eliminating $\overset{\circ}{\mathbf{t}}{}^{*}$ between (11.90) and (11.89) and rearranging, we obtain

$$
\begin{aligned}
\mathbf{L}^{*}\cdot\overset{\circ}{\mathbf{e}} &= \mathbf{B}\cdot(\overset{\nabla}{\sigma}{}^{*}-\overset{\nabla}{\sigma}{}^{\mathrm{ext}})-\mathbf{Z}\cdot\mathbf{D}\\
&= (\mathbf{B}\cdot\bar{\mathbf{L}}^{*}-\mathbf{Z})\cdot\mathbf{D}\\
&= (\mathbf{B}\cdot\bar{\mathbf{L}}^{*}-\mathbf{Z})\cdot\mathbf{A}^{-1}\cdot\overset{\circ}{\mathbf{e}}.
\end{aligned}
\tag{11.91}
$$

This must hold for arbitrary variations in $\overset{\circ}{\mathbf{e}}$, which implies

$$\mathbf{L}^* = (\mathbf{B} \cdot \bar{\mathbf{L}}^* - \mathbf{Z}) \cdot \mathbf{A}^{-1} . \tag{11.92}$$

This completes the derivation of the transformation rules (11.28).

References

1. Asaro, R.J. (1983) Micromechanics of crystals and polycrystals. *Advances in Applied Mechanics, Vol.* 23, pp. 1-115. Acad. Press, New York.
2. Asaro, R.J. and Rice, J.R. (1977) Strain localization in ductile single crystals. *J. Mech. Phys. Solids* **25**, 309-338.
3. Bažant, Z.P. and Kim, S.S. (1979) Plastic-fracturing theory of concrete. *ASCE I. Engng. Mech. Div.* **105**, 407-428.
4. Bigoni, D. and Zaccaria, D. (1992) Loss of strong ellipticity in non-associative elastoplasticity. *J. Mech. Phys. Solids* **40**, 1313-1331.
5. Carol, I. and Prat, P.C. (1995) A multicrack model based on the theory of multisurface plasticity and two fracture energies. In *Computational Plasticity*, ed. D.R.J. Owen, E. Oñate and E. Hinton, pp. 1583-1594. Pineridge Press, Swansea.
6. Cherkaoui, M. Berveiller, M. and Sabar, H. (1998) Micromechanical modeling of martensitic transformation induced plasticity (TRIP) in austenitic single crystals. *Int. J. Plasticity* **14**, 597-626.
7. Cottle, R.W., Pang, J.-S. and Stone, R.E. (1992) *The Linear Complementarity Problem.* Acad. Press, San Diego.
8. Dragon, A. and Mróz, Z. (1979) A continuum model for plastic-brittle behaviour of rock and concrete. *Int. J. Engng. Sci.* **17**, 121-137.
9. Eshelby, J.D. (1957) The determination of the elastic field of an ellipsoidal inclusion, and related problems. *Proc. Roy. Soc. London* **A 241**, 376-396.
10. Fischer, F.D., Sun, Q.-P. and Tanaka, K. (1996) Transformation induced plasticity (TRIP). *Appl. Mech. Rev.* **49**, no 6, 317-364.
11. Hansen, N.R. and Schreyer, H.L. (1994) A thermodynamically consistent framework for theories of elastoplasticity coupled with damage. *Int. J. Solids Structures* **31**, 359-389.
12. Havner, K.S. (1992) *Finite Plastic Deformation of Crystalline Solids*, University Press, Cambridge.
13. Hill, R. (1962) Acceleration waves in solids. *J. Mech. Phys. Solids* **10**, 1-16.
14. Hill (1965) Continuum micro-mechanics of elastoplastic polycrystals. *J. Mech. Phys. Solids* **13**, 89-101.
15. Hill, R. (1966) Generalized constitutive relations for incremental deformation of metal crystals by multislip, *J. Mech. Phys. Solids* **14**, 95-102.
16. Hill, R. (1967a) On the classical constitutive laws for elastic/plastic solids. In: *Recent Progress in Applied Mechanics, The Folke Odkvist Volume*, ed. B. Broberg *at. al.*, pp. 241-249. Almqvist & Wiksell, Stockholm.
17. Hill, R. (1967b) Eigenmodal deformations in elastic/plastic continua. *J. Mech. Phys. Solids* **15**, 371-386.
18. Hill, R. (1968) On constitutive inequalities for simple materials, *J. Mech. Phys. Solids* **16**, 229-242, 315-322.
19. Hill, R. (1978) Aspects of invariance in solids mechanics, *Advances in Applied Mechanics, Vol.* 18, pp. 1-75. Acad. Press, New York.
20. Hill, R. (1983) On intrinsic eigenstates in plasticity with generalized variables. *Math. Proc. Camb. Phil. Soc.* **93**, 177-189.

21. Hill, R. and Rice, J.R. (1972) Constitutive analysis of elastic- plastic crystals at arbitrary strain, *J. Mech. Phys. Solids* **20**, 401-413.
22. Huang, M. and Brinson, L.C. (1998) A multivariant model for single crystal shape memory alloy behavior. *J. Mech. Phys. Solids* **46**, 1379-1409.
23. Klisinski, M. (1998) On constitutive relations for arbitrary stress-strain control in multi-surface plasticity. *Int. J. Solids Structures* **35**, 2655-2678.
24. Koiter, W.T. (1953) Stress-strain relations, uniqueness and variational theorems for elastic-plastic materials with a singular yield surface, *Quart. Appl. Math.* **11**, 350-353.
25. Leblond, J.B., Mottet, G. and Devaux, J.C. (1986) A theoretical and numerical approach to the plastic behaviour of steels during phase transformations – I. Derivation of general relations. *J. Mech. Phys. Solids* **34**, 395-409.
26. Maier, G. (1969) "Linear" flow-laws of elastoplasticity: a unified general approach, *Rend. Acc. Naz. Lincei, Ser. VIII*, Vol. XLVII, 266-276.
27. Mandel, J. (1965) Generalisation de la théorie de plasticité de W.T. Koiter, *Int. J. Solids Structures* **1**, 273-295.
28. Nemat-Nasser, S., Sumi, Y. and Keer, L.M. (1980) Unstable growth of tension cracks in brittle solids: stable and unstable bifurcations, snap-through, and imperfection sensitivity. *Int. J. Solids Structures* **16**, 1017-1035.
29. Nguyen, Q.S. (1994) Bifurcation and stability in dissipative media (plasticity, friction, fracture). *Appl. Mech. Rev.* **47**, no 1, Part 1, 1-31.
30. Nguyen, Q.S., Stolz, C. and Debruyne, G. (1990) Energy methods in fracture mechanics: stability, bifurcation and second variations. *Eur. J. Mech. A/Solids* **9**, 157-173.
31. Ogden, R.W. (1984) On Eulerian and Lagrangean objectivity in continuum mechanics. *Arch. Mech.* **36**, 207-218.
32. Patoor, E., Eberhardt, A. and Berveiller, M. (1988) Thermomechanical behaviour of shape memory alloys. *Arch. Mech.* **40**, 775-794.
33. Patoor, E. and Berveiller, M. (1997) Micromechanical modelling of the thermomechanical behavior of shape memory alloys. In *CISM Lecture-Notes* 368, ed. M. Berveiller and F.D. Fischer. Springer, Wien - New York.
34. Peirce, D. (1983) Shear band bifurcations in ductile single crystals. *J. Mech. Phys. Solids* **31**, 133-153.
35. Petryk, H. (1989) On constitutive inequalities and bifurcation in elastic-plastic solids with a yield-surface vertex. *J. Mech. Phys. Solids* **37**, 265-291.
36. Petryk, H. (1992) Material instability and strain-rate discontinuities in incrementally nonlinear continua. *J. Mech. Phys. Solids* **40**, 1227-1250.
37. Petryk, H. (1997) Plastic instability: criteria and computational approaches. *Arch. Comp. Meth. Engng.* **4**, 111-151.
38. Prat, P.C. and Bažant, Z.P. (1997) Tangential stiffness of elastic materials with systems of growing or closing cracks. *J. Mech. Phys. Solids* **45**, 611-636.
39. Raniecki, B. and Bruhns, O.T. (1981) Bounds to bifurcation stresses in solids with non-associated plastic flow law at finite strain. *J. Mech. Phys. Solids* **29**, 153-172.
40. Rice, J.R. (1971) Inelastic constitutive relations for solids: an internal-variable theory and its application to metal plasticity. *J. Mech. Phys. Solids* **19**, 433-455.
41. Rice, J.R. (1975) Continuum mechanics and thermodynamics of plasticity in relation to microscale deformation mechanisms. In *Constitutive Equations in Plasticity*, ed. A.S. Argon, pp.23-79. MIT Press, Cambridge Mass.
42. Rice, J.R. (1977) The localization of plastic deformation. In: *Theoretical and Applied Mechanics* , ed. W.T. Koiter, pp. 207-220. North- Holland, Amsterdam.
43. Rice, J.R. and Rudnicki, J.W. (1980) A note on some features of the theory of localization of deformation. *Int. J. Solids Structures* **16**, 597-605.
44. Rizzi, E., Maier, G. and Willam, K. (1996) On failure indicators in multi-dissipative materials. *Int. J. Solids Structures* **33**, 3187-3214.
45. Sewell, M.J. (1967) On configuration-dependent loading. *Arch. Rat. Mech. Anal.* **23**, 327-351.
46. Sewell, M.J. (1972) A survey of plastic buckling, in *Stability*, ed. H.H.E. Leipholz,, pp. 85-197. Univ. of Waterloo Press, Ontario.

47. Steinmann, P. (1996) On localization analysis in multisurface hyperelasto-plasticity. *J. Mech. Phys. Solids* **44**, 1691-1713.
48. Sun, Q.-P. and Lexcellent, C. (1996) On the unified micromechanics constitutive description of one-way and two-way shape memory effects. *Journal de Physique IV* **6**, C1-367-375.
49. Xiao, H., Bruhns, O.T. and Meyers, A. (1998) Objective corotational rates and unified work-conjugacy relation between Eulerian and Lagrangean strain and stress measures. *Arch. Mech.* **50**, 1015-1045.
50. Yu, J.W. (1989) On energy-based coupled elastoplastic damage theories: constitutive modeling and computational aspects. *Int. J. Solids Structures* **25**, 803-833.

Chapter 12
Incremental Energy Minimization in Dissipative Solids

Abstract The incremental energy minimization is examined as a method of determining solution paths for time-independent dissipative solids. Isothermal quasi-static deformations are considered, and the deformation work is locally decomposed into the increments in free energy and intrinsic dissipation. General conditions necessary for applicability of the minimization procedure are derived and discussed.

Keywords: Solids and structures; Dissipative materials; Plasticity; Energy; Path stability

12.1 Introduction

This paper is concerned with a variational method of determining the response of an inelastic body to slowly varying loading. It is based on the concept of minimization of *incremental* energy supplied to the examined *system*, apparently first developed by Petryk [1, 2, 3] for a class of time-independent problems at finite deformation. Essentially the same idea can also be found in more recent papers by other authors, e.g. [4, 5, 6]. An advantage of the incremental energy minimization method is that it provides a criterion of choice if non-unique solutions exist. Justification of that approach follows from its consistency with Hill's bifurcation theory [7, 8], as shown in [2], and with the concept of instability of plastic deformation processes or paths [3]. The non-convex minimization of an incremental energy supply enables crossing multiple bifurcation points with automatic selection of the post-critical deformation path in finite element simulations [9, 10]. A general condition of applicability is that the global tangent stiffness matrix must be symmetric.

The deformation work, which is the principal part of the minimized energy functional, is usually expressed as an integral of the stress power. According to the classical thermodynamics, the work supplied in isothermal quasi-static deformation

[0] Reprinted from: Comptes Rendus Mécanique 331, 469-474, 2003. H.Petryk, Incremental energy minimization in dissipative solids. Copyright © 2003, with permission from Académie des sciences.

H. Petryk, *Theory of Plastic Instability*,
https://doi.org/10.1007/978-3-031-98223-1_12

can be decomposed into the increments in free energy and intrinsic dissipation. With this decomposition adopted, the aim of this paper is to examine applicability of the incremental energy minimization to time-independent dissipative solids when the increments in internal state variables are treated as independent unknowns. In distinction to the closely related earlier works dealing with path stability, see [11, 4] for comparison, the main task here is to characterize the minimization procedure itself along with its prerequisities. In particular, a symmetry restriction on the dissipation function is derived as a condition necessary for the intrinsic consistency between the first- and second-order minimization. This symmetry restriction has not been mentioned in some recent works dealing with numerical applications of the minimization procedure.

12.2 Minimization of Incremental Energy

Following the approach originated in [1, 2], consider the minimization problem

$$\Delta E \longrightarrow \min \qquad \text{subject to kinematical constrains} \tag{12.1}$$

where ΔE is the increment in energy to be supplied from external sources to the mechanical *system*, consisting of the deformed body and the loading device, in order to produce quasi-statically a deformation increment, generally with the help of additional perturbing forces. The prefix Δ denotes a virtual increment from a given state, corresponding to an increment $\Delta\lambda$ of a loading parameter λ used to define changes in external constrains. For inelastic solids with path-dependent energy consumption, (12.1) is fundamentally different from the usual energy minimization at *fixed* λ.

Suppose that the minimum in (1) exists, and denote it by ΔE^0. Then (12.1) is rewritten as

$$\Delta E^0 \leq \Delta E \qquad \text{subject to kinematical constrains.} \tag{12.2}$$

Under the restriction to isothermal processes, ΔE is defined as follows

$$\Delta E = \Delta W + \Delta \Omega, \tag{12.3}$$

where ΔW is the total work increment supplied to the deforming body, and $\Delta\Omega$ is a potential energy increment of the loading device, assuming that the incremental loading is *conservative* in an overall sense. A further insight into the meaning of ΔE is gained if the mechanical system is imagined to be placed in a heat reservoir. Then ΔE can be identified with a change in the internal energy of the resulting compound thermodynamic system {deforming body + loading device + heat reservoir} [11].

The mathematical expression for the basic physical quantity ΔW may depend on the problem. In particular, for a continuous body that in a fixed reference configuration occupies a domain B of volume V, the deformation work increment ΔW can be split at constant temperature as follows

$$\Delta W = \int_B \Delta w \, dV = \int_B (\Delta\phi + \Delta\mathcal{D}) \, dV = \int \dot{W} \, dt, \quad \dot{W} = \int_B \dot{w} \, dV, \quad \dot{w} = \dot{\phi} + D, \tag{12.4}$$

where a superimposed dot denotes the forward rate (one-sided material time derivative) with respect to time t, ϕ is the Helmholtz free energy density, and $\mathcal{D}$ is the intrinsic dissipation density, being a path-dependent integral of a local dissipation function D. The constitutive law used to determine ΔW is here time-independent and local. The analysis can be extended to global internal variables [4].

Let a local material state $\mathcal{G}$ be characterized in terms of the deformation gradient $\mathbf{F}$ and internal state variables considered in the reference configuration, denoted collectively by α, so that $\mathcal{G} := (\mathbf{F}, \alpha)$ at a fixed temperature. The two functions ϕ and D needed to determine ΔW from (12.4) are assumed to be given. For a smooth free energy density function ϕ, denoted for simplicity by the same symbol as its value, we have the usual formulae

$$\phi = \phi(\mathbf{F}, \alpha), \qquad \phi_{,\mathbf{F}} = \mathbf{S}, \qquad -\phi_{,\alpha} = \boldsymbol{A}, \tag{12.5}$$

where the comma as a subscript denotes the partial Fréchet derivative, $\mathbf{S}$ is the transposed nominal stress, and $\boldsymbol{A}$ stands for the set of the thermodynamic driving forces conjugate to α. A continuous dissipation function D for a time-independent material is assumed to satisfy

$$D = D(\dot{\alpha}, \alpha) \geq 0, \qquad D(r\dot{\alpha}, \alpha) = rD(\dot{\alpha}, \alpha) \quad \text{for all } r > 0. \tag{12.6}$$

The condition of thermodynamic compatibility

$$D = \boldsymbol{A} \cdot \dot{\alpha} \tag{12.7}$$

equivalent to $\dot{w} = \mathbf{S} \cdot \dot{\mathbf{F}}$, is necessary for $\dot{\alpha}$ to be compatible with the current thermodynamic forces $\boldsymbol{A}(\mathcal{G})$, while (12.7) fails for other *virtual* $\dot{\alpha}$. A central dot denotes full contraction (or a scalar product). It will be convenient to rewrite the condition (12.7) in a more mathematical form as follows

$$\dot{\alpha} \in \mathcal{L}(\mathcal{G}), \qquad \mathcal{L}(\mathcal{G}) = \{\dot{\alpha} : \ D(\dot{\alpha}, \alpha) = \boldsymbol{A}(\mathcal{G}) \cdot \dot{\alpha}\}, \tag{12.8}$$

$\mathcal{L}(\mathcal{G})$ being thus defined as a cone in $\dot{\alpha}$-space within which the dissipation function D yields the actual value of the dissipation rate. Note that D is necessarily linear in $\dot{\alpha}$ within that cone.

The question arises in which circumstances the minimizer in (12.2) is a solution to the incremental boundary-value problem, satisfying locally the constitutive law. Whatever might be an approximate sense of a solution in small but finite increments, the minimization procedure (12.2) should yield an exact solution in the first-order rates in the limit as the magnitude of all increments tends to zero. To examine this, applicability of the minimization rule (12.1) or (12.2) must be checked for ΔE developed up to the *second*-order terms.

Denote by $\mathcal{G}_0 = (\mathbf{F}_0, \alpha_0)$ a fixed initial state from which any increment is initiated. Assuming that the function ϕ is twice Fréchet differentiable, we obtain

$$\Delta\phi = \Delta_1\phi + \Delta_2\phi + o(\rho^2)\,, \qquad \rho = a\|\Delta\mathbf{F}\| + \|\Delta\alpha\|\,, \qquad 0 < a = \text{const}, \quad (12.9)$$

where $o(\rho^2)/\rho^2 \to 0$ as $\rho \to 0$, $\|\mathbf{x}\| = (\mathbf{x}\cdot\mathbf{x})^{1/2}$ denotes a norm in an appropriate (Hilbert) space, and

$$\Delta_1\phi = \mathbf{S}(\mathcal{G}_0)\cdot\Delta\mathbf{F} - \boldsymbol{A}(\mathcal{G}_0)\cdot\Delta\alpha\,, \qquad \Delta_2\phi = \tfrac{1}{2}\Delta\mathbf{S}\cdot\Delta\mathbf{F} - \tfrac{1}{2}\Delta\boldsymbol{A}\cdot\Delta\alpha\,. \quad (12.10)$$

Under the assumption that the function $D(\dot\alpha, \alpha)$ possesses the Fréchet derivative $D_{,\alpha}$ with respect to α, we have

$$D(\dot\alpha, \alpha_0+\Delta\alpha) = D(\dot\alpha, \alpha_0)+D_1(\dot\alpha, \Delta\alpha)+\|\dot\alpha\|\, o(\|\Delta\alpha\|),\ D_1(\dot\alpha, \hat\alpha) \equiv \hat\alpha\cdot D_{,\alpha}(\dot\alpha, \alpha_0), \quad (12.11)$$

where $D_1(r\dot\alpha, \hat\alpha) = rD_1(\dot\alpha, \hat\alpha)$ for every $r > 0$ and $D_1(\dot\alpha, \hat\alpha)$ is linear with respect to $\hat\alpha$. It follows that

$$\Delta\mathcal{D} = \Delta_1\mathcal{D} + \Delta_2\mathcal{D} + o(\|\Delta\alpha\|^2) \quad (12.12)$$

on a radial path of constant direction of the tangent ($\dot\alpha$), where

$$\Delta_1\mathcal{D} = D(\Delta\alpha, \alpha_0)\,, \qquad \Delta_2\mathcal{D} = \tfrac{1}{2}D_1(\Delta\alpha, \Delta\alpha). \quad (12.13)$$

In contrast to (12.10), the formulae (12.13) do not apply to arbitrary circuitous paths in α-space, cf. [12]. In this paper we restrict attention to smooth paths of bounded curvature that start from $\mathcal{G}_0$ and are kept fixed as $\|\Delta\alpha\| \to 0$. Under the simplifying assumption that D is Gateaux differentiable with respect to $\dot\alpha$ at $\dot\alpha \neq \mathbf{0}$ (but *not* at $\dot\alpha = \mathbf{0}$), and using $D = D_{,\dot\alpha}\cdot\dot\alpha$ as the well-known consequence of $(12.6)_2$, validity of (12.13) is extended to such "direct" paths. The case when D is not differentiable with respect to $\dot\alpha$ can be examined separately.

12.3 Conditions for Applicability of the Incremental Energy Minimization

12.3.1 First-Order Minimization

We have $\Delta E = \dot E \Delta t$ to first order, where from (12.3), (12.4) and (12.5),

$$\dot E = \int_B (D(\dot\alpha, \alpha_0) - \boldsymbol{A}(\mathcal{G}_0)\cdot\dot\alpha)\, dV + \int_B \mathbf{S}(\mathcal{G}_0)\cdot\dot{\mathbf{F}}\, dV + \dot\Omega. \quad (12.14)$$

By the principle of virtual power, and excluding unilateral constraints for simplicity, the current global state is in mechanical equilibrium if and only if the sum of the last two terms in (12.14) has a constant value for all virtual (kinematically admissible) velocity fields. This is required for applicability of the minimization (12.1) to first order, along with another necessary condition

$$D(\dot\alpha, \alpha_0) - \boldsymbol{A}(\mathcal{G}_0)\cdot\dot\alpha \geq 0 \qquad \text{for all virtual } \dot\alpha \quad (12.15)$$

which must hold almost everywhere in B, since otherwise $\dot{E}$ would be unbounded from below.

If the above two conditions imposed on the initial *state* of incremental energy minimization are satisfied then, from (12.8), $\dot{E} \geq \dot{E}^0$ if and only if the minimizing rate $\dot{\alpha}^0 \in \mathcal{L}(\mathcal{G}_0)$ almost everywhere in B. If, moreover, also $\dot{\alpha} \in \mathcal{L}(\mathcal{G}_0)$ in B then $\dot{E} = \dot{E}^0$. Clerly, to select $\dot{\alpha}^0$, the energy minimization (12.2) must be extended to the second-order terms, which is examined in Subsection 12.3.2.

Remark 1 In general, convexity of the dissipation function D with respect to $\dot{\alpha}$ need not be assumed in advance when applying the incremental energy minimization. However, fulfillment of (12.15) is related to *convexification* (i.e. finding a convex envelope) of D; this problem is not pursued further here.

Denote by superscript 0 the quantities evaluated along a path determined by applying the minimization procedure (12.1) successively. The condition (12.15) must hold along that path at the beginning of a next time step, since otherwise no minimum of $\dot{E}$ could be found at that instant and the minimization procedure would break down. Therefore, by treating the left-hand expression in (12.15) as a function of time at variable $\mathcal{G}_0$ but fixed $\dot{\alpha}$, its forward increment must be non-negative if the expression vanishes initially, which is the case if $\dot{\alpha} \in \mathcal{G}_0$. For a time step arbitrarily small, this yields

$$D_1(\dot{\alpha}, \dot{\alpha}^0) - \dot{A}^0 \cdot \dot{\alpha} \geq 0 \qquad \text{for all } \dot{\alpha} \in \mathcal{L}(\mathcal{G}_0). \tag{12.16}$$

The above argument and result (12.16) are, in essence, the same as those established first by Nguyen (cf. [13]); however, his starting point was the assumed maximum dissipation principle, and not the minimization (12.1) as here.

12.3.2 Second-Order Local Minimization With Respect To $\dot{\alpha}$

We begin with the remark that in energy calculations with accuracy to the second-order terms, the leading terms like in $\Delta_1\mathcal{D}$ or $\Delta_1\phi$, and hence $\Delta\alpha$ or $\Delta\mathbf{F}$ therein, must be also correct to second order, e.g. on substituting $\Delta\alpha = \dot{\alpha}\,\Delta t + \frac{1}{2}\ddot{\alpha}\,(\Delta t)^2$. In turn, in the second-order term $\Delta_2 w$ it suffices to substitute $\Delta\alpha = \dot{\alpha}\,\Delta t$. Here and below, all derivatives are evaluated in $\mathcal{G}_0$.

Suppose first that (12.2) holds to second order for $\dot{\alpha} = \dot{\alpha}^0 \in \mathcal{L}(\mathcal{G}_0)$ and the only difference between both sides of (12.2) is due to $\ddot{\alpha} \neq \ddot{\alpha}^0$. This can be shown to imply locally

$$D(\Delta\alpha^0, \alpha_0) - A(\mathcal{G}_0) \cdot \Delta\alpha^0 = o(\|\Delta\alpha^0\|^2). \tag{12.17}$$

Next, on using (12.12) and (12.9), taking $\ddot{\alpha} = \mathbf{0}$ and substituting (12.17), and rearranging with the help of (12.13), (12.10), and (12.8), it is shown that the fulfillment of (12.2) to second order implies [11]

$$\lim_{\Delta t \to 0} (\Delta_2 w - \Delta_2 w^0)/(\Delta t)^2 \geq 0 \text{ at } \dot{\mathbf{F}} = \dot{\mathbf{F}}^0 \iff J(\dot{\alpha}) \geq J(\dot{\alpha}^0) \text{ for all } \dot{\alpha} \in \mathcal{L}(\mathcal{G}_0) \tag{12.18}$$

almost everywhere in B, where $\dot{\mathbf{F}}^0$ is fixed and

$$J(\dot{\alpha}) = \tfrac{1}{2}\dot{\alpha} \cdot \phi_{,\alpha\alpha} \cdot \dot{\alpha} + \tfrac{1}{2}D_1(\dot{\alpha}, \dot{\alpha}) + \dot{\mathbf{F}}^0 \cdot \phi_{,\mathbf{F}\alpha} \cdot \dot{\alpha}. \tag{12.19}$$

We note that fulfillment of (12.18) requires $dJ(r\dot{\alpha}^0)/dr = 0$ at $r = 1$, which yields

$$D_1(\dot{\alpha}^0, \dot{\alpha}^0) - \dot{\alpha}^0 \cdot \dot{\mathbf{A}}^0 = 0 \tag{12.20}$$

as a consequence of (12.2), where $\dot{\mathbf{A}}^0$ is related to $(\dot{\mathbf{F}}^0, \dot{\alpha}^0)$ through (12.5). Hence, $J(\dot{\alpha}^0) = \frac{1}{2}\dot{\mathbf{F}}^0 \cdot \phi_{,\mathbf{F}\alpha} \cdot \dot{\alpha}^0$ and $\Delta_2 w^0 = \frac{1}{2}\Delta\mathbf{S}^0 \cdot \Delta\mathbf{F}^0 = \frac{1}{2}\dot{\mathbf{S}}^0 \cdot \dot{\mathbf{F}}^0\,(\Delta t)^2$, to second order.

For *existence* of a minimum in (12.18) it is necessary that $J(\dot{\alpha})$ be bounded from below, which requires

$$\dot{\alpha} \cdot \phi_{,\alpha\alpha} \cdot \dot{\alpha} + D_1(\dot{\alpha}, \dot{\alpha}) \geq 0 \qquad \text{for all } \dot{\alpha} \in \mathcal{L}(\mathcal{G}_0). \tag{12.21}$$

Condition (12.21) is imposed on the current *state* and related to directional stability of equilibrium, cf. [13, 11].

12.3.3 The Symmetry Condition

The first variation of $J(\dot{\alpha})$ at $\dot{\alpha} = \dot{\alpha}^0$ reads

$$\delta J(\dot{\alpha}^0) = \tfrac{1}{2}D_1(\dot{\alpha}^0 + \delta\dot{\alpha}, \dot{\alpha}^0) + \tfrac{1}{2}D_1(\dot{\alpha}^0, \dot{\alpha}^0 + \delta\dot{\alpha}) - D_1(\dot{\alpha}^0, \dot{\alpha}^0) - \dot{\mathbf{A}}^0 \cdot \delta\dot{\alpha} \tag{12.22}$$

for an infinitesimal $\delta\dot{\alpha}$. More rigorously, the variation (12.22) represents the Gateaux differential of J at $\dot{\alpha}^0$ in the direction of $\delta\dot{\alpha}$. Clearly, if (12.18) holds then the *one-sided* variation $\delta J(\dot{\alpha}^0) \geq 0$ whenever $\dot{\alpha}^0 + \delta\dot{\alpha} \in \mathcal{L}(\mathcal{G}_0)$. If the minimizer $\dot{\alpha}^0$ is an interior point of $\mathcal{L}(\mathcal{G}_0)$ with respect to the direction of $\delta\dot{\alpha}$, in the sense that the equivalence:

$$(\,\dot{\alpha}^0 + \delta\dot{\alpha} \in \mathcal{L}(\mathcal{G}_0) \iff \dot{\alpha}^0 - \delta\dot{\alpha} \in \mathcal{L}(\mathcal{G}_0)\,) \qquad \text{holds for infinitesimal } \delta\dot{\alpha}, \tag{12.23}$$

then $\delta J(\dot{\alpha}^0) = 0$ for $\delta\dot{\alpha}$ satisfying (12.23). Moreover, a similar variation of the left-hand expression in (12.16) must also vanish at $\dot{\alpha} = \dot{\alpha}^0$, as a consequence of (12.16) and (12.20). Jointly, this requires

$$D_1(\dot{\alpha}^0, \dot{\alpha}^0 + \delta\dot{\alpha}) = D_1(\dot{\alpha}^0 + \delta\dot{\alpha}, \dot{\alpha}^0) \qquad \text{subject to (12.23)}. \tag{12.24}$$

This is the symmetry restriction imposed on the function D_1, that is, on the dissipation function D itself. In fact, the condition (12.24) should be satisfied *in advance*, being necessary for applicability of the incremental energy minimization (12.1) to the first- and second-order simultaneously. Other forms of (12.24) can be specified, e.g. if $D_1(\dot{\alpha}, \hat{\alpha})$ linear with respect to $\dot{\alpha} \in \mathcal{L}(\mathcal{G}_0)$ then (12.24) holds for any $\dot{\alpha}^0 \in \mathcal{L}(\mathcal{G}_0)$

if and only if

$$D_1(\dot{\alpha}, \hat{\alpha}) = D_1(\hat{\alpha}, \dot{\alpha}) \qquad \text{for all } \dot{\alpha}, \hat{\alpha} \in \mathcal{L}(\mathcal{G}_0)\,. \tag{12.25}$$

Remark 2 In the special case of classical (single-mode) plasticity, the cone $\mathcal{L}(\mathcal{G}_0)$ reduces to a single ray in $\dot{\alpha}$-space, and the symmetry condition (12.24) is satisfied automatically. Another special case when (12.24) holds trivially is when $D(\dot{\alpha})$ is state-independent. In general, (12.24) represents an additional restriction imposed on the constitutive law.

Consider a minimizer $\dot{\alpha}^0$ of $J(\dot{\alpha})$ associated with some $\dot{\mathbf{F}}^0$ and another one $\dot{\alpha}^0 + \delta\dot{\alpha}^0$ associated with $\dot{\mathbf{F}}^0 + \delta\dot{\mathbf{F}}^0$ for an infinitesimal $\delta\dot{\mathbf{F}}^0$. Since each minimizer lies within $\mathcal{L}(\mathcal{G}_0)$ and satisfies (12.20), after some transformations with the use of (12.20), (12.22) and (12.5) it follows that

$$\delta\dot{\alpha}^0 \cdot \dot{\mathbf{A}}^0 = \dot{\alpha}^0 \cdot \delta\dot{\mathbf{A}}^0 \qquad \text{and} \qquad \delta\dot{\alpha}^0 \cdot \phi_{,\alpha\mathbf{F}} \cdot \dot{\mathbf{F}}^0 = \dot{\alpha}^0 \cdot \phi_{,\alpha\mathbf{F}} \cdot \delta\dot{\mathbf{F}}^0. \tag{12.26}$$

From now on, the superscript 0 is omitted for brevity. If a minimizer $\dot{\alpha}$, related to $\dot{\mathbf{S}}$, is a single valued, continuous and piecewise differentiable (but never fully linear) function of $\dot{\mathbf{F}}$ then, from (12.26), (12.5) and by the time-independence of the material, we obtain

$$\delta\dot{\mathbf{S}} \cdot \dot{\mathbf{F}} = \dot{\mathbf{S}} \cdot \delta\dot{\mathbf{F}} \qquad \text{and} \qquad \dot{\mathbf{S}} = U_{,\dot{\mathbf{F}}}(\dot{\mathbf{F}}, \mathcal{G}_0)\,, \qquad U = \tfrac{1}{2}\dot{\mathbf{S}}\cdot\dot{\mathbf{F}}. \tag{12.27}$$

This form of the constitutive rate relationship, postulated long ago by Hill [7], is shown here to follow from the minimization (12.18) applied to finding $\dot{\alpha}$ as a function of $\dot{\mathbf{F}}$.

Remark 3 Note that the symmetry requirement (12.24) has *not* been used when deriving another symmetry condition (12.27). However, as shown above, the symmetry restriction (12.24) is required for the intrinsic consistency between the first- and second-order incremental energy minimization.

12.3.4 Second-Order Global Minimization

Applicability of (12.2) up to second-order terms within the incremental constitutive framework with $\dot{\mathbf{S}}$ as a function of $\dot{\mathbf{F}}$ was thoroughly examined [2, 3]. Here we only recall that the minimizer is shown to satisfy automatically the rate form of the principle of virtual power if and only if the potential form (12.27) holds, and then the second-order minimization (12.1) is shown to reduce to

$$\int_B U(\Delta\mathbf{F}, \mathcal{G}_0)\, dV \;+ \cdots = (\Delta t)^2 \int_B U(\dot{\mathbf{F}}, \mathcal{G}_0)\, dV \;+ \cdots \to \min \tag{12.28}$$

subject to kinematical constrains. In (12.28), only the leading term (corresponding to $\Delta_2 W$) is displayed, and the second-order terms added in case of configuration-dependent loading as well as all higher-order terms are dotted.

It should be noted that the minimizer determined from (12.28) under assumption (12.27) represents an exact solution in velocities, in the limit as $\Delta t \to 0$. Finite increments found from (12.28) are merely approximate unless further constitutive assumptions are introduced. Recent examples of the use of (12.28) can be found in several references listed in Introduction.

References

1. H. Petryk, A consistent energy approach to defining stability of plastic deformation processes, in: F.H. Schroeder (Ed.), Stability in the Mechanics of Continua, Proc. IUTAM Symp. Nümbrecht 1981, Springer, Berlin - Heidelberg 1982, pp. 262-272.
2. H. Petryk, On energy criteria of plastic instability, in: Plastic Instability, Proc. Considère Memorial, Ecole Nat. Ponts Chauss., Paris 1985, pp. 215-226.
3. H. Petryk, The energy criteria of instability in time-independent inelastic solids. Archives of Mechanics **43** (1991) 519-545.
4. B. Fedelich, A. Ehrlaher, An analysis of stability of equilibrium and of quasi-static transformations on the basis of the dissipation function, Eur. J. Mech. A/Solids **16** (1997) 833-855.
5. M. Ortiz, E.A. Repetto, Nonconvex energy minimization and dislocation structures in ductile single crystals, J. Mech. Phys. Solids **36** (1999) 286-351.
6. C. Miehe, J. Schotte, M. Lambrecht, Homogenization of inelastic solid materials at finite strains based on incremental minimization principles. Application to the texture analysis of polycrystals, J. Mech. Phys. Solids **50** (2002) 2123-2167.
7. R. Hill, Some basic principles in the mechanics of solids without a natural time, J. Mech. Phys. Solids **7** (1959) 209-225.
8. R. Hill, Aspects of invariance in solids mechanics, Advances in Applied Mechanics, Vol. 18, Acad. Press, New York 1978, pp. 1-75.
9. H. Petryk, K. Thermann, On discretized plasticity problems with bifurcations, Int. J. Solids Structures **29** (1992) 745-765.
10. H. Petryk, K. Thermann, Post-critical plastic deformation in incrementally nonlinear materials, J. Mech. Phys. Solids **50** (2002) 925-954.
11. H. Petryk, Stability and constitutive inequalities in plasticity, in: Muschik W. (Ed.), CISM Courses and Lectures No. 336, Springer, Wien - New York 1993, pp. 259-329.
12. H. Petryk, Second-order work and dissipation on indirect paths, C.R. Mecanique **330** (2002) 121-126.
13. Q.S. Nguyen, Stability and Nonlinear Solid Mechanics, Wiley, Chichester 2000.

Chapter 13
A Quasi-Extremal Energy Principle for Non-Potential Problems in Rate-Independent Plasticity

Abstract The rate-problem of continuing equilibrium is examined for a general class of rate-independent elastoplastic solids, without assuming the normality flow rule or symmetry of the tangent stiffness matrix. Accordingly, the problem addressed is of non-potential type, for which the usual stationarity or minimum principles for a governing potential do not apply. It is shown that the rate-problem can nevertheless be formulated as a quasi-extremal energy principle. It is characterized by explicit dependence of the minimized energy function or functional not only on variables undergoing variations but also, although only in a particular way, on an unknown solution as a parameter. To enable transparent and mathematically simple presentation of the main concept, the energy function is defined in a finite-dimensional setting for a spatially discretized material body with generalized velocities and a number of plastic multipliers as unknowns. If a solution is non-unique then incrementally stable solutions can be selected using the quasi-extremal principle in which the minimized energy function includes the second-order terms. Examples and extensions concern an elastic-plastic continuum obeying a non-associative plastic flow rule, without or with a higher-order gradient term in the loading function. The issue of selection of active slip-systems in a single crystal of a non-symmetric slip-system interaction matrix is also addressed.

Keywords: Non-associative plasticity; Crystal plasticity; Rate problem; Free energy potential; Dissipation; Internal variables; Thermodynamic basis; Variational formulation; Incremental energy minimization; Non-uniqueness; Path stability; Gradient theory

---

[0] Reprinted from: Journal of the Mechanics and Physics of Solids 136, 10369, 2020. H.Petryk, A quasi-extremal energy principle for non-potential problems in rate-independent plasticity. Copyright © 2019 The Author, open access.

H. Petryk, *Theory of Plastic Instability*,
https://doi.org/10.1007/978-3-031-98223-1_13

13.1 Introduction

The aim of this paper is to develop an energy approach to the *non-potential* problems in rate-independent plasticity through the proposed 'quasi-extremal' principle when ordinary stationarity or minimum principles fail. The ultimate purpose is not just another variational formulation of the problem, but also the introduction of a criterion of choice when a rate-solution to the problem in nonuniqe.

There is a long-standing difficulty in solving incremental problems in rate-independent plasticity which is related to non-uniqueness of solutions. There are commonly known branching phenomena like buckling, necking, shear banding or subgrain formation. A rate-independent theory of finite deformation plasticity may thus be regarded as insufficiently general if a possibility of branching of a deformation path is ruled out in advance. Hill [29, 31] laid down the bifurcation theory for elastic-plastic solids obeying an incremental constitutive potential and provided a variational formulation in velocities. The question of selection of the post-bifurcation path was left open in that theory since the criterion of stability of *equilibrium* was insufficient for that. An answer was proposed later for incremental problems of potential type by using the incremental energy minimization which will be addressed below in more detail.

However, not every incremental problem in plasticity admits a minimum principle for an energy-based potential. In a general theory of nonlinear operators, an operator is called a potential operator if it is the gradient of a functional, and the potentiality test is based on the symmetry of the Gateaux derivative of the operator [64]. The notions of potentiality and symmetry are thus strictly related and can be used interchangeably. In physically congruent finite-dimensional problems, an incremental problem is of non-potential type if the basic matrix (e.g., a constitutive matrix or a global tangent stiffness matrix) is non-symmetric. The rate-problem of continuing equilibrium is considered here for a general class of rate-independent inelastic solids, without assuming normality or symmetry properties of the plastic response. The problem addressed is thus of non-potential type, for which the usual variational principles, formulated as stationarity or extremal principles for a governing potential, do not apply. In this paper, we restrict attention to the rate-problems of continuing equilibrium posed on terms of velocities and plastic multipliers as unknowns.

The primary purpose of the present paper is to extend the energy-based variational formulation to a broad class of rate-problems of *non-potential* type in rate-independent plasticity. The focus is on the approach that

(i) is energy-based and thermodynamically consistent, possessing thus a physical interpretation;
(ii) is able to generate an exact solution to the original rate-problem, if it exists;
(iii) provides a criterion of choice if non-unique solutions exist.

As typical advantages, a variational approach may (a) remain valid under conditions for which differential equations need not make sense, (b) provide a weak formulation of the problem, (c) be useful in numerical calculations, and (d) provide a criterion of choice between physically realizable and unstable solutions. A classical example of (d) is the minimum potential energy principle for fully conservative systems which excludes unstable saddle-point equilibrium states.

Variational approaches to rate-independent plasticity have a long-standing and rich history. It is definitely beyond the scope of the present paper to provide a comprehensive review of the existing vast literature related to this topic. The introduction is limited to a brief overview of original concepts that are most relevant to the points (i)–(iii) above. Further references to the literature will be given in later sections where specific aspects of the present approach will be discussed in more detail.

As far as (i) is concerned, the principle of maximum dissipation rate, which for perfectly plastic solids can be traced back to maximizing the plastic work with respect to stresses bounded by a yield surface [28, 6], has been widely used in different versions. Following Rice [55], it can be formulated more generally in terms of scalar *thermodynamic driving forces* bounded in the rate-independent limit by their critical values that must be reached to induce inelastic structural rearrangements. If the bounds do not depend on the state of stress (or *total* strain) then a remarkable normality structure of constitutive relations is obtained for a variety of the corresponding microscale processes [56]. By using the concept of a subdifferential [39, 57] of a convex dissipation function generated from the maximum dissipation principle, the theory of generalized standard materials was formulated [26] in terms of two potentials, of free energy and dissipation. In the same spirit but more generally, the rate-problem with velocities and rates of internal state variables as unknowns has been formulated in terms of a (generally non-symmetric) variational inequality, which includes the first and second derivatives of the dissipation and free energy potentials, respectively, with respect to state variables [40]. The approach based on the principle of maximum plastic dissipation in its rate-independent form has been applied and developed in a number of other works, not reviewed here. Variational approaches that involve a (rate-dependent or viscous) dissipation function different from a plastic potential, cf. [65] and discussion in [25], are not considered here.

Another energy approach that leads to variational formulation of the incremental problem of rate-independent plasticity is based on the incremental energy minimization. Early works, e.g., [58],[11], dealt with minimization of the incremental plastic work in rigid-plastic crystals. The concept of incremental minimization of energy supplied to the examined elastic-plastic system at finite deformation was developed later, as an extension of Hill's bifurcation theory [29, 30, 31]. Under the general assumptions of Hill's theory that lead to a symmetric tangent stiffness matrix, Petryk [45, 46, 47] formulated and analysed the minimum principle for a non-convex energy functional that encompasses the deformation work and external loading potential up to second-order terms. Subsequently, the principle was given a thermodynamic interpretation [48, 19] by the standard split of the isothermal deformation work into the sum of incremental free energy and dissipation. In order to satisfy goal (ii) by the incremental energy minimization, the dissipation function was subjected to a symmetry condition [49]. In a time-discretized framework focused on formation of dislocation structures in ductile metal crystals, Ortiz et al. [42, 43] formulated the boundary value problem of finite crystal plasticity as a sequence of incremental minimization problems for a pseudoelastic energy function that encompassed dissipation. A time-incremental energy minimization problem is also involved in a mathematically rigorous and derivative-free theory developed by Mielke and Roubíček and co-workers in a series of papers and summarized in the monograph [38]. Other mathematical approaches to variational evolution, including the theory of minimizing movements [15], related to the incremental energy minimization in irregular cases, are presented in the book [7] and references therein.

Referring to point (ii), the rate problem of continuing equilibrium in rate-independent plasticity has been differently formulated as: a system of equations and inequalities, a variational inequality, a differential inclusion, or a linear complementarity problem in a discretized case. For non-potential problems, it cannot be straightforwardly reduced to the energy minimization problem. For instance, a direct incremental minimization of the energy supply including dissipation, while addressing goals (i), (ii) and (iii) for potential problems, does not provide an exact solution to a general rate-problem when the plastic flow rule is non-associative. The reason for that will be discussed in more detail in Sect. 13.4.3. Actually, the existing variational formulations in non-associative elastoplasticity are typically not given the form of an ordinary extremal principle. In a discretized version they can have the form of a non-symmetric linear complementarity problem (LCP), which is associated with a constrained minimization problem but only of a special kind [34, 35]. While every LCP is associated with that special minimization problem [13], in the non-symmetric case it does not reduce to the incremental energy minimization and does not meet the goal (iii). There are other variational approaches to general non-potential problems, cf. [59] and [61], in particular in non-associative plasticity [60, 9], but without the energy interpretation desired in (i) and associated physical meaning. In a mathematically rigorous theory of evolutionary problems in rate-independent plasticity in a finite time interval, a solution to the rate-problem need not exist, cf. [14, 38] and references therein. In the engineering literature, the rate-problem is the basic one in plasticity for formulation of incremental problems solved numerically.

The author was unable to find in the literature a general approach that would satisfy the goals (i), (ii) and (iii) simultaneously for non-potential problems, which has motivated the present work. In view of failure of the usual minimum principles for non-potential problems, a generic rate-problem of continuing equilibrium in rate-independent plasticity is formulated in the present paper in an apparently novel form as a *quasi-extremal principle*. In this principle, the minimized function or functional is explicitly dependent not only on variables undergoing variations but also on an unknown solution as a parameter.

There is an analogy to the well-known mathematical concept of a quasi-variational inequality that is a variational inequality in which the constraint set depends on the variable, which has motivated the present name. The quasi-extremal energy principle represents a special case of a fixed point equation, where the mapping involves the incremental energy minimization and possesses thus a physical meaning.

In the mechanical context, the present concept can be compared with the approaches used in frictional contact mechanics [33, 1, 16], although a close equivalent to the present theory has not been found in the literature. Similar 'incomplete' minimization has appeared in a finite-step incremental problem with a state-dependent dissipation function, cf. [63].

The specific form of the minimized non-convex energy function is defined below in a finite-dimensional setting to enable transparent and mathematically simple presentation of the main concept. The needed mathematical background can be found in the book [13] on the related linear complementarity problem. Extensions to a broad class of rate-independent continuous systems can be relatively straightforward if limited to a formal treatment without entering the mathematical subtleties of functional analysis. Selected examples and extensions are presented in Sect. 13.5.

The recent emergence of a number of gradient theories of plasticity, for which variational approaches are of particular interest in view of non-standard boundary conditions, has provided another motivation for this work. Such variational formulations of various gradient theories of plasticity have been developed, for instance, in [20, 21], [53, 54], [37], [2], [41] and [17]. Not all of gradient-enhanced incremental constitutive laws admit a potential, and application of the concept of a quasi-extremal principle developed here may be of interest. This direction of research is only briefly addressed in the present work in Sect. 13.5.3 and is expected to be continued in future work.

13.2 Rate-Independent Plasticity: Preliminaries

13.2.1 Notation

A natural time does not appear in the considerations limited to rate-independent models, and its role is played by a time-like non-dimensional parameter t called 'time' for simplicity. Throughout the paper, a 'rate' is defined as the material time derivative taken always in the *one-sided*, forward (or right-hand) sense $\mathrm{d}/\mathrm{d}t^+$ and assumed to exist. The actual rate that appears in the rate-problem formulation or solution is denoted by a superimposed dot, while a virtual rate, which will play an essential role in what follows, will be denoted either by a superimposed tilde or by another symbol. The distinction is fundamental for understanding the variational formulations that will appear in this paper.

$\mathcal{B}$ denotes a material body identified with the domain in a physical Euclidean space $\mathbb{R}^3$ it occupies in a fixed reference configuration. $\mathbf{n}$ is a unit outward normal to a piecewise smooth boundary $\partial\mathcal{B}$ of $\mathcal{B}$, and $\mathrm{d}V$ and $\mathrm{d}S$ denote infinitesimal volume and area elements, of $\mathcal{B}$ and $\partial\mathcal{B}$, respectively. The standard absolute notation is used for vectors or tensors in the physical space which are denoted respectively by lower- and upper-case boldface letters. A central dot denotes a scalar product of vectors, a central boldface dot a double contraction of tensors, in the sense $\mathbf{A}\boldsymbol{\cdot}\mathbf{B} = \sum_{i,j} A_{ij}B_{ij}$, direct juxtaposition of two tensors (or tensor and vector) indicates simple contraction, and $\otimes$ a tensor product.

Independently, the standard matrix notation is used in other finite-dimensional spaces like $\mathbb{R}^n$, the Euclidean space of real n-tuples $x = (x_1, \ldots, x_n)$, except the scalar (or inner) product $x^{\mathrm{T}}z$ in $\mathbb{R}^n$ that is denoted here by a central dot, $x \cdot z = x_1 z_1 + \cdots + x_n z_n$. The same symbol A is used for a square $n \times n$ real matrix and the corresponding linear transformation $A : \mathbb{R}^n \to \mathbb{R}^n, x \mapsto Ax$. A nonnegative orthant of $\mathbb{R}^n$ is denoted by $\mathbb{R}^n_+$ which is the set of all vectors $x \in \mathbb{R}^n$ such that $x \geq 0$, where the vector inequality is to be read componentwise as $x_i \geq 0\ \forall i = 1, \ldots, n$ ($\forall$ means 'for all'). If $z, x \in \mathbb{R}^n$ and $p < n$ then a statement like $z \geq x$ in $\mathbb{R}^p$ means that both z and x are restricted to $\mathbb{R}^p$. The Cartesian product of Euclidean spaces or their subsets is denoted by $\times$, i.e. $(q, v) \in \mathbb{R}^m \times \mathbb{R}^n_+ \iff q \in \mathbb{R}^m$ and $v \in \mathbb{R}^n_+$.

13.2.2 Rate-Problem of Classical Elastoplasticity

Due to inherent path-dependence of an elastic-plastic material, the basic problem to be solved in a time-continuous setting is formulated in the rate form. As a starting point, consider conventional constitutive rate-equations for isothermal elastoplasticity at finite deformation, reduced to the following form, cf. [5],

$$\dot{\mathbf{S}} = \mathbb{C}^{\mathrm{e}} \,.\, \nabla\dot{\mathbf{u}} - \mathbf{M}\dot{\gamma}\,, \qquad \dot{f} = \mathbf{N} \,.\, \nabla\dot{\mathbf{u}} - g\dot{\gamma}\,, \tag{13.1}$$

where $\dot{\mathbf{S}}$ is the rate of Piola stress $\mathbf{S}$ (the first Piola-Kirchhoff stress), $\mathbb{C}^{\mathrm{e}}$ the fourth-order elastic stiffness tensor that relates $\dot{\mathbf{S}}$ to $\dot{\mathbf{F}} = \nabla\dot{\mathbf{u}}$, the rate of deformation gradient equal to the velocity gradient $\dot{\mathbf{u}}$ in a fixed reference configuration, $\mathbf{M}$ defines the direction of plastic stress-rate $\dot{\mathbf{S}}^{\mathrm{p}} = \dot{\mathbf{S}} - \mathbb{C}^{\mathrm{e}} \,.\, \nabla\dot{\mathbf{u}}$, $\dot{\gamma}$ is a scalar plastic multiplier, $\mathbf{N}$ a yield-surface normal transformed to $\mathbb{R}^{3\times 3}$ space, $\dot{f}$ the rate of a yield function f, and $g > 0$ a coefficient that includes strain hardening/softening dominated by elastic-plastic interaction. After suitable normalization that is taken as granted, $\mathbf{M} = \mathbf{N}$ corresponds to the normality flow rule typical for metals, and $\mathbf{M} \neq \mathbf{N}$ to a non-associative flow rule typical for frictional materials, see [5] for relevant references and examples. All constitutive relations are subject to the basic requirement of material frame indifference, which is well known and thus not discussed here.

The basic rule of plastic activity is written down in the form

$$\dot{\gamma} \geq 0, \quad f \leq 0, \quad f\dot{\gamma} = 0\,. \tag{13.2}$$

An unknown plastic multiplier $\dot{\gamma}$ in the plastic region $\mathcal{B}^{\mathrm{p}}$ of $\mathcal{B}$ where $f = 0$ (otherwise $f < 0$ and $\dot{\gamma} = 0$) is to be found from the familiar consistency conditions

$$f = 0\,, \quad \dot{f} \leq 0, \quad \dot{f}\dot{\gamma} = 0\,. \tag{13.3}$$

In a given current state of $\mathcal{B}$, *the rate problem* is defined by the partial differential equation of continuing equilibrium, $\mathrm{Div}\,\dot{\mathbf{S}} = \mathbf{0}$ in the absence of body forces, along with the prescribed boundary conditions of a specified type. To fix attention, we will assume the standard boundary conditions, such that a continuous velocity field from an appropriate function class $\mathcal{V}$ takes values $\bar{\mathbf{v}}$ prescribed on a part $\partial\mathcal{B}_u$ of the body boundary, while on the remaining part $\partial\mathcal{B}_T = \partial\mathcal{B} \setminus \partial\mathcal{B}_u$ the nominal traction rate $\dot{\mathbf{T}}$ is prescribed. Thus, in the assumed absence of other external forces,

$$\mathrm{Div}\,\dot{\mathbf{S}} = \mathbf{0} \quad \text{in } \mathcal{B}, \qquad \dot{\mathbf{u}} = \bar{\mathbf{v}} \quad \text{on } \partial\mathcal{B}_u, \qquad \dot{\mathbf{S}}\mathbf{n} = \dot{\mathbf{T}} \quad \text{on } \partial\mathcal{B}_T. \tag{13.4}$$

The rate-problem to be solved is defined by Eqs. (13.1), (13.2), (13.3) and (13.4). In a weak formulation, a solution is searched among virtual velocity fields $\underline{\mathbf{v}} \in \mathcal{V}$ and plastic multiplier fields $\underline{\nu} \in \mathcal{R}_+$, where an underline is used to distinguish a field over $\mathcal{B}$ from its value at a material point. A suitable space $\mathcal{V}$ and half-space $\mathcal{R}_+$ are such that $\underline{\mathbf{v}} \in \mathcal{V}$ is kinematically admissible and $\underline{\dot{\gamma}} \in \mathcal{R}_+$ satisfies locally, i.e. at a material point, the constraint $\dot{\gamma} \geq 0$. An extension to a gradient-enhanced model will be discussed in Sect. 13.5.3.

In the non-gradient case, if $g > 0$ then $\dot{\gamma}$ can be eliminated between Eqs. (13.1) in a standard way with the help of conditions (13.2) and (13.3), which yields $\dot{\mathbf{S}}$ as a *nonlinear* function $\hat{\mathbf{S}}$ of $\nabla\dot{\mathbf{u}}$ in a given state, viz.

$$\dot{\mathbf{S}} = \hat{\mathbf{S}}(\nabla\dot{\mathbf{u}}), \quad \hat{\mathbf{S}}(\nabla\mathbf{v}) = (\mathbb{C}^e - \frac{r}{g}\mathbf{M}\otimes\mathbf{N}) \,.\, \nabla\mathbf{v}, \quad r = \begin{cases} 1 & \text{if } f = 0 \text{ and } \mathbf{N} \,.\, \nabla\mathbf{v} > 0, \\ 0 & \text{otherwise.} \end{cases} \tag{13.5}$$

A conventional rate-problem to be solved is now defined by Eqs. (13.1), (13.4) and (13.5).

It is evident that the stationarity principle $\delta\mathcal{J} = 0$ at $\underline{\dot{\mathbf{u}}}$, due to [29], where

$$\mathcal{J}(\underline{\mathbf{v}}) = \int_{\mathcal{B}} \frac{1}{2}\hat{\mathbf{S}}(\nabla\mathbf{v}) \,.\, \nabla\mathbf{v}\,\mathrm{d}V - \int_{\partial\mathcal{B}_T} \dot{\mathbf{T}} \cdot \mathbf{v}\,\mathrm{d}S \tag{13.6}$$

is Hill's velocity functional, does not apply if $\mathbf{M} \neq \mathbf{N}$ after normalization, that is, if the plastic flow rule is not associative with respect to the yield surface. It implies that a minimum principle for the velocity functional $\mathcal{J}$ must also fail in non-associative plasticity.

In case $\mathbf{M} = \mathbf{N}$ that corresponds to the normality flow rule, a solution field $\underline{\dot{\mathbf{u}}}$ for the rate-problem can be found from the minimum principle

$$\underline{\dot{\mathbf{u}}} = \arg\min_{\underline{\mathbf{v}} \in \mathcal{V}} \mathcal{J}(\underline{\mathbf{v}}), \tag{13.7}$$

This was proven in [29] under the assumption that the functional $\mathcal{J}$ is strictly convex in $\underline{\mathbf{v}}$. Later it was shown that the extremal principle (13.7) is also applicable without assuming convexity of $\mathcal{J}$ but then with a different justification coming from the energy criterion of path stability. We return to minimization (13.7) as a special case of the quasi-extremal principle, see Sect. 13.5.1.

13.2.3 Multiple Deformation Mechanisms in Crystal Plasticity

In crystal plasticity, or more generally in case of multiple inelastic deformation mechanisms ot rate-independent type, Eqs. (13.1) are generalized by replacing a scalar $\dot{\gamma}$ with a vector $\dot{\gamma} = (\dot{\gamma}_1, \ldots, \dot{\gamma}_s)$ of s plastic multipliers associated with vector-valued yield function $f = (f_1, \ldots, f_s)$; see Sect. 13.2.1 for the adopted notation convention. Constitutive rate-equations, originally formulated in [32] in a generic objective form, can be transformed to non-symmetric variables in analogy to Eqs. (13.1) as follows

$$\dot{\mathbf{S}} = \mathbb{C}^{\mathrm{e}} \,.\, \nabla\dot{\mathbf{u}} - \sum_{\alpha=1}^{s} \mathbf{M}_\alpha \dot{\gamma}_\alpha \,, \quad \dot{f}_\alpha = \nabla\dot{\mathbf{u}} \,.\, \mathbf{N}_\alpha - \sum_{\beta=1}^{s} g_{\alpha\beta}\dot{\gamma}_\beta \quad \text{for } \alpha = 1, \ldots, s \,, \tag{13.8}$$

while Eqs. (13.2), (13.3) are replaced with

$$\dot{\gamma}_\alpha \geq 0, \quad f_\alpha \leq 0, \quad f_\alpha \dot{\gamma}_\alpha = 0 \qquad \text{for } \alpha = 1, \ldots, s \,, \tag{13.9}$$

$$f_\alpha = 0, \quad \dot{f}_\alpha \leq 0, \quad \dot{f}_\alpha \dot{\gamma}_\alpha = 0 \qquad \text{for } \alpha = 1, \ldots, s_p \,. \tag{13.10}$$

Here, the plastic multipliers are assumed to be numbered so that the first s_p of them form a vector corresponding to all plastic deformation mechanisms currently stressed to the yield point, $f = 0$ in $\mathbb{R}^{s_p}$. By using the matrix notation for operations in $\mathbb{R}^s$ in addition to operations on tensors in $\mathbb{R}^{3\times 3}$, see Sect. 13.2.1, Eqs. (13.8)–(13.10) are equivalently rewritten in a more compact form as follows:

Constitutive rate-equations

$$\dot{\mathbf{S}} = \mathbb{C}^{\mathrm{e}} \,.\, \nabla\dot{\mathbf{u}} - \mathbf{M}\dot{\gamma} \quad \in \mathbb{R}^{3\times 3}\,, \qquad \dot{f} = \mathbf{N}^{\mathrm{T}} \,.\, \nabla\dot{\mathbf{u}} - g\dot{\gamma} \quad \in \mathbb{R}^s \,. \tag{13.11}$$

Plastic activity rule

$$\dot{\gamma} \geq 0, \quad f \leq 0, \quad f \cdot \dot{\gamma} = 0 \qquad \text{in } \mathbb{R}^s \,. \tag{13.12}$$

Consistency conditions:

$$f = 0, \quad \dot{f} \leq 0, \quad \dot{f} \cdot \dot{\gamma} = 0 \qquad \text{in } \mathbb{R}^{s_p} \,. \tag{13.13}$$

Recall that a central dot at $\dot{\gamma}$ indicates a scalar product in $\mathbb{R}^s$ (or restricted to $\mathbb{R}^{s_p} \subset \mathbb{R}^s$), and an inequality of vectors is to be read componentwise, cf. Sect. 13.2.1.

The rate-problem to be solved is defined by Eqs. (13.11), (13.12), (13.13) and (13.4).

In contrast to the preceding Sect. 13.2.2, for a given velocity gradient $\nabla\dot{\mathbf{u}}$ there is in general no closed-form expression for vector $\dot{\gamma}$ of plastic multipliers. Rather, to determine $\dot{\gamma}$ locally in terms of any (virtual) velocity gradient $\nabla\mathbf{v}$ (except in the elastic region within which $f < 0$ and $\dot{\gamma} = 0$ in $\mathbb{R}^s$), *a linear complementarity problem* (LCP) is to be solved that is defined by consistency conditions (13.13) subject to $\dot{\gamma} \geq 0$ and Eq. $(13.11)_2$, viz.

$$\dot{\gamma} \geq 0, \quad -\dot{f} \geq 0, \quad \dot{f} \cdot \dot{\gamma} = 0 \qquad \text{in } \mathbb{R}^{S_P}\,. \tag{13.14}$$

A number of useful statements concerning LCP can be found in the comprehensive mathematical monograph [13], including the conditions for existence and uniqueness of a solution in a general case, and the reduction to the minimum principle (13.15) below in the symmetric case.

In the literature on rate-independent plasticity, consistency conditions (13.13) are frequently called the Kuhn-Tucker conditions. However, to justify this name, they should represent optimality conditions for some function which is not always the case. If, at a material point, g is a *symmetric* matrix then a solution $\dot{\gamma}$ (if it exists) to the following minimum principle

$$\dot{\gamma} = \arg\min_{\nu \geq 0} J(\nu, \nabla\dot{\mathbf{u}})\,, \tag{13.15}$$

where J is a quadratic function of virtual slip-rates ν defined by

$$J(\nu, \nabla\mathbf{v}) = \frac{1}{2}\nu \cdot g\nu - \nabla\mathbf{v}\,.\,\mathbf{N}\nu\,, \qquad \nu \in \mathbb{R}^{S_P}\,, \tag{13.16}$$

must satisfy the Kuhn-Tucker conditions that indeed coincide with the consistency conditions (13.13). However, if $g \neq g^{\mathrm{T}}$ then the coincidence is no longer true in general since

$$J_{,\nu}(\dot{\gamma}, \nabla\mathbf{v}) = \frac{1}{2}(g + g^{\mathrm{T}})\dot{\gamma} - \nabla\mathbf{v}\,.\,\mathbf{N} \qquad \neq \qquad -\dot{f} = g\dot{\gamma} - \nabla\mathbf{v}\,.\,\mathbf{N}\,, \tag{13.17}$$

where $J_{,\nu} = \partial J/\partial\nu$. It means that $\dot{\gamma}$ that satisfies Eq. (13.15) for $g \neq g^{\mathrm{T}}$ is *not* a solution to LCP (13.14) in general. For realistic models of crystal plasticity at finite deformation, the matrix g need not be symmetric [32, 44].

If LCP (13.14) is uniquely solvable but g is not symmetric then the constitutive relationship $\dot{\mathbf{S}}(\nabla\dot{\mathbf{u}})$ that results from Eq. (13.1) generally does not admit a potential [31], so that the minimum principle (13.7) breaks then down even if $\mathbf{M}_\alpha \equiv \mathbf{N}_\alpha$.

In Sect. 13.5.2, the minimum principle (13.15) will be extended to the quasi-extremal principle applicable to an arbitrary non-symmetric matrix g.

13.2.4 A Discretized Rate-Problem of Multi-Mode Plasticity

Consider now a heterogeneous material body $\mathcal{B}$ as a *spatially discretized* system. Then a kinematically admissible velocity field $\underline{\dot{\mathbf{u}}}$ from a suitable class $\mathcal{V}$ is represented by a finite-dimensional vector of generalized velocities $\dot{q} \in \mathbb{R}^m$, and a field of plastic multipliers is represented by a vector $\dot{\mu} \in \mathbb{R}^n$. In gradient plasticity, Eqs. (13.1) or (13.11) may be enhanced by terms dependent on higher-order gradients of $\dot{\mathbf{u}}$ or $\dot{\gamma}$, and then it is assumed that the spatial discretization (e.g., by the finite element method) provides estimates of the involved gradients of $\dot{\mathbf{u}}$ and $\dot{\gamma}$ in terms of $\dot{q}$ and $\dot{\mu}$, respectively (which need not be straightforward computationally).

Suppose that constitutive rate-equations (13.1) or (13.11) can be transformed to a global form for a spatially discretized system,

$$\dot{Q}_i = \sum_{j=1}^{m} K_{ij}\dot{q}_j - \sum_{\kappa=1}^{n} M_{i\kappa}\dot{\mu}_\kappa\,, \qquad \dot{F}_\kappa = \sum_{i=1}^{m} N_{i\kappa}\dot{q}_i - \sum_{\iota=1}^{n} G_{\kappa\iota}\dot{\mu}_\iota\,, \qquad (13.18)$$

where Q_i are generalized internal forces and $\dot{Q}_i$ their rates, q_j generalized displacements and $\dot{q}_j$ generalized velocities, K_{ij} coefficients of the elastic stiffness matrix, $\dot{\mu}_\kappa$ global plastic multipliers, $(M_{i\kappa})$ a global counterpart to local $\mathbf{M}_\alpha$ and $(N_{i\kappa})$ a counterpart to $\mathbf{N}_\alpha$ above, and $(G_{\kappa\iota})$ an interaction matrix between inelastic deformation mechanisms indexed by κ as a global counterpart to a local matrix $(g_{\alpha\beta})$ at a material point. F_k are global loading functions of state and $\dot{F}_k$ their rates, where the order of κ in a given state is taken such that $F_\kappa = 0$ for $\kappa = 1, \ldots, p$, while $F_\kappa < 0$ for $\kappa = p + 1, \ldots, p + n$.

In the matrix notation which is more transparent, it reads

$$\dot{Q} = K\dot{q} - M\dot{\mu} \quad \in \mathbb{R}^m\,, \qquad \dot{F} = N^{\mathrm{T}}\dot{q} - G\dot{\mu} \quad \in \mathbb{R}^n\,, \qquad (13.19)$$

where M and N are rectangular $m \times n$ matrices, a square $n \times n$ matrix G need not be symmetric, and K^{e} is a square $m \times m$ symmetric matrix as the elastic tangent stiffness matrix. All matrices in Eq. (13.19) are dependent in general on the current state. Eqs. (13.19) can be introduced directly as the assumption that the rates $\dot{Q}$ of internal forces and the rates $\dot{F}$ of loading functions are collectively regarded as state-dependent linear functions of generalized velocities $\dot{q}$ and global plastic multipliers $\dot{\mu}$.

The equation of continuing equilibrium and standard boundary conditions mentioned above impose conditions on $\dot{q}$ and $\dot{Q}$ in the form

$$\dot{Q}_i = \dot{P}_i \quad \text{for } i = 1, \ldots, m \qquad \text{and} \qquad \dot{q}_i = \dot{\bar{q}}_i \quad \text{for } i > m\,, \qquad (13.20)$$

where P_i are external generalized forces and $\bar{q}_i$ external displacements, prescribed as given functions of time in a complementary manner. To simplify the formulae below, it is assumed that $\mathbf{v} = \mathbf{0}$ on $\partial\mathcal{B}_{\mathrm{u}}$ and that the corresponding zero variables $\dot{q}_i = \dot{\bar{q}}_i = 0$ for $i > m$ have already been eliminated from Eqs. (13.19).[1] Then, the imposed conditions of equilibrium and of continuing equilibrium take the simplest form

$$Q = P\,, \qquad \dot{Q} = \dot{P} \quad \text{in } \mathbb{R}^m\,, \qquad \dot{q}_i = 0 \quad \text{for } i > m\,. \qquad (13.21)$$

Suppose that each component pair $(F_\kappa, \dot{\mu}_\kappa)$ of the vectors $(F, \dot{\mu})$ of global loading functions and plastic multipliers corresponds to a specific mode of inelastic deformation. Accordingly, the basic conditions (13.9) for activity of inelastic deformation

[1] It is not difficult to remove this simplification by adding a given term $\hat{K}\dot{\bar{q}}$ to the expression for $\dot{Q}$, with $\hat{K}$ as an extension of K to kinematically controlled degrees of freedom. Elastic stiffness of the loading device could also be easily included by taking $\dot{P}$ to be linearly dependent on $\dot{q}$ and modifying K accordingly.

mechanisms are extended to[2]

$$\dot{\mu}_\kappa \geq 0, \quad F_\kappa \leq 0, \quad F_\kappa \dot{\mu}_\kappa = 0 \qquad \text{for } \kappa = 1, \ldots, n\,. \tag{13.22}$$

Once the spatial discretization scheme and the entire discrete set of unknowns are prescribed, the conditions (13.22), similarly as Eqs. (13.21), are required to be satisfied exactly. In this sense we are looking for an exact solution to the spatially discretized rate-problem.

Using the vector notation in $\mathbb{R}^n$, the inelastic activity conditions (13.22) take the form

$$\dot{\mu} \geq 0, \quad F \leq 0, \quad F \cdot \dot{\mu} = 0 \qquad \text{in } \mathbb{R}^n\,. \tag{13.23}$$

In a subspace $\mathbb{R}^p \subset \mathbb{R}^n$ such that $F = 0$ in $\mathbb{R}^p$, by taking a forward time derivative of Eqs. $(13.23)_{2,3}$ we obtain the consistency conditions

$$F = 0, \quad \dot{F} \leq 0, \quad \dot{F} \cdot \dot{\mu} = 0 \qquad \text{in } \mathbb{R}^p\,. \tag{13.24}$$

We have arrived at the following discretized rate-problem to be solved in a given state:

Rate-Problem 1. Solve system $\{(13.19), (13.21), (13.23), (13.24)\}$ for unknown $(\dot{q}, \dot{\mu}) \in \mathbb{R}^{m+n}$. (13.25)

Obviously, Rate-Problem 1 has no solution in the current state if $Q \neq P$ or if some $F_\kappa > 0$, and is non-linear due to the inequality constraints involved.

Similar problems have been addressed in the literature, following [34], although usually not within the energy framework described below in Sect. 13.3. A linear subproblem for *active* inelastic mechanisms can be simply formulated as follows

$$\dot{\mu} > 0 \text{ in } \mathbb{R}^p \quad \Longrightarrow \quad Ax = b \quad \text{in } \mathbb{R}^{m+p}\,, \tag{13.26}$$

where[3]

$$x = \begin{bmatrix} \dot{q} \\ \dot{\mu} \end{bmatrix}, \quad A = \begin{bmatrix} K & -M \\ -N^{\mathrm{T}} & G \end{bmatrix}, \quad b = \begin{bmatrix} \dot{P} \\ 0 \end{bmatrix}. \tag{13.27}$$

[2] This extension need not be immediate. It imposes a restriction on the spatial discretization scheme so that $F_\kappa, \dot{\mu}_\kappa$ may be treated as finite-dimensional counterparts of the spatial fields of $f_\alpha, \dot{\gamma}_\alpha$, respectively.

[3] Within the vector b of external force-rate, there is a visible room for extending it to the rates of external thermodynamic forces if needed, for instance, in a gradient theory.

13.3 Thermodynamic Background and Energy Approach

13.3.1 Total Internal Energy of a Compound Thermodynamic System

The discussion below is consistent with the classical thermodynamics [8, 62]. It is conceptually useful to consider a compound thermodynamic system $\mathcal{A}$ that, in the absence of disturbances caused by an external agency, can be treated as *isolated* if a control parameter λ is kept constant. Such a system is shown schematically in Fig. 13.1. It consists of the material body $\mathcal{B}$ (of internal energy U and entropy S), mechanical loading device (assumed conservative, of potential energy Ω) and heat reservoir (of internal energy E_h, entropy S_h and constant absolute temperature $\bar{T}$), cf. [48, 50].[4] The total internal energy of $\mathcal{A}$ is

$$E = U + \Omega + E_h \,. \tag{13.28}$$

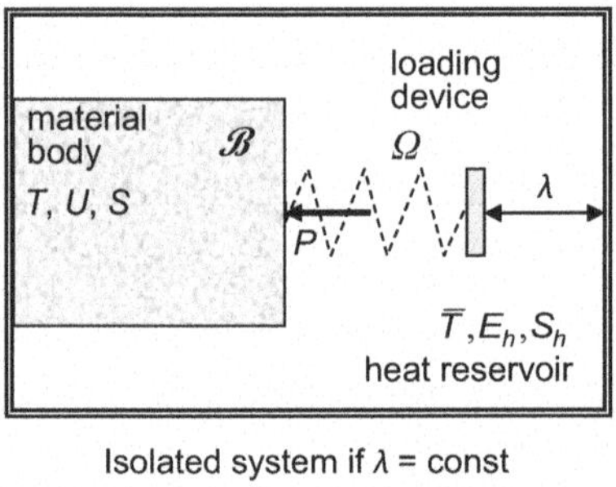

Fig. 13.1 The compound thermodynamic system $\mathcal{A}$ under consideration

Throughout this paper, the considerations are limited to quasi-static isothermal processes, $T \equiv \bar{T} = \text{const}$. On substituting $\dot{U} = \dot{\Phi} + \bar{T}\dot{S}$ and $\dot{E}_\text{h} = \bar{T}\dot{S}_\text{h}$, the rate $\dot{E}$ of change in the total internal energy E of the compound system $\mathcal{A}$ in an *unperturbed* process takes the form

$$\dot{E} = \dot{\Phi} + \dot{\mathcal{D}} + \dot{\Omega} \qquad \text{at } T \equiv \bar{T} \,, \tag{13.29}$$

where $\Phi = U - \bar{T}S$ is the Helmholtz free energy of $\mathcal{B}$, and $\dot{\mathcal{D}} = \bar{T}(\dot{S} + \dot{S}_h)$ [5] the dissipation rate in $\mathcal{B}$.

Remark 3.1 The right-hand expression (13.29) has been widely used and understood in two equivalent ways. First, as a sum of the internal work rate $\dot{W} = \dot{\Phi} + \dot{\mathcal{D}}$ and the

[4] The body $\mathcal{B}$ can exchange work (but not heat) with the loading device, and can exchange heat (but not work) with the heat reservoir. The heat exchange cannot be neglected for isothermal processes in $\mathcal{B}$ since they are not adiabatic.

[5] There is a difference in notation in the source article: $-\dot{S}_h$ (inflow to $\mathcal{B}$) instead of the outflow $+\dot{S}_h$ here and in [48, 50]. This has no effect on the resulting formula (13.29).

rate $\dot{\Omega}$ of external potential Ω. Second, as a sum of the rate of total potential $(\Phi + \Omega)$ and the dissipation rate $\dot{\mathcal{D}}$. The aforementioned third interpretation as the single rate $\dot{E}$ of total *internal* energy of the compound system does not seem to be common, although it has a clear meaning in the framework of classical thermodynamics.

Let a state of mechanical equilibrium of $\mathcal{B}$ be characterized by a pair (q, α) of a generalized displacement vector $q \in \mathbb{R}^m$ and a vector $\alpha \in \mathbb{H}$ of internal variables of any kind. Vector α describes structural rearrangements in the material body and need not be finite dimensional ($\mathbb{H}$ may be a Hilbert space, but is treated below as Euclidean space of unspecified dimension). An important step is that not α itself but its *rate* is characterized by discrete variables, here plastic multipliers $\dot{\mu}_K \geq 0$, whose time integrals $\int \dot{\mu}_K \mathrm{d}t$, in contrast to α, need not have a physical meaning [55].[6] Accordingly, it is assumed here that

$$\dot{\alpha} = \sum_{K=1}^{n} \boldsymbol{\eta}_K \dot{\mu}_K = \boldsymbol{\eta}(q, \alpha)\dot{\mu}\,, \tag{13.30}$$

where $\boldsymbol{\eta} : \mathbb{R}^m \times \mathbb{H} \to \mathbb{H} \otimes \mathbb{R}^n$ is a given (matrix-valued and Fréchet differentiable) function of state parameters (q, α). $\boldsymbol{\eta}$ defines the state-dependent set of admissible directions of current changes in α, while $\dot{\mu} \geq 0$ specifies the actual direction and magnitude of $\dot{\alpha}$.

The Helmholtz free energy Φ of body $\mathcal{B}$ is assumed to be a given function (twice Fréchet differentiable) of external variables q and internal variables α taken as *global* state parameters. In the discretized *unperturbed* system, the rate of Φ reads

$$\dot{\Phi} = \frac{\mathrm{d}}{\mathrm{d}t^+}\Phi(q, \alpha) = \Phi_{,q} \cdot \dot{q} + \Phi_{,\alpha} \cdot \dot{\alpha} = Q \cdot \dot{q} - \Pi \cdot \dot{\mu}\,, \qquad \Pi = -\boldsymbol{\eta}^{\mathrm{T}} \Phi_{,\alpha}\,, \tag{13.31}$$

where a comma as a subscript denotes the partial Fréchet derivative with respect to the quantity that follows the comma. Π is defined as a vector of thermodynamic driving forces that is conjugate to $\dot{\mu}$. In analogy to [55], the yield functions are defined through the difference between the current (Π) and critical (Π^{c}) driving forces,

$$F_K = \Pi_K - \Pi_K^{\mathrm{c}} \quad \forall K \qquad \longleftrightarrow \qquad F = \Pi - \Pi^{\mathrm{c}}\,, \tag{13.32}$$

with the distinction that Π^{c} can depend here not only on α but also on the current state of stress through q. The critical forces Π^{c} are discussed in the next subsection; at the moment we formally assume that $\Pi^{\mathrm{c}} : \mathbb{R}^m \times \mathbb{H} \to \mathbb{R}^n$ is a given (vector-valued and Fréchet differentiable) function of state parameters (q, α).

Accordingly, all state-dependent quantities that appear in Eq. (13.19) are expressed through partial derivatives of Φ and F as follows

$$Q = \Phi_{,q}\,, \quad K = \Phi_{,qq}\,, \quad M = -\Phi_{,q\alpha}\boldsymbol{\eta}\,, \quad N^{\mathrm{T}} = F_{,q}\,, \quad G = -F_{,\alpha}\boldsymbol{\eta}\,. \tag{13.33}$$

[6] The presence of $\boldsymbol{\eta}$ in Eq. (13.30) is thus enforced by the fact that the time integrals of $\dot{\mu}_K$ cannot themselves be regarded as state variables in general.

Clearly, $Q_{,\alpha}\dot{\alpha} = -M\dot{\mu}$ is the plastic part of the rate $\dot{Q}$ of vector Q of generalized internal forces.

Remark 3.2 In the special case when both Π^{c} and η are independent of q, we have $N^{\mathrm{T}} = \Pi_{,q} = -\eta^{\mathrm{T}}\Phi_{,\alpha q} = M^{\mathrm{T}}$, and the normality structure in [55, 56] is recovered. The same conclusion follows from $\Pi^{\mathrm{c}}_{,q} = 0$ even if $\eta_{,q} \neq \mathbf{0}$ provided $\eta^{\mathrm{T}}_{,q}\Phi_{,\alpha} = 0$. On the other hand, symmetry of G can be destroyed by $\eta_{,\alpha} \neq \mathbf{0}$, even at constant Π^{c}. These conclusions show the important role that is played by the state-dependent matrix link η between $\dot{\alpha}$ and $\dot{\mu}$ in deciding whether the final rate-problem is of potential type or not.

Under the simplifying assumption (13.21) that external forces P_i are prescribed for $i = 1, \ldots, m$ and complementary velocities $\dot{q}_i = 0$ for $i > m$, the potential energy Ω of the loading device and the rate of work supplied to $\mathcal{B}$, called the external power $\mathcal{P}^{\mathrm{ext}}$, take the standard form

$$\Omega(q, \lambda) = -P(\lambda) \cdot q\,, \qquad \mathcal{P}^{\mathrm{ext}} = P(\lambda) \cdot \dot{q} = -\Omega_{,q}\dot{q}\,, \tag{13.34}$$

where a scalar control parameter λ varies in time t in a prescribed, possibly non-monotonic manner.[7]

For a quasi-static ($Q = P$), isothermal ($T \equiv \bar{T}$) *and unperturbed* process in $\mathcal{B}$, the actual dissipation rate reads

$$\dot{\mathcal{D}} = \mathcal{P}^{\mathrm{ext}} - \dot{\Phi} = Q \cdot \dot{q} - \dot{\Phi} = \Pi \cdot \dot{\mu} \geq 0\,. \tag{13.35}$$

The first equality above follows from the first law of thermodynamics for $\mathcal{B}$ at $T \equiv \bar{T}$, the second from $Q = P$, the third from Eq. (13.31), and the inequality from the second law of thermodynamics. Of course, the dissipation rate should not be confused with the plastic work-rate.

13.3.2 Virtual Dissipation Rate

In a variational formulation of the rate-problem, *virtual* rates are considered that can be associated with imagined *perturbations* of the inelastic deformation process (still assumed isothermal). Virtual rates, $\tilde{q}, \tilde{\mu}$, etc., are distinguished from the actual rates in an unperturbed process, $\dot{q}, \dot{\mu}$, etc., by a superimposed tilde instead of a dot. The actual dissipation rate $\dot{\mathcal{D}}$ in an unperturbed process is also to be replaced with a virtual dissipation rate $\tilde{\mathcal{D}}$ in a perturbed process.

To define $\tilde{\mathcal{D}}$, we will use the concept of a *dissipation function*, which is nowadays a classical notion that in rate-independent plasticity coincides with a dissipation

[7] It is thus assumed that the *external* power $\mathcal{P}^{\mathrm{ext}}$ is independent of $\dot{\mu}$, which is possible also in gradient plasticity under appropriate boundary conditions imposed on the body surface, cf. Sect. 13.5.3.

potential, cf. [39, 26, 22, 18, 12, 10] and many others. The value of a dissipation function is identified with a *virtual* dissipation rate.

It is assumed that the dissipation function $D : \mathbb{R}^n \to \mathbb{R}$ satisfies the following requirements:

- in a given state, D is a non-negative function of virtual plastic multipliers $\tilde{\mu}$ that is positively homogeneous of degree one, i.e. $D(r\tilde{\mu}) = rD(\tilde{\mu}) \geq 0$ if $r \in \mathbb{R}_+$;
- function D itself can be state-dependent, i.e. $D_{,q} \neq 0$ or $D_{,\alpha} \neq 0$;
- D is compatible with the actual dissipation rate for any solution $\dot{\mu}$, i.e. $D(\dot{\mu}) = \dot{\mathcal{D}}$.

A (convex) dissipation function D that satisfies these three requirements can be given the simplest form for system $\mathcal{A}$ as follows [48, 50]

$$D(\tilde{\mu}) = \Pi^{\mathrm{c}} \cdot \tilde{\mu}\,, \qquad \Pi^{\mathrm{c}}(q, \alpha) \geq 0\,, \qquad \tilde{\mu} \geq 0\,. \tag{13.36}$$

The function $D : \mathbb{R}^n_+ \to \mathbb{R}_+$ is defined here on the non-negative orthant $\mathbb{R}^n_+$ of $\mathbb{R}^n$, but it can be extended to the whole space $\mathbb{R}^n$ by setting $D(\tilde{\mu}) = +\infty$ if some of $\tilde{\mu}_\kappa$ is negative.

This dissipation function D acts as a *pseudopotential* [39, 57] *for driving forces* Π, that is

$$\Pi \in \partial D(\dot{\mu}) \qquad \Longleftrightarrow \qquad \dot{\mu} = \arg \min_{\tilde{\mu} \in \mathbb{R}^n_+} (D(\tilde{\mu}) - \Pi \cdot \tilde{\mu})\,. \tag{13.37}$$

Here, $\partial D(\dot{\mu})$ denotes subdifferential of D at $\dot{\mu}$ in a given state, and the equivalence above follows from the definition of a subdifferential. As a consequence, the minimum value of $D(\tilde{\mu}) - \Pi \cdot \tilde{\mu}$, if attained at $\dot{\mu}$, must be zero, implying $D(\dot{\mu}) = \Pi \cdot \dot{\mu}$ as required earlier. On substituting Eqs. $(13.36)_1$ and (13.32), the minimized function, $D(\tilde{\mu}) - \Pi \cdot \tilde{\mu}$, coincides with $(-F \cdot \tilde{\mu})$, so that

$$D(\tilde{\mu}) - \Pi \cdot \tilde{\mu} = -F \cdot \tilde{\mu} \geq -F \cdot \dot{\mu} = D(\dot{\mu}) - \Pi \cdot \dot{\mu} = 0 \qquad \forall \tilde{\mu} \geq 0\,. \tag{13.38}$$

Consequently, property (13.37) is equivalent to the inelastic activity rule (13.23), as expected.

In a dual formulation, function D^* conjugate to D is the indicator function of the closed set in Π-space consisting of an elastic domain $(F < 0)$ and its boundary as the yield surface ($F_\kappa = 0$ for some κ). The obtained classical normality structure, limited here to the relation between Π and $\dot{\mu}$ defined by dissipation function D, is illustrated in Fig. 13.2.

Remark 3.3 It is remarkable that the simple normality relation between Π and $\dot{\mu}$, implied by Eq. (13.37) and illustrated in Fig. 13.2, does *not* imply normality in the commonly understood sense. In particular, normals N to the yield surface in q-space can depend on $\Pi^{\mathrm{c}}_{,q} \neq 0$ and hence differ from the directions M of plastic response that are independent of $\Pi^{\mathrm{c}}_{,q}$. Furthermore, the direction of $\dot{\alpha}$ can depend on stress through $\eta(q, \alpha)$ in a manner assumed independently of $\Pi^{\mathrm{c}}(q, \alpha)$, and thus need not be normal to the yield surface in the space of thermodynamic forces defined as $-\Phi_{,\alpha}$.

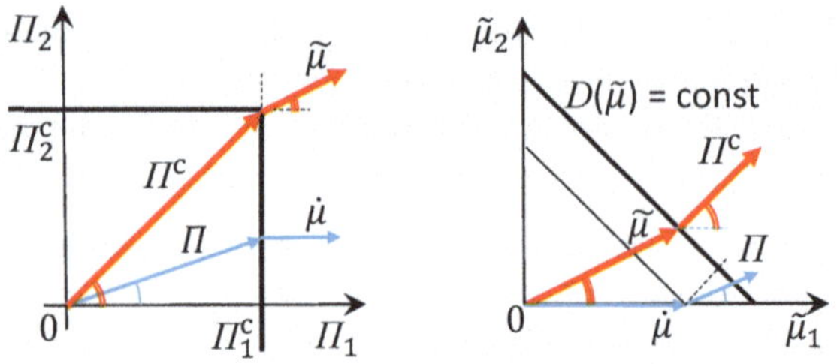

Fig. 13.2 Illustration of the inelastic activity rule generated from dissipation function D of virtual plastic multipliers $\tilde{\mu}$

13.3.3 Minimization of the Energy Rate

The difference between $\Pi^c(q, \alpha)$ and $\Pi(q, \alpha)$ may be compensated by an extra set of perturbing forces $\delta\Pi$, applied by a disturbing agency and needed to induce a virtual $\tilde{\mu}$. The difference $D(\tilde{\mu}) - \Pi \cdot \tilde{\mu} = (\Pi^c - \Pi) \cdot \tilde{\mu} = -F \cdot \tilde{\mu}$ is thus the power expended by those perturbing forces identified as $\delta\Pi = -F$. Other perturbing forces $\delta P = Q - P$ may be imagined to compensate a deviation from mechanical equilibrium.

A virtual rate of change in the internal energy E of the compound system $\mathcal{A}$ in a perturbed process is no longer given by Eq. (13.29) but, using consistently a tilde to denote any virtual rate, can be expressed as follows

$$\tilde{E} = \tilde{\Phi} + \tilde{\mathcal{D}} + \tilde{\Omega} \qquad \text{at } T \equiv \bar{T} . \tag{13.39}$$

Since Φ and Ω are state-dependent functions, their virtual rates are obtained straightforwardly. The virtual dissipation rate $\tilde{\mathcal{D}}$ is identified with $D(\tilde{\mu})$ defined by Eq. (13.36). From Eqs. (13.31), (13.34) and (13.36) it follows that

$$\begin{aligned} \tilde{E}(\tilde{q}, \tilde{\mu}) &= \tilde{\Phi}(\tilde{q}, \tilde{\mu}) + \tilde{\Omega}(\tilde{q}, \dot{\lambda}) + D(\tilde{\mu}) \\ &= (Q - P) \cdot \tilde{q} - F \cdot \tilde{\mu} - \dot{P} \cdot q \end{aligned} \tag{13.40}$$

for any admissible virtual rates ($\tilde{q}$ arbitrary and $\tilde{\mu} \geq 0$) in a given state.

For $(\dot{q}, \dot{\mu})$ to be a solution to Rate Problem 1, a prerequisite is that

$$Q = P \quad \text{in } \mathbb{R}^m, \qquad \dot{\mu} \geq 0 \quad F \leq 0 \quad F \cdot \dot{\mu} = 0 \quad \text{in } \mathbb{R}^n. \tag{13.41}$$

Besides the two restrictions on admissible states, i.e. mechanical equilibrium ($Q = P$) and preserved bounds on driving forces ($F \leq 0$), conditions (13.41) impose also two restrictions on the *cone* of admissible $\dot{\mu}$, which is the nonnegative orthant $\mathbb{R}_+^p$ in $\mathbb{R}^p \subset \mathbb{R}^n$ such that $F = 0$ when restricted to $\mathbb{R}^p$. From conditions $(13.41)_{1,4}$ it follows that $\tilde{E}(\dot{q}, \dot{\mu})$ takes a given value $\dot{E} = -\dot{P} \cdot q$.

In Sect. 13.3.2 it has been noted that the last three conditions (13.41) are equivalent to the minimum principle (13.38). The first condition $(13.41)_1$ is equivalent to a trivial variational equality $(Q - P) \cdot \tilde{q} = (Q - P) \cdot \dot{q} = 0 \ \forall \tilde{q} \in \mathbb{R}^m$. By combining these conclusions, it becomes evident that the conditions (13.41) jointly are equivalent to the variational inequality

$$(\tilde{q} - \dot{q}) \cdot (Q - P) - (\tilde{\mu} - \dot{\mu}) \cdot F \geq 0 \qquad \forall (\tilde{q}, \tilde{\mu}) \in \mathbb{Z} = \mathbb{R}^m \times \mathbb{R}^n_+ \tag{13.42}$$

and thus equivalent to the extremal energy-rate principle

$$\tilde{E}(\dot{q}, \dot{\mu}) \leq \tilde{E}(\tilde{q}, \tilde{\mu}) \qquad \forall (\tilde{q}, \tilde{\mu}) \in \mathbb{Z}. \tag{13.43}$$

In an equilibrium state, $\tilde{E}(\tilde{q}, \tilde{\mu})$ is equal to $\dot{E}(\dot{q}, \dot{\mu}) = -\dot{P} \cdot q$, the actual rate of E along a solution path, if and only if $F \cdot \tilde{\mu} = 0$. Inequality (13.43) can be rewritten as

$$\dot{E}(x) \leq \tilde{E}(z) \qquad \text{for all perturbing forces,} \tag{13.44}$$

where x is any rate-solution and z any virtual rate. This formulation of principle (13.43), when referred to the original expression (13.28) for total internal energy E of system $\mathcal{A}$, is free of all specific assumptions made in Sect. 13.2.4 and thus more general. Principle (13.44), derived here in a general case of non-normality and non-symmetry, appears to be widely accepted nowadays in one form or another.

Principle (13.44) states that under the restriction to isothermal processes and rate-independent dissipation, any kinematically admissible perturbation of a quasi-static solution path requires the rate of total energy supply to the system $\mathcal{A}$ to be not smaller than that along the solution path. It may be regarded as the first-order energy condition for stability of a solution *path* [45]. Since the last term $(-\dot{P} \cdot q)$ in Eq. (13.40) is independent of virtual rates, inequality (13.44) reduces to $0 \leq \tilde{E}(z)$ with $\dot{P} = 0$, which is the first-order energy condition for stability of the current *state*. The coincidence is no longer true for the *second-order* energy conditions for stability of a quasi-static solution path and of an equilibrium state.

Obviously, the extremal principle (13.44) or (13.43), when applied in a single given state (q, α) only, is insufficient to determine an unknown inelastic rate-solution $(\dot{q}, \dot{\mu})$. Approaches based on a postulated extension of inequality (13.44) to the second-order rate or to a finite increment of E have been developed in the literature, as reviewed briefly in Introduction. However, those approaches do not generate an exact solution to Rate Problem 1 in a general non-symmetric case, which is discussed in Sect. 13.4.3.

In this work, another possibility is explored, namely, by applying the extremal principle (13.43) at a next time instant along a solution path initiated from the current state (q, α) in direction $(\dot{q}, \dot{\mu})$. This is analysed in Sect. 13.4.2, after introducing in the next Sect. 13.4.1 the needed concept of a *quasi-extremal* principle.

13.4 A Quasi-Extremal Energy Principle

13.4.1 The Concept of a Quasi-Extremal Principle

A general form of a *quasi-extremal principle* (QEP) is introduced by

$$x = \arg\min_{z \in \mathbb{Z}} \varphi(z, x)\,, \tag{13.45}$$

where a closed set $\mathbb{Z}$ in a suitable vector space $\mathbb{X}$ and a real-valued function $\varphi : \mathbb{Z} \times \mathbb{Z} \to \mathbb{R}$ are given. $\mathbb{X}$ may be an infinite-dimensional function space (preferably a Hilbert space to have a scalar product) in a continuum setting, or a finite-dimensional Euclidean space for a discretized problem examined below, while φ is specified later as a function derived from the energy function E. In contrast to a 'true' extremal principle corresponding to a typical optimization problem, the objective function φ in QEP (13.45) depends not only on variable z but also on an unknown optimal solution x. Such a function φ will be called *a quasi-potential.*

Eq. (13.45) can be equivalently expressed by the following statement

$$\exists y: \quad \varphi(y,x) \le \varphi(z,x) \quad \forall z \qquad \text{and} \qquad y = x \qquad (\text{all in } \mathbb{Z})\,. \tag{13.46}$$

It can be seen that Eq. (13.45) encompasses as special cases the usual minimum principles (if φ is independent on x) and variational inequalities (if φ is linear in z). A minimum can trivially be converted to a maximum by reversing the sign of φ.

Eq. (13.45) represents a special case of a fixed point equation $x = \mathcal{T}(x)$, where the mapping $\mathcal{T} : \mathbb{Z} \to \mathbb{Z}$ has a particular form of finding a minimum point (dependent on x) of a function $\varphi(\cdot, x)$. To the author's knowledge, a fixed point problem in the form of a quasi-extremal principle (13.45) has not been investigated so far in the literature concerning rate-independent plasticity. It is worth emphasizing that the number of unknowns in the proposed QEP is not increased since x is a particular value of z.

The form of QEP (13.45) suggests a possible iteration scheme defined by

$$x^{k+1} = \arg\min_{z \in \mathbb{Z}} \varphi(z, x^k)\,. \tag{13.47}$$

This need not be, however, an effective computational procedure. A simple counterexample is shown in Fig. 13.3 for a nearly-quadratic convex function $\phi(\cdot, x)$ in $\mathbb{R}^2$ without constraints. Details are omitted as they are inessential for purely illustrative purposes. What is essential is the slow convergence of the iterative procedure (13.47) to the final solution as marked in Fig. 13.3a,b by a blue line. For some other parameters there was no convergence at all.

Another mapping $\mathcal{T}$ can be used to generate a fixed point iteration

$$x^{k+1} = \mathcal{T}(x^k)$$

under the condition that $x = \mathcal{T}(x)$ implies (13.45). In the same illustrative example without constraints, the standard Newton iteration procedure behaved much better, which is marked in Fig. 13.3b by a red line.

Numerical procedures of solving QEP (13.45) in more complex cases are beyond the scope of this paper. The next question is how to construct an appropriate function φ of physical meaning.

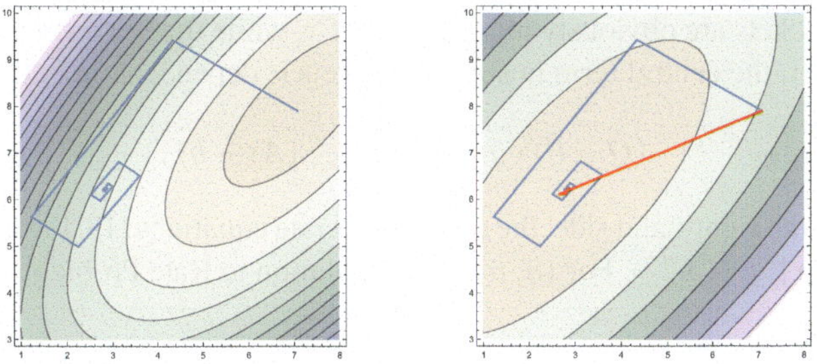

Fig. 13.3 Illustration of two iterative procedures of solving QEP for a nearly-quadratic convex function $\varphi(\cdot, x)$ in $\mathbb{R}^2$ superimposed on contours of $\varphi(\cdot, x)$ (a) for initial $x = 0$ and (b) for final $x = \arg\min_{z\in\mathbb{R}^2} \varphi(z, x)$

13.4.2 Quasi-Extremal Energy Principle of First Order

The attention is now focused on a spatially discretized rate-problem formulated in Sect. 13.2.4, addressed by using the energy approach presented in Sect. 13.3.

Recall that a solution $x \equiv (\dot{q}, \dot{\mu})$ to Rate Problem 1 in a given state of the compound thermodynamic system $\mathcal{A}$ satisfies inequality (13.43), rewritten for later reference as follows

$$x = \arg\min_{z\in\mathbb{Z}} \tilde{E}(z)\,, \qquad \mathbb{Z} = \mathbb{R}^m \times \mathbb{R}^n_+\,, \tag{13.48}$$

where

$$\tilde{E}(z) = \tilde{E}(\tilde{q}, \tilde{\mu}) = (Q - P)\cdot\tilde{q} - F\cdot\tilde{\mu} - \dot{P}\cdot q \tag{13.49}$$

is the virtual rate of total internal energy E (13.28) that is supplied to system $\mathcal{A}$ undergoing transformation with a virtual rate $z = (\tilde{q}, \tilde{\mu})$.

Suppose that inequality (13.43) holds true in a given state at time $t = 0$. Apply now inequality (13.43) after a small time increment $\tau > 0$ along a solution path, viz.

$$\tilde{E}(x)|_\tau \leq \tilde{E}(z)|_\tau = (Q_\tau - P_\tau)\cdot\tilde{q} - F_\tau\cdot\tilde{\mu} - \dot{P}_\tau\cdot q_\tau\,, \tag{13.50}$$

where an index τ indicates a quantity evaluated at time τ and assumed to vary sufficiently smoothly in time. There is a certain analogy between the rather simple approach based on inequality (13.50) and a more general mathematical technique of minimizing movements [15]. However, a result in the form of the quasi-extremal principle QEP1 in Theorem 1 below has not been found in the literature.

On evaluating $\tilde{E}(z)|_\tau$ to first order in τ in direction of x, we obtain

$$\tilde{E}_\tau(z,x) = \big[(Q-P)\cdot\tilde{q} - F\cdot\tilde{\mu}\big] + \tau\big[(\dot{Q}-\dot{P})\cdot\tilde{q} - \dot{F}\cdot\tilde{\mu}\big] - \big[\dot{P}_\tau\cdot q + \tau\dot{P}\cdot\dot{q}\big]\,. \tag{13.51}$$

The key point is that the above expression depends not only on virtual rate-variables $z = (\tilde{q}, \tilde{\mu})$ but also on an unknown rate-solution $x = (\dot{q}, \dot{\mu})$. The function $\tilde{E}_\tau$ depends also on a time distance τ from a state given at $t = 0$. The quantities in the first term in square brackets are obviously independent of x, while the last term is independent of z. However, the central term in square brackets is dependent on both z and x, viz.

$$(\dot{Q} - \dot{P}) \cdot \tilde{q} - \dot{F} \cdot \tilde{\mu} = z \cdot (Ax - b), \tag{13.52}$$

on using, on the right-hand side, the constitutive rate equations (13.19) in the notation introduced in Eq. (13.27). For $(\dot{q}, \dot{\mu})$ to be a solution to Rate Problem 1 (13.25), the conditions

$$\dot{Q} = \dot{P} \quad \text{in } \mathbb{R}^m, \qquad F = 0, \quad \dot{F} \leq 0, \quad \dot{F} \cdot \dot{\mu} = 0 \quad \text{in } \mathbb{R}^p \tag{13.53}$$

are to be satisfied in addition to conditions (13.41).

The first two terms in square brackets in Eq. (13.51) have an analogous form. It has been shown, cf. inequalities (13.42) and (13.43), that $z = x \in \mathbb{Z}$ is a minimizer of the first term if and only if x satisfies the conditions (13.41). This suggests that if the first term is zero then minimization of the second term with respect to z can imply fulfilment of the remaining conditions (13.53), and conversely.

This is indeed so but within the set of z smaller than $\mathbb{Z}$, which corresponds to the set of $\tilde{\mu}$ smaller than $\mathbb{R}^n_+$ in order to exclude $F_\kappa \tilde{\mu}_\kappa \neq 0$ in advance. Such a set of z can easily be constructed in two ways: first, as the set $\mathbb{Z}^{\mathrm{p}} = \mathbb{R}^m \times \mathbb{R}^p_+$ of reduced dimension $m + p \leq m + n$, and second, as a subset $\mathbb{Z}_{\mathrm{p}} = \{z \in \mathbb{Z} : F \cdot \tilde{\mu} = 0\}$ of the same dimension of $\mathbb{Z}$. The only difference is that every $\tilde{\mu} \in \mathbb{R}^n_+$ within a pair $(\tilde{q}, \tilde{\mu}) \in \mathbb{Z}_{\mathrm{p}}$ is an extension of equivalent $\tilde{\mu} \in \mathbb{R}^p_+$ within a pair $(\tilde{q}, \tilde{\mu}) \in \mathbb{Z}^{\mathrm{p}}$ by a zero vector in $\mathbb{R}^{n-p}$. In short-hand notation, $\mathbb{Z}_{\mathrm{p}} = \mathbb{Z}^{\mathrm{p}} \times \{0\}$, where $0 \in \mathbb{R}^{n-p}$.

The following simple statement has a mathematical character, therefore it is formulated as a lemma.

Lemma 1 $x = (\dot{q}, \dot{\mu}) \in \mathbb{Z}^p = \mathbb{R}^m \times \mathbb{R}^p_+$ *satisfies the following* quasi-extremal principle, *with expression (13.52) as a quasi-potential,*

$$x = \arg \min_{z \in \mathbb{Z}^p} \big(z \cdot (Ax - b)\big) \tag{13.54}$$

if and only if x satisfies the conditions (13.53).

Proof. Suppose that QEP (13.54) is satisfied by $x \in \mathbb{Z}^{\mathrm{p}}$. Since $\tilde{q} \in \mathbb{R}^m$ is unrestricted, it follows that $\dot{Q} = \dot{P}$ and

$$\dot{\mu} = \arg \min_{\tilde{\mu} \in \mathbb{R}^p_+} (-\dot{F} \cdot \tilde{\mu}). \tag{13.55}$$

We have $F = 0$ and $\tilde{\mu} \geq 0$ in $\mathbb{R}^p_+$ by definition, and the remaining two conditions (13.53) follow as the Kuhn-Tucker conditions for minimization (13.55).

Conversely, suppose that $x \in \mathbb{Z}^{\mathrm{p}}$ satisfies the conditions (13.53). By $\dot{Q} = \dot{P}$, expression (13.52) reduces to $(-\dot{F} \cdot \tilde{\mu})$ which attains among $\tilde{\mu} \in \mathbb{R}^p_+$ a minimum value $(-\dot{F} \cdot \dot{\mu}) = 0$ at $\tilde{\mu} = \dot{\mu}$ on account of $\dot{F} \leq 0$ and $\tilde{\mu} \geq 0$. This completes the proof of Lemma 1. □

It is not difficult to show that Lemma 1 remains true if set $\mathbb{Z}^{\mathrm{p}}$ is interchanged with $\mathbb{Z}_{\mathrm{p}}$ defined above. Eq. (13.54) can also be straightfowardly reformulated as a variational inequality

$$(z - x) \cdot (Ax - b) = (\tilde{q} - \dot{q}) \cdot (\dot{Q} - \dot{P}) - (\tilde{\mu} - \dot{\mu}) \cdot \dot{F} \geq 0 \tag{13.56}$$

for either all $z \in \mathbb{Z}^{\mathrm{p}}$ or all $z \in \mathbb{Z}_{\mathrm{p}}$ since $\dot{F}$ is defined in $\mathbb{R}^n$.[8]

The following theorem[9] establishes the relation between the entire Rate Problem 1 (13.25) to be solved and the quasi-extremal principle (13.45) specified for the energy function $\tilde{E}_\tau$, of first order in τ, as a quasi-potential φ.

Theorem 1 (Energy formulation of Rate-Problem 1) *Let* $x = (\dot{q}, \dot{\mu}) \in \mathbb{R}^{m+n}$, $z = (\tilde{q}, \tilde{\mu}) \in \mathbb{Z} = \mathbb{R}^m \times \mathbb{R}^n_+$, *and function* $\tilde{E}_\tau(z, x)$ *in a given state be defined by Eq. (13.51). The following statements are equivalent:*

(a) *x satisfies the quasi-extremal energy principle of first order (QEP1):*

$$x = \arg\min_{z \in \mathbb{Z}} \tilde{E}_\tau(z, x) \qquad \textit{for all } \tau \geq 0 \textit{ sufficiently small}\,. \tag{13.57}$$

(b) *x is a solution to Rate-Problem 1 (13.25).*

Proof. We have already shown that x is a solution to *Rate-Problem 1* (13.25) if and only if Eqs. (13.48) and (13.54) are satisfied jointly. It remains to prove that Eqs. (13.48) and (13.54) are satisfied if and only if QEP1 (13.57) is satisfied.

Since the last term in square brackets in Eq. (13.51) is independent of z, it can be disregarded in the proof. It follows that if x is a minimizer of $\tilde{E}_\tau(z, x)$ with respect to $z \in \mathbb{Z}$ then (13.57) $\Leftrightarrow$ (13.48) if $\tau = 0$, while (13.57) $\Rightarrow$ (13.54) if $\tau > 0$ since $z \in \mathbb{Z}^{\mathrm{p}} \subset \mathbb{Z}$ implies that the first term in square brackets in Eq. (13.51) vanishes. Hence, (a)$\Rightarrow$(b).

Conversely, fulfilment of Eqs. (13.48) and (13.54) implies that function (13.51) attains at $z = x$ a minimum value, equal to the last term in square brackets with minus sign, among $z \in \mathbb{Z}_{\mathrm{p}} = \{z \in \mathbb{Z} : F \cdot \tilde{\mu} = 0\}$. If $z \notin \mathbb{Z}_{\mathrm{p}}$ then $(-\dot{F} \cdot \tilde{\mu})$ is the only additional term that can be negative, but then $(-F \cdot \tilde{\mu}) > 0$ and such $\tilde{\mu}$ must have components $\tilde{\mu}_\kappa > 0$ conjugate to $F_\kappa < 0$ outside $\mathbb{R}^p$. Taking these terms jointly, the respective term in expression (13.51) reads $-(F + \tau\dot{F}) \cdot \tilde{\mu}$, where $\tilde{\mu} \geq 0$ and $-(F + \tau\dot{F}) \geq 0$ for all $\tau > 0$ sufficiently small. Thus, QEP1 (13.57) holds true if Eqs. (13.48) and (13.54) are satisfied. Hence, (b)$\Rightarrow$(a) which completes the proof of Theorem 1. □

Remark 4.1 By Theorem 1, existence of a solution to QEP1 (13.57) and to Rate Problem 1 (13.25) requires identical conditions. A similar observation concerns

[8] Note that if $Q = P$ then the first term in square brackets in Eq. (13.51) vanishes if and only if $z \in \mathbb{Z}_{\mathrm{p}}$, however, in this case $\mathbb{Z}_{\mathrm{p}}$ cannot be interchanged with $\mathbb{Z}^{\mathrm{p}}$.

[9] It is named 'theorem' *not* because of its mathematical value, since the proof requires only elementary mathematical tools, but because of a novel energy formulation of non-potential Rate Problem 1.

uniqueness. Hence, all statements in the literature concerning problems of the form of Rate Problem 1 and dealing with existence or uniqueness of their solutions apply also to QEP1. In reference to Lemma 1, this concerns in particular the theorems on the linear complementarity problem and variational inequlities. A review is not attempted here; however, see the end of Sect. 13.4.4.

Remark 4.2 The energy formulation of Rate-Problem 1 as the quasi-extremal energy principle QEP1 appears to be new. Since the underlying energy function $\tilde{E}_\tau$ has been derived as a specification of the first-order increment of a global energy function E (13.28) evaluated at an instant τ, Theorem 1 opens perspectives towards formulation of other rate-independent problems starting from other expressions for $\tilde{E}$ and applying a similar energy approach.

Another possible extension is by modifying the function φ in QEP (13.45) to obtain a criterion of choice between non-unique solutions. This concept is developed in Sect. 13.4.4 after introducing in the next Sect. 13.4.3 a more general expression for the increment of energy E on arbitrary paths.

13.4.3 Incremental Energy Supply on Arbitrary Paths

We have defined by Eq. (13.40) the rate of energy E for arbitrary rate-variables $(\tilde{q}, \tilde{\mu})$ subject to $\tilde{\mu} \geq 0$. A time integral of expression (13.40) gives an energy increment $\Delta E_\tau = E_\tau - E_0$ along an arbitrary path defined by virtual rate-variables $(\tilde{q}_t, \tilde{\mu}_t)$ that are uniformly bounded and vary in an interval $[0, \tau]$ of time t,

$$\Delta E_\tau = \int_0^\tau \Big((Q_t - P_t) \cdot \tilde{q}_t - F_t \cdot \tilde{\mu}_t - \dot{P}_t \cdot q_t \Big)\, dt\,, \tag{13.58}$$

where an index t denotes a quantity evaluated at time t. Quantities $P_t, \dot{P}_t$ are given at every t, including $P = P_0 = 0$ at $t = 0$, since $P_t = P(\lambda(t))$ has been prescribed. Quantities q_t, μ_t are defined by integrals

$$q_t = \int_0^t \tilde{q}_\theta \, \mathrm{d}\theta\,, \qquad \mu_t = \int_0^t \tilde{\mu}_\theta \, \mathrm{d}\theta\,. \tag{13.59}$$

Recall that quantity μ_t is not a state variable and need not have any physical meaning, and note that there is no loss of generality in taking $q = q_0 = 0$ at $t = 0$.

In turn, time integration of constitutive rate-equations Eq. (13.19) applied to rate-variables $(\tilde{q}_\theta, \tilde{\mu}_\theta)$ gives

$$\begin{bmatrix} Q_t \\ -F_t \end{bmatrix} = \begin{bmatrix} Q \\ -F \end{bmatrix} + \underbrace{\begin{bmatrix} K & -M \\ -N^{\mathrm{T}} & G \end{bmatrix}}_{A} \begin{bmatrix} q_t \\ \mu_t \end{bmatrix} + o(t) \tag{13.60}$$

under a mild assumption that matrix A varies continuously in time. In consistency with the notation used in earlier sections, $Q_0 = Q$, $F_0 = F$ and $A_0 = A$ at $t = 0$.

On substituting expressions (13.59) and (13.60) into Eq. (13.58) and calculating the integral using integration by parts (of a *half* of an integral) and symmetry of the elasticity matrix $K = K^{\mathrm{T}}$, after some transformations not described here in detail the following expression is obtained

$$\Delta E_\tau = \underbrace{\begin{bmatrix} Q-P \\ -F \end{bmatrix}}_{a_0} \cdot \underbrace{\begin{bmatrix} q_\tau \\ \mu_\tau \end{bmatrix}}_{y_\tau} - \underbrace{\begin{bmatrix} P_\tau - P \\ 0 \end{bmatrix}}_{b_\tau} \cdot \underbrace{\begin{bmatrix} q_\tau \\ \mu_\tau \end{bmatrix}}_{y_\tau} + \frac{1}{2} \underbrace{\begin{bmatrix} q_\tau \\ \mu_\tau \end{bmatrix}}_{y_\tau} \cdot \underbrace{\begin{bmatrix} K & -M \\ -N^{\mathrm{T}} & G \end{bmatrix}}_{A} \underbrace{\begin{bmatrix} q_\tau \\ \mu_\tau \end{bmatrix}}_{y_\tau}$$
$$+\frac{1}{2}\int_0^\tau \underbrace{\begin{bmatrix} \tilde{q}_t \\ \tilde{\mu}_t \end{bmatrix}}_{\tilde{y}_t} \cdot \underbrace{\begin{bmatrix} 0 & (N-M) \\ (M-N)^{\mathrm{T}} & (G-G^{\mathrm{T}}) \end{bmatrix}}_{2C=A-A^{\mathrm{T}}} \underbrace{\begin{bmatrix} q_t \\ \mu_t \end{bmatrix}}_{y_t} dt + o(\tau^2)\,. \tag{13.61}$$

If $A = \text{const}$ then the formula displayed above is *exact* for an arbitrary history of rate-variables $(\tilde{q}_t, \tilde{\mu}_t)$ (subject to $\tilde{\mu}_t \geq 0$) for arbitrary $\tau > 0$, the term $o(\tau^2)$ being then precisely zero. ΔE_τ expressed by Eq. (13.61) has a physical meaning of the total energy supplied to the compound thermodynamic system $\mathcal{A}$ in an isothermal quasi-static process along a path of (q_t, μ_t) in interval $t \in [0, \tau]$. Obviously, $y_\tau \cdot A y_\tau = y_\tau \cdot B y_\tau$ for $B = \frac{1}{2}(A + A^{\mathrm{T}})$. The integral in Eq. (13.61) that involves a skew-symmetric part C of A cannot be eliminated in general and represents the effect of the dependence of energy consumption on the path in (q_t, μ_t)-space. For continuous systems, the scalar products in Eq. (13.61) can be replaced by spatial integrals of the products of respective local quantities or their gradients, see Sect. 13.5.

Remark 4.3 If A is varying continuously in time then τ acts as a small parameter. The first term $(a_0 \cdot y_\tau)$ is of the order of τ and complies with expression (13.40) since $q = q_0 = 0$ here. The remaining terms that are displayed, including the path-dependent integral, are of the second order in τ. The limiting passage as $\tau \to 0$ is subtle since a path in (q_t, μ_t)-space need *not* be held fixed when approaching the limit. Formula (13.61) remains valid for arbitrarily circuitous paths whose complexity can be preserved as their length tends to zero, for instance, by scaling down a path proportionally to its length.

Keeping the above remark in mind, suppose from now on that $A = \text{const}$, so that τ may be arbitrary. In fact, in view of rate-independence of the material, the time scale can be changed without loss of generality, leaving the value of ΔE unaffected. Let the time interval $[0, \tau]$ be rescaled to $[0, 1]$, accompanied by $\lambda(\tau) \mapsto \lambda(1)$. Using the shortened notation defined in Eq. (13.61), the energy increment ΔE can be rewritten in the final compact form as the functional

$$\Delta E(y) = \frac{1}{2} y_1 \cdot B y_1 + (a_0 - b_1) \cdot y_1 + \int_0^1 \tilde{y}_t \cdot C y_t \, dt\,, \quad B = \frac{1}{2}(A + A^{\mathrm{T}})\,,\ C = \frac{1}{2}(A - A^{\mathrm{T}})\,, \tag{13.62}$$

of a function $y : [0, 1] \to \mathbb{Z}$, $y(t) = y_t$, such that the *one-sided*, forward time derivative $\dot{y}$ at t coincides with a virtual rate $\tilde{y}_t \in \mathbb{Z}$ which is assumed to exist and to be uniformly bounded on $[0, 1)$. Denote the class of such functions y by $\mathcal{Y}$. Path $y \in \mathcal{Y}$ is continuous and right-hand differentiable but may exhibit any number of sharp corners, and formula (13.62) is exact for all such paths. In general, $\Delta E(y)$ is a functional of path y in interval $[0, 1]$ and not only of the initial ($y(0) = 0$) and final ($y(1) = y_1$) values of y.

On a straight path $y : y_t \propto y_1$, the integral in Eq. (13.62) vanishes by antisymmetry od C, and ΔE becomes a function of the end point y_1, influenced only by a symmetric part B of matrix A,

$$\Delta E(y_1) = \frac{1}{2} y_1 \cdot B y_1 + (a_0 - b_1) \cdot y_1 \qquad \text{on a straight path } y : y_t \propto y_1 \,. \tag{13.63}$$

Formula (13.63) is also precisely valid for arbitrary paths y if $A = A^{\mathrm{T}} = \text{const}$.

The question arises whether direct minimization of ΔE can generate an exact solution to the original rate-problem. The answer is affirmative if $A = A^{\mathrm{T}}$, which will be discussed in Sect. 13.4.5.

If skew-symmetric matrix $C \neq 0$ then the answer is negative in general, and only in a special case it can be affirmative. This can be shown by examining a *one-sided* directional derivative δE (Gateaux differential) of ΔE taken at argument y in direction δy,

$$\delta E(y) = \delta y_1 \cdot B y_1 + (a_0 - b_1) \cdot \delta y_1 + \int_0^1 (\delta \tilde{y}_t \cdot C y_t + \tilde{y}_t \cdot C \delta y_t)\, \mathrm{d}t \,, \tag{13.64}$$

where admissible δy are subject to the constraint $\tilde{y}_t + \delta \tilde{y}_t \in \mathbb{Z}\ \forall t \in [0, 1)$. For $y = x$ to be a minimizer of $\Delta E(y)$ it is necessary that $\delta E(x) \geq 0$ for all admissible δy. The inequality constraint $\dot{\mu}_K + \delta \tilde{\mu}_K \geq 0$ is not relevant to $\delta \tilde{\mu}_K$ small enough whenever $\dot{\mu}_K > 0$ along path x. Hence, the inequality $\delta E(x) \geq 0$ implies that $\delta E(x) = 0$ if $\delta \tilde{\mu}_K = 0$ whenever $\dot{\mu}_K = 0$, for variations δy being otherwise arbitrary.

Denote by $\hat{C}_t$ the matrix obtained from C by zeroing the rows and columns of indexes κ_t such that $\dot{\mu}_{\kappa_t} = 0$ at instant t along path x. Taking $\delta y_1 = 0$ and using $C = -C^{\mathrm{T}}$, by the Euler-Lagrange equation for functional (13.62) or by integration by parts of the integrand in Eq. (13.64), it follows that

$$\hat{C}_t \dot{x}_t = 0 \qquad \text{along path } x = \arg\min_{y \in \mathcal{Y}} \Delta E(y) \,. \tag{13.65}$$

In view of its implications,[10] Eq. (13.65) is also derived on another route, by applying an end-point variation $\delta y_1 \neq 0$ to arbitrarily short segments of a path. Consider a segment x of a general path $y \in \mathcal{Y}$ in a vanishingly short time interval $[0, \tau]$, which after rescaling again $\tau \mapsto 1$ corresponds in the limit to a proportional path $x : x_t = t x_1$, associated with $P_t = P + t\dot{P}$ and $b_1 = 1b = b$ for given $\dot{P}$. We may

[10] Condition (13.65) is comparable to the symmetry condition imposed on a dissipation function in order to validate the incremental energy minimization with respect to internal variables [49].

take a path variation δy that is also proportional to t, $\delta y_t = t\delta y_1$, $\delta\tilde{y}_t = \delta y_1 \in \mathbb{Z}$. Suppose that $Q = P$ at $t = 0$ and $F \cdot \delta\mu_1 = 0$. Then $a_0 \cdot \delta y_1 = 0$ and Eq. (13.64) reduces to

$$\delta E(x_1) = \delta y_1 \cdot (Bx_1 - b)\,, \tag{13.66}$$

which obviously follows also directly from Eq. (13.63). For x_1 to be a minimizer of ΔE expressed by Eq. (13.63), it is necessary that $\delta E(x_1) \geq 0$ for all admissible δy_1 such that $x_1 + \delta y_1 \in \mathbb{Z}$. Upon substituting $x_1 = x$ and $\delta y_1 = z - x$, inequality $\delta E(x_1) \geq 0$ becomes visibly inconsistent, due to $A \neq B$, with the variational inequality (13.56) for a rate-solution x. The inconsistency can be eliminated only if $\hat{C}x = 0$, where $\hat{C}$ is the restriction of C to the rows and columns of indexes κ such that $\dot{\mu}_\kappa > 0$ in solution x. This gives condition (13.65) again, now for $\dot{x}_t = x$ at $t = 0$ which is, however, an arbitrary instant on a solution path.

Remark 4.4 We have established a general condition $(13.65)_1$ whose violation implies that minimization of ΔE does not provide an exact solution to Rate Problem 1. The derived condition $(13.65)_1$ imposes serious limitation on the class of problems whose rate-solutions can be obtained by direct minimization of the energy increment ΔE. Nevertheless, Eq. $(13.65)_1$ does *not* imply that the *entire* matrix $C \in \mathbb{R}^{m+n} \times \mathbb{R}^{m+n}$ must vanish, see an example of crystal plasticity discussed in Sect. 13.5.2. Clearly, Eq. $(13.65)_1$ is not applicable to irregular solution paths along which $\dot{x}_t$ does not exist, cf. the derivative-free theory of [38].

13.4.4 Quasi-Extremal Energy Principle of Second Order

With reference to expression (13.62) for an energy increment ΔE in time interval $[0, 1]$, consider a special 'dog-leg' path consisting of two short straight segments, $y : 0 \twoheadrightarrow x \twoheadrightarrow z$, such that $\tilde{y}_t = \text{const}$ on each segment $0 \twoheadrightarrow x$ and $x \twoheadrightarrow z$ separately. We take here the liberty to identify a rate with an increment on the time scale $[0, 1]$, and substitute into Eq. (13.62) the end-point value $y_1 = 1z = z$. The first segment can be traversed, for instance, with constant rate x/θ for $t \in [0, \theta)$ and for a given $\theta \in (0, 1)$. Then, on account of antisymmetry of matrix C, the integral in Eq. (13.62) is equal to 0 over the first segment and to $z \cdot Cx$ over the second segment. The expression (13.62) reduces to

$$\Delta E|_{0 \twoheadrightarrow x \twoheadrightarrow z} = \frac{1}{2} z \cdot Bz + (a_0 - b_1) \cdot z + z \cdot Cx\,. \tag{13.67}$$

Throughout Sect. 13.4.4, we will assume that a given initial state at $t = 0$ is in mechanical equilibrium ($P = Q$), and restrict attention to short paths y that leave the initially unloaded inelastic mechanisms $F_\kappa < 0$ inactive. It follows that $F \cdot \tilde{\mu}_t \equiv 0$ and $a_0 \cdot z = 0$. Furthermore, on short paths we can take $\dot{P}$ constant so that $b_1 = 1b = b$. This prompts us to introduce the following *second-order energy function* of virtual rate z with an unknown rate-solution x as a parameter,

$$\mathcal{E}(z,x) = \frac{1}{2} z \cdot Bz - z \cdot b + z \cdot Cx \,, \qquad x, z \in \mathbb{Z}_\mathrm{p} = \{z \in \mathbb{Z} : F \cdot \tilde{\mu} = 0\} \,. \quad (13.68)$$

From now on, $\mathcal{E}$ is interpreted as an energy function defined formally by Eq. (13.68) and treated in its own merit for *arbitrary* arguments $x, z \in \mathbb{Z}_\mathrm{p}$, thus disregarding the temporary assumption ($z \geq x$) used implicitly when deriving Eq. (13.67) from Eq. (13.62).

For better transparency and later reference, Eq. (13.68) is rewritten as

$$\mathcal{E}(z,x) = \mathcal{E}_\mathrm{sym}(z) + z \cdot Cx \,, \qquad (13.69)$$

where in the expanded matrix form

$$\mathcal{E}_\mathrm{sym}(z) = \frac{1}{2} z \cdot Bz - b \cdot z = \frac{1}{2} \begin{bmatrix} \tilde{q} \\ \tilde{\mu} \end{bmatrix} \cdot \begin{bmatrix} K & -M \\ -N^\mathrm{T} & G \end{bmatrix} \begin{bmatrix} \tilde{q} \\ \tilde{\mu} \end{bmatrix} - \begin{bmatrix} \dot{P} \\ 0 \end{bmatrix} \cdot \begin{bmatrix} \tilde{q} \\ \tilde{\mu} \end{bmatrix} . \qquad (13.70)$$

The term $z \cdot Cx$ is expanded as follows

$$z \cdot Cx = \begin{bmatrix} \tilde{q} \\ \tilde{\mu} \end{bmatrix} \cdot \begin{bmatrix} 0 & -\frac{1}{2}(M - N) \\ \frac{1}{2}(M - N)^\mathrm{T} & \frac{1}{2}(G - G^\mathrm{T}) \end{bmatrix} \begin{bmatrix} \dot{q} \\ \dot{\mu} \end{bmatrix} . \qquad (13.71)$$

Clearly, the one-sided partial derivative $\mathcal{E}_{,z}$ of the quadratic function $\mathcal{E}$ defined by Eq. (13.68) reads

$$\mathcal{E}_{,z}(z,x) = Bz - b + Cx \,, \qquad \mathcal{E}_{,z}(x,x) = Ax - b \,. \qquad (13.72)$$

The second-order Taylor series decomposition

$$\mathcal{E}(z,x) = \mathcal{E}(x,x) + \mathcal{E}_1(z,x) + \mathcal{E}_2(z,x) \qquad (13.73)$$

is exact, with

$$\mathcal{E}(x,x) = \mathcal{E}_\mathrm{sym}(x) \,, \quad \mathcal{E}_1(z,x) = (z-x) \cdot (Ax-b) \,, \quad \mathcal{E}_2(z,x) = \frac{1}{2}(z-x) \cdot B(z-x) \,. \qquad (13.74)$$

On substituting the above expressions into Eq. (13.73), after elementary transformations we recover Eq. (13.68). The term $\mathcal{E}_1(z,x)$ is recognized as the left-hand term in variational equality (13.56).

It becomes evident that energy function $\mathcal{E}$ may be used as a quasi-potential, since the minimization of its first-order term $\mathcal{E}_1$ with respect to z at x provides a solution to variational inequality (13.56) and QEP (13.54). However, $\mathcal{E}$ includes not only $\mathcal{E}_1$ but also the quadratic term $\mathcal{E}_2$ which is of the second order with respect to an original time increment τ, and also of the second order with respect to $z - x = \delta x$ if regarded as a perturbation. Therefore, an extra condition is expected to be imposed on solution x obtained by quasi-minimization of $\mathcal{E}(\cdot, x)$. This is specified by the following theorem.

Theorem 2 (Energy criterion of stability of a solution path) *Let* $x = (\dot{q}, \dot{\mu}) \in \mathbb{R}^{m+n}$ *and assume that the conditions (13.41) are satisfied. Let* $z = (\tilde{q}, \tilde{\mu})$ *be arbitrary within the set* $\mathbb{Z}_p = \{z \in \mathbb{Z} : F \cdot \tilde{\mu} = 0\}$. *Then a solution* x *to the quasi-extremal energy principle of second order (QEP2):*

$$x = \arg\min_{z \in \mathbb{Z}_p} \mathcal{E}(z, x) \tag{13.75}$$

solves Rate Problem 1 (13.25) and satisfies additionally the inequality

$$(z - x) \cdot B(z - x) \geq 0 \qquad \forall z \in \mathbb{Z} \quad \text{subject to } F \cdot \tilde{\mu} = 0 \text{ and } \dot{F} \cdot \tilde{\mu} = 0 . \tag{13.76}$$

Proof. Suppose that x satisfies QEP2 (13.75). From decomposition (13.73) in the limit as $z \to x$ it follows that the first-order term $\mathcal{E}_1(z, x)$ must be nonnegative, which implies the variational inequality (13.56) since the magnitude of $(z - x)$ is inessential there. In turn, variational inequality (13.56) is equivalent to QEP (13.54). By Lemma 1 this implies fulfilment of the conditions (13.53), which jointly with assumed conditions (13.41) imply that x is a solution to Rate Problem 1.

Since x is a solution then the condition $\dot{F} \cdot \tilde{\mu} = 0$ at inequality (13.76) implies $\mathcal{E}_1(z, x) = 0$ by Eq. (13.52). Under this condition, fulfilment of QEP2 (13.75) implies $\mathcal{E}_2(z, x) \geq 0$, which coincides with inequality (13.76). This completes the proof of Theorem 2. □

Inequality (13.76) can be compared with the second-order optimality condition in the non-convex quadratic programming. It is emphasized that condition (13.76) imposes a restriction only on a principal submatrix $\hat{A}(x)$ of A that is restricted to the rows and columns of indexes κ such that that $F_\kappa = 0$ and $\dot{F}_\kappa = 0$ in solution x. This is discussed further below.

Condition (13.76) can be given a physical meaning. Suppose that $z = x + \delta x$ represents a solution to a *perturbed* rate-problem, analogous to Rate-Problem 1 defined by (13.25), but with b replaced with $b + \delta b = A(x + \delta x)$. Then

$$\delta b \cdot \delta x = (z - x) \cdot (Ax - b) + (z - x) \cdot B(z - x) . \tag{13.77}$$

In the above expression, the first right-hand term is non-negative on account of variational inequality (13.56). Hence,

$$\delta b \cdot \delta x \geq (z - x) \cdot B(z - x) \qquad \forall z \in \mathbb{Z}_\mathrm{p} . \tag{13.78}$$

Moreover, for solution x we have $x \cdot (Ax - b) = 0$ as the complementarity condition, and also

$$z \cdot (Ax - b) = \tilde{q} \cdot (\dot{Q} - \dot{P}) + \tilde{\mu} \cdot \dot{F} = 0 \qquad \text{if } \dot{F} \cdot \tilde{\mu} = 0 . \tag{13.79}$$

Condition (13.76) is thus given the following interpretation

$$\delta b \cdot \delta x = (z - x) \cdot B(z - x) \geq 0 \qquad \text{if } F \cdot \tilde{\mu} \text{ and } \dot{F} \cdot \tilde{\mu} = 0 . \tag{13.80}$$

It means that perturbing forces δb make a non-negative scalar product with a deviation δx they induce. This scalar product may, in turn, be interpreted as the second-order work supplied by a disturbing agency to the material body $\mathcal{B}$ that is *deforming*[11] along a solution path of the current rate x. Its non-negativeness provides a criterion of path stability in the energy sense.

If there are multiple solutions then fulfilment of QEP2 can be used as a *criterion of choice* among them. Namely, the associated condition (13.80), imposed on matrix $\hat{A}(x)$ dependent on x, can fail for one solution x and be satisfied for another, especially when the latter corresponds to a smaller dimension of vector $\dot{\mu} > 0$. In that case, the solution that satisfies QEP2 and hence condition (13.80) can be considered as preferable from the point of view of stability of the associated solution path, at least in the above energy sense.

However, a word of caution is needed. Unlike for rate-independent problems of potential type, for which the path instability in the energy sense has a definite meaning, see Sect. 13.4.5, consequences of failure of inequality (13.76) are much less clear for non-potential problems examined here. The condition (13.80) may thus be over-sufficient for acceptance of certain solutions. This shortcoming can become less essential the smaller the magnitude of C.

A prerequisite for investigating QEP2 is that the involved minimization subproblem has a solution. It always does if the second-order energy condition for *directional* stability of an equilibrium state (q, α) at constant loads ($\dot{P} = 0$) is satisfied. On account of Eq. (13.63), this condition limited to straight paths of departure from equilibrium takes the form, cf. [50],

$$z \cdot Bz = \tilde{q} \cdot K\tilde{q} - \tilde{q} \cdot (M + N)\tilde{\mu} + \tilde{\mu} \cdot G\tilde{\mu} > 0 \qquad \forall z = (\tilde{q}, \tilde{\mu}) \in \mathbb{Z}_{\rm p}\,,\ z \neq 0\,. \tag{13.81}$$

Then the quadratic function $\mathcal{E}(\cdot, x)$ is bounded from below and, by the Frank-Wolfe theorem of the quadratic programming theory, attains its infimum on $\mathbb{Z}_{\rm p}$.

Inequality (13.81) has the following immediate consequences:

(a) K is positive definite, i.e. $v \cdot Kv > 0\ \forall v \in \mathbb{R}^m$, $v \neq 0$ (purely elastic stability for $\tilde{\nu} \equiv 0$).

(b) G is strictly copositive in $\mathbb{R}^p$, i.e. $\nu \cdot G\nu > 0\ \forall \nu \in \mathbb{R}^p_+$, $\nu \neq 0$ (intrinsic plastic stability for $\tilde{q} \equiv 0$).

(c) The reduced $p \times p$ matrix $\hat{B}_{\rm q}$ is strictly copositive,

$$\nu \cdot \hat{B}_{\rm q}\nu > 0\ \ \forall \nu \in \mathbb{R}^p_+\,,\ \nu \neq 0, \quad \hat{B}_{\rm q} = \frac{1}{2}(G + G^{\rm T}) - \frac{1}{4}(N + M)^{\rm T}K^{-1}(N + M)\,. \tag{13.82}$$

If K is positive definite then conditions (13.81) and (13.82) are equivalent. This follows easily by minimizing $z \cdot Bz$ with respect to $\tilde{q}$ for any fixed $\tilde{\mu}$, which gives $\tilde{q}(\tilde{\mu}) = \frac{1}{2}K^{-1}(M + N)\tilde{\mu}$ as a linear function of $\tilde{\mu}$. The equivalence is obtained on substituting $\tilde{q}(\tilde{\mu})$ into inequality (13.81).

---

[11] This makes a substantial difference with respect to the concept of stability of *equilibrium* when the second-order work is considered to be supplied to the body being *at rest* when unperturbed [29]; cf. also condition (13.81) that is distinct from condition (13.80).

In a similar way, on introducing the assumption (a) of purely elastic stability, minimization of $\mathcal{E}(\cdot, x)$ with respect to $z = (\tilde{q}, \tilde{\mu})$ can be straightforwardly transformed to minimization of a reduced quadratic function of the leading term $\tilde{\mu} \cdot \hat{B}_q \tilde{\mu}$ with respect to $\tilde{\mu}$. Matrix $\hat{B}_q$ in inequality $(13.82)_1$ can still be indefinite, so that the quadratic energy function $\mathcal{E}(\cdot, x)$ is *non-convex* in general.

Uniqueness of a solution can be proven under the stronger condition that entire A (or B, equivalently) is positive definite, cf. [13, 27]. The special cases of existence of a unique solution are of less interest for potential applications of Theorem 2, whose major advantage may be in providing a criterion of choice between non-unique solutions which are expected in rate-independent problems of finite deformation plasticity. It can be noted that Theorems 1 and 2 are valid for arbitrary non-symmetric indefinite matrices A without guarantee that a solution does exist.

13.4.5 Incremental Energy Minimization in Symmetric Case

Let $A = \text{const}$ so that expression (13.62) is exact for arbitrary paths. In the special case of normality ($M = N$) and symmetry ($G = G^{\mathrm{T}}$), the matrix A is symmetric and $C = 0$. The energy functional (13.61) becomes then path-independent and, on substituting $y_\tau = \tau z$, reduces to a one-parameter quadratic energy function

$$\Delta E_\tau(z) = a_0 \cdot z\tau + \mathcal{E}_{\text{sym}}(z)\tau^2 \qquad z \in \mathbb{Z}, \qquad A = A^{\mathrm{T}}, \tag{13.83}$$

where $\mathcal{E}_{\text{sym}} : \mathbb{Z}_{\text{p}} \to \mathbb{R}$ is defined by Eq. (13.70), i.e.

$$\mathcal{E}_{\text{sym}}(z) = \frac{1}{2} z \cdot Az - b \cdot z, \qquad z \in \mathbb{Z}_{\text{p}}, \qquad A = A^{\mathrm{T}}. \tag{13.84}$$

This is in agreement with formula (13.63) for straight paths. After time rescaling $\tau \mapsto 1$, z can be regarded either as a virtual rate $z = (\tilde{q}, \tilde{\mu})$ at $t = 0$ or a virtual increment at $t = 1$. In formula (13.84) it is again assumed that $\dot{P} = \text{const}$, so that $b_\tau = \tau b$.

In the symmetric case above, an incremental solution can be found by minimizing $\Delta E_\tau(z)$. A prerequisite is, of course, that $Q = P$ and $F \leq 0$ at the initial instant $t = 0$ of the time increment. Then, after restricting the search of rate-solutions at $t = 0$ to the set of $z \in \mathbb{Z}$ such that $F \cdot \tilde{\mu} = 0$, i.e. to $z \in \mathbb{Z}_{\text{p}}$, it follows that $a_0 \cdot z = 0$. Hence, both the minimization of $\Delta E_\tau(z)$ over $\mathbb{Z}_{\text{p}}$ and QEP2 (13.75) reduce to minimization of $\mathcal{E}_{\text{sym}}(z)$,

$$x = \arg \min_{z \in \mathbb{Z}_{\text{p}}} \mathcal{E}_{\text{sym}}(z), \tag{13.85}$$

which corresponds also to minimization of the second time derivative $\ddot{E}$ of E. The point is that a minimizer, if exists, is automatically a solution, while the converse is not true in general since $\mathcal{E}_{\text{sym}}(\cdot)$ need not be convex.

A minimizer does exist (by the Frank-Wolfe theorem of the quadratic programming theory) if $\mathcal{E}_{\text{sym}}$ is bounded from below. This is so if the condition (13.81) of stability of equilibrium holds, which in the case of purely elastic stability and $N = M$ is transformed to the strict copositivity condition, cf. Eq. (13.82),

$$\nu \cdot (G - N^{\mathrm{T}} K^{-1} N)\nu > 0 \qquad \forall \nu \in \mathbb{R}^{p}_{+},\ \nu \neq 0\,. \tag{13.86}$$

Note that the matrix in parentheses is of reduced dimension $p \times p$ and can still be indefinite, in consistency with non-convexity of $\mathcal{E}_{\mathrm{sym}}$.

An important consequence is that along a solution path such that $\dot{x}_t = x$ satisfies Rate Problem 1 but violates Eq. (13.85), while equilibrium is stable according to condition (13.86), at every instant there exists another rate-solution that satisfies Eq. (13.85) and initiates thus an energetically preferable branching path. This is the essence of path instability associated with violation of Eq. (13.85) in the potential case; see [47, 50] for the theory and [52, 51] and references therein for a number of numerical applications to automatic branch switching.

13.5 Examples and Extensions of the Quasi-Extremal Principle to Elastoplastic Continuum

13.5.1 A Classical Non-Associative Elastoplastic Continuum

Consider the rate-boundary value problem formulated in Sect. 13.2.2 for an unknown velocity field $\underline{\dot{\mathbf{u}}} \in \mathcal{V}$ and a plastic multiplier field $\underline{\dot{\gamma}} \in \mathcal{R}_+$ over a reference domain B of a continuous material body. Denote by $\underline{\mathbf{v}} \in \mathcal{V}$ and $\underline{\nu} \in \mathcal{R}_+$ the respective virtual fields, assuming for simplicity that $\mathbf{v} \equiv \mathbf{0}$ on $\partial\mathcal{B}_{\mathrm{u}}$ and $\underline{\mathbf{u}} = \underline{\mathbf{0}}$ at $t = 0$. A virtual rate (13.39) of energy E becomes now a functional of virtual fields $(\underline{\mathbf{v}}, \underline{\nu})$,

$$\tilde{E}(\underline{\mathbf{v}}, \underline{\nu}) = \int_{\mathcal{B}} (\mathbf{S} \,.\, \nabla\mathbf{v} - f\nu)\, \mathrm{d}V - \int_{\partial\mathcal{B}_{\mathrm{T}}} \mathbf{T} \cdot \mathbf{v}\, \mathrm{d}S\,. \tag{13.87}$$

The first integrand is the virtual rate of internal work density, $\tilde{w} = \mathbf{S} \,.\, \nabla\mathbf{v} - f\nu$, with the second term as the power expended by local perturbing forces $\delta\pi = \pi^{\mathrm{c}} - \pi = -f$, cf. the first paragraph of Sect. 13.3.3.

What follows is a straightforward adaptation of the lines established in Sect. 13.4, although only in a formal manner, without addressing the mathematical problem of existence of minima of the examined functionals in suitable function spaces. The first-order energy function $\tilde{E}_\tau$ (13.51) becomes a functional of a virtual pair $(\underline{\mathbf{v}}, \underline{\nu}) \leftrightarrow z$ and a solution pair $(\underline{\dot{\mathbf{u}}}, \underline{\dot{\gamma}}) \leftrightarrow x$ associated with fields $(\underline{\dot{\mathbf{S}}}, \underline{\dot{f}})$,

$$\mathcal{E}_\tau(\underline{\mathbf{v}}, \underline{\nu}; \underline{\dot{\mathbf{u}}}, \underline{\dot{\gamma}}) = \tilde{E}(\underline{\mathbf{v}}, \underline{\nu}) + \tau \int_{\mathcal{B}} (\dot{\mathbf{S}} \,.\, \nabla\mathbf{v} - \dot{f}\nu)\, \mathrm{d}V - \tau \int_{\partial\mathcal{B}_{\mathrm{T}}} (\dot{\mathbf{T}} \cdot \mathbf{v} + \dot{\mathbf{T}} \cdot \dot{\mathbf{u}})\, \mathrm{d}S\,. \tag{13.88}$$

Energy formulation of the rate-boundary value problem takes the form of a quasi-extremal energy principle of first order analogous to QEP1,

$$(\underline{\dot{\mathbf{u}}}, \underline{\dot{\gamma}}) = \arg \min_{\underline{\mathbf{v}} \in \mathcal{V}} \min_{\underline{\nu} \in \mathcal{R}_+} \mathcal{E}_\tau(\underline{\mathbf{v}}, \underline{\nu}; \underline{\dot{\mathbf{u}}}, \underline{\dot{\gamma}}) \quad \text{for all } \tau \geq 0 \text{ sufficiently small}. \tag{13.89}$$

The second-order energy functional corresponding to $\mathcal{E}$ in Eq. (13.68) is now defined by

$$\mathcal{E}(\underline{\mathbf{v}}, \underline{\nu}; \underline{\dot{\mathbf{u}}}, \underline{\dot{\gamma}}) = \int_{\mathcal{B}} \varepsilon(\nabla\mathbf{v}, \nu; \nabla\dot{\mathbf{u}}, \dot{\gamma})\, \mathrm{d}V - \int_{\partial\mathcal{B}_\mathrm{T}} \dot{\mathbf{T}} \cdot \mathbf{v}\, \mathrm{d}S, \tag{13.90}$$

with the integrand (denoting $\tilde{\mathbf{F}} = \nabla\mathbf{v}$)

$$\varepsilon(\tilde{\mathbf{F}}, \nu; \nabla\dot{\mathbf{u}}, \dot{\gamma}) = \frac{1}{2}(\tilde{\mathbf{F}}.\mathbb{C}^\mathrm{e}.\tilde{\mathbf{F}} - \tilde{\mathbf{F}}.(\mathbf{M}+\mathbf{N})\nu + g\nu^2) - \tilde{\mathbf{F}}.\frac{\mathbf{M}-\mathbf{N}}{2}\dot{\gamma} + \nabla\dot{\mathbf{u}}.\frac{\mathbf{M}-\mathbf{N}}{2}\nu \tag{13.91}$$

constructed by using locally the form of Eqs. (13.68), (13.70) and (13.71), with the last term simplified since matrix G corresponds here to a scalar g.

The energy formulation of the rate-boundary value problem that includes a criterion of stability of a solution path takes the form of a quasi-extremal energy principle of second order analogous to QEP2,

$$(\underline{\dot{\mathbf{u}}}, \underline{\dot{\gamma}}) = \arg \min_{\underline{\mathbf{v}} \in \mathcal{V}} \min_{\underline{\nu} \in \mathcal{R}_+^\mathrm{p}} \mathcal{E}(\underline{\mathbf{v}}, \underline{\nu}; \underline{\dot{\mathbf{u}}}, \underline{\dot{\gamma}}), \tag{13.92}$$

where $\mathcal{R}_+^\mathrm{p} \subseteq \mathcal{R}_+$ denotes the set of fields of $\nu \geq 0$ over $\mathcal{B}$, such that $\nu = 0$ in the elastic domain where $f < 0$, and $\nu \geq 0$ in the plastic domain $\mathcal{B}^\mathrm{p}$ where $f = 0$.

Since ν is a scalar and $g > 0$, the minimization with respect to $\underline{\nu}$ can be done directly, by performing local minimization of $\varepsilon(\tilde{\mathbf{F}}, \cdot; \nabla\dot{\mathbf{u}}, \dot{\gamma})$. This is pursued further here in the case of associative plasticity. If $\mathbf{M} = \mathbf{N}$ then quasi-extremal principle (13.92) reduces to a minimum principle analogous to Eq. (13.85),

$$(\underline{\dot{\mathbf{u}}}, \underline{\dot{\gamma}}) = \arg \min_{\underline{\mathbf{v}} \in \mathcal{V}} \min_{\underline{\nu} \in \mathcal{R}_+} \mathcal{E}_\mathrm{sym}(\underline{\mathbf{v}}, \underline{\nu}), \tag{13.93}$$

where

$$\mathcal{E}_\mathrm{sym}(\underline{\mathbf{v}}, \underline{\nu}) = \int_{\mathcal{B}} \frac{1}{2}(\tilde{\mathbf{F}} . \mathbb{C}^\mathrm{e} . \tilde{\mathbf{F}} - 2\tilde{\mathbf{F}} . \mathbf{N}\nu + g\nu^2)\, \mathrm{d}V - \int_{\partial\mathcal{B}_\mathrm{T}} \dot{\mathbf{T}} \cdot \mathbf{v}\, \mathrm{d}S. \tag{13.94}$$

If $g > 0$ then after eliminating $\underline{\nu}$ as in Eq. (13.5), quasi-extremal principle (13.92) reduces further to the minimum principle (13.7) for Hill's velocity functional $\mathcal{J}(\underline{\mathbf{v}})$. In contrast to [29], uniqueness of a solution is not required to formulate the minimum principle for Hill's velocity functional which may be non-convex here. The key point is that the minimum principle (13.7), when derived from inequality (13.93), includes not only the stationarity condition, valid for every solution in the symmetric case, but also the energy condition of path stability. The latter provides the criterion of choice between velocity solutions that selects an energetically favourable branching path expected to have a physical meaning, cf. [47] for more details.

13.5.2 Single Crystals Deformed Plastically by Multislip

The local rate-problem for single crystals presented in Sect. 13.2.3 is now revisited in more detail. At finite deformation of gradient $\mathbf{F}$, the Helmholtz free energy density ϕ, in isothermal case and per unit reference volume, is frequently assumed in a separable form

$$\phi(\mathbf{F}^*, \boldsymbol{\xi}) = \phi^{\mathrm{e}}(\mathbf{U}^{\mathrm{e}}) + \phi^{\mathrm{p}}(\boldsymbol{\xi}), \qquad \mathbf{F} = \mathbf{F}^*\mathbf{F}^{\mathrm{p}}, \quad \mathbf{F}^* = \mathbf{R}^*\mathbf{U}^{\mathrm{e}}, \tag{13.95}$$

where $\mathbf{F}^{\mathrm{p}}$ is the plastic deformation gradient, $\mathbf{F}^*$ is a contraction of lattice rotation $\mathbf{R}^*$ and elastic stretch tensor $\mathbf{U}^{\mathrm{e}}$ with respect to the stress-free lattice configuration, and $\boldsymbol{\xi}$ denotes local internal variables of unspecified type. Since $\mathbf{U}^{\mathrm{e}}$ depends not only on $\mathbf{F}$ but also on $\mathbf{F}^{\mathrm{p}}$, in correspondence to $\boldsymbol{\alpha}$ in Sect. 13.3.1 that included all variables except q, we add $\mathbf{F}^{\mathrm{p}}$ to the list $\boldsymbol{\alpha}_{\mathrm{s}} = (\mathbf{F}^{\mathrm{p}}, \boldsymbol{\xi})$ of local internal variables. Following the conventional, non-gradient theory of single crystals, we adopt the standard evolution equation for $\mathbf{F}^{\mathrm{p}}$, complemented with a (less standard) evolution equation for $\boldsymbol{\xi}$,

$$\dot{\mathbf{F}}^{\mathrm{p}}(\mathbf{F}^{\mathrm{p}})^{-1} = \sum_\beta \mathbf{m}_\beta \otimes \mathbf{n}_\beta \dot{\gamma}_\beta, \qquad \dot{\boldsymbol{\xi}} = \sum_\beta \boldsymbol{\zeta}_\beta(\boldsymbol{\alpha}_{\mathrm{s}})\dot{\gamma}_\beta. \tag{13.96}$$

Here, $\dot{\gamma}_\beta$ represents the slip-rate on the β-th slip-system and $(\mathbf{m}_\beta, \mathbf{n}_\beta)$ a pair of orthogonal vectors[12] defining slip direction and slip-plane normal, respectively, in the intermediate, stress-free configuration of the crystallographic lattice, assumed unaffected by changes in $\boldsymbol{\alpha}_{\mathrm{s}}$. Accordingly, as a local counterpart to Eq. (13.30) we obtain

$$\dot{\boldsymbol{\alpha}}_{\mathrm{s}} = (\dot{\mathbf{F}}^{\mathrm{p}}, \dot{\boldsymbol{\xi}}) = \sum_\beta \boldsymbol{\eta}_\beta \dot{\gamma}_\beta = \boldsymbol{\eta}_{\mathrm{s}} \cdot \dot{\boldsymbol{\gamma}}, \quad \boldsymbol{\eta}_{\mathrm{s}} = (\boldsymbol{\eta}_\beta), \quad \boldsymbol{\eta}_\beta(\mathbf{F}^{\mathrm{p}}, \boldsymbol{\xi}) = (\mathbf{m}_\beta \otimes \mathbf{F}^{\mathrm{p}\,\mathrm{T}}\mathbf{n}_\beta, \boldsymbol{\zeta}_\beta). \tag{13.97}$$

In a given state where current values of all variables in Eq. (13.95) are known, the rate $\dot{\phi}$ of ϕ is a function of either $(\dot{\mathbf{F}}^*, \dot{\boldsymbol{\xi}})$ or $(\dot{\mathbf{F}}, \dot{\boldsymbol{\gamma}})$, denoted for simplicity by the same symbol,

$$\dot{\phi} = \frac{\mathrm{d}}{\mathrm{d}t^+}\phi(\dot{\mathbf{F}}^*, \dot{\boldsymbol{\xi}}) = \mathbf{S}^* \cdot \dot{\mathbf{F}}^* + \sum_\beta p_\beta \dot{\gamma}_\beta = \frac{\mathrm{d}}{\mathrm{d}t^+}\phi(\dot{\mathbf{F}}, \dot{\boldsymbol{\gamma}}) = \mathbf{S} \cdot \dot{\mathbf{F}} - \sum_\beta \pi_\beta \dot{\gamma}_\beta, \tag{13.98}$$

where

$$\mathbf{S}^* = \frac{\partial \phi(\mathbf{F}^*, \boldsymbol{\xi})}{\partial \mathbf{F}^*}, \quad p_\beta = \frac{\partial \phi(\mathbf{F}^*, \boldsymbol{\xi})}{\partial \boldsymbol{\xi}} \cdot \boldsymbol{\zeta}_\beta, \quad \mathbf{S} = \frac{\partial \phi(\mathbf{F}, \boldsymbol{\alpha}_{\mathrm{s}})}{\partial \mathbf{F}}, \quad \pi_\beta = \frac{\partial \phi(\mathbf{F}, \boldsymbol{\alpha}_{\mathrm{s}})}{\partial \boldsymbol{\alpha}_{\mathrm{s}}} \cdot \boldsymbol{\eta}_\beta. \tag{13.99}$$

After some transformations which are omitted here, the generalized Schmid stress τ_β, that is work-conjugate to $\dot{\gamma}_\beta$ per unit reference volume, appears in the final expression for π_β, cf. [55] and [51],

[12] It implies that the plastic deformation is isochoric, by using Jacobi's formula for the rate of det $\mathbf{F}^{\mathrm{p}}$ and substituting $\mathbf{m}_\beta \cdot \mathbf{n}_\beta = 0$.

$$\pi_\beta = \tau_\beta - p_\beta\,, \qquad \tau_\beta = \mathbf{\Pi}\,.\,(\mathbf{m}_\beta \otimes \mathbf{n}_\beta)\,, \tag{13.100}$$

where $\mathbf{\Pi}$ is the Mandel stress [36] related to $\mathbf{S}$ (the Piola stress work-conjugate to $\mathbf{F}$) through $\mathbf{\Pi} = \mathbf{F}^{*\mathrm{T}}\mathbf{S}\mathbf{F}^{\mathrm{pT}}$.

In comparison with Eq. (13.31), there is the correspondence $q \leftrightarrow \mathbf{F}$, $\dot{\mu} \leftrightarrow \dot{\gamma} = (\dot{\gamma}_\beta)$, $Q \leftrightarrow \mathbf{S}$ and $\Pi \leftrightarrow \pi = (\pi_\beta)$. Further considerations at a material point level can follow closely those in Sect. 13.3 and 13.4. To avoid repetitions, only the major conclusions are re-stated as illustrations.

By imposing upper bounds on either π_β or τ_β, the yield function f_β for β-th slip-system is defined by

$$f_\beta = \pi_\beta - \pi_\beta^{\mathrm{c}} = \tau_\beta - \tau_\beta^{\mathrm{c}}\,, \qquad \tau_\beta^{\mathrm{c}} - \pi_\beta^{\mathrm{c}} = p_\beta\,. \tag{13.101}$$

The dissipation function of vector $\nu = (\nu_\beta)$ of virtual slip-rates,

$$D(\nu) = \sum_\beta \pi_\beta^{\mathrm{c}} \nu_\beta = \pi^{\mathrm{c}} \cdot \nu\,, \qquad \pi^{\mathrm{c}}(\mathbf{F}, \alpha_{\mathrm{s}}) \geq 0\,, \qquad \nu \geq 0\,, \tag{13.102}$$

generates the normality relation between π and $\dot{\gamma}$ fully analogous to that illustrated in Fig. 13.2.

According to Remarks 3.2 and 3.3, as long as the dependence of π^{c} on $\mathbf{F}$, and hence on $\mathbf{F}^*$ and consequently on stress, has not been excluded, we can have $\mathbf{M} \neq \mathbf{N}$ as in Sect. 13.2.3. In other words, the normality relation between $\dot{\gamma}$ and π does *not* imply normality of the plastic strain rate to the yield surface in stress space. It is clear that we can also have $g \neq g^{\mathrm{T}}$.

As far as the local rate-problem (13.14) is concerned, which is repeated here for convenience as

$$\dot{f} = \mathbf{N}^{\mathrm{T}}\,.\,\dot{\mathbf{F}} - g\dot{\gamma}\,, \qquad \dot{\gamma} \geq 0, \qquad -\dot{f} \geq 0, \qquad \dot{f} \cdot \dot{\gamma} = 0 \qquad \text{in } \mathbb{R}^{S_p} \tag{13.103}$$

for a given $\dot{\mathbf{F}}$, normality is inessential since $\mathbf{M}$ is not involved. The crucial point is that, as already remarked, in realistic models of crystal plasticity at finite deformation, the matrix $g = (g_{\alpha\beta})$ need not be symmetric in general [32, 44, 51]. Therefore, in view of insufficiency of the minimum principle (13.15) discussed in Sect. 13.2.3, and as a specification of the quadratic energy function $\mathcal{E}$ (13.68), consider a local quadratic function of virtual slip-rates $\nu = (\nu_\beta)$ with an unknown solution $\dot{\gamma}$ as a parameter,

$$\varepsilon(\nu, \dot{\gamma}) = \frac{1}{2}\nu \cdot g\nu - \dot{\mathbf{F}}\,.\,\mathbf{N}\nu + \frac{1}{2}\nu \cdot (g - g^{\mathrm{T}})\dot{\gamma}\,, \quad \nu, \dot{\gamma} \in \mathbb{R}_+^{S_p}\,, \quad \dot{\mathbf{F}} \text{ given}\,, \tag{13.104}$$

where $\mathbb{R}_+^{S_p}$ denotes the set of non-negative slip-rates for potentially active slip-systems for which $f_\beta = 0$.

From Theorem 2 it follows that a solution to the quasi-extremal principle for function (13.104), viz.

$$\dot{\gamma} = \arg \min_{\nu \in \mathbb{R}_+^{S_p}} \varepsilon(\nu, \dot{\gamma}) \tag{13.105}$$

solves rate-problem (13.103) and satisfies additionally the path-stability condition

$$\delta\dot{\mathbf{F}}\,.\,\mathbf{N}\delta\dot{\gamma} = (\nu - \dot{\gamma}) \cdot g(\nu - \dot{\gamma}) \geq 0 \qquad \text{subject to } f \cdot \nu = 0 \text{ and } \dot{f} \cdot \nu = 0\,. \quad (13.106)$$

Remark 5.1 QEP (13.105) provides through condition (13.106) a criterion of selection of active slip-systems for any given velocity gradient $\dot{\mathbf{F}}$. Condition (13.106) implies positive semi-definiteness of the slip-system interaction submatrix $\hat{g}(\dot{\gamma}) = (g_{\alpha\beta})$ with the rows and column restricted to *active slip-systems* for which $\dot{\gamma}_\beta > 0$. On physical grounds, when latent hardening exceeds self-hardening of slip systems, this is expected only for linearly independent active slip-systems whose number does not exceed 5 [3]. Therefore, QEP (13.105) can be used to eliminate solution paths with more than 5 slip systems active simultaneously at one material point.

QEP (13.105) represents an extension of the minimum principle (13.15) for symmetric g to an arbitrary non-symmetric g. It is worth mentioning, however, that the minimum principle (13.15) can still yield a correct solution $\dot{\gamma}$, even if $g \neq g^{\mathrm{T}}$ when considered in the whole space $\mathbb{R}^{s\times s}$, provided the restriction $\hat{g}(\dot{\gamma})$ of matrix g to the slip-systems that are *active in the solution* ($\dot{\gamma}_\beta > 0$) is a symmetric submatrix. This is an example of a more general condition (13.65). Under this relaxed assumption of selective symmetrization, minimization (13.15) has been used in [51] to calculate energetically favourable solutions with ≤ 5 active slip-systems for a large number of different $\dot{\mathbf{F}}$.

13.5.3 Gradient Crystal Plasticity in Single-Slip Case

The expression (13.30) for evolution of internal variables may include the effect of spatial gradient $\nabla\dot{\gamma}$ of a plastic multiplier. In the continuum framework of single-slip plasticity with a field of one plastic multiplier $\dot{\gamma}$, this corresponds to the evolution equation for local internal variables $\boldsymbol{\alpha}_{\mathrm{sg}}$, composed of the previous variables $\boldsymbol{\alpha}_{\mathrm{s}}$ and their gradient enhancement $\boldsymbol{\alpha}_{\mathrm{g}}$, of the form

$$\dot{\boldsymbol{\alpha}}_{\mathrm{sg}} = (\dot{\boldsymbol{\alpha}}_{\mathrm{s}}, \dot{\boldsymbol{\alpha}}_{\mathrm{g}}), \qquad \dot{\boldsymbol{\alpha}}_{\mathrm{s}} = \boldsymbol{\eta}_{\mathrm{s}}\dot{\gamma}\,, \qquad \dot{\boldsymbol{\alpha}}_{\mathrm{g}} = \boldsymbol{\zeta}_{\mathrm{g}} \cdot \nabla\dot{\gamma}\,, \qquad (13.107)$$

where $\boldsymbol{\eta}_{\mathrm{s}}$ and $\boldsymbol{\zeta}_{\mathrm{g}}$ are state-dependent matrix parameters. The rate $\dot{\phi}$ of the Helmholtz free energy density $\phi(\mathbf{F}, \boldsymbol{\alpha}_{\mathrm{sg}})$ then reads, in correspondence to Eq. (13.31),

$$\dot{\phi} = \mathbf{S}\,.\,\dot{\mathbf{F}} - \pi\dot{\gamma} - \boldsymbol{\xi}_{\mathrm{g}} \cdot \nabla\dot{\gamma} \qquad \pi = -\phi_{,\boldsymbol{\alpha}_{\mathrm{s}}} \cdot \boldsymbol{\eta}_{\mathrm{s}}\,, \qquad \boldsymbol{\xi}_{\mathrm{g}} = -\boldsymbol{\zeta}_{\mathrm{g}}^{\mathrm{T}}\phi_{,\boldsymbol{\alpha}_{\mathrm{g}}}\,. \qquad (13.108)$$

The non-gradient part of crystal plasticity has been described in Sect. 13.5.2. Only one simple version of the slip-rate gradient enhancement is presented here; much more details on slip-gradient theories can be found in [23, 24] and other works. $\boldsymbol{\xi}_{\mathrm{g}}$ is here an *energetic* microforce vector conjugate to the projection of plastic slip-rate gradient $\nabla\dot{\gamma}$ on the slip plane. With this constitutive change as the only change with respect to Sect. 13.5.1, and following the general lines of Sect. 13.3, the virtual first-order rate of the energy functional $\tilde{E}$ takes the form

$$\tilde{E}(\underline{\nu}, \underline{\mathbf{v}}) = \int_{\mathcal{B}} (\mathbf{S} \,.\, \nabla\mathbf{v}) - f\nu - \boldsymbol{\xi}_{\rm g} \cdot \nabla\nu)\, \mathrm{d}V - \int_{\partial\mathcal{B}_{\rm T}} \mathbf{T} \cdot \mathbf{v}\, \mathrm{d}S\,. \tag{13.109}$$

For simplicity, we assume the boundary conditions for ν to be of either micro-hard or micro-soft type, so that

$$\boldsymbol{\xi}_{\rm g} \cdot \mathbf{n}\nu = 0 \quad \text{over } \partial\mathcal{B}. \tag{13.110}$$

Then, by using the divergence theorem for a continuous field of $\boldsymbol{\xi}_{\rm g}$, from the extremal energy-rate principle (13.44) we easily deduce[13] the equilibrium conditions in $\mathcal{B}$ and on $\partial\mathcal{B}_{\rm T}$, along with the gradient-enhanced condition of plastic flow,

$$\dot{\gamma} \geq 0\,, \qquad f_{\rm g} \leq 0\,, \qquad f_{\rm g}\dot{\gamma} = 0\,, \qquad f_{\rm g} = f + \nabla \cdot \boldsymbol{\xi}_{\rm g} \qquad \text{in } \mathcal{B}. \tag{13.111}$$

Assume that $\dot{\pi}$ and $\dot{\pi}_{\rm c}$ are independent of $\nabla\dot{\gamma}$, and $\dot{\boldsymbol{\xi}}_{\rm g}$ is independent of $\dot{\gamma}$. The rate of the yield function then reads

$$\dot{f}^{\rm g} = \dot{f} + \nabla \cdot \dot{\boldsymbol{\xi}}_{\rm g}\,, \qquad \dot{\boldsymbol{\xi}}_{\rm g} = \boldsymbol{\theta} \cdot \nabla\dot{\gamma}\,, \qquad \boldsymbol{\theta} = -(\boldsymbol{\zeta}_{\rm g}^{\rm T}\phi_{,\alpha_{\rm g}})_{,\alpha_{\rm g}}\boldsymbol{\zeta}_{\rm g}\,. \tag{13.112}$$

We allow for deviation ($\mathbf{M} \neq \mathbf{N}$) from the normality flow rule to account for non-Schmid effects, whose physical background and implications for shear band formation have been discussed in [4]. What follows is along the lines presented in the preceding sections. In particular, in full analogy to Eq. (13.88) and with the only difference in the use of the yield-function $f_{\rm g}$ in place of f, the first-order energy function $\mathcal{E}_\tau(\underline{\mathbf{v}}, \underline{\nu}; \underline{\dot{\mathbf{u}}}, \underline{\dot{\gamma}})$ now reads

$$\mathcal{E}_\tau(\underline{\mathbf{v}}, \underline{\nu}; \underline{\dot{\mathbf{u}}}, \underline{\dot{\gamma}}) = \tilde{E}(\underline{\mathbf{v}}, \underline{\nu}) + \tau \int_{\mathcal{B}} (\dot{\mathbf{S}} \,.\, \nabla\mathbf{v} - \dot{f}_{\rm g}\nu)\, \mathrm{d}V - \tau \int_{\partial\mathcal{B}_{\rm T}} (\dot{\mathbf{T}} \cdot \mathbf{v} + \dot{\mathbf{T}} \cdot \dot{\mathbf{u}})\, \mathrm{d}S\,. \tag{13.113}$$

With this change, the quasi-extremal principle Eq. (13.89) can also be applied in the gradient case. As before, QEP (13.89) in an equilibrium state with $f_{\rm g}$ non-negative is formally equivalent to the weak formulation of the following rate problem for $(\underline{\dot{\gamma}}, \underline{\dot{\mathbf{u}}})$ subject to $\dot{\mathbf{u}} = \mathbf{0}$ on $\partial\mathcal{B}_{\rm u}$ and $\dot{\gamma} = 0$ whenever $f_{\rm g} < 0$:

$$\mathrm{Div}\,\dot{\mathbf{S}} = \mathbf{0} \ \text{ in } \mathcal{B}\,, \quad \dot{\mathbf{S}}\mathbf{n} = \dot{\mathbf{T}} \ \text{ on } \partial\mathcal{B}_{\rm T}\,, \quad \dot{\gamma} \geq 0, \ \ \dot{f}_{\rm g} \leq 0, \ \ \dot{f}_{\rm g}\dot{\gamma} = 0, \ \ f_{\rm g} = 0 \ \text{ in } \mathcal{B}^{\rm p}. \tag{13.114}$$

Analysis of QEP2 is omitted here. The aim of this section was to illustrate the way of implementing the results of finite-dimensional analysis given in Sect. 13.4 in a gradient-enhanced theory for a continuum. A detailed comparison with the

[13] Moreover, if the virtual internal power is defined by $\mathcal{P}^{\rm int} = \tilde{\Phi} + \tilde{\mathcal{D}}$ and external power by $\mathcal{P}^{\rm ext} = P \cdot \tilde{q}$, which corresponds to $\tilde{E}(\underline{\nu}, \underline{\mathbf{v}}) = \mathcal{P}^{\rm int} - \mathcal{P}^{\rm ext}$ in Eq. (13.109) under assumption (13.110), then the *virtual power inequality* $\mathcal{P}^{\rm int} \geq \mathcal{P}^{\rm ext}$ is obtained from principle (13.44). The equality in the frequently used 'principle of virtual power' is recovered if $\nu \geq 0$ is not arbitrary but subject to $f_{\rm g}\nu = 0$.

variational formulations of gradient plasticity theories listed in the introduction is beyond the scope of this paper.

13.6 Conclusion

An energy-based variational formulation has been proposed for a broad class of rate problems of non-potential type in rate-independent plasticity. It encompasses pressure-sensitive plastic yielding, but also other rate-independent inelastic phenomena that do not require symmetry of the resulting tangent stiffness matrix. A spatially discretized problem has been considered for better transparency of the analysis and results. The major feature of this problem is the lack of symmetry of a global matrix A that specifies the relation between generalized velocities and non-negative plastic multipliers as unknown rate-variables, and the rates of generalized forces and yield functions as quantities present in the system of equations and inequalities that define the global evolution rule.

It has been shown that, in a general case of non-symmetric A, an exact solution to the rate-problem posed is not provided by a direct incremental energy minimization. The restrictive condition under which it is possible has been established in Sect. 13.4.3. This result might not be evident in advance in view of the extremal energy-*rate* principle (13.44) that is valid more universally. However, this principle applied at a given instant generates only a cone within which a rate-solution is to be found. When applied at a next time instant, it has lead to a *quasi-extremal* energy principle of first order (QEP1). The difference between extremal and quasi-extremal principles is that in the latter the minimized function depends on an unknown solution itself. Equivalence of a solution to QEP1, if it exists, to a solution to the basic rate-problem has been established and formulated as Theorem 1. This result has a general character and can be formally extended in a rather straightforward manner to quasi-extremal principles for a broad class of rate-independent continuous systems, also for gradient-enhanced theories of plasticity; examples have been provided in Sect. 13.5.

The reformulation of the incremental problem as a quasi-extremal energy principle, in view of the reference to the total internal energy of the system under consideration, provides a basis for possible extensions. One of them has been given the form of Theorem 2, where the minimization involved in the quasi-extremal principle QEP2 concerns a quadratic energy function. In effect, a solution to QEP2 is not only a solution to the basic rate-problem but satisfies additionally an extra condition that may be used as a criterion of choice between non-uniquely defined solutions. The condition has been given a physical interpretation as non-negativeness of the second-order work expended by small perturbing forces on the conjugate deviations they induce *during* the deformation process. This is a form of the energy condition of path stability, the concept developed earlier by the present author and associated with the incremental energy minimization approach to problems of potential type. QEP2 may be regarded as an extension of that concept to non-potential rate-problems.

However, consequences of violation of the path stability condition for non-potential problems are not yet fully clear.

The results obtained have a theoretical character. In view of a series of earlier papers with applications of the second-order incremental energy minimization to automatic branch switching in numerical calculations, there are similar perspectives of applicability of a version of the quasi-extremal energy principle to a broader class of problems of non-potential type. This constitutes a challenging task for the future.

References

1. Alart, P., Curnier, A., 1991. A mixed formulation for frictional contact problems prone to Newton like solution methods. Comp. Meth. Appl. Mech. Engng. 92, 353–375.
2. Alessi, R., Marigo, J.-J., Vidoli, S., 2015. Gradient damage models coupled with plasticity: Variational formulation and main properties. Mech. Mater. 80, 351–367.
3. Asaro, R. J., 1983. Micromechanics of crystals and polycrystals. Adv. Appl. Mech. 23, 1–115.
4. Asaro, R. J., Rice, J. R., 1977. Strain localization in ductile single crystals. J. Mech. Phys. Solids 25 (5), 309–338.
5. Bigoni, D., 2012. Nonlinear Solid Mechanics: Bifurcation Theory and Material Instability. Cambridge University Press.
6. Bishop, J. F. W., Hill, R., 1951. A theory of the plastic distortion of a polycrystalline aggregate under combined stresses. Phil. Mag. 42 (327), 414–427.
7. Braides, A., 2014. Local minimization, variational evolution and Γ-convergence. Springer Verlag, Berlin.
8. Callen, H. B., 1960. Thermodynamics. Wiley and Sons, New York.
9. Carini, A., Genna, F., 2001. Variational formulations in the non-associated flow theory of plasticity. Int. J. Engng Sci. 39, 1765–1784.
10. Carstensen, C., Hackl, K., Mielke, A., 2002. Non-convex potentials and microstructures in finite-strain plasticity. Proc. R. Soc. Lond. A 458, 299–317.
11. Chin, G. Y., Wonsiewicz, B. C., 1969. Deformation banding and stability of (100)-(111) fiber textures of fcc metals. Trans. Metall. Soc. AIME 245 (4), 871–872.
12. Collins, I., Houlsby, G., 1997. Application of thermomechanical principles to the modelling of geotechnical materials. Proc. R. Soc. A 453, 1975–2001.
13. Cottle, R. W., Pang, J.-S., Stone, R. E., 1992. The Linear Complementarity Problem. Academic Press, Boston, MA.
14. Dal Maso, G., DeSimone, A., Solombrino, F., 2011. Quasistatic evolution for Cam-Clay plasticity: a weak formulation via viscoplastic regularization and time rescaling. Calc. Var. 40, 125–181.
15. De Giorgi, E., 1993. New problems on minimizing movements. In: Lions, J.-L., Baiocchi, C. (Eds.), Boundary Value Problems for Partial Differential Equations and Applications. Masson, Paris, pp. 81–98.
16. De Saxcé, G., Feng, Z.-Q., 1998. The bipotential method: a constructive approach to design the complete contact law with friction and improved numerical algorithms. Mathl. Comput. Modelling 28, 225–245.
17. Ebobisse, F., Neff, P., Forest, S., 2018. Well-posedness for the microcurl model in both single and polycrystal gradient plasticity. Int. J. Plast. 107, 1–26.
18. Eve, R. A., Reddy, B. D., Rockafellar, R. T., 1990. An internal variable theory of elastoplasticity based on the maximum plastic work inequality. Quart. Appl. Math. 48, 59–83.
19. Fedelich, B., Ehrlaher, A., 1997. An analysis of stability of equilibrium and of quasi-static transformations on the basis of the dissipation function. Eur. J. Mech. A/Solids 16, 833–855.
20. Fleck, N. A., Willis, J. R., 2009a. A mathematical basis for strain-gradient plasticity theory–Part I: Scalar plastic multiplier. J. Mech. Phys. Solids 57, 161–177.

21. Fleck, N. A., Willis, J. R., 2009b. A mathematical basis for strain-gradient plasticity theory. Part II: Tensorial plastic multiplier. J. Mech. Phys. Solids 57, 1045–1057.
22. Germain, P., Nguyen, Q. S., Suquet, P., 1983. Continuum thermodynamics. Trans. ASME J. Appl. Mech. 50, 1010–1020.
23. Gurtin, M. E., 2000. On the plasticity of single crystals: free energy, microforces, plastic-strain gradients. J. Mech. Phys. Solids 48, 989–1036.
24. Gurtin, M. E., Anand, L., Lele, S. P., 2007. Gradient single-crystal plasticity with free energy dependent on dislocation densities. J. Mech. Phys. Solids 55, 1853–1878.
25. Hackl, K., Fischer, F. D., 2008. On the relation between the principle of maximum dissipation and inelastic evolution given by dissipation potentials. Proc. R. Soc. Lond. A 464, 117–132.
26. Halphen, B., Nguyen, Q. S., 1975. Sur les matériaux standards généralisés. J. Mécanique 14 (1), 39–63.
27. Harker, P. T., Pang, J.-S., 1990. Finite-dimensional variational inequality and nonlinear complementarity problems: a survey of theory, algorithms and applications. Math. Programming 48, 161–220.
28. Hill, R., 1948. A variational principle of maximum plastic work in classical plasticity. J. Mech. Phys. Solids 1, 18–28.
29. Hill, R., 1958. A general theory of uniqueness and stability in elastic-plastic solids. J. Mech. Phys. Solids 6, 236–249.
30. Hill, R., 1959. Some basic principles in the mechanics of solids without a natural time. J. Mech. Phys. Solids 7, 209–225.
31. Hill, R., 1978. Aspects of invariance in solid mechanics. In: Advances in Applied Mechanics. Vol. 18. Academic Press, New York, pp. 1–75.
32. Hill, R., Rice, J. R., 1972. Constitutive analysis of elastic-plastic crystals at arbitrary strain. J. Mech. Phys. Solids 20, 401–413.
33. Kikuchi, N., Oden, J. T., 1988. Contact Problems in Elasticity. SIAM Publishing Co., Philadelphia.
34. Maier, G., 1969. "Linear" flow-laws of elastoplasticity: a unified general approach. Rend. Acc. Naz. Lincei, Ser. VIII Vol. XLVII, 266–276.
35. Maier, G., 1970. A minimum principle for incremental elastoplasticity with non-associated flow laws. J. Mech. Phys. Solids 18, 319–330.
36. Mandel, J., 1971. Plasticité classique et viscoplasticité. Vol. 97 of CISM Courses and Lectures. Springer-Verlag, Wien-New York.
37. Miehe, C., Mauthe, S., Hildebrand, F. E., 2014. Variational gradient plasticity at finite strains. Part III: Local–global updates and regularization techniques in multiplicative plasticity for single crystals. Comp. Meth. Appl. Mech. Engng. 268, 735–762.
38. Mielke, A., Roubíček, T., 2015. Rate-independent Systems. Springer, New York.
39. Moreau, J. J., 1970. Sur le lois de frottement, de plasticité et de viscosité. C. R. Acad. Sci Paris A 271, 608–611.
40. Nguyen, Q. S., 1990. Stabilité et bifurcation des systèmes dissipatifs standards à comportement indépendant du temps physique. C. R. Acad. Sci Paris 310, Séries II, 1375–1380.
41. Nguyen, Q. S., 2016. Quasi-static responses and variational principles in gradient plasticity. J. Mech. Phys. Solids 97, 156–167.
42. Ortiz, M., Repetto, E. A., 1999. Nonconvex energy minimization and dislocation structures in ductile single crystals. J. Mech. Phys. Solids 47, 286–351.
43. Ortiz, M., Repetto, E. A., Stainier, L., 2000. A theory of subgrain dislocation structures. J. Mech. Phys. Solids 48, 2077–2114.
44. Peirce, D., Asaro, R. J., Needleman, A., 1982. An analysis of nonuniform and localized deformation in ductile single crystals. Acta Metall. 30, 1087–1119.
45. Petryk, H., 1982. A consistent energy approach to defining stability of plastic deformation processes. In: Schroeder, F. H. (Ed.), IUTAM Symposium on Stability in the mechanics of continua, Nümbrecht, 1981. Springer, Berlin, pp. 262–272.
46. Petryk, H., 1985. On energy criteria of plastic instability. In: Plastic Instability, Proc. Considère Memorial. Ecole Nat. Ponts Chauss., Paris, pp. 215–226.

47. Petryk, H., 1991. The energy criteria of instability in time-independent inelastic solids. Arch. Mech. 43 (4), 519–545.
48. Petryk, H., 1993. Stability and constitutive inequalities in plasticity. In: Muschik, W. (Ed.), Non-equilibrium thermodynamics with application to solids. Vol. 336 of CISM Courses and Lectures. Springer, Wien, New York, pp. 259–329.
49. Petryk, H., 2003. Incremental energy minimization in dissipative solids. C. R. Mecanique 331, 469–474.
50. Petryk, H., 2005. Thermodynamic conditions for stability in materials with rate-independent dissipation. Phil. Trans. R. Soc. Lond. A 363, 2479–2515.
51. Petryk, H., Kursa, M., 2015. Incremental work minimization algorithm for rate-independent plasticity of single crystals. Int. J. Num. Meth. Engng. 104, 157–184.
52. Petryk, H., Thermann, K., 2002. Post-critical plastic deformation in incrementally nonlinear materials. J. Mech. Phys. Solids 50, 925–954.
53. Reddy, B. D., 2011a. The role of dissipation and defect energy in variational formulations of problems in strain-gradient plasticity. Part 1: polycrystalline plasticity. Continuum Mech. Thermodyn. 23, 527–549.
54. Reddy, B. D., 2011b. The role of dissipation and defect energy in variational formulations of problems in strain-gradient plasticity. Part 2: single-crystal plasticity. Continuum Mech. Thermodyn. 23, 551–572.
55. Rice, J. R., 1971. Inelastic constitutive relations for solids: an internal-variable theory and its application to metal plasticity. J. Mech. Phys. Solids 19, 433–455.
56. Rice, J. R., 1975. Continuum mechanics and thermodynamics of plasticity in relation to microscale deformation mechanisms. In: Argon, A. S. (Ed.), Constitutive Equations in Plasticity. MIT Press, Cambridge, Mass., pp. 23–79.
57. Rockafellar, R. T., 1970. Convex Analysis. Princeton University Press, Princeton.
58. Taylor, G. I., 1938. Plastic strain in metals. J. Inst. Metals 62, 307–324.
59. Telega, J. J., 1979. On variational formulations for non-linear, non-potential operators. J. Inst. Maths Appl. 24, 175–195.
60. Telega, J. J., 1980. Variational principles for rate boundary-value problems in non-associated plasticity. ZAMM 60, 71–82.
61. Tonti, E., 1984. Variational formulation for every nonlinear problem. Int. J. Engng Sci. 22, 1343–1371.
62. Truesdell, C., Toupin, R. A., 1960. The Classical Field Theories. In: Flugge, S. (Ed.), Handbuch der Physik. Vol. III/I. Springer-Verlag, Berlin, pp. 226–793.
63. Tůma, K., Stupkiewicz, S., Petryk, H., 2016. Size effects in martensitic microstructures: finite-strain phase field model versus sharp interface approach. J. Mech. Phys. Solids 95, 284–307.
64. Vainberg, M. M., 1956. Variational Methods for the Study of Nonlinear Operators. Government Publishing House for Technical-Theoretical Literature, Moscow.
65. Ziegler, H., 1963. Some extremum principles in irreversible thermodynamics with application to continuum mechanics. In: Sneddon, I. N., Hill, R. (Eds.), Progress in Solid Mechanics. Vol. IV. North-Holland, Amsterdam, pp. 91–193.

The manufacturer's authorised representative in the EU is Springer Nature Customer Service Centre GmbH, Europaplatz 3, 69115 Heidelberg, Germany. If you have any concerns regarding our products, please contact ProductSafety@springernature.com

Printed and bound by CPI Group (UK) Ltd, Croydon, CR0 4YY
01/10/2026
02225971-0015